# Surface Finishing Theory and New Technology

Shengqiang Yang · Wenhui Li

# Surface Finishing Theory and New Technology

国防工业出版社
National Defense Industry Press

Springer

Shengqiang Yang
College of Mechanical Engineering
Taiyuan University of Technology
Taiyuan
China

Wenhui Li
College of Mechanical Engineering
Taiyuan University of Technology
Taiyuan
China

ISBN 978-3-662-54131-9 ISBN 978-3-662-54133-3 (eBook)
https://doi.org/10.1007/978-3-662-54133-3

Jointly published with National Defense Industry Press, Beijing, China

The print edition is not for sale in China Mainland. Customers from China Mainland please order the print book from: National Defense Industry Press.

Library of Congress Control Number: 2017956322

Translation from the Chinese language edition: 《表面光整加工理论与新技术》 *Biao Mian Guang Zheng Jia Gong Lilun Yu Xin Jishu* by 杨胜强 Yang Shengqiang, © National Defense Industry Press 2011. All Rights Reserved.


Printed on acid-free paper

This Springer imprint is published by Springer Nature
The registered company is Springer-Verlag GmbH, DE
The registered company address is: Heidelberger Platz 3, 14197 Berlin, Germany

# Preface

Taiyuan University of Technology (TYUT) has been researching and spreading the application of surface finishing technology for about 30 years and is one of the early domestic research institutes carrying out the free abrasive finishing technology. TYUT has obtained a number of achievements and patents. One of the projects was listed in the national key project of scientific and technological achievements. TYUT was awarded one prize for Science and Technology Progress of China and two prizes for Science and Technology Progress of Shanxi province. Based on the former research results, Professor Shichun Yang published a monograph "surface quality and finishing technology" supported by high-standard writing publication fund of China Machine Press in 2000. This monograph won extensive recognition from all sectors of society such as colleges, research institutes, and enterprises. Some colleges are using this book as the textbook for postgraduate or undergraduate students. Some manufacturers have used the new processes and equipment introduced in this monograph. It is certain that the editor's original intention has been realized.

In the past decade, the Finishing Technology Research Institute of TYUT has been committed to the technical challenge in production. There are four doctoral theses, respectively, discussing many new theories, technologies, and processes related to the barrel finishing for shaft parts with large sizes, magnetic abrasive finishing, fluid magnetic abrasive finishing, and two-phase flow finishing. These theses are supported by National Natural Science Foundation of China, Innovation Funds for Small and Medium Corporation from Nation's Technology Ministry, Shanxi Natural Science Foundation, Shanxi Scientific and Technological Project, Shanxi Science and Technology Research and Development Program, and Shanxi Foundation for Returned Scholars. Our institute has been cooperating with Langfang branch of China North Industries Group Corporation (Langfang Beifang Tianyu Mechanical and Electrical Technology Co., Ltd.) on developing many new finishing equipments adapting to market demand. The research achievements have been successfully used in military-industrial enterprises, civil enterprises, and foreign-funded enterprises. These enterprises achieved better economic benefits and social benefits. Based on the necessity in practice of teaching, scientific research, and manufacturing, the monograph "surface finishing theory and new technology" was published in 2011, which was supported by publishing fund of National Defense Science and Technology Books. The principle of this monograph is that the monograph focuses on the research results and achievements of authors. It also emphasizes the systematicness and integrity of theory, experiment, simulation, and application and summarizes the practical finishing methods at home and abroad. This monograph is intended for the practicing engineers who engaged in machinery manufacturing and related engineering field, as well as for scientific personnel and university teachers and students.

In this revision, authors have made some modifications and improvements for every chapter. Chapter 1 supplemented some concepts and explanation. Chapter 2 added simulation and analysis of centrifugal barrel finishing process; types of vibratory barrel finishing process are improved, preparation process of the media are added, and some application cases are updated. Chapter 3 increased one preparation method of magnetic abrasive using the high-energy ball milling. Chapter 4 increased Sect. 6, and the surface modification of FMA is put forward to change the surface chemical properties of the particles in this section. Chapter 5 increased Sect. 6, and this section explored the finishing principle, fluent field simulation, and finishing effects of the two-phase compulsive circulation flows finishing.

This monograph is comprised of 5 chapters. Chapter 1 introduces the concept of surface quality and surface finishing technology, and the development of finishing technology. Chapters 2–5 explore the finishing mechanism, equipment, influence factors, finishing effects and applications of barrel finishing, magnetic abrasive finishing, fluid magnetic abrasive finishing, and two-phase flow finishing, respectively.

This monograph is written by Sheng-qiang Yang, Wen-hui Li, et al. This book is written in the form of brainstorm and division writing. The breakdown of this monograph is Chap. 1 is written by Sheng-qiang Yang, Wen-hui Li, Feng-feng Wu; Chap. 2 is written by Wen-hui Li; Chap. 3 is written by Hong-ling Chen, Yong-gang Li, Zhi-yan Hou; Chap. 4 is written by Wei-dong Li and Huan-wu Sun; Chap. 5 is written by Xiu-hong Li. The book is compiled by Wen-hui Li and Sheng-qiang Yang. English proofreading is done by Jian-yan Tian, and Yan-qing Wang, Xiu-zhi Wang, Li-ping Liu in our institution took part in the translation. The authors express great appreciation to their supports. The authors are grateful to the National Defense Science and Technology Publishing Press for the support.

The authors are limited, and this monograph may have some mistakes. We welcome readers to correct and make valuable comments!

Taiyuan, China
March 2017

Shengqiang Yang
Wenhui Li

# Contents

# Chapter 1
# Surface Quality and Finishing Technology

## 1.1 Introduction

In the machining processes of the mechanical part from part blanks to finished products, including machining, casting, forging, punching, welding, there are usually different levels of roughness and various defects on the surfaces of the machined parts, such as uneven surface, broken edge, burr, scratch, micro-cracks. These have influenced not only on the quality and usability of these parts, for instance wearability, fatigue strength, corrosion resistance, but also on the entire machine assembly accuracy, serviceable performance, and life after assembly of the part, such as the working failure of electric appliance, the wear speeding up of mechanical transmission system, the changing of fitting property, the increasing of vibration and noise, the sealing failure, the mechanical efficiency falling, the weakened binding force of electroplating protection layer. The reason occurring roughnesses and defects on the surface of the machined part, involving the material, size, structural shape, machining methods, and other factors, is an intricate collaboration among material microstructure, machining deflection, physicochemical reaction, etc. To significantly overcome these surface defects and meet the requirements for use, the countermeasures from many aspects must be considered [1–5].

The surface qualities of parts not only have significant influcnces on their usability, life, and reliability, but also have a close relationship with the entire machine performance and life. With the development of scientific technology and modern production, many types of machinery are developing to be large-scale (or miniaturization), high-speed (or high-efficiency), high-accuracy, specialty (flexibility), comprehensive, integration, modularization, automation, intelligence, digitization, networking, greenization, and so on, which deliver a higher demand for the surface qualities of parts (including the surface geometric characteristics, physical and mechanical properties).

The surface finishing technology, as effective tactics for promoting the surface qualities of parts, has become one of the important ways to improve the serviceable

 
S. Yang and W. Li, *Surface Finishing Theory and New Technology*,
https://doi.org/10.1007/978-3-662-54133-3_1

performance of parts and products, increase reliability, and prolong life span. Taking 17 categories of parts with friction pairs on the LL480QB diesel engine as an example, the surface qualities of parts are improved by adopting surface finishing process, according to the part structure, material, surface condition, and finishing requirements. After comparing the test results of the whole device performance, it is discovered that the performance has been improved. According to running-in indicators, the running-in time of the entire machine leaving factory has been reduced by 50%, the fuel consumption rate at the commissioning stage has been reduced by 3–4 g/kW·h, and the noise has been reduced by 1 dB [6].

## 1.2 Surface Quality of Part

### *1.2.1 Product Quality*

Along with the scientific and technological progress and social development, the growing demands for mechanical product's performance, quality, and price are arising. These demands prompt enterprises to innovate and strive incessantly, to hone the product value in the fierce market competition to expand the market and gain profit, and to ensure the development and survival of the enterprises. The product value is determined by its function, quality, market, profit, after-sales service, and other factors, among which function and quality are the most important. In the product design and manufacture, the conception to drive value improvement, which can obtain the desired functionality at low cost to generate more profit, should be set up for improving the product competitiveness in both domestic and foreign markets.

The product functionality that is explained the product's purpose and the product quality that is reflected in the degree to meet the user's expectations of product functionality are the essential criteria for the evaluation of the products. For general mechanical products, there are two kinds of evaluation indexes: ① The performance index of the mechanical product, including product function parameters, precision indexes, and other non-physical indexes. For a regular lathe, its function parameters include maximum machining diameter and length, rotating speed, feed, power, automation. The function parameters express the fundamental performance of the product and explain the application range. The precision indexes of the mechanical product, including the accuracy of the spindle, guide ways, transmitting train, reflect the satisfaction degree of its functionality. The non-physical indexes, including the request to appearance, pattern, color, reflect the aesthetic function of the product and other requirements. ② The endurance index of the mechanical product, which means the period for keeping a mechanical product in good condition and its performance without trouble. The endurance index is the reflection of ability to maintain the product performance and quality, usually appraised by reliability. For example, the endurance index of a machine tool is

7 years to 15 years, and it can provide a basis to users for maintenance and depreciation [1].

### 1.2.2 Part Quality

The product performance and endurance indexes are determined by product design and guaranteed by the manufacturing and assembly quality of parts. The modern mechanical product generally consists of mechanics, hydraulics, pneumatics, electrics, and other components, so their performance and endurance directly affect those of the products. To ensure the quality of the mechanical product, strict requirements for the quality of mechanical parts and the reliability of hydraulic, pneumatic, electric, and electronic components must be raised.

In the manufacturing process of mechanical parts, the blank is made from the raw materials through thermoplastic deforming methods such as casting, forging; the cold plastic deforming methods such as cold-rolling, punching, rolling; and welding method. Then, the sizes, shapes, and performances of the blank surface which fail to satisfy the design requirement are changed through machining, heat treating, and other methods. At last, the part which can meet the design requirements is manufactured successfully.

The part quality is the conformance degree between the manufactured part and design requirements, in the material qualitative, size, shape, surface status, and other aspects. Therefore, by the blank manufacturing, heat treating, and machining, the part quality is guaranteed. For the blank quality and heat treatment quality, they should be evaluated according to the design drawings and corresponding process documents. For the machining quality of the part, it is assessed according to machining precision and surface quality in the two areas. Among these, the evaluation index of machining precision, which is the accuracy of the size, geometrical shape, and relative positions, represents the demand for part macro-level parameters. The surface quality expresses the request for microscopic geometric characteristics, physical–mechanical performance on the part surface.

### 1.2.3 Surface Quality of Part

The surface quality of the part is one of the essential criteria for the evaluation of part quality and has a significant effect on the usability of the part, especially on the reliability of part running at high speed, high temperature, and high pressure.

The part surface in the way of forming surface can be classified into two main types: ① The blank surface. Its size, shape, and performance are formed during blank manufacturing to reach the design pattern requirement without any mechanical machining processes. Its surface quality is determined by the corresponding blank manufacturing method and evaluated according to its standard.

② The machined surface. The blank surface is changed to achieve the design pattern requirement of sizes, shapes, and performances through mechanical machining and other machining methods. Its surface quality is the quality of the very thin surface layer formed in the machining process, and the geometrical, physical, chemical, or other engineering performance condition of the part surface layer obtained correlates significantly with the machining technique and processing parameters. Thus, these characteristics and influence factors of the surface layer must be researched when studying surface quality of the machined part, and then, the corresponding content of surface quality evaluation is put forward based on the characteristics of the surface layer.

1. Characteristics of surface layer

The very thin surface layer is formed on the part surface after the mechanical machining, and its characteristics are very different from those of the internal base of the part, the surface obtained through planning is taken as an example to analyze and study, and the forming process of the cutting surface is shown in Fig. 1.1.

As the workpiece is gradually close to the cutting edge with velocity $v$, compressive and shear metal deformation occurs in the cutting layer, and the chip is generated through shearing slip along the OM-direction on the shear plane at the end. At this point, on the machined surface at the separation between the chip and metal, not the entire cutting thickness $a_p$ is cut off by tool. However, there is a thin layer about $\Delta a$ thick of metal that will not yield slipping deformation along the OM-direction, to continue moving forward by following the workpiece, and be extruded under Point O of the cutting edge. Then, the cutting layer metal is separated into two sections at Point O: The section of Point O above becomes the chip and flows out along rack face, and the section of Point O below remains on the machined surface after being extruded by the cutting edge. This section of metal is obtained extrusion and friction from the edge BC on the flank surface after undergoing through Point B on the cutting-edge roundness, and the severe friction has caused shear effect on the surface metal of workpiece. As the workpiece continues to move forward, the machined surface restores elasticity along the flank face of tool. During the restore process, the surface metal continues to be friction with the flank face on the edge CD. When the restoration reaches the ultimate value $\Delta h$, the whole cutting process is finished. Therefore, the workpiece surface is ever

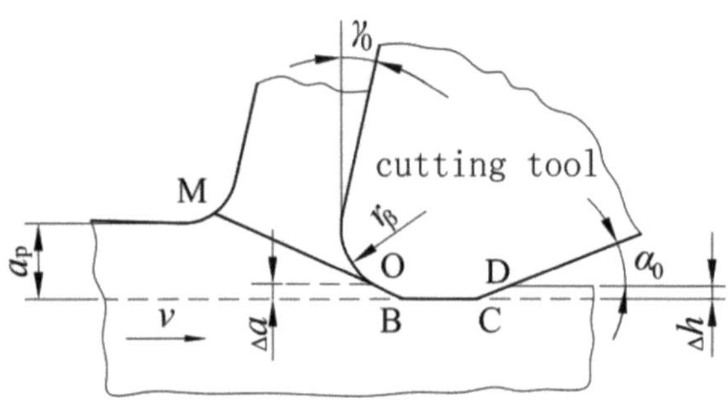


**Fig. 1.1** The forming process of machined surface

in the complex stress state of wedging, extrusion, breaking, and friction during the whole cutting process for elastic and plastic deformations. Based upon the interaction of the cutting force, cutting heat, and ambient medium, the original geometric characteristics and physical–mechanical performance of the workpiece surface are changed. The other cutting methods are roughly the same as the planning except for the force state and the degree of deformation. And the condition of surface layer machined is also changing.

The surface layer composition of the metal machined can be represented in Fig. 1.2.

(1) The bulk metal (also known as the metal base) has a structure that depends on the composition and processing history of the metal.
(2) As shown in Fig. 1.2, the surface layer is separated into inner and outer layer above the bulk metal. The outer surface layer is primarily composed of adsorption layer and oxidation layer. The inner surface layer is a deformation layer deformed plastically and work-hardened to a greater extent.
(3) The depth and properties of the deformation layer depend on such factors as the processing method used and how much frictional sliding the surface encountered, and the metal in the deformation layer can cause the change of residual stress, hardness, and metal tissue. For example, if the surface is produced by machining with a dull and worn tool or the surface is grounded by a dull grinding wheel, the work-hardened layer will be relatively thick and usually will also have residual stresses.
(4) Unless the metal is processed and kept in an inert (oxygen-free) environment or is a noble metal (such as gold or platinum), an oxide layer forms over the deformation layer. The oxide on a metal surface is generally much harder than the base metal; hence, it is more abrasive. As a result, it has important effects on friction, wear, and lubrication. The thickness of surface oxidation layer is at least 1 nm; some can reach 20–30 nm or be thicker. As the thickness of

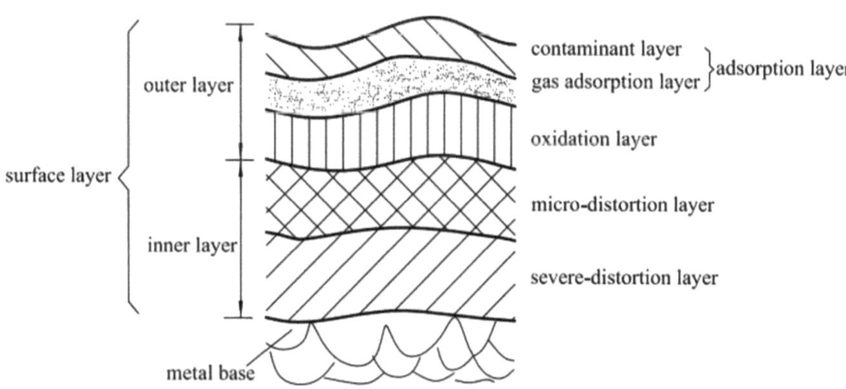


**Fig. 1.2** Surface layer composition of metal machined

oxidation layer exceeds 40 nm, the intrinsic color of oxide becomes apparent. For example, iron has an oxide structure with FeO adjacent to the bulk metal, followed by a layer of $Fe_3O_4$ and then a layer of $Fe_2O_3$ (which is exposed to the environment); aluminum has a dense, amorphous (with any crystalline structure) layer of $Al_2O_3$ with a thick, porous, and hydrated aluminum oxide layer over it; copper has a bright, shiny surface when freshly scratched or machined. Soon after, however, it develops a $Cu_2O$ layer, which then is covered with a layer of CuO (The layer gives copper its somewhat dull color); stainless steels are "stainless" because they develop a protective layer chromium oxide (by passivation).

(5) It is difficult to determine the depth of the surface layer that varies according to different workpiece materials, manufacturing methods, and treatment methods. For example, the depth of residual stress layer in the rough grinded surface can reach 0.318 mm, the depth of hardness layer can be 0.254 mm, the depth of plastic deformation can be 0.089 mm, but the depth of deformation layer in the shot-blasting surface can reach 0.5–1.5 mm.

2. Implications of surface quality

The surface layer condition of any machined part cannot be perfectly ideal face; a certain micro-geometry deviation is always there. The surface layer metal usually gets the intrinsic physical–mechanical performance changed, under the influence of cutting force and cutting heat. Therefore, in mechanical machining, "surface quality" is used to evaluate the geometrical, physical, chemical, or other engineering performances of part surface layer obtained by one or several machining and processing methods and the degree of conformity with the technical request of the part. What is expressed mostly is two contents:

(1) Geometric characteristics of machined surface: such as surface roughness, surface texture, and machining defects on the surface. There are many kinds of defects, such as the burr, which is the most frequent defect generated in the cutting process. Burr is the unnecessary tiny sharp-edged asperities derived on the edge of machined surface during cutting process, due to plastic deformation, surface tearing, etc.

(2) Physical–mechanical performances of machined surface layer: such as the performances used to reflect plastic deformation and work-hardening of the surface layer, residual stress of the surface layer, microstructure changes of the surface layer, and other special performances used to reflect the requirement of surface corrosion, optical quality, etc.

3. Impact of surface quality on parts usability

Surface quality reflects only surface geometric characteristics and surface layer performances, but it has certain effects on the usability of mechanical parts such as wearability, fit quality, fatigue strength, corrosion resistance, contact stiffness.

(1) Impact on parts' wearability

The life span of mechanical parts depends largely on parts' wearability. The parts' wearability is mainly related to the friction pair material, heat treating, and lubricating conditions. In the case where these conditions have been determined, the surface qualities of parts play a decisive role.

The wear process of parts has been generally classified into three stages (Fig. 1.3):

(1) The Initial Worn Stage
It is also known as running-in stage; when two parts surfaces begin to contact each other, only the profile peaks of rough surface contact each other, the real contact area is very small, and pressure per unit area is very large. Without forming the lubricating oil film, the local dry friction is generated at contact area, and the surface contact stress might exceed the yield limit and strength limit of the metal. With the relative motion of the friction pairs, metal in the contact section is torn, broken, and cut off; the profile peaks are rolled out; and worn speed is faster, as shown in Fig. 1.3 part I.
(2) The Normal Worn Stage
As running-in process continues, the value of surface roughness decreases gradually, the real contact area increases, and the pressure per unit area decreases. Good lubrication condition appears between friction surfaces, and wear continues at a slow rate, as shown in Fig. 1.3 part II.
(3) The Acutely Worn Stage
After a very long period, as wear continues, the surface roughness value decreases further, causing the lubricating oil to extrude from the contact area. Semi-dry friction and even dry friction formed in the contact surface, which makes friction damping increase and wear intensified. Meanwhile, the adsorption forces among contact surface molecules is enhanced by increasing the real contact area, and surface gnawing solder caused by the bonding among surface molecules could easily come up. As the movement continues, the surface that comes up gnawing solder is torn. Sometimes, for certain quantity of friction heat to be produced, the appearance of the high temperature on the

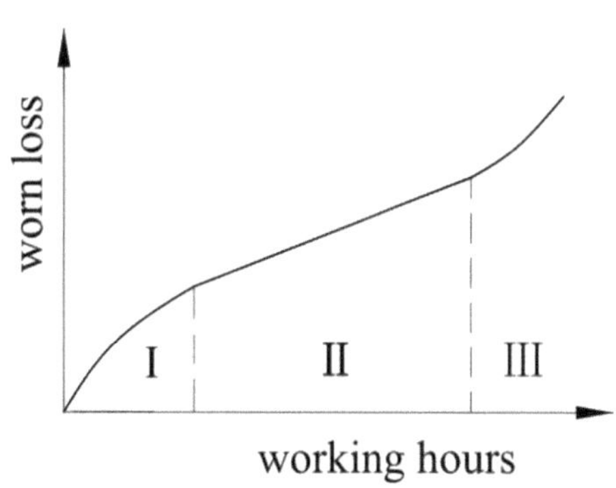


**Fig. 1.3** Basic rules in wear process

contact surfaces changes microstructure, decreases surface hardness, and even can make the local contact surfaces fusion welded together and make wear increase dramatically, as shown in Fig. 1.3 part III.

As the two factors of surface quality, geometric characteristics and physical–mechanical performances have certain impacts on parts' wearability. The parts' initial worn loss has close relation with the value of surface roughness. Figure 1.4 shows that there can be an optimal value of surface roughness corresponding to the minimum initial worn loss under a certain condition. With the increase of the load, the curve shifts upper right, and the optimal value of surface roughness shifts to $O_2$ from $O_1$. In general case, when the range of surface roughness *Ra* value is 0.4–0.8 μm, the initial worn loss is the lowest.

The profile shape of surface roughness and the orientation of surface texture have a significant impact on parts' wearability. For example, the surface roughness values of various surfaces shown in Fig. 1.5 are the same, but the parts' wearabilities are very different, because the real contact area and the remaining lubricating oil distribution between parts vary dramatically with differences among the profile shapes of surfaces. Comparing with the same group of surfaces with profile shapes (a), (c), (e) as shown in Fig. 1.5, surfaces with profile shapes (b), (d), (f) have better wear resistance.

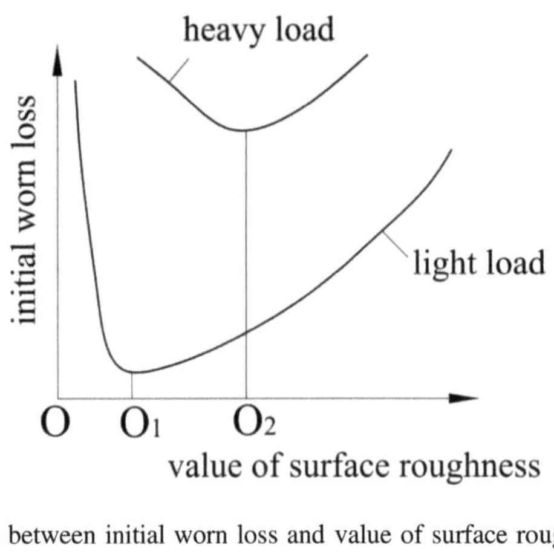


**Fig. 1.4** Relationship between initial worn loss and value of surface roughness

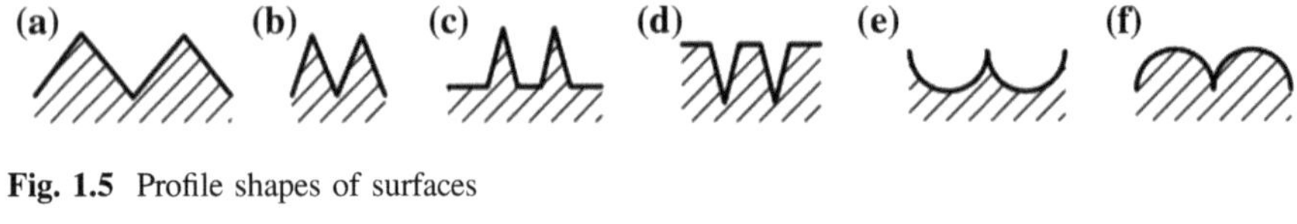


**Fig. 1.5** Profile shapes of surfaces

For example, in order to assure that lubricating oil between cylinder liner and piston ring would not be extruded from the contact area, hydrodynamic lubricating oil film is generated among surfaces in relative motion to improve the lubricating state between cylinder liner and piston ring. A large number of mutual independent, mutual non-penetrating honeycomb-like micro-pits with identical length based on a certain distribution law are arranged on the working surface of the cylinder liner. The mass bench test results and automotive operations show that the life span of cylinder increases 80–200%, lube oil consumption decreases 30–60%, environment pollution decreases (such as HC emission decreases 10–20% than using other cylinders), and running-in period is shortened. It can avoid the cylinder scoring to improve the life span of engine.

The correlation between the orientation of surface texture and the direction of relative movement on the parts surface has a significant impact on parts' worn loss, as shown in Fig. 1.6. Under light load, worn loss is the lowest when the texture direction and the direction of relative movement are the same. Worn loss becomes the highest when they are perpendicular. Under heavy load, worn loss is the lowest when the texture directions of two surfaces are perpendicular and the direction of movement is parallel to the texture directions of lower surface. Also, worn loss is the highest because holding-on often occurs when the texture directions of two surfaces and the direction of movement are the same.

The relationship between surface micro-hardness and wearability is shown in Fig. 1.7. After cold plastic deformation, the surface micro-hardness of parts machined increases, thus reducing the likelihood of the surface plastic deformation and gnawing solder of the metal on the surface and improving wearability. For example, the hardness of carbon steel 40 can improve 15–45% after cold-drawing. Wear test result shows that the worn loss can decrease 20–30%. However, when work-hardening increases to some extent, further cold work-hardening causes the excessive porosity of metal organization, and intensified wear and even the emergence of crack and pealing will decrease wearability. For example, when the surface

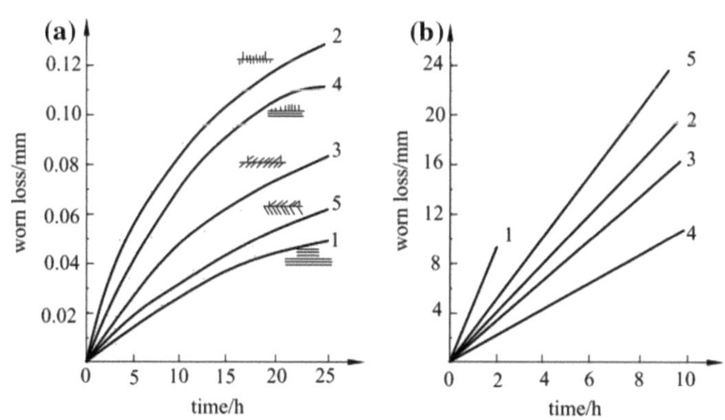


**Fig. 1.6** Relationship between worn loss and directions of surface texture. **a** Under light load; **b** under heavy load

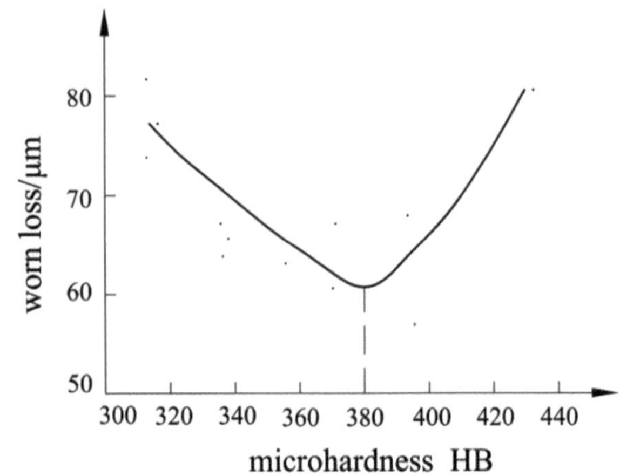


**Fig. 1.7** Relationship between surface micro-hardness and wearability

micro-hardness of carbon tool steel T7A raises to 380HB as shown in Fig. 1.7, the worn loss achieves lowest value, and the further improvement of micro-hardness can decrease wearability.

If cooling is abnormal in the grinding process, the surface layer temperature will rise, and surface burning occurs, which causes the microstructure changes of surface layer metal and the changes of original hardness that has influences on part wearability. For steel parts, superficial hardening is realized by heat treatment to improve their wearability, but it is necessary to prevent surface burning that can reduce the hardness and influence on wearability.

(2) Impact on fit quality

The surface roughness of parts has an obvious influence on fit accuracy of mating surfaces. Tests have shown that the rough profile peaks on the surface can be rolled out during assembly for interference fits. Even if the interference values of mating parts are the same, different joint strengths can be obtained for different surface roughness. The joint strength is comparatively weak, while the roughness value is high. For clearance fit, the initial worn is serious with high surface roughness value. So fit clearance increases rapidly and fit accuracy decreases.

In design, the surface roughness of parts should be selected according to the requirement of corresponding fit accuracy. If the requirement to fit accuracy is high, the surface roughness of corresponding mating surfaces should be low to assure the fit quality of the part and the working reliability.

The mechanical processing method that includes turning, boring, drilling, milling, planning, grinding, broaching, thread machining, and gear cutting often results in different degrees of machining burr on part surface, in particular for edge burr, and thus leads to surface defects. These burrs can cause parts to assemble ineffectively, the noise to increase during the machine operation, the abnormal worn to generate, and even the jammed phenomenon to occur on the precision matching surfaces.

(3) Impact on fatigue strength

Under alternative loading, the stress concentration caused by rough surface, scratch, and crack can cause the stress at the valley of the surface profile to exceed the fatigue limit easily, which forms and extends fatigue cracks and produces the fatigue failure. The stress concentration factor, which is employed in the calculation of fatigue strength, increases along with the increase of surface roughness value. When the surface roughness values of grinding surface are 0.4, 0.8, 1 μm, values of the stress concentration factor are 1.2, 1.24, and 1.48, respectively. The sensitivity of different materials to stress concentration is different. The smaller the material grain size and the denser the material microstructure is, the more sensitive is to stress concentration, and the more serious the surface roughness effect on fatigue strength is. Table 1.1 shows the relative fatigue strength on the surface obtained by different machining methods with different strength steel. As shown in Table 1.1, the higher the ultimate strength and the bigger the surface roughness value is, the more serious the decrease of the fatigue strength is. The experiment results show that the fatigue strength can improve 30% to 40% with the continuous decrease of the surface roughness value for parts under alternate loads.

Work-hardening also has an effect upon the fatigue strength. Appropriate work-hardening caused by tensile and compression can reinforce surface metal and then prevent formation of fatigue cracks and relieve expansion of cracks. It is conducive to the improvement of the fatigue strength. However, excessive hardening can cause formation of larger brittle cracks, so it reduces the fatigue strength.

The surface residual stress has a considerable impact on fatigue strength. When the tensile stress appears on the surface layer, the formation and expansion of fatigue cracks are much accelerated, and the fatigue strength decreases. For example, if the nickel-plated or chromed parts produce residual tensile stress under 400 $N/mm^2$ on the surface layer, the fatigue strength will drop 20–50% even more dramatically. When the compressive stress appears, it can counteract partly the tensile stress caused by the alternate loading. Then, the fatigue strength of the parts increases. Figure 1.8 shows the relationship between the residual compressive stress of 40Cr steel sample and the fatigue strength.

**Table 1.1** Relative fatigue strength obtained by different finishing ways

| Machining methods | Ultimate strength of steel $\sigma_B$ ($N/mm^2$) | | |
|---|---|---|---|
| | 470 | 930 | 1420 |
| | Relative fatigue strength (%) | | |
| Precise polishing or precision grinding | 100 | 100 | 100 |
| Polishing or grinding | 95 | 93 | 90 |
| Precision grinding or finish turning | 93 | 90 | 85 |
| Coarse grinding or rough turning | 90 | 90 | 70 |
| Using hot-formed steel | 70 | 50 | 35 |

Fig. 1.8 Relationship between residual compressive stress of 40Cr steel sample and fatigue strength

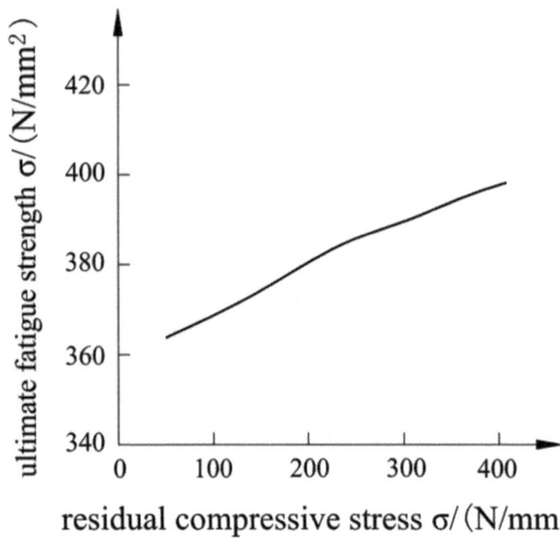


The microstructure changes of the metal on the parts surface layer also have an effect upon the fatigue strength. For example, because of grinding and burning, the fatigue strength of 45Cr hardening steel parts drops from 450 to 330 $N/mm^2$. If grinding cracks appear in the grinding, the effects become more severe.

(4) Impact on corrosion resistance

When the parts work in the wet moist air or the caustic medium, the corroding medium will deposit in the valleys and the cracks on the surface, and it will become a chemical corrosion or electrochemical corrosion. If the valleys are deeper and sharper, especially if there are cracks, the corrosive effects are stronger. It is practical way to decrease the surface roughness values, reduce the surface cracks, form the residual compressive stress, and produce the proper work-hardening for improving the corrosion resistance of parts.

(5) Impact on others

The foregoing is only the impacts of parts surface quality on the general properties of products. For products with special requirements, the impact on adhesive strength, painting performance, plating quality, flow resistance, electrical appliances shortcut, cleaning condition, glossiness, food hygiene, and appearance quality should be considered. For example, for products with high appearance quality requirement, along with the surface roughness, the impact of surface texture, scratch, crack, and luster on product properties need to be considered. For hydraulic and pneumatic components, the impact of surface quality, especially burr and flashing, and other surface defects on the sealability and flow resistance needs to be considered. For engine products, the impact of surface quality, especially burr and dust sticking to the surface on products' cleanliness, needs also to be considered. For electrical products, such as electric motors, transformers, commutators, and relays, the burr impact on electrical appliances shortcut should be considered to avoid product failure and even accidents.

## 1.3 Surface Finishing Technology

Many machining methods are used in forming parts surface, and each method would allow the workpieces to get not only a certain shape and dimensional accuracy, but also relevant surface quality. In many machining methods for ensuring the surface qualities of parts, some focuses on improving the geometrical characteristics of parts surface, such as reducing surface roughness value and removing burrs and scratches. Other focuses on improving the physical–mechanical performance of parts surface layer, such as improving surface hardness and changing stress status. There are also some machining methods that can be used to balance the improvement of the geometrical characteristics and physical–mechanical performance to obtain a good result. When the focus tends to reduce the surface roughness value, the traditional finishing methods such as grinding, superfinishing, and polishing would typically be used. When the focus tends to improve the physical–mechanical performance of parts surface layer, chipless machining such as rolling and diamond press polishing would be used as the last manufacturing procedure. When these processes are adopted, the plastic deformation is yielded on the surface layer of the metal, the surface hardness can be increased, and there exist residual compressive stresses, so the wear and fatigue resistant property of parts are enhanced.

In recent years, surface finishing technology and deburring technique have gained rapid development, and many new techniques and technologies have emerged because of rising quality requirements on the parts surface. These processes not only effectively reduce the surface roughness value and improve surface stress state, but it can also remove burrs and do chamfering and rounding. Thus, the surface qualities of parts can be improved.

### *1.3.1 Connotation*

In order to improve the appearance quality of the parts surface, enhance the wear resistance of parts surface, achieve their full potential function of every part, and guarantee the performance of products, these operations such as removing burrs, flashes, grooves, reducing the surface roughness value, improving surface stress state, eliminating all defects left on the parts surface layer should be performed as needed, when the surface qualities have not yet met the requirements after the given dimension accuracy and geometric accuracy of parts are obtained. So all kinds of processing methods and processing technologies to improve the surface qualities of parts are known in mechanical machining as surface finishing technology, which is called finishing technology for short.

The corresponding surface finishing is often used on the blank surface and machined surface as required. As for the surface finishing on blank surface, it is necessary to be processed because there are also lots of defects such as surface oxide faults, wrinkling, sand burnings, and remaining weld spatters on the blank parts surface, although the specified blank sizes and geometrical shapes have been obtained after casting, stamping, and welding. These blank parts must be treated by sand removal and electroplating pre-treatment before entering into the next machining procedure. This kind of surface finishing technologies is called blank surface finishing technology with the goal of guaranteeing the quality of blank parts surface.

The process of chamfering, deburring, eliminating micro-crack, reducing the surface roughness value, and improving physical–mechanical performance, which is operated after parts are machined, is called the final surface finishing technology, also known as precision surface finishing technology.

Noteworthy is that the analysis and study of the blank part's surface quality have been ignored or not paid much attention, even confused with parts surface quality after machined. So the corresponding technological methods are less targeted and have no obvious effects.

With the advance of science and technology, the demand for parts surface quality is much more strict. So parts surface finishing technology has developed rapidly and gradually formed its own independent disciplinary system; in particular the surface finishing and the deburring have gained rapid development as final manufacturing procedure. It is a multi-discipline-crossing edge subject, which integrates machining, materialogy, physics, chemistry, electromagnetism, etc., and it has got progressively attention.

Scientific and technical personnel in many colleges, universities, and research institutes are dedicated to the researching and developing of this technology and energetically carry out academic discussion and experience exchange. Many new processes, new technologies, and new products developed in the finishing technology field lower the labor intensity for workers, reduce environmental pollution, and have brought significant economic and social benefits.

## *1.3.2 Function and Characteristics*

1. Main functions of finishing technology

The finishing is designed to improve the surface quality of parts. Its primary functions are as follows.

(1) To reduce the surface roughness values; to remove surface defects, such as scratches and micro-crack; to improve the surface qualities of parts.

(2) To enhance physical–mechanical performances of the part's surface layer and to improve surface stress state of parts.
(3) To remove the edge burr, round, and chamfer for ensuring a smooth transition between surfaces and to improve assembly ability of parts.
(4) To improve luster and brightness on the parts surface and to improve the cleanliness of parts.

2. Main characteristics of finishing technology

Whether traditional finishing methods or new techniques appearing recently all have the following characteristics:

(1) The machining allowance of finishing processes is small and only a fraction of tolerance bandwidth of the last process in general. The technology, as a general rule, can only improve the surface quality with no impact on machining accuracy. If the machining allowance is oversized, the production efficiency will drop, and the original machining precision of workpiece will decrease sometimes.
(2) The equipment used in finishing processes needs not full-precision forming motion, but the relative motion between abrasive tools and workpieces should be complicated as far as possible. The finishing process is a micro-grinding process that uses fine-grained abrasives to roll, rub, and score the workpiece surfaces, just be sure that there are enough random contacts between abrasive tools and workpiece surfaces; the surface error will be gradually homogenized and eliminated in the end. As a result, high surface quality can be obtained.
(3) During finishing, the position of the abrasive tool relative to the workpiece is not confirmed, so the shape and position error of the surface cannot be corrected in general. The guarantee of the workpiece surface precision has relied on the former process.

## 1.3.3 Classification

In order to guarantee and improve the surface quality, the finishing methods that are a very necessary and effective measure should be adopted. At the moment, there are various processing methods for the implementation of the finishing technology, and the finishing technology can be classified in different ways.

1. Classified by main function of finishing

(1) The finishing processes which mainly aim at decreasing surface roughness value, such as grinding, honing, polishing.
(2) The finishing processes which mainly aim at improving the physical–mechanical properties of the part surface, such as rolling, shot-peening strengthening, diamond press polishing, extruding holes.

(3) The finishing processes which mainly aim at deburring, deflashing, and chamfering, such as abrasive blasting, high-temperature blasting, barrel finishing, power brush finishing.

2. Classified by finishing mechanism

(1) The finishing processes that abrasive particles roll, rub, and score the workpiece surfaces under the action of certain pressure and relative velocity, such as grinding, honing, barrel finishing, magnetic abrasive finishing, liquid-magnetic abrasive tool finishing, jet machining.
(2) The finishing processes that the burr and flash on the workpiece dissolve rapidly under the action of chemical reaction, such as chemical polishing and electrochemical polishing.
(3) The finishing processes that the heat generated by the high temperature makes the workpiece material locally melt down, gasify, and even evaporate, such as high-temperature deburring, ion beam deburring, laser deburring.

3. Classified by supply pattern of energy during finishing

The finishing technology can be classified into mechanical method, chemical and electrochemical method, and heat energy method according to this classification [7].

For mechanical method, it can be classified into two categories, free abrasive tool finishing and non-free abrasive tool finishing according to the different states of abrasives during processing. And beyond that, the combined finishing methods have emerged, which combine several finishing methods and let them complement each other. The mechanical method has around 30 different finishing methods, as shown in Fig. 1.9.

The following section briefly introduces the non-free abrasive tool finishing and the free abrasive tool finishing.

(1) Non-free abrasive tool finishing

When the abrasive or the non-abrasive tools keep a relatively definite location with the workpiece, and work on the workpiece surface with certain cutting parameters to improve surface quality, this kind of finishing is called the non-free abrasive tool finishing, such as the grinding process with the grinding wheel on the workpiece surface.

(1) Main functions of the non-free abrasive tool finishing are to the reduce surface roughness value, improve the surface quality; retain or even improve the precision degree; increase the physical–mechanical properties of the part's surface layer; improve the stress state of the part's surface layer; remove edge burrs, round, and chamfer; realize smooth transition between different surfaces; improve the assembly ability of parts.
(2) Main characteristics of the non-free abrasive tool finishing includes significant effect on reducing the surface roughness value, very little effect on improving precision, and small machining allowance.

- Surface finishing
  - Mechanical method
    - Non - free abrasive tool finishing
      - Grinding
      - Super finishing
      - Polishing
        - Belt polishing
        - Elastic wheel polishing
      - Honing
      - Power brush finishing
      - Chipless finishing
        - Rolling
        - Diamond press polishing
        - Extruding holes
      - Ultrasonic polishing
    - Free abrasive tool finishing
      - Grinding
      - Barrel finishing
      - Magnetic abrasive finishing
      - Liquid - magnetic abrasive tool finishing
      - Gas - particles or liquid - particles two - phase flow finishing
      - Shot peening
  - Chemical and electrochemical method
    - Electrochemical polishing
    - Chemical polishing
  - Heat energy method
    - Laser - beam polishing
    - Ion - beam polishing
    - Hot - wire machining
    - High - temperature deburring
  - Combined finishing method
    - Chemical - mechanical polishing
    - Electrochemical - mechanical finishing
    - Magnetic - electrochemical finishing
    - Electrical discharge - ultrasonic grinding
    - Electrochemical - ultrasonic grinding
    - Electrical discharge - electrochemical polishing
    - Other combined finishing processes

**Fig. 1.9** Surface finishing methods classified by different patterns of supply energy

(2) Free abrasive tool finishing

When the abrasives do not keep a definite relative location with the workpiece, they are in a loose status during processing, and work on the surface of the workpiece with certain relative speed and acting pressure to improve surface quality, this kind of finishing processes is called the free abrasive tool finishing, such as the barrel finishing process of the abrasive blocks on the workpiece surface.

During the free abrasive tool finishing processes, free abrasives always roll, score, and micro-grind the workpiece surfaces randomly under the action of certain pressure and certain motion to realize the finishing of the workpiece surface. This kind of machining method contrasts strikingly with the grinding of the grinding wheel to the workpiece surface because of that the abrasive particle fixed on the surface of grinding wheel has constant cutting parameters in grinding and changes directly the precision and quality of the workpiece surface.

Free abrasive tool finishing technology develops rapidly, and it includes a wide range of items, but all feasible finishing methods follow one principle: The abrasive tool (or abrasives) provides certain acting pressure and relative speed in some movement mode to realize the rolling, scoring, and micro-grinding of workpiece surface and achieve the goal of improving workpiece surface quality.

To this kind of free abrasive tool finishing method belong grinding, barrel finishing, magnetic abrasive finishing, extrusion honing, etc. As there are striking differences between different abrasive tools in the aspects of constitutional form, action and machining state, machining mechanism, technological parameters, and technological characteristics also have their own unique feature. It is very urgent and important to develop constantly various new processes of free abrasive tool finishing and apply to the production practice and acquire economic benefit for scientific academia and industrial enterprise.

(1) Main functions of the free abrasive tool finishing

① To remove edge burr, chamfer, and smooth edge and improve the edge quality of the workpiece.
The main functions of such finishing are to remove edge burr and flash for the edge with good quality, such as edge fillet and chamfering. This can provide adequate assurance of the smooth transition and connection between different surfaces. After this processing, the assembly ability of these parts can be improved to realize the automatization of part assembly.

② To reduce the surface roughness value of parts and improve the surface quality of parts.
Abrasives participate directly to the rolling, scoring, and micro-grinding under the action of certain pressure. This method can remove the surface oxidation layers of forging/casting/stampings, the surface oxidation films after heat treatment, and the surface defects of colliding/scratch/micro-crack on the cutting surface. It can effectively reduce the surface roughness value and thus improve the surface quality.
After finishing, the surfaces of every part with assembly relations have better matching property and leak tightness and have better electric heating conductivity and force conductivity between different parts. This kind of finishing processes can effectively shorten the time for the initial worn stage of the part after completing the whole assembly, give full play to the functions of every part, and fully reflect the overall performance and service life of the machine.

③ To improve the physical–mechanical properties of part surface layer.
The abrasives blasting and shot-peening of the casting, forging, or punching parts belong to this kind of finishing processes. In the barrel

finishing process on the workpiece surface, the abrasives have the impact of rolling, scoring, and colliding on the surface to effectively enhance the stress distribution of part surface layer and thus further improve the physical–mechanical properties of surface layer. Finally, this can prolong the service life of the part itself and the entire machine.

④ To improve and change the surface luster and brightness of the part and increase parts' cleanliness.
With the cutting and the chemical effects of the abrasive tool (abrasives), this kind of finishing methods can change the surface luster and increase the brightness and cleanliness of these parts. Their finishing effects relate directly to parts material, surface condition, abrasive material, and chemical properties of grinding compounds. The luster of the part surface is changed, the surface is brightened, and the overall surface of the part is cleaned (by integrated finishing effects such as the rolling and micro-grinding of abrasives and the chemical reaction between compounds and workpiece surface). The kind of finishing methods is usually used in the pre-treatment of parts or the final machining of parts in order for the need of luster finishing.

(2) Main characteristics of free abrasive tool finishing

① The kind of finishing methods can keep the original size and shape accuracy of parts and put their emphasis on the improvement of surface quality.
Generally, the main purpose of the free abrasive tool finishing is to improve the surface quality. They focus on improving the surface quality, such as surface luster, surface roughness, surface hardness, surface stress state, and edge deburring without breaking the size and shape accuracy of the former process.

② The abrasives achieve the process of workpiece surface by the aid of some carrier.
The abrasives must depend on a gaseous, liquid, or solid form of the carrier to constitute abrasive tool and randomly realize the rolling, scoring, and micro-grinding machining in a certain moving manner and at a certain pressure by the aid of the abrasive tool.

③ Finishing equipment is simple and does not need to have a higher precision, and it has a lower cost.
Comparing with the cutting machines, finishing equipment does not need to have a higher motion precision. This is because there are not any requirements to improve the size and shape accuracy of the original parts during the finishing process. The kind of finishing method only needs some form of motion, such as rotational motion and reciprocating motion, to realize the relative motion of free abrasive tool and workpiece surface.

④ Finishing technology has many kinds of matched accessory devices, and most of them are nonstandard.

In the course of finishing, the finishing equipment still needs to be equipped with various accessory devices for the joint participation of abrasive tool, compounds, and work carrier (gas, liquid, or solid). These devices need specialized design and manufacture due to different finishing technological methods.

### *1.3.4 Finishing Effects*

Finishing method includes a wide range of methods. The production applications are wide, but there is none a single finishing method that can replace all of the other finishing methods. A method can be successful in solving certain quality problem of the surface, but it can be a failure when it solves the other quality problem of the surface. Although many free abrasive tool finishing methods are already available, it is still being studied and developed constantly.

Various finishing methods have their own characteristics, and they also have drawbacks. With the development of science and technology, and the continuous improvement of manufacture, finishing technology continues improvement, development, and innovation. Facing many complex problems of part surface quality and various finishing methods, the first important problem that needs to be solved is the proper choice of feasible, effective finishing method.

The analysis and comparison on machining quality, productivity, and economics of finishing methods have been shown below.

1. Machining quality

After the finishing of the part surface, the main contents of measuring the surface quality include the decreasing (attainable value) of the surface roughness *Ra* value, the improvement of the removing burr and edge quality, the improvement of surface physical–mechanical property (the increasing of surface hardness, the improvement of the surface stress state, the strength of metal layer included), and others (such as the changing of surface luster, the improving of contour perception, and the overall cleanliness of part). The machining quantities of various finishing methods are shown in Table 1.2.

2. Machining efficiency

For any kind of advanced technology, it should be advanced not only in machining quality but also in machining efficiency. To judge whether some kind of finishing technology has any advancement or not, the assessment on machining efficiency is a necessary factor that should be taken into consideration. If one kind of finishing technology has advantages in machining quality, but without a certain

**Table 1.2** Machining characteristics and application range of various surface finishing

| Finishing methods | Basic principle of finishing | | | | | Burr size | Machining state and range | Machining time | Superficial effect after machining | | | Characteristics of machining technology | Applicable parts |
|---|---|---|---|---|---|---|---|---|---|---|---|---|---|
| | Grinding | Cutting | Chemistry | Thermal energy | Electrolysis | | | | Size | *Ra*/μm | Chamfering /mm | | |
| Manual processing with tools | | | | | | Large/ small | Workpiece highly contacted With surface, manual control processing | 1–20 min | | | 0.05 | Worker operates various tools to machine, various tools have been commercialized | All kinds of parts |
| Rotary barrel finishing | | | | | | Middle/ small | Designed to machine outer surface of workpiece; difficult to machine inner surface, deburring | 1 min ~ 20 h | <5 μm | 0.05–0.8 | | Machining microelements is effective; low productivity; unsuitable for machining soft, easily deformed workpieces that easily bump | Sheet metal, stamping, medical device, hardware |
| Sand blasting | | | | | | Middle/ small | Machined workpieces is easy to be deformed; designed to machine outer and inner surface | 1–2 min | 10 μm | 1–8 | 0.08–0.25 | Jetting surface of electronic part by walnut shell; jetting surface of metallic die by glass beads | Camshaft, gear, box, box cover |
| Belt grinding | | | | | | Large/ small | Can machine plane and cylindrical surface; cannot machine concave surface | 30 s ~ 3 min | <10 μm | 0.8–1.6 | | Remove stamping burrs effectively; could machine castings by NC control device | Castings and forgings, stamping, shaft parts |
| Brush grinding | | | | | | Large/ small | Brush can work according to workpieces contour, thoroughly eliminate burr | 1–5 min | 0.08 mm | | 0.1 | Machine large parts by wire brush; processing small parts by animals and plants fiber brush | Connecting rod, chain wheel, rotor, impeller, valve |

(continued)

**Table 1.2** (continued)

| Finishing methods | Basic principle of finishing | | | | | Burr size | Machining state and range | Machining time | Superficial effect after machining | | | Characteristics of machining technology | Applicable parts |
|---|---|---|---|---|---|---|---|---|---|---|---|---|---|
| | Grinding | Cutting | Chemistry | Thermal energy | Electrolysis | | | | Size | $Ra$/μm | Chamfering /mm | | |
| Electrolytic grinding | | | | | | Small | Unsuitable for machining castings and Si, S, C containing alloy parts | 30 s ~ 10 min | | 0.05–0.4 | 0.08 | Designed to machine stainless steel material | Screen mesh, filter screen |
| Liquid honing | | | | | | Small | Designed to machine workpieces with complex shape | 1–3 min | <3 μm | 0.4 | 0.002–0.13 | Can machine burrs of metallic and nonmetallic parts; also can machine burrs of precision parts | Screen mesh, filter screen |
| Spindle barrel finishing | | | | | | Middle/small | Designed to machine precision parts without collision | 10 s ~ 2 min | <5 μm | 0.4–1 | 0.005 | Can realize automatic operation if the equipment is equipped with automatic loading and unloading device | Gear, camshaft, precision part |
| Vibratory barrel finishing | | | | | | Middle/small | Designed to machine workpieces with complex shape, difficult to machine inner surface | 30 min ~ 2 h | <5 μm | 0.2–0.8 | 0.08–0.25 | Barrel with periodic vibrating movement; Separation workpieces from abrasive automatically | Air compressor parts, piston, shaft parts |
| Chemical vibratory grinding | | | | | | Small | Batch processing in bulk; can machine 10 ~ 20 pieces at a time | 20 min ~ 1 h | 0.08–0.15 mm | 0.1–0.5 | <0.5 | Remove burrs and corrosion effectively with vibration device and chemical methods | Microelement, special-shaped parts |
| Ultrasonic machining | | | | | | Small | Designed to machine ultra-small miniature parts and workpieces without collision | | 0.05–0.08 mm | 0.3 | 0.05–0.13 | | Shaft bearing, gear, instrument parts |

(continued)

**Table 1.2** (continued)

| Finishing methods | Basic principle of finishing | | | | | Burr size | Machining state and range | Machining time | Superficial effect after machining | | | Characteristics of machining technology | Applicable parts |
|---|---|---|---|---|---|---|---|---|---|---|---|---|---|
| | Grinding | Cutting | Chemistry | Thermal energy | Electrolysis | | | | Size | $Ra$/μm | Chamfering /mm | | |
| Centrifugal barrel finishing | | | | | | Middle/small | Designed to machine various hardware | 5–20 min | <5 μm | 0.05–0.8 | 0.12–0.25 | Tumbling barrels with rotation, revolution, high machining efficiency, and symmetrical machining quality | Shaft bearing, gear, instrument parts |
| Chemical energy processing | | | | | | Small | Designed to machine steel and Cu–Al alloy without the limits of part shape | 15–40 s | 10–80 μm | 0.15–0.8 | 0.03–0.2 | Designed to machine thin-walled parts, for thick burrs, using this method firstly, then use barrel finishing | Threaded workpieces, stamping |
| Chlorine processing | | | | | | Small | | 20 s ~ 1 min | 0.01–0.25 mm | Around 0.2 | | Have oxidizing effect on metal containing iron; at the stage of research and development due of harmful emissions | |
| Electrochemical machining | | | | | | Large/small | Remove burrs of conductive material and inner surface | 12 s ~ 3 min | | | 0.13–0.25 | Remove burrs of cross-hole and inner surface; electrolytic deburring has been commercialized | Gear, shaft parts, gearbox, gearbox cover |
| Electrochemical vibratory grinding | | | | | | Small | | around 1 min | 2 μm ~ 0.1 mm | Around 2 | 0.05–0.10 | Mediated by spherical graphite for improving conductivity | |
| Processing by flame | | | | | | Middle | Designed to machine large size castings | 0.8–1.7 s | | | Around 0.13 | Melt burrs by the combustions of oxyacetylene, suitable for deburring of large castings | |
| Processing by thermal shock | | | | | | Small | Designed to machine parts of low thermal conductivity such as castings and Zn–Al alloy parts | 20 s | 3 μm | Around 0.8 | 0.5–1.5 | Burning hydroxide gas mixture, melt burrs momentarily at high temperature | Zn, Cu, brass parts, vaporization parts, valve |

(continued)

**Table 1.2** (continued)

| Finishing methods | Basic principle of finishing | | | | | Burr size | Machining state and range | Machining time | Superficial effect after machining | | | Characteristics of machining technology | Applicable parts |
|---|---|---|---|---|---|---|---|---|---|---|---|---|---|
| | Grinding | Cutting | Chemistry | Thermal energy | Electrolysis | | | | Size | *Ra*/μm | Chamfering /mm | | |
| Abrasive flow machining | | | | | | Small | Design fixture for workpieces shape, designed to machine inner surface of complex shape | 1–5 min | <0.5 mm | 0.05–0.4 | 0.04 | Suitable for fine machining of precision parts, deburring of cross-hole and complex surface | Engine parts, hydraulic valve parts |
| High pressure water jetting | | | | | | Small | Designed to machine micro-parts, deburr, especially suit for brittle materials | 3–10 min | | | 0.05–0.25 | Nozzle diameter: 0.1–0.5 mm; water pressure: 1.5–7.5 MPa | Gearbox, gearbox cover, shaft gear |

degree of machining efficiency, a modern production pace, and a high efficiency of production, the finishing technology looks incomplete and unsuccessful.

In the fundamental sense of the finishing technology, the goal is to liberate the productive forces. That is to say that hand labor (old-fashioned, out-dated, and heavy) should be replaced by modern advanced technique to change the production environment of the worker thoroughly and free the worker from heavy hand labor. Therefore, taking productivity level as a criterion to judge finishing technology performance is inevitable.

In the finishing process of parts, increasing productivity means that the technology should remove more metal of the surface and more flash and burr, help the surface to look smooth and refining, and chamfer within limited machining time. Only then, all of the requirements of the surface quality have been satisfied.

Usually, machining efficiency is expressed by the material removal rate of the surface layer, and the unit is "μm/h," which means the removal thickness of metallic layer per hour. It can also be expressed by the corner radius of the edge fillet after removing edge burr, and the unit is "μm/h," which means the corner radius of the fillet with the edge burr removed per hour.

In brief, the quantification of the productivity of every free abrasive tool finishing method from different angles can conduct a better comprehensive analysis and comparison to choose better finishing methods.

3. Economical evaluation of machining

Stable and reliable machining quality, high productivity to adapt to modern production rhythm, high economic benefits with less investment and quick results are three aspects of the basic requirements for finishing technology. None of manufacturer's deciders can make a decision only considering the machining quality and productivity, not considering the production economy requirements when selecting and determining one finishing method or finishing process.

The responsible deciders can give a conclusion after the detailed calculation of everything from engineering implementation, workshop layout, equipment installation to trial production; from loading, unloading, machining, inspection of parts to finished product; from the configuration of abrasive tools, compounds, and assistive devices to power utilization, water utilization, and labor-hour utilization, and so on. Moreover, they should make a comparison between what happened before and after the engineering improvement or a comparison between various engineering schemes. Even if a finishing engineering has been defined to implement, it still often needs to be realized recurrent cost accounting. The result of cost accounting and the market feedback is combined and then endlessly offered for the decision-making department to make deciders easier to discuss about whatever situation is ongoing, in order to improve and perfect the identified production engineering again.

So we can say that the economic analyses of finishing engineering need pass through the beginning and end of the production engineering operations.

Finishing method includes a wide range of methods, and the corresponding calculation methods to the cost accounting are also abundant. Companies and manufacturers usually keep their department engineering account private, so it is difficult to collect complete information. A famous American professor and an expert in finishing technology named Gillespic has collected some cost accounting methods of the finishing ways from some US manufacturers, which has certain reference value. Table 1.3 gives the cost accounting methods of one piece for various finishing methods, which can learn reference for relevant professionals.

## 1.4 Evaluation of Surface Quality and Selection of Finishing Methods

### *1.4.1 Surface Quality Evaluation*

With the increasing advance of requirements of part's surface quantity by the machinery, people have got to develop study on how to evaluate and control the surface quantity. M.·Filed and J.·Kahles of American Association for the Study of Metal Cutting first proposed the concept of surface integrity in 1964; K.·Tagazawa put forward the concept of surface integrity and conducted an appropriate research in 1989 and has made a certain progress.

Surface integrity is inherent or enhanced condition of a surface produced by machining processes or other surface generation operations. That is, it is a technical indicator, and it describes, identifies, and controls how the machining processes of parts affect machined surface layer. Surface integrity of cutting parts requires that surface layer is kept intact, also requires that physical and mechanical properties of surface layer and microstructure can meet the service demand, as well as ensures that the parts have potential usable life after machining. The content included in the surface integrity is more comprehensive and specific than that is included in the conventional surface quality. Surface integrity has been used in production by industrially developed country to evaluate and control surface condition and performance of crucial parts during manufacturing process and fully analyze the effect of surface quality on product's properties [8, 9].

1. Evaluation indexes

Surface integrity can evaluate and control surface quality in many directions including geometry texture of the machined surface and physical, chemical, mechanical properties of disturbed surface material section. It can be classified into five categories:

(1) Surface topography: including surface roughness, surface waviness, and surface texture direction.
(2) Surface defect: including machining burr, flash, macro-crack, surface tearing, buckling, and other defects.

**Table 1.3** Cost accounting methods of one piece for various finishing methods

| Finishing methods | Calculation formulas of cost per unit | Comment |
|---|---|---|
| Barrel finishing | $C = [C_D + C_M + C_L(1 + D_O) + WC_P + C_A + C_E + C_C + C_W]\frac{1}{N}$ | $C$ is processing cost of single part; $N$ is the number of machining parts per hour; $C_D$ is equipment depreciation cost per hour; $n$ is the number of machining parts per production cycle; $C_M$ is minimum cost per hour of using equipment; $t$ is the time per production cycle ($h$); $C_L$ is wage per hour of operator of equipment; $C_G$ is gas cost per production cycle; $C_A$ is cleaning charge per hour; $N_P$ is the total number of parts; $C_P$ is electricity charge of starting equipment[yuan/(kW·h)]; $C_B$ is the cost of brushes; $C_E$ is the cost per hour of grinding block; $N_{P1}$ is the number of machining parts in finite life of a tool; $C_C$ is the cost per hour of chemical agent; $C_T$ is the cost of all kinds of tools; $C_S$ is the cost of all kinds of grinding fluid; $D_O$ is percentage of indirect cost of rents; $W$ is power consumption (kW) |
| Power brush finishing | $C = [C_D + C_M + C_L(1 + D_O) + WC_P + C_A]\frac{1}{N} + \frac{C_B}{N_{P1}}$ | |
| Manual finishing | $C = [C_L(1 + D_O) + C_A]\frac{1}{N} + \frac{C_T}{N_P}$ | |
| Mechanical tool finishing | $C = [C_D + C_M + C_L(1 + D_O) + WC_P + C_A]\frac{1}{N} + \frac{C_S}{N_P}$ | |
| Chemical finishing | $C = [C_D + C_M + C_L(1 + D_O) + WC_P + C_A]\frac{1}{N} + \frac{C_S}{N_P}$ | |
| Electrolytic finishing | $C = [C_D + C_M + C_L(1 + D_O) + WC_P + C_A]\frac{1}{N} + \frac{C_T}{N_P} + \frac{C_S}{N_{P1}}$ | |
| Electrolytic grinding | $C = [C_D + C_M + C_L(1 + D_O) + WC_P + C_A]\frac{1}{N} + \frac{C_T}{N_P} + \frac{C_S}{N_{P1}}$ | |
| Thermal explosion finishing | $C = [C_D + C_M + C_L(1 - D_O) + WC_P + C_A]\frac{1}{N} + \frac{C_g}{n} + \frac{C_T}{N_P}$ | |
| Flame combustion | $C = [C_D + C_M + C_L(1 + D_O) + WC_P + C_A]\frac{1}{N} + \frac{C_g}{n}$ | |

(3) Microstructure, surface metallurgy, and surface chemistry: including metallographic structure, micro-crack, and surface chemical properties.
(4) Surface mechanical properties: including work-hardening, surface depth, and magnitude direction and distribution of residual stress.
(5) Other surface engineering traits: including electronic performance changes (electric conductivity, magnetic variation, and resistance variation) and optical property variation (refection, such as radiance).

For a specific part, partial evaluation content should be selected as the individual evaluation indicators based on specific requirement. Three different groups of data prescribed in Table 1.4 can be used for evaluating in general.

In the three different groups of data prescribed in Table 1.4, the latter group of data is the development of the previous data group. Basic data group is the low limit of data group; reference data group is used for more critical part; generalized data group expands the content of testing including mechanical property testing and other particular technical requirements for the special demands of surface quality at design time based on basic data group.

This has been valued and used in the industry such as the aerospace industry, automobile industry, atomic power generation, and machine tool industry, presenting that surface quality of crucial parts is evaluated and controlled by applying the surface integrity. The materials used in the crucial parts and the working method about final process can be analyzed through the studies of surface integrity. Doing different experiments in a range of limited technical parameters and studying the strength properties such as microstructures, residual stress, and fatigue character of the surface layer after processing help us determine their sensitivity to the process conditions and technical parameters. For example, the data of the machined surface integrity turbine disks made of Ni-based alloy suggests that this material is not sensitive to the effect of fatigue strength when surface roughness $Ra = 0.38$–$6.1\ \mu m$. According to the results of this experiment, this surface roughness can be relaxed and be chosen as $Ra = 3.2\ \mu m$ instead of the original design $Ra = 1.6\ \mu m$. Parts' processing hour brings down the time approximately 10% because of the increase of surface roughness value, and the production costs are reduced without affecting its performance.

In order to control processing method and technological parameter using surface integrity, a series of experiments commonly need to be conducted and a large amount of data need to be tested. This will lead to an increase in costs and a decrease in productivity. In order to guarantee the surface quality of key parts and lower overall production costs, it can only be used to selectively apply policies to crucial part or key areas of the part and loosen the requirement of non-key parts. Surface quality evaluation is conducted with some parameters of basic data groups on parts of common product based on the technical requirement of parts.

2. Surface geometrical characteristics evaluation

Surface geometrical characteristic of the part is the general term that describes surface roughness, surface waviness, surface burr, and other defects. It not only has

**Table 1.4** Evaluating data groups

| Data group | Content | Remarks |
|---|---|---|
| Basic data group | Surface roughness and surface texture | |
| | Macrostructure | Observable machining burr, flash, macro-crack, and macro-corrosion found at 10 times or lower times magnification |
| | Structure | Micro-crack, plastic deformation, phase change, intercrystalline corrosion, pocking mark, tearing, buckling, built-up edge, melting and redeposition layer, selective corrosion |
| | Micro-hardness | |
| Reference data group | Basic data group | |
| | Fatigue strength test | |
| | Stress corrosion test | |
| | Residual stress and distortion | |
| Generalized data group | Basic data group | |
| | Extensive fatigue test | Such as low-cycle fatigue test, used to obtain the information to satisfy the needs of design |
| | Appended mechanical property test | Tests such as stretching, stress cracking, creeping |
| | Other special performance | Such as friction characteristic, corrosion characteristic, Optical characteristic, electrical characteristic |

a great impact on use property of the part but also relates to the machining cost of the part. It is sensitive to the tool used in machining, the machining conditions, material properties, and other parameters, so it is an important gist for control processing. Thus, it has received extensive attention in engineering area to study and evaluate surface geometrical characteristics. This book only explains the relevant contents of the surface roughness and surface defect. Surface waviness relates to the problem of vibration in machining, but this has been discussed by many works before and need not to be repeated here.

(1) Surface roughness evaluation

The surface geometry obtained by every method used in machining can never be a complete perfect geometry, and it is the surface with profile peaks and profile valleys. This geometric feature of the surface can be described by 3D surface topography shown in Fig. 1.10. As shown in Fig. 1.11, it is a 2D surface topography obtained from one normal section of the part, and it only reflects the relations between the height and the horizontal distance, which expresses 2D geometric features of the part. It can be regarded that this surface profile is composed of three-order wave with different wave front and wavelength if analyzed by the content of harmonic wave. The shape A shown in the figure is the first order of the harmonic wave, it reflects the objective unevenness, and it belongs to the studies of

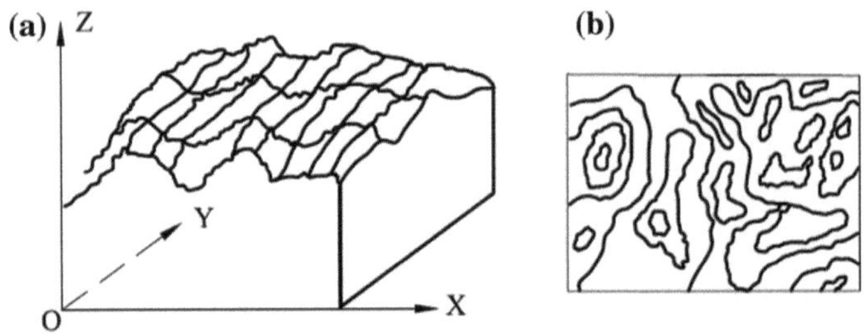


**Fig. 1.10** 3D surface topography. **a** Stereogram represents; **b** geomorphological map represents

**Fig. 1.11** Sketch of 2D surface topography

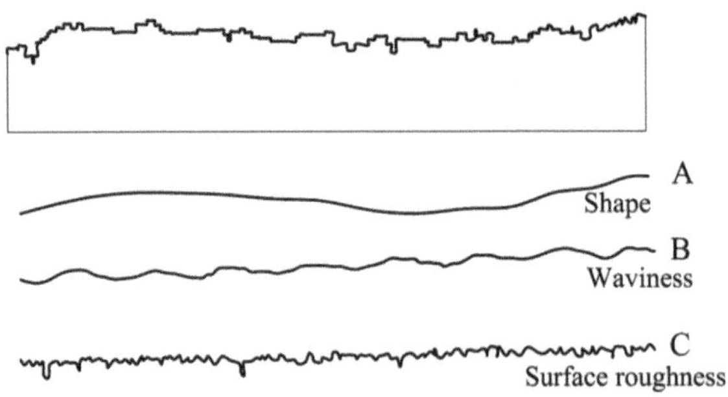


machining precision, which is caused by the flexibility of machine tool and workpiece or the error of the guide rail of a machine way. The shape B shown in the figure is the second order of the harmonic wave, and it reflects surface waviness, and the vibration in the machining, the inhomogeneous materials, and the transmission error that cause it. The period of surface waviness is not only obvious but also fairly large in terms of the pitch wave. The shape C shown in the figure is the third order of the harmonic wave, it reflects the surface roughness, it is the microscopic shape characteristic consisting of valleys with a smaller periodic distance on the machined surface, and it relates to the machining method such as the surface roughness obtained from machining, revolution mark, and plastic deformation of the tool, tear performance of the materials.

The classification method of flatness, surface waviness, and surface roughness in general is shown in Table 1.5.

**Table 1.5** Classification of flatness, surface waviness, and surface roughness

| Surface shape error classification | Ratio of wave height $h$ to wave distance $l$ | Gauge probe head radius |
|---|---|---|
| Surface roughness | 1/1–1/50 | 2–10 |
| Surface waviness | 1/50–1/1000 | 60–200 |
| Planeness | <1/1000 | >1000 |

The definition of surface and its parameters related to surface roughness are dictated in national standards GB/T 3505-2009 (Geometrical Product Specifications—Surface texture: Profile method—Terms, definitions, and surface texture parameters) [10]. The general rule of evaluating the parameters and its magnitude of surface roughness is dictated in national standards GB/T 1031-2009 (Geometrical Product Specifications—Surface texture: Profile method—Surface roughness parameters and their values) [11]. These standards apply to evaluate the surface roughness of manufactured goods.

The surface roughness is evaluated by mean line system based on two-dimension contour in standard at the present. This is an accounting system based on the mean line, and it is considered as datum line for the evaluation of contour. The parameters for the evaluation of the surface roughness are as follows:

(1) The parameters of surface roughness related to the range characteristics of micro-irregularities. The parameter of height of surface roughness is selected from *Ra* and *Rz*, and this is dictated in national standards GB/T 1031-2009.

The arithmetical mean deviation of the profile *Ra* is the mathematic average for the absolute value of vertical coordinates *Z*(*x*) within the sampling length *lr*. As shown in Fig. 1.12, the formula is given by

$$Ra = \frac{1}{lr}\int_0^{lr} |Z(x)|\,\mathrm{d}x \tag{1.1}$$

It is approximated as

$$Ra = \frac{1}{n}\sum_{i=1}^{n} |Z_i| \tag{1.2}$$

*Ra* is the first suggested parameter to evaluate the surface roughness; it is also a parameter used by almost all international countries. The values of *Ra* ruled by international standards are presented in Table 1.6. The value of *Ra* should be chosen firstly when *Ra* = 0.025 6.3 μm and *Rz* = 0.1–25 μm. The main concern is using simple, high measuring precision and simple stylus instrument for measurement and using standard comparison specimen for comparative measurement.

The maximum height of the profile *Rz* is the sum of maximum height of peak crest in the outline *Rp* (the maximum *Rp* within a sampling length) and maximum depth of profile valley *Rv* (the maximum *Rv* within a sampling length) within

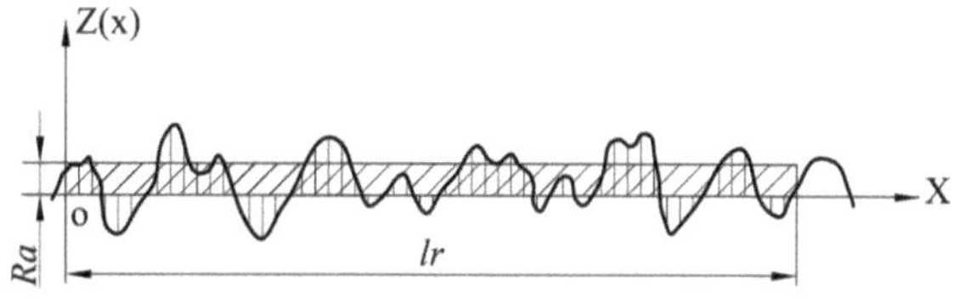


**Fig. 1.12** Arithmetical mean deviation of profile *Ra*

**Table 1.6** Profile arithmetical mean deviation value of *Ra*

| | | | | |
|---|---|---|---|---|
| *Ra* | 0.012 | 0.2 | 3.2 | 50 |
| | 0.025 | 0.4 | 6.3 | 100 |
| | 0.05 | 0.8 | 12.5 | |
| | 0.1 | 1.6 | 25 | |

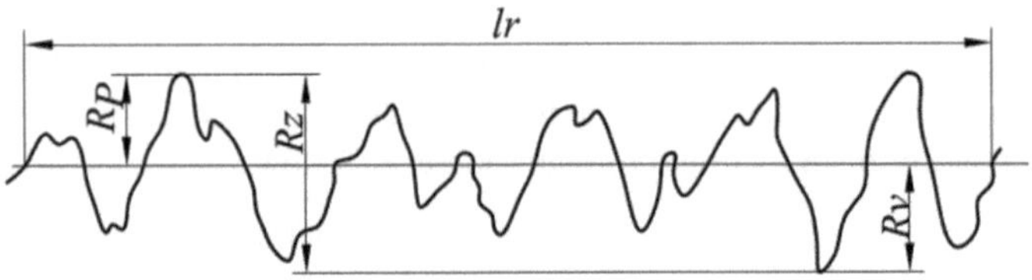


**Fig. 1.13** Maximum profile height

sampling length *lr*, as shown in Fig. 1.13. *Rp* is the maximum height of peak crest in the outline, it represents the distance between the top line of profile peaks and the middle line of profile peaks within sampling length, and it relates to the wear resistant and contact stiffness of the surface. *Rv* is the maximum depth of profile valley, it represents the distance between the bottom line of profile valley and the middle line of profile valley within sampling length, and it relates to the fatigue strength and corrosion resistance of workpiece. So *Rz* should be selected except for selecting *Ra* for the surface with stress concentration and easily leading to fatigue failure, to ensure that the maximum height of profile is not exceed acceptable limit.

*Rz* is used to evaluate small surface, curved surface, or the surface of workpiece with the requirement of fatigue strength. The values of *Rz* are shown in Table 1.7 based on the national standards.

Based on some related datum, the relationship between *Ra* and *Rz* is:

$$Rz = (4 - 15)Ra \tag{1.3}$$

(2) Surface roughness parameters relate to the micro-space of irregularities.

With the development of industry, it is discovered that the transverse spacing is an effective role on some occasion besides the altitude characteristic of the micro-surface irregularities, which has an effect on workpiece performance. When painting performance, vibration resistance, corrosion, reducing the friction drag of

**Table 1.7** Maximum profile height value of *Rz*

| | | | | | |
|---|---|---|---|---|---|
| *Rz* | 0.025 | 0.4 | 6.3 | 100 | 1600 |
| | 0.05 | 0.8 | 12.5 | 200 | |
| | 0.1 | 1.6 | 25 | 400 | |
| | 0.2 | 3.2 | 50 | 800 | |

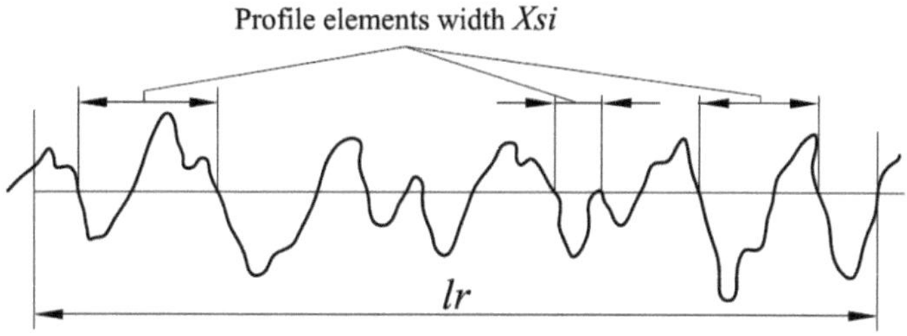


**Fig. 1.14** Profile elements width

the fluid flow, and preventing the formation of cracks in shock forming need to be reflected, the spacing parameters should be measured to evaluate the fineness of the surface texture, except evaluating the micro-surface irregularities.

The mean width of the profile elements *Rsm* is an average value of element width with a sampling length *lr*, as shown in Fig. 1.14; its expression is

$$Rsm = \frac{1}{m}\sum_{i=1}^{m} Xs_i \qquad (1.4)$$

This parameter is needed to determine the width and space of the profile elements. If there are no special provisions, the default height resolution should be chosen as 10% of *Rz*, and the default horizontal spacing resolution should be chosen as 1% of the sampling length. The above two conditions should be satisfied in the meantime; *Rsm* stated in national standards is shown in Table 1.8.

Through test analysis of specimens with different *Ra* processed by some general machining method, it can be concluded that the smaller the *Ra* is, the smaller the *Rsm* is in most cases. Moreover, through statistics of the test results for different processing method including grinding, turning, planning, end milling, and plain milling, the range of *Rsm* corresponding to *Ra* is shown in Table 1.9.

(3) Relevant parameters

It is unable to explain the entire feature of surface roughness with only parameters of height and transverse space. In some occasion, it also cannot reflect the effect on the part function surface. For example, the height and transverse space of profile irregularity of two surfaces are the same, but the shape of contour curves between peaks and valleys is different, and this will bring a different concern to the wearing quality of parts. That's why the characteristic parameter related to the shape of irregularities should be represented.

**Table 1.8** Profile elements width value of *Rsm*

| *Rsm* | 0.006 | 0.1 | 1.6 |
|---|---|---|---|
| | 0.0125 | 0.2 | 3.2 |
| | 0.025 | 0.4 | 6.3 |
| | 0.05 | 0.8 | 12.5 |

**Table 1.9** Correspondence between *Ra* and *Rsm*

| Machining methods | *Ra*/μm | | | | | | | |
|---|---|---|---|---|---|---|---|---|
| | 0.16 | 0.32 | 0.63 | 1.25 | 2.5 | 5 | 10 | 20 |
| | *Rsm*/ mm | | | | | | | |
| Grinding | 19 ~ 62 | 19 ~ 87 | 32 ~ 153 | 44 ~ 263 | 64 ~ 300 | – | – | – |
| Lathe turning | – | 27 ~ 51 | 20 ~ 78 | 32 ~ 307 | 70 ~ 307 | 125 ~ 666 | 153 ~ 600 | – |
| Planning | – | – | – | 88 ~ 226 | 105 ~ 330 | 114 ~ 413 | 236 ~ 666 | ~ 582 |
| Vertical milling | – | – | – | 54 ~ 363 | 129 ~ 800 | 171 ~ 1562 | 223 ~ 1041 | – |
| Plain milling | – | – | – | 126 ~ 780 | 302 ~ 1400 | 593 ~ 2200 | ~ 1250 | ~ 2300 |

Contour entity material length at height *c* of the horizontal section is the sum of individual intercept of a line parallel to the X-axis and profile element at a given height *c* of horizontal section. As shown in Fig. 1.15, the formula is expressed as

$$Ml(c) = Ml_1 + Ml_2 + \cdots + Ml_i + \cdots + Ml_n. \quad (1.5)$$

Profile bearing length rate *Rmr*(*c*) is the ratio of the contour entity material length *Ml*(*c*) to the evaluation length *ln* at a given height *c* of horizontal section; its formula is

$$Rmr(c) = \frac{Ml(c)}{ln} \quad (1.6)$$

*Rmr*(*c*) is given at different heights of horizontal section, so there has to be given the values of height *c* of the contour horizontal section when selecting the parameter of profile bearing length rate. The value of *c* is represented by micrometer or the percentage of *Rz*; the percentage series of *Rz* are 5, 10, 15, 20, 25, 30, 40, 50, 60, 70, 80, and 90%.

The values of *Rmr*(*c*) in national regulatory standards are shown in Table 1.10.

The values of *Rmr*(*c*) can directly reflect workpiece surface wearability; it is important for increasing the carrying capacity. But the value of *Rmr*(*c*) at one height *c* of horizontal section cannot completely reflect the performance of workpiece surface; it is more comprehensive for having the change of *Rmr*(*c*) at different heights *c* of horizontal section. Profile bearing length rate curve is the relationship between the profile bearing length rate and the height *c* of horizontal section, as shown in Fig. 1.16.

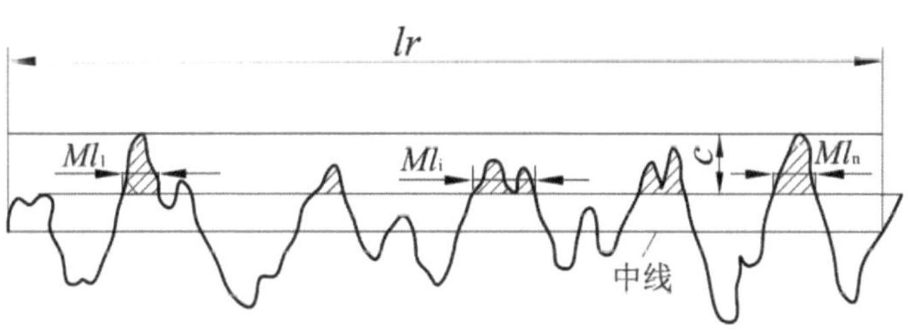


**Fig. 1.15** Contour entity material length

**Table 1.10** Values of profile bearing length rate *Rmr*(*c*)

| *Rmr*(*c*) | 5 | 10 | 15 | 20 | 25 | 30 | 40 | 50 | 60 | 70 | 80 | 90 |
|---|---|---|---|---|---|---|---|---|---|---|---|---|

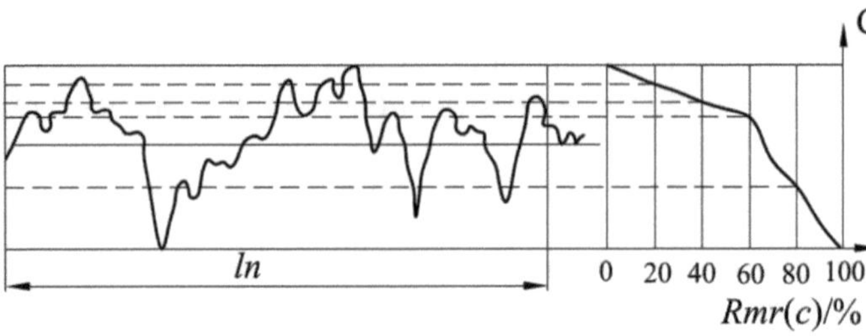


**Fig. 1.16** Profile bearing length rate curve

(4) Sampling length and evaluation length

The sampling length is used to evaluate the length of the irregularity of roughness profile along X-coordinate axis.

The machined surface profile obtained by any technique is composite profile containing in the surface roughness, the surface waviness, and the geometric form error. For different sampling lengths, different surface roughness *Ra* is obtained when evaluating surface roughness, as shown in Fig. 1.17. In order to make measurement results to fully reflect the properties of the surface roughness, the influence of waviness and the geometric form error on the measurement results must be restricted and weakened, so the appropriate sampling length should be selected from the series of sampling length in national standard shown in Table 1.11 when measuring.

The recommended sampling lengths corresponding to the parameter values of *Ra* and *Rz* are given in Tables 1.12 and 1.13.

Evaluation length *ln* is used to evaluate the length parallel to the X-axis of the roughness profile. When the sampling length is determined, the parameters of surface roughness measured in this sampling length may be close to or very different from that measured in near sampling length, due to the uneven surface irregularities after processing. Therefore, an appropriate minimum surface length should be defined as evaluation length, in order to ensure that the average parameters of surface roughness obtained in evaluation length should be representative and that the characteristics of surface roughness on the surface section should be reliably reflected. Evaluation length contains one or several sampling lengths, and it is recommended that evaluation length is five times the sampling length in general. The exact numerical values are shown in Tables 1.12 and 1.13.

**Fig. 1.17** Sampling length and surface roughness

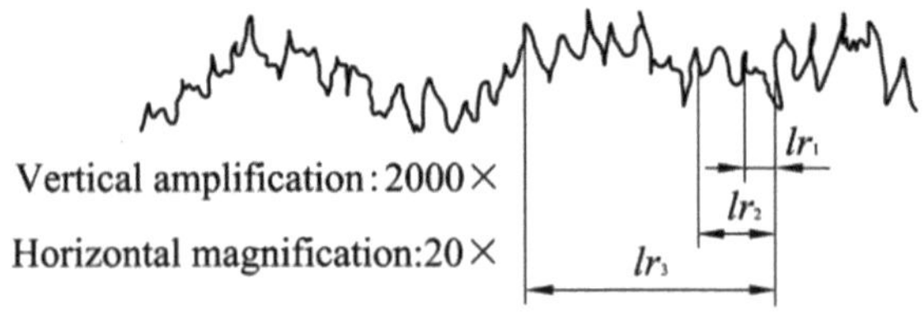

**Table 1.11** Value of sampling length *lr*

| *lr* | 0.08 | 0.25 | 0.8 | 2.5 | 8 | 25 |
|---|---|---|---|---|---|---|

**Table 1.12** Preferred values of sampling length *lr* and evaluation length *ln* corresponding to *Ra*

| *Ra*/μm | *lr*/mm | *ln*/mm (*ln* = 5 × *lr*) |
|---|---|---|
| ≥0.008–0.02 | 0.08 | 0.4 |
| >0.02–0.1 | 0.25 | 1.25 |
| >0.1–2.0 | 0.8 | 4.0 |
| >2.0–10.0 | 2.5 | 12.5 |
| >10.0–80.0 | 8.0 | 40.0 |

**Table 1.13** Preferred values of sampling length *lr* and evaluation length *ln* corresponding to *Rz*

| *Rz*/μm | *lr*/mm | *ln*/mm (*ln* = 5 × *lr*) |
|---|---|---|
| ≥0.025–0.10 | 0.08 | 0.4 |
| >0.10–0.50 | 0.25 | 1.25 |
| >0.50–10.0 | 0.8 | 4.0 |
| >10.0–50.0 | 2.5 | 12.5 |
| >50–320 | 8.0 | 40.0 |

(5) Surface texture feature

Surface texture means the major direction of parts surface microstructure, and it represents the category and direction of machine marks that were formed in machining. It is the important content for evaluating parts with the need for controlling the surface texture. Seven kinds of common texture direction symbols are defined in the standard of GB/T 131-2006 (Geometrical Product Specifications—Indication of Surface texture in technical product documentation) [12]. Symbols of surface texture and their understanding definitions are shown in Table 1.14.

(2) Surface defect evaluation

Related characteristics, types of surface defect, and other parameters are defined in the standard of GB/T 15757-2002 (Geometrical Product Specifications—Surface imperfections—Terms, definitions, and parameters) [13]. The surface defect is defined as an element, grouped elements, and irregularity of actual surface formed by unintentional or accident during machining, storing, or using process. These kinds of grouped elements and irregularity are obviously different from those forms of an uneven surface. It is noted that the actual surface with defects is not applicable during the evaluation of the defect, the acceptability of the defect depends on the purpose or functionality, and it is determined by appropriate item such as length, width, depth, height, number of defects per unit area.

The surface defect is classified into four kinds containing dent defects, punch defects, combined surface defects, and region and apparent defects in the national standard. These include the following:

**Table 1.14** Symbols and understanding definitions of surface texture

| Symbol | Explanations and example | |
|---|---|---|
| = | The texture is parallel to the projection plane of the view | 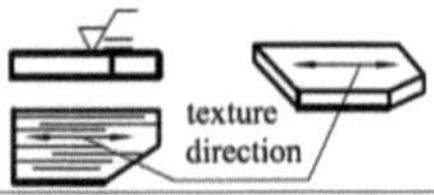 |
| ⊥ | The texture is perpendicular to the projection plane of the view | 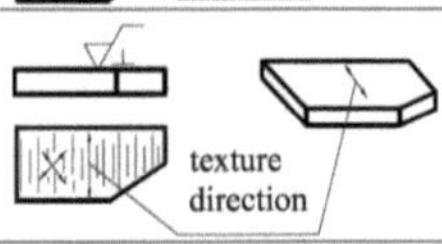 |
| × | The texture arranged into two-diagonal crossing intersects the projection plane of the view | 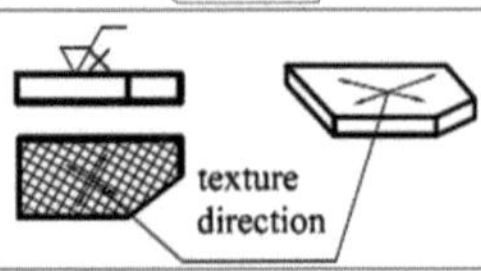 |
| M | The texture forms multi-direction | 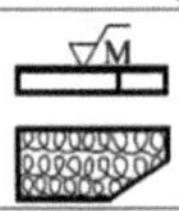 |
| C | The texture is approximately concentric circles, and the circle centers are related to the surface center |  |
| R | The texture is approximately radial pattern and related to the surface center | 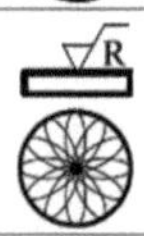 |
| P | The texture is microsome, bulge, and without direction | 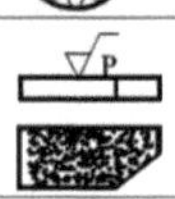 |

(1) Dent defect. It means inward defect, such as groove, scratch, rupture, pore, blister, shrinkage, crack, gap, fissure, vitium, concave buckling, and crypt, as shown in Table 1.15.

(2) Punch defect.
It is outward defects, such as burl, scar, convex buckling, oxide skin, inclusions, flash, joint fin, and attachment, as shown in Table 1.16.

(3) Combined surface defect.
It is the surface defect that is outward partly and inward partly, such as crater, folding, scratch, and cutting residue, as shown in Table 1.17.

(4) Region defect and apparent defect.
They scattered around the outermost surface, they have no sharp outline in general, and they have no actual depth and height measured, such as slide mark, abrasive wear, corrosion, mottling, crack, spot, stripe, fading, fringe, split, scale, as shown in Table 1.18.

**Table 1.15** Dent defect types and their definition

| Types | Definition | Sample pictures |
|---|---|---|
| Groove | Dent defect with a certain length, a crater, or flat bottom | 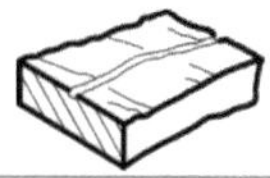 |
| Scratch | Dent defect with irregular shape and uncertain direction | 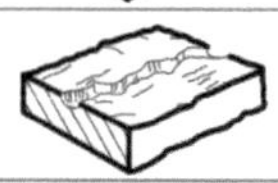 |
| Rupture | Sliver defect with sharp bottom caused by the integrity damage of the surface and the basal body |  |
| Pore | Small-sized cavity with very steep sloping wall and sharp edge, and the upper edge of the cavity is not higher than the tangent plane of the reference surface |  |
| Blister | Surface defect in the form of a single dent defect results from the influence of the foreign particle losing, corrosion, or gas |  |
| Shrinkage | Dent defect caused by differential shrinkage during the casting or weld metal solidification process |  |
| Crack, gap, fissure | Cusp-shaped sliver dent defect with shallow irregular openings |  |
| Vitium | Arc-shaped defect at the intersection of the two workpiece surfaces |  |
| Concave buckling | Dent defect caused by the local bending of the plate surface |  |
| Crypt | Concave without swell, dent defect caused by plastic deformation generated by impressing or beating | 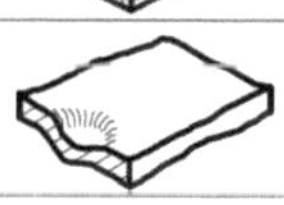 |

The characteristics and parameters of surface defect can be expressed by using defect length, width, depth, height, area (the area projected the individual defect onto the reference plane), total area (the sum total of individual surface defect area in the agreed judgement limit), the number of surface defect (the total number of surface defect on the actual surface in the agreed judgement limit), and the number of surface defect per unit area (the number of surface defect in the given assess area of A).

(3) Burr defect evaluation

The burr formed in part machining process is one of the most common surface defects, and it belongs to the combined surface defect. Although the burr is small, it does great damage. The existence of burr would have effects on the precision, assembly, performance, and working life of products; it may even cause the

**Table 1.16** Punch defect types and their definition

| Types | Definition | Sample pictures |
|---|---|---|
| Burl | Ridge or mound bulge with small size and finite length |  |
| Scar | Local bulge caused by the gases or liquids contained below the surface | 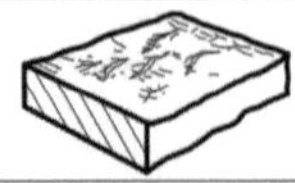 |
| Convex buckling | Arch defect caused by the local bending of the plate surface |  |
| Oxide skin | Partial detached thin flake-shaped bulge formed when the surface layer different from basis material composition peels off |  |
| Inclusions | Clutter embedded within workpiece material | 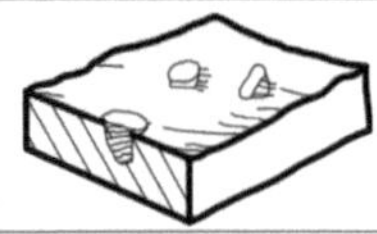 |
| Flash | Sharp bulge on the periphery of the surface, the defect often appears on the corresponding side |  |
| Joint fin | Ridge bulge of workpiece material formed in vertical direction of compression face, when the material is pushed out of the mold gap during die-casting forming or die-forging forming, or the two surfaces are welded by resistance (e.g., resistance butt welding, fusion welding) | 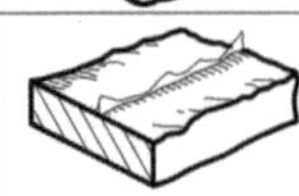 |
| Deposits | Clutter piled up on the workpiece or the material of another workpiece |  |

performance failure of the shipment, the short circuit of the electric elements, and the cause of major accident such as equipment damage. Therefore, more and more attention has been given to study the mechanism and characteristics of the burr formation, explore the method of curbing and reducing the burr formation, lay down the criteria for burr, and ways improve the technology of deburring.

The machining technology on parts is usually classified into three types: removal cutting, deformation process, and addition process in the machining process. The mechanism, shape, size, position, and characteristics of the burr formation (e.g., stiffness, hardness, and strength) are different depending on different machining methods. Table 1.19 lists the burrs formed by applying different machining methods. The burr is formed due to plastic deformation during cutting process of material; trimming can be formed because of the overflowing of materials in deformation such as cast and forge; the cuspidal edge is formed on the machining

**Table 1.17** Combined surface defect types and their definition

| Types | Definition | Sample pictures |
|---|---|---|
| Crater | Circular peripheral bulge, similar to the pit of the crater, and its circum is higher than reference surface |  |
| Folding | Tiny thickness of snakelike bulge, in the shape of wrinkle in general, formed when the material is pressed against the skin layer by folding during rolling or forging |  |
| Scratch | Continuous concave–convex defect formed by eliding or extruding the workpiece surface material, due to the moving of the foreign material | 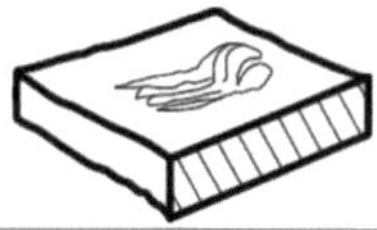 |
| Cutting residue | Banding bulge caused by bad chip removal |  |

parts in addition process such as welding and electroplating. It will not form burr by using the machining methods such as electrochemical or electrolytic machining, corrosion machining, chemical milling, or other. The size and location of the formed burrs are different since the degree of plastic deformation is different, which is dependent on the machining conditions (such as tool, workpiece material, cutting data), and the smallest burr is formed by using the wear machining.

The shape and size of the burrs formed by the use of general machining are shown in Fig. 1.18. The characteristics of the shape and size on the burrs are represented by height, thickness, and root radius. $h$ is the height, $b$ is the thickness, and $r$ is the root radius in Fig. 1.18. Additionally, there are problems such as location, stiffness, length, and errors about the burrs.

At present, the standards related to the characteristics of burrs and the marked method have not yet been built in China. Usually, the designers only propose the standard of deburring on the design of some important precision components and do not involve the standard of quantitative requirements of deburring. Some countries have begun to research in this field; for example, the S. A. E. (Society of American Engineer) proposed the concept of "edge quality" in the 1970s. Subsequently, Germany and Japan introduced the representation method of edge quality to give a helpful discussion on setting the standards related to the burrs, but the standards have not been used in engineering because of incompleteness.

In order to restrain and reduce the formation of the burrs, it is better to choose the appropriate part structure and materials with good hardening exponent and stretching rate in the design on the premise that meeting the requirements of parts performance. Many methods can be adopted to restrain and reduce the formation of

**Table 1.18** Region defect and apparent defect types and their definition

| Types | Definition | Sample pictures |
|---|---|---|
| Slide mark | Appears on the discontinuous area of the surface due to intermittent overload, for example the fog surface damage formed on the ball bearing and the bearing race | 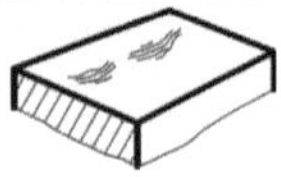 |
| Abrasive wear | Surface damage caused by the physical destruction or the wear |  |
| Corrosion | Surface damage caused by the chemical ablation | 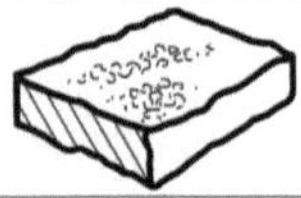 |
| Mottling | Deep concave spot and tiny hole-shaped defect distributed on a large area of the surface | 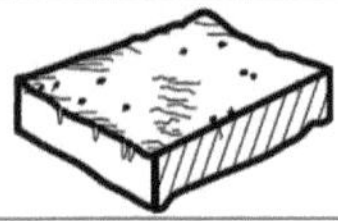 |
| Crack | Broken defect presented a net construction on the surface | 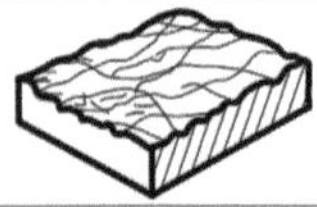 |
| Spot, stripe | Area with different appearances from the adjacent surface |  |
| Fading | Area with discoloration or color dodge | 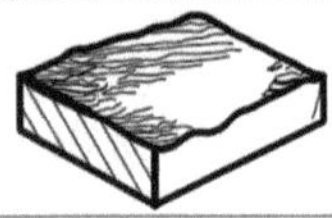 |
| Fringe | Shallow depth of recessed area presented banding pattern, or area with different surface structure | 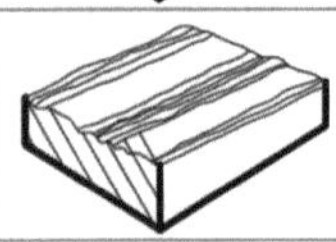 |
| Split, scale | Defect formed by the separation of the partial workpiece surface layer | 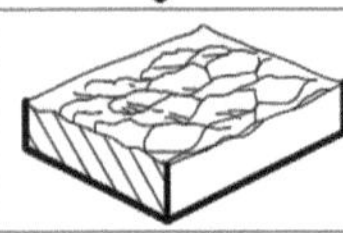 |

the burrs in making process, for example, using the machining method with burr-free, changing the process sequence of parts, changing machining parameters, using appropriate tool material and geometrical terms.

3. Physical and mechanical properties evaluation of the surface

The changes of mechanical, chemical, and thermal energy always occur in machining process of the parts, this will lead to the change of the properties of metallurgy, chemistry, physics, and mechanics on the surface layer of the parts, and that will result in the formation of work-hardening, residual stress, and

**Table 1.19** Burr formed by different machining methods

| Classification | | Machining methods | Instructions |
|---|---|---|---|
| Material removal machining | Cutting burrs | Cutting<br>Grinding | Burrs resulted from the cutting process are small, can be used to remove big burr resulted from other processes; burrs resulted from the cutting process are smaller |
| | Sheared burr | Punch & Shear | Generating abrasive tool clearance and wear |
| | Burr-free | Electrochemical machining, electrolytic grinding, chemical corrosion | Can be used to remove burrs |
| Deforming machining | Melt solidification burr | Electrical discharge machining, laser machining, gas cutting | Generating melting burr of all sizes |
| | | Casting, die-casting, plastics forming, sintering | Generated by raw material spillage |
| | Plastic deformation burr | Die forging, gear rolling | Raw material rolling spillage |
| Adding machining | Melt solidification burr | Electrowelding, soldering | Welding bead, bumps around the welding point |
| | Adhesive burr | Electroplating, metal spraying | Cuspidal edge, Small bumps |

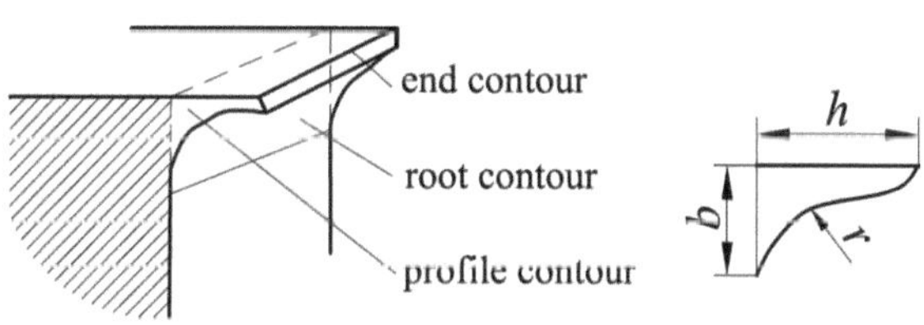


**Fig. 1.18** Shape and size of burr. **a** Burr appearance; **b** burr section size of burr. *h*—height; *b*—thickness; *r*—root radius

metallographic change on the surface layer, which can further affect the abrasion resistance, carrying capacity, fatigue strength, contact stiffness, and the fit quality of the parts.

(1) Work-hardening of surface layer

The lattice of the microstructure of metal material will have shearing slide, serious distort, traction, fibrillation, and fracture during the cold plastic deformation of the surface layer. In turn, it will lead to the reduction of surface plastic degree and significant improvement of the strength and stiffness, and this phenomenon is called work-hardening.

The electron movements are partly disrupted, which will lead to larger surface resistance and lower electrical conductivity because of larger lattice distortion and dislocation density in the process of work-hardening. Meanwhile, it is easy to start the chemical reaction with the adjacent medium because of the larger dislocation density and enhanced chemical of the crystal at high power, and this will reduce corrosion resistance and accelerate the corrosion rate.

In the metal cutting process, the hardness of the surface layer will generally increase 20–40% and the depth of the hardened layer will reach 20–100 μm. The level of work-hardening depends on the force that produces the plastic deformation, the deformation velocity, and the deformation temperature. The plastic deformation and the level of work-hardening increase with the increase in the force; the level of work-hardening reduces with the increase in the deformation velocity because of the insufficient plastic deformation; the deformation temperature $T$ not only affects the level of plastic deformation, it will also affect the recovery degree of metallographic structure after deformation. When $T$ is among the range of $(0.25–0.3)T_m$ ($T_m$ is absolute temperature at the time of melting), the phenomenon about the recovery of metallographic structure will appear and partly eliminate work-hardening.

The evaluation index of surface layer work-hardening includes the following kinds:

(1) The surface micro-hardness $HV$.
(2) The work-hardening capacity $N$.

$$N = \frac{HV - HV_0}{HV_0} \times 100\% \tag{1.7}$$

where $HV$ is the micro-hardness of the material before processing; $HV_0$ is the micro-hardness of the surface layer after processing.

(3) The depth of work-hardening layer $h$.
(4) The gradient of work-hardening $\mu$.

$$\mu = \frac{HV - HV_0}{h} \tag{1.8}$$

(2) Change of surface layer microstructure.

The microstructure will change in the machining process due to the surface layer of parts being affected by cutting heat. The affecting factors include workpiece materials, temperatures, temperature gradients, and cooling velocities. For example, the average temperature of ground surface can reach 300–400 °C, while the transient temperature can reach 800–1000 °C.

If the workpiece material is the quenched steel, when the temperature of grinding zone over martensite starting temperature (that of medium carbon steel is

250–300 °C) and below its critical temperature of the phase transformation (that of the carbon steel is 720 °C), the primary martensitic structure on the workpiece surface will convert into the softer tempered structure including sorbite and troostite, and this is called grinding temper burn. When the temperature of grinding zone was over the critical temperature of the phase transformation, a thin and brittle surface (the outer layer of the surface) will form under sharp quenching by cooling liquid, this outer layer of the surface is the secondary quenching martensitic structure with the higher hardness than the primary tempered martensite, its lower layer will convert into the softer tempered structure because the temperature is cooler and the cooling rate is also relatively low, and this is called grinding quenched burn. If the grinding process happened without the cooling liquid and the temperature of grinding zone was over the critical temperature of the phase transformation, the metal on the surface layer will form annealed structure because the cooling rate is decreased, the hardness and strength will have a sharp fall, and this is called grinding annealing burn.

This condition, the grinding burn, will influence the surface properties of ground parts. For example, it may bring soft spots to the part surface or bring uneven surface hardness, so to reduce its abrasion resistance. Therefore, the primary metallographic structure should be maintained by taking measures during the machining process.

(3) Residual stress

Residual stress refers to the residual force remaining inside the object when the external cause of stress is eliminated. The shape and structure of the metal on the surface layer will change in the machining process because of the effects of cutting force and cutting heat, and then, the balanced elastic stresses produce on the surface layer and the interface between the surface and matrix.

The main causes of the residual stresses in the machining process include the following aspects:

(1) The effects of cold-state plastic deformation.

The volume of machined surface layer will change because of severe plastic deformation under the actions of cutting forces, while the inner metal is in elastic deformation; the inner metal will have trends to recover when the cutting force disappears, but it cannot recover to initial status because it is restricted by the plastic deformation surface layer, and thus, the residual stresses appear. In general case, when the effects of extrusion and friction on the surface layer are relatively large (e.g., peening process), the stretching plastic deformation will produce on the surface layer, and the surface area will increase; so the residual compressive stress will produce on the surface layer with the action of inner elastic recovery, and the residual tensile stress will produce on the inner layer.

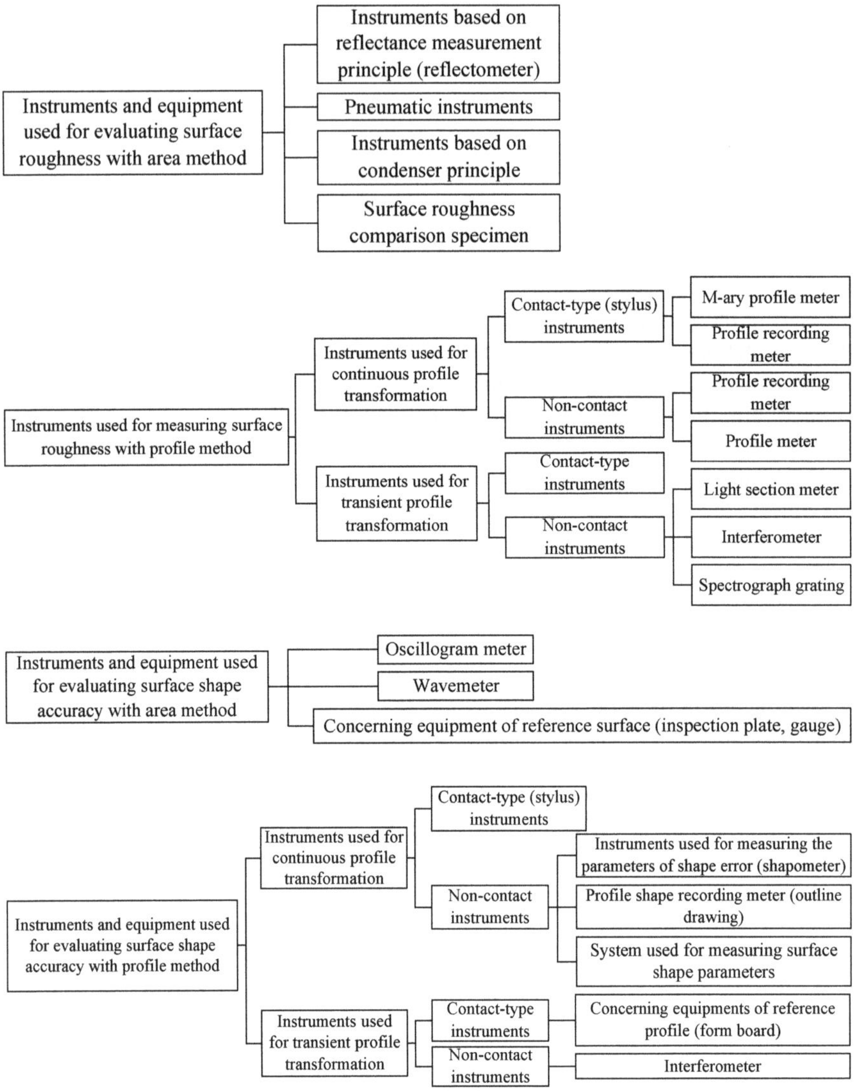


**Fig. 1.20** Concrete classification of instruments and equipment

immediate, not a gradual and consecutive process, such as light section microscope. When measuring, the instrument and measured surface have no direct contact, but the measuring process is completed by non-touching measurement based on light pattern. After the instrument is set up and the power source is turned on, the profile presents it immediately on the measuring graticule template, and the surface roughness is measured immediately with the immediate profile transformation.

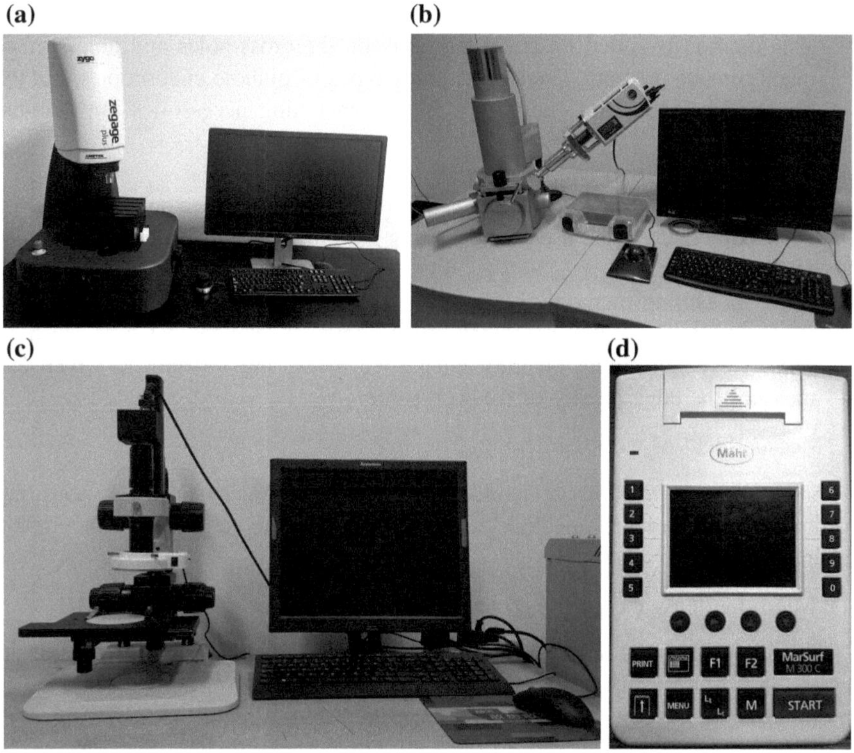


**Fig. 1.21** Measurement instruments of surface geometric parameters. **a** ZeGage 8000Plus surface profilers; **b** TSC-CSM100X scanning electron microscope; **c** MIV-IN 3D microscope; **d** perthometer M3 surface texture measuring instruments

Figure 1.21 shows some measurement instruments of surface geometric parameters.

2. Measurement of burr defects

Measurement methods of the burr which are widely used in China are as follows:

(1) Observation.
This method is suited to detect big burr on the outside of the parts. When the size of the burr is greater than 0.1 mm, the measurement can be completed by human eyes and hand touching and can also be completed by direct measurement with all kinds of measuring implement, such as vernier caliper.
(2) Microscope method.
This method is suited to detect small burr on the outside of the parts. When the size of the burr is less than 0.1 mm, the used equipment has all kinds of metallography microscope, stereoscopic microscope, tool microscope, optical projector, etc.

(3) Fiber-optical pinhole endoscopy.
This method is suited to detect the burrs on the cross-holes and hole surface quality inside the part. There are a many types of pinhole endoscopies, but the fiber-optical industrial endoscopy and its matching accessories are adopted currently.
(4) Electronic industrial endoscopy.
This method has the same application field as the fiber-optical industrial endoscopy and can be used for detecting the burrs throughout the inside holes of the part.
(5) Other measuring methods.
Some special burr detecting instruments have developed in China, such as servo-valve machining quantity integrative measuring instrument, which can measure the edge quality of the workpiece.

3. Measurement of work-hardening

The measuring methods of work-hardening include micro-hardness measuring method, X-ray organic method, recrystallization method, and metallographic method.

(1) Micro-hardness measuring method

The micro-hardness of the surface layer can be measured by the micro-hardness tester, and it can reflect the hardening degree of the surface finishing. The micro-hardness variety in depth can be measured by means of the following three ways.

(1) Peel measurement
The micro-hardness of the machined surface is measured firstly by the micro-hardness tester. Then, a thin layer of metal is removed from the surface by means of mechanical methods, electropolishing, or etching methods. The thickness of the removed metal layer is measured using the dial micrometer, and then, the micro-hardness of newly emerging surface layer can be measured using the micro-hardness tester. The measurement is carried out over and over again by removing metal layers, until the measured micro-hardness equals the original material hardness. The micro-hardness varieties in depth direction can be obtained. This way has a large amount of labor and cannot measure very thin work-hardening layer.
(2) Cross-sectional measurement
The transverse section perpendicular to the machined surface needs to be made into metallographic specimens during measurement, and then, the micro-hardness is measured every 50–100 μm from the surface layer to the inner layer on the metallographic specimens until the metal matrix appears. This way is simple, but only suitable for measuring thicker work-hardening layer.

(3) Oblique-sectional measurement
The tilted section, at 1°–3° to the machined surface, needs to be made into metallographic specimens during measurement, and then, the micro-hardness is measured every 50–100 μm from the surface layer to the inner layer on the metallographic specimens until the metal matrix appears. Finally, the depth of the hardening layer can be calculated in accordance with the measured length of the hardening layer and the tilt angle. Since the tilt angle is small, the measurable length is larger when the tilting plane passes through the metal layer, and this way can significantly enlarge the measuring length (about 30 times ~60 times). This way can measure thinner work-hardening layer.

(2) X-ray method

The fundamental principle is that the polycrystalline metal surface is shined by X-ray beam, and the interference rings can be obtained on the photographic plate due to the reflector from crystalline atomic surface. The interference rings reflect the structural change during the plastic deformation such as the change of the crystal lattices, the breaking of the crystal grains. The distance between atomic surfaces can be worked out in accordance with the diameter size of the interference ring and the wavelength of X-ray on the X-ray image, which reflects the plastic deformation.

To measure the depth of work-hardening layer, a metal layer with 10 μm is removed by means of electropolishing and is shined by X-ray until the reflected image is the same as the reflected image prior to deformation. At the moment, all deformed layers have been removed, the thickness of every removed layer is counted up, and the depth of the work-hardening layer can be obtained. This way has a long work time and a large amount of labor.

(3) Recrystallization method

The principle is that the new metal structure is generated during plastic deformation, and the size of the recrystallized crystal grain is different from that of the initial crystal grain. The size of the recrystallized crystal grain is either magnified many times, or minified many times. The hardening layer is judged by the size of the recrystallized crystal grain.

The machined specimen is heated up to annealing temperature (above the crystallization temperature), and the side surface is made into metallographic specimens which is corroded by the reagent during measurement. The depth of the hardening layer can be obtained according to the crystal grain size within the extent of certain depth, which is observed on a microscope.

(4) Metallographic method

The metallographic method is that the side surface of the machined specimen is made into metallographic specimens, and corroding the surface of the metallographic specimens is used to reveal microstructure. The depth and the hardening degree of the surface hardening layer are evaluated in accordance with the disintegrating degree and the shape distortion degree of the crystal grain. This way is simple and suitable for the qualitative analysis of surface layer status.

4. Measurement of residual stress

There are many ways to measure residual stresses, and these ways can be divided into two kinds: mechanical measurement and physical measurement.

(1) Mechanical measurement
The principle of this method is that the local separation/division (such as removing layer with corrosive method) is carried out on the specimens with residual stresses by using some technical methods, and the residual stresses are locally released. The corresponding deformation before and after the residual stresses released is measured, and the values representing residual stress can be calculated by using relevant elastic formulas. The residual stresses distribution of different depths can be obtained by successive separation and measurement.
(2) Physical measurement
The principle of this method is that the specimens are measured by adopting the methods of X-ray, magnetism, ultrasonic, and optics, and the X-ray method has a wider application. The X-ray method is a kind of nondestructive testing method, and it can measure the surface stress within 10–35 μm. In addition, it can measure the surface stress in a small area and the surface stress of every phase in the multi-phase alloys.

5. Measurement of other performances

There are many methods and instruments to measure the surface brightness. For example, it can be measured by UV-240 ultraviolet–visible spectrophotometer, expressed by reflectivity. The wavelength range is 600–800 nm taking the mirror surface as the reference, and the reflectivity of the mirror surface is 100%. The measuring results can be directly obtained by the DFW broadband reflectivity measuring instrument, which can measure multiple points, and the average is taken as the final result. Its measuring accuracy can reach 0.05%.

The corrosion resistance is measured by corrosion test of mass method and saline spray method, and it includes the following:

(1) Mass method.
The specimen is cleaned up, weighed, and soaked by different corrosive mediums. After some time, the corroded specimen is fetched out, cleaned, and weighed. The mass change of the specimen before and after the soaking is the corrosion quantity. When the corrosion products on the metal surface are easy to remove and the metal matrix is not damaged, the weight-loss method is the most common method; when the corrosion products attach tightly to the metal surface, the weight-increment method is adopted.

$$V^{\pm} = \frac{M_0 - M_1}{S_0 t} \tag{1.9}$$

where $V^{\pm}$ are, respectively, the corrosion rate of the weight increment and the weight loss after soaking; $M_0$ is the mass before soaking; $M_1$ is the mass after soaking; $S_0$ is the area; $t$ is the soaking time.

(2) Saline spray method.
Various types of saline spray corrosion test chambers can be used, such as YQ-250 salt spray test chamber, which cycles every 24 h and sprays 8 h continuously with a 16-h interval, and the test temperature is 25–30 °C. The pressurized spray is neutral 3–5% NaCl solution. The specimen is hung in the salt spray chamber and observed every several spray cycles. The corrosive states of the surface, which include the change of the brightness, the generation condition, and degree of rust spot, are recorded. The salt-fog resistance property is evaluated by comparison. In general, the specimen with bright surface and without any rust spot after 600-h salt spray test is deemed to have good corrosion resistance.

### 1.4.3 The Choice of Finishing Methods

Finishing technology is a new discipline; it involves metal materials, cutting, physical, chemical, electromagnetic, and other basic theories. In terms of some sort of finishing method, it may be feasible in production, but its theoretical basis may not be perfect. When choosing a correct finishing method, it is sufficient to consider the actual production status and to analyze and compare the quality of the machined surface, productivity, and economy based on trial machining process of several kinds of finishing techniques. From the result, it can be determined which is the more appropriate.

1. Choosing criteria of finishing methods

Finishing is the final process of pre-treatment for plating after a series of complex processes. When determining a finishing method, the basic situation of the objects (manufactured parts) must be analyzed.

The influencing factors of finishing method selection include the appearance of the parts, size, material, surface conditions, edge burrs, surface hardness, production batch. The first step is to do the detailed analysis of the objects for determining a finishing method.

(1) Comprehensive analysis of influence factors

Finishing methods will be affected and restricted by various factors, especially the appearance of the parts, size, finished positions, parts material, the conditions of the former procedure, the surface quality requirements after finishing, and the finishing engineering requirements (such as productivity, economy, environmental pollution). Even if a particular finishing method is adopted, it may not be static. The problem is found by trial processing, and certain modifications will be done in the future to make sure the best solution is put forward.

The following problems are carefully considered for choosing one solution:

① What is the purpose of the finishing method? Is it for improving the surface quantity? Is it for increasing productivity? Is it for reducing the production cost? Is it for improving the working conditions or environmental pollution?

② The function is analyzed carefully for the part in the machine, and technical requirements, especially surface quality, are also analyzed carefully for the part in the technical drawings.

③ Master parts in sophistication, size, material, weight, and surface quality problems.

④ Does the adopted method meet the technical requirements in degree of distortion, glossiness, roughness of surface quality, and edge quality?

⑤ Compare the equipment costs, mold costs, labor costs, processing costs, water, and electricity fee and then determine which is the best solution based on the economy.

⑥ Do the automation and productivity of all kinds of finishing methods meet the request?

⑦ Compare the adaptability and alterability of the mode of production of finishing methods for the multiple types and small quantity of products.

⑧ Which stage is introduced for the finishing process? Is it the first? Secondary? or multiple? The problem of surface quality, whether it is a satisfactory solution or not?

⑨ Whether will there use the manual mode when the finishing method selection cannot meet the surface quality?

⑩ With the use of new technologies, a few problems need to be considered. How to arrange the technical staff studying for the technique and operational training? How to deal with the machining parts for transportation and storage after adopting the new technology? What facilities and tools are added? How to deal with the processing of chips, waste liquid, and waste residues? What facilities should be added? What kind of arrangement for the cleaning, drying, and screening of the parts and facilities should be added?

(2) Available range

Determining finishing methods for the free abrasive finishing methods or non-free abrasive finishing methods, experts, scholars, and engineering and technical personnel have accumulated valuable experiences in practice. These experiences and achievements are valuable for finishing technologies.

Table 1.20 summarizes the comprehensive evaluation indexes of various finishing methods. Table 1.21 is the available range of finishing equipment (“O” is

**Table 1.20** Evaluation indexes for selecting finishing methods

| Evaluation items | | Machining methods | | | | | | |
|---|---|---|---|---|---|---|---|---|
| | | Belt grinding | Barreling | Power brush machining | Injection | Electrolysis | Ultrasonic | Press-working |
| 1 | Machined edge quality | Δ | ○ | ○ | ○ | ○ | ○ | × |
| 2 | Original precision retaining ability after machining | ○ | Δ | Δ | Δ | × | Δ | × |
| 3 | Installation cost | ○ | Δ | Δ | Δ | × | × | Δ |
| 4 | Batches production | ○ | ○ | ○ | ○ | ○ | × | ○ |
| 5 | Ability to give stable processing conditions | ○ | ○ | ○ | ○ | ○ | Δ | ○ |
| 6 | Work-operating cost | ○ | ○ | ○ | ○ | Δ | × | ○ |
| 7 | Equipment maintenance cost | ○ | ○ | ○ | ○ | ○ | ○ | ○ |
| 8 | Production automation | ○ | ○ | ○ | ○ | Δ | × | Δ |
| 9 | Ability to setup operating cycle | ○ | ○ | ○ | ○ | ○ | ○ | Δ |
| 10 | Environmental pollution condition | Δ | ○ | Δ | ○ | ○ | ○ | ○ |
| 11 | Cleaning project settings | Δ | Δ | Δ | Δ | Δ | Δ | ○ |
| 12 | Safety of production | ○ | ○ | ○ | ○ | Δ | ○ | ○ |
| Identification | Tolerable item numbers ○ | 9 | 9 | 8 | 9 | 6 | 5 | 7 |
| | Number of items being considered Δ | 3 | 3 | 4 | 3 | 4 | 3 | 3 |
| | Impossible item numbers × | | | | | 2 | 4 | 2 |

**Table 1.21** Available range of various finishing equipment and installations

| Equipment and installations name | Available range | | | | | | | | | |
|---|---|---|---|---|---|---|---|---|---|---|
| | Finishing cross-hole edge | Finishing complex surface | Finishing inner hole surface | Finishing uneven edge | Finishing continuous surface | Need sharp edge | Need continuous operation | Need batches processing | No need follow-up processing | Indicative price of equipment/ten thousand yuan |
| Gear chamfering machine | | ○ | | ○ | ○ | | ○ | | ○ | 1.8–30 |
| Belt grinding machine | | | | ○ | ○ | ○ | ○ | | ○ | 6.0–60 |
| Cylindrical polishing machine | | | | | ○ | ○ | ○ | | ○ | 3.0–30 |
| Tube plate chamfering machine | | | | | ○ | | | | ○ | 60–120 |
| Rotary barreling machine | | ○ | ○ | ○ | ○ | | | ○ | | 3.0–12 |
| Vibrating barreling machine | | ○ | ○ | ○ | ○ | | | ○ | | 3.0–42 |
| Centrifugal barreling machine | | ○ | ○ | | ○ | | | ○ | | 6.0–30 |
| Reciprocating barreling machine | | ○ | ○ | | ○ | | | ○ | | 3.0–18 |
| Spindle barreling machine | | ○ | ○ | ○ | ○ | | | ○ | | 18–60 |

(continued)

**Table 1.21** (continued)

| Equipment and installations name | Available range | | | | | | | | | Indicative price of equipment/ten thousand yuan |
|---|---|---|---|---|---|---|---|---|---|---|
| | Finishing cross-hole edge | Finishing complex surface | Finishing inner hole surface | Finishing uneven edge | Finishing continuous surface | Need sharp edge | Need continuous operation | Need batches processing | No need follow-up processing | |
| Electrolytic barreling machine | | ○ | ○ | | ○ | | | ○ | | 30–60 |
| Liquid honing installation | ○ | ○ | ○ | | ○ | | ○ | ○ | | 3.0–30 |
| Dry honing installation | ○ | ○ | ○ | | ○ | | ○ | ○ | ○ | 3.0–18 |
| Sandblasting polishing installation | | ○ | | | ○ | | | ○ | ○ | 3.0–12 |
| Shot-blasting polishing installation | ○ | ○ | | | ○ | | | ○ | ○ | 12–18 |
| Thermal energy machining installation | ○ | ○ | ○ | | ○ | | | ○ | | Above 300 |
| Electrochemistry machining installation | ○ | ○ | ○ | | ○ | | ○ | | | 30–60 |
| High-pressure water jet installation | ○ | ○ | | | ○ | | ○ | | | Above 120 |
| Chemistry abrasive device | ○ | ○ | ○ | | ○ | | | ○ | | 12–60 |

**Table 1.22** Available range of various finishing equipment

| Equipment names | Select features | | | | | | | | | | | | | | | | | | | | | | | | | | | | | |
|---|---|---|---|---|---|---|---|---|---|---|---|---|---|---|---|---|---|---|---|---|---|---|---|---|---|---|---|---|---|---|
| | Parts surfaces condition before final machining | | | | | | | | | | | | | | | | | | | | | | | Machining restricted factors | | | | | | |
| | Formation source of parts surfaces | | | | | | | Burrs | | Location of burrs | | | | | | | | Part material | | | | Size dimension | | Production pattern | | | | Surface quality | | |
| | Cutting | Punching | Casting and forging | Sintering | Welding | Pressure casting | Plastics forming | 0~0.1 mm | 0.1~0.5 mm | Straight line location | Outer circle location | Inner circle location | Cross hole location | Complex surfaces | Outside surface | Internal surface | Minimum size of inner boremm | Steel, cast iron | Copper aluminium alloy | Stainless steel | Plastic | 0mm~100mm | 100mm~500mm | One-piece | Batch | Mass | Various mixing | Chamfering | Decrease in size | Surface roughness |
| Abrasive belt grinder | ○ | ○ | ○ | 0○ | ○ | | | ○ | ○ | ○ | ○ | | | | ○ | | 3 | ○ | ○ | ○ | | ○ | ○ | ○ | ○ | ○ | | ○ | Δ | Δ |
| Rotary barrelling machine | ○ | ○ | | Δ | | ○ | ○ | ○ | | ○ | ○ | | | ○ | ○ | ○ | 3 | ○ | ○ | ○ | ○ | ○ | | | ○ | | ○ | ○ | Δ | Δ |
| Vibrating barrelling machine | ○ | ○ | | Δ | | ○ | ○ | ○ | | ○ | ○ | | | ○ | ○○ | ○ | 3 | ○ | ○ | ○ | ○ | ○ | ○ | | ○ | ○ | ○ | ○ | Δ | Δ |
| Centrifugal barrelling machine | ○ | ○ | | Δ | | ○ | ○ | ○ | | ○ | ○ | | | ○ | ○ | ○ | 3 | ○ | ○ | ○ | ○ | ○ | | | ○ | Δ | ○ | ○ | Δ | Δ |
| Reciprocating barrelling machine | ○ | ○ | | Δ | | | | ○ | | ○ | ○ | | | ○ | ○ | | 3 | ○ | ○ | ○ | | ○ | ○ | Δ | ○ | ○ | | ○ | Δ | Δ |
| Spindle barrelling machine | ○ | ○ | | Δ | | | | ○ | | ○ | ○ | | | ○ | ○ | | 3 | ○ | ○ | ○ | | ○ | | Δ | ○ | | | ○ | Δ | Δ |
| Electrolytic barrelling equipment | ○ | ○ | | | | | | ○ | | ○ | ○ | ○ | ○ | ○ | ○ | ○ | 1 | ○ | Δ | Δ | | ○○ | ○ | | ○ | ○ | | ○ | Δ | Δ |
| Liquid honing equipment | ○ | ○ | | Δ | | ○ | ○ | ○ | | ○ | ○ | | Δ | ○ | ○ | ○ | 2 | ○ | ○ | ○ | ○ | ○ | ○○ | ○ | ○ | ○ | ○ | ○ | Δ | Δ |
| Dry honing equipment | ○ | ○ | | ○ | ○ | ○ | ○ | ○ | | ○ | ○ | | Δ | ○ | ○ | ○ | 3 | ○ | ○ | ○ | ○ | ○ | ○ | ○ | ○ | ○ | ○ | ○ | Δ | Δ |
| Sand blasting polishing equipment | ○ | ○ | ○ | ○ | ○ | ○ | Δ | ○ | ○ | ○ | ○ | | | ○ | ○ | | 5 | ○ | ○ | ○ | | ○ | ○ | | ○ | ○ | | ○ | Δ | Δ |
| Shot blasting polishing equipment | ○ | ○ | ○ | ○ | ○ | ○ | Δ | ○ | ○ | ○ | ○ | | | ○ | ○ | | 5 | ○ | ○ | ○ | ○ | ○○ | ○○ | ○ | ○ | | | ○ | Δ | Δ |
| Abrasive flow machining equipment | ○ | ○ | | | | | | ○ | | ○ | ○ | ○ | ○ | ○ | ○ | ○ | 2 | ○ | ○ | ○ | | ○ | ○ | ○ | ○ | | | ○ | Δ | Δ |
| Thermal energy machining equipment | ○ | ○ | Δ | | | ○ | | ○ | | ○ | ○ | ○ | ○ | ○ | ○ | ○ | 2 | ○○ | ○ | ○ | | ○ | ○ | ○ | ○ | | | ○ | Δ | Δ |
| Electrochemistry equipment | ○ | ○ | | | | | | ○ | | ○ | ○ | ○ | ○ | ○ | ○ | ○ | 5 | ○ | Δ | Δ | | ○ | ○ | | ○ | ○ | ○ | ○ | Δ | Δ |
| Chemistry abrasive device | ○ | ○ | | | | | | ○ | | ○ | ○ | ○ | ○ | ○ | ○ | ○ | 2 | ○ | Δ | Δ | | ○ | | ○ | ○ | | | ○ | Δ | Δ |
| Liquid gas machining equipment | ○ | ○ | | | | | | ○ | | ○ | ○ | | | ○ | ○ | | 0.5 | ○ | Δ | ○ | ○ | ○ | | | ○ | | | ○ | ○ | ○ |

Ok). Table 1.22 is the finishing factors of processing equipment ("O" is Ok, and "Δ" is considered and discussed). The basic principles, processing characteristics, and application range of various types of finishing methods are listed in Table 1.2.

The above tables are derived from the production practice for choice reference; the data is obtained from processing parts, machining process characteristics, application field, equipment cost, and cost accounting.

2. The selection procedure of finishing methods

The selection procedure is as follows:

(1) According to consumers requirements or finishing problems of certain parts, the relevant technological materials are collected extensively in various forms, such as drawings, technical process, and technical data of the likewise products. Then, diagnose, analyze, and discuss the existing problems.
(2) In order to improve technician's technical quality, the current status of factory is investigated, such as production capacity, technological level, and production of peripheral conditions; the technical level and professional quality of the workers are also investigated.
(3) It should have related processes when proposing a finishing method. The test data are collected by experiments. Machining technologies and engineering proposals are proposed with certain induction and analysis.
(4) Customer comments should be adopted and respected, such as factory existing conditions, project costing, working conditions, and environmental pollution.

Figure 1.22 shows the analysis program of the finishing methods. Figure 1.23 shows the analytical procedure of the finishing methods.

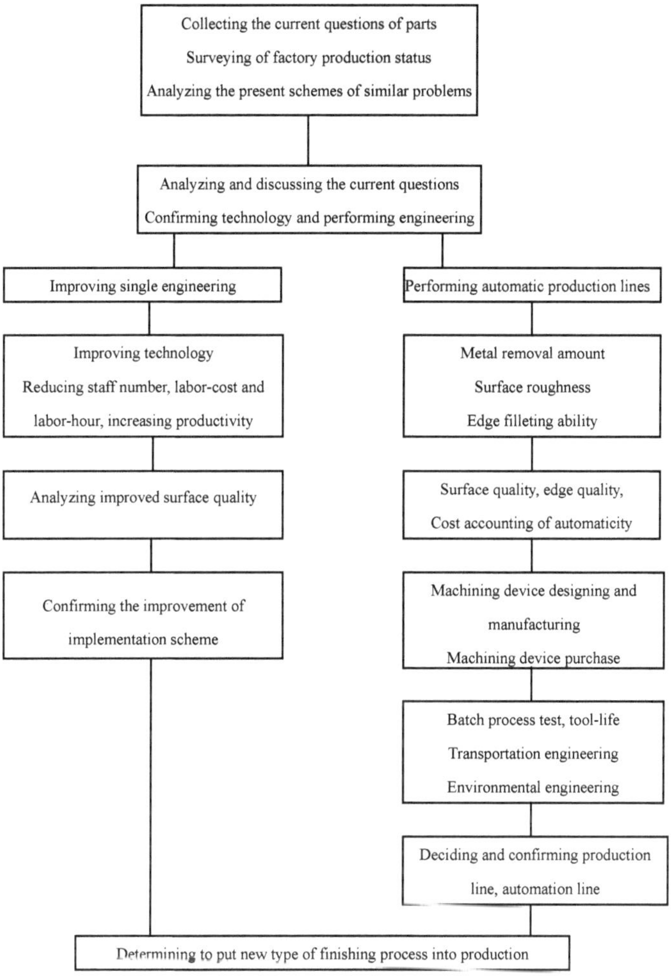


**Fig. 1.22** Analysis program of finishing methods

# 1.5 Current Situation and Trend of Finishing Technologies

## *1.5.1 Brief History and Present Situation*

Since the 1970s, precision surface finishing and deburring technologies have generally been of great importance to each industrially developed country, and the concepts of "Burr Engineering" and "Burr Technology" are proposed. The real awareness about the importance of burr technologies started in 1974. Mr. LarruxK. Gillepie of Bendix Company used the technical term at first during the meeting of

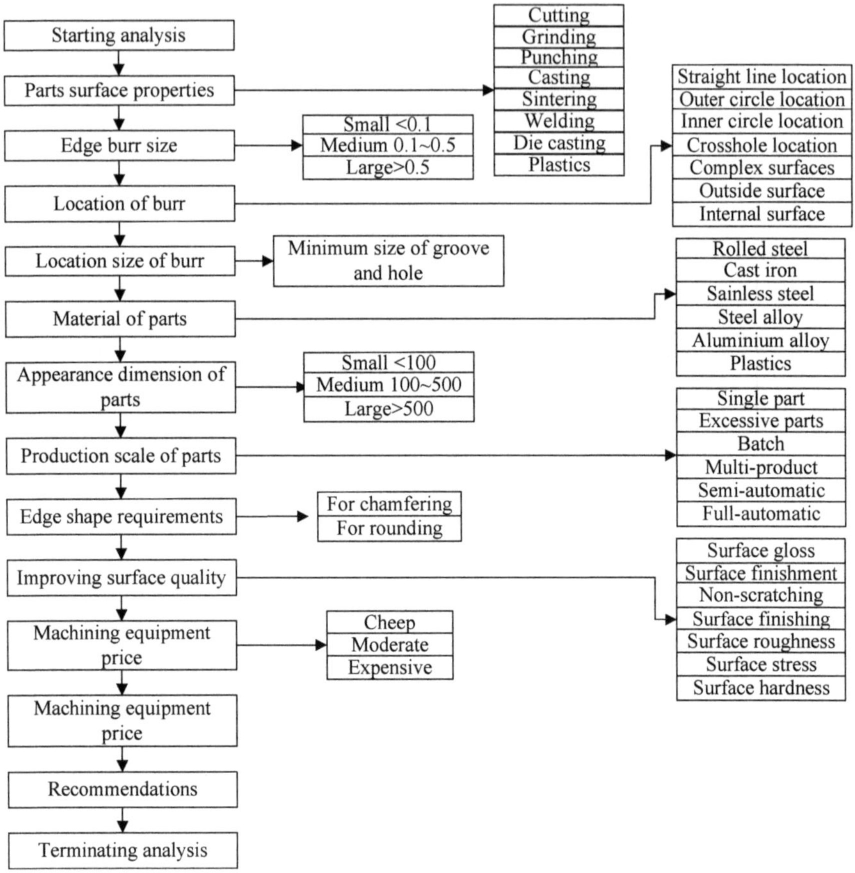


**Fig. 1.23** Analytical procedure of the finishing methods

the Burr Technology Division of the Society of Manufacturing Engineers, and then, the term has been used ever since. As time went on, this technology always brings Edge and Surface Finishing Technology together to communicate and explore; therefore, the meeting has officially changed its name to Burr Edge and Surface Finishing Technique, abbreviated to BEST. The first International Academic Conference on Burr and Flash organized by International Organization for Standardization was held in 1975. Afterward, the USA, West German, Japan, Russia, and other countries have successively established many specialized research institutions and product development departments. They have done a great deal of research work and developed many practical deburring and finishing processes and corresponding equipment. Above all, International BEST Organization established in 1997 played an active role in popularizing and applying the strategic technology around the world. Then, the International Academy for Production Engineering published a series of papers about deburring and finishing at their academic annual meeting. In the 1990s, the technology has occupied an important position in machinery industry and attached more attention of various countries.

In China, the development of deburring and finishing technologies started later. The first book of deburring technology was published in the early 1980s, and there are only about ten academic books till now. In the early 1990s, "Finishing Technology Research Institute of Chinese Instrument Technology Society" and "Finishing Technology Research Institute of Chinese Mechanical Engineering Society" were organized successively. In 1999, Production Engineering Branch of Chinese Mechanical Engineering Society added "Finishing Professional Board" in its organization. Under the proposal and organization of other societies or enterprises, the seminars on industrial (hydraulic industry, internal combustion engine industry, bearing industry, textile machinery industry included) promoting finishing technology have been held many times in recent years. For now, the mechanism that holds International Conference on Finishing Technology and Surface Engineering every two years has been formed in China, many new ideas, methods, processes, and equipment continue to spring up, and the research and application of finishing technology all reveal a good growth trend.

The traditional finishing technologies based on bonded abrasive and loose abrasive finishing have been widely used in manufacturing and greatly improve the surface quality of parts. However, it is difficult or even impossible to machine shaped surfaces, free-form surfaces, and inside surfaces of the cavity using the traditional finishing technologies. Although manual grinding and polishing are able to accomplish the finishing of plane, rotary surface, free-form surface, mold cavity, and others, it is difficult to be popularized in large scale, due to high labor intensity, unstable quality caused by various technical grades and technical proficiency of operators, and low production rate. In order to overcome the limitation in application of traditional finishing technologies, researchers from different countries aim at studying not only the improvement of the surface quality, the machining precision, and the production efficiency of traditional finishing, but also non-traditional finishing technologies and combined finishing technologies, which draw on the complementary strengths of electrical energy, magnetic energy, acoustic energy, luminous energy, chemical energy, thermal energy, and compound actions of various machining methods to achieve the finishing. At the moment, researchers and companies place high importance on these technologies, and lots of achievements have been obtained and been widely applied in some fields. Chemical polishing, electrochemical polishing, magnetic abrasive finishing, electrochemical–mechanical finishing, ultrasonic finishing, barrel finishing, and other finishing ways have been successfully applied in practice and yielded favorable benefits [14].

### *1.5.2 Development Trends*

Finishing technologies have been widely applied in China. However, many companies only treat finishing like polishing, which have not introduced finishing processes into the manufacturing process, due to the development time of finishing technologies not being very long. This leads to the low quality of parts, components, and products and decreases in the use performance, reliability, and the working life of products. Meanwhile, systematic and in-depth research into finishing mechanism is seldom conducted, which hinders the further development of finishing technologies.

With the higher-level development of machinery manufacturing, new demands are set for finishing technologies. The new demands are mainly reflected in three aspects: Firstly, it is to solve the unresolved conundrum of finishing large- and medium-sized parts, complicated surfaces, inner cavities, and inner walls of tiny pipe with high quality and efficiency (like ultra-large crankshafts for ship applications, ultra-large bearings for windmill applications). Secondly, it is to solve the finishing conundrum of difficult-to-machine materials and fragile materials (like ceramic materials, composite materials). Lastly, it is to require developing finishing technology to higher level and to achieve technological progress from micrometer-scale finishing to nanoscale finishing. In addition, what is also needed is to make the parts surface by finishing process appearances more perfect, the physical–mechanical properties more excellent, and the machining process more economical and convenient.

Each finishing method has their own optimal application scope and also has some problems and certain limitations more or less. The problem of how to give full play to their strengths has become an important objective for researchers from different countries to explore industriously. Not only did this significantly reduce surface roughness value and achieve high-quality surface layer, it also improved the manufacturing efficiency and realized the finishing of large- and medium-sized parts, complicated surfaces, and difficult-to-machine materials. While researchers from different countries continue to aim at studying how to improve the surface quality, the machining precision, and the production efficiency of traditional finishing, they are also working hard for new-type finishing technologies and finishing processes. They take greater interests in the research of non-traditional finishing technologies and combined finishing technologies and propose many new finishing technologies and processes. In recent years, elastic emission machining (also known as soft particle polishing), hydrodynamic polishing, magnetic fluid polishing, extrusion lapping and polishing, abrasive flow machining, barrel finishing, magnetic abrasive finishing, rolling polishing, tow-phase flow finishing, jet machining, and others are highly regarded by related research institutes and researchers. New theories and processes continue to rise.

In order to overcome the limitation of the various finishing methods, researchers are now putting their eyes on the studying of combined finishing technologies and processes recently. Combined finishing methods, a mode of combining various machining processes based on the concept of complementary advantages and supplement each other, take full advantage of different processes, circumvent their respective weaknesses, and have obtained certain achievements. For example, electrochemical–mechanical finishing uses simple equipment, is low-cost process, combines the advantages of mechanical finishing precision controllability and electrochemical polishing efficiency, and successfully achieves the compounding of electrochemical polishing and mechanical finishing. Pulse electrochemical–mechanical finishing technology synthesizes the advantages of both pulse electrochemical machining and electrochemical–mechanical finishing, gives full play to its own advantages, further improves machining precision and production efficiency, and mainly is used for the finishing of the comparatively rough surfaces. Furthermore, combined finishing methods also include magnetic electrochemical finishing technology which adds magnetic field into the electrochemical polishing directly, electrochemical magnetic abrasive finishing technology which introduces electrochemical finishing into magnetic abrasive finishing, the finishing technology which introduces ultrasonic-assisted machining into magnetic abrasive finishing, hole surface finishing technology which introduces magnetic field into abrasive jet finishing technology, barrel finishing technology which combines two different types of barrel finishing. The advent and application of combined finishing technology, which takes full advantage of different processes, proves that combined finishing technology is feasible.

With extended application of finishing technologies in production, new demands are set for finishing automation in order to further enhance product quality, increase efficiency, reduce processing cost, save labor, and improve the working condition. In recent years, many companies have invested a large amount of manpower and resources in researching and developing finishing technology. They have also developed several different forms of automatic finishing equipment and manufacturing systems, such as the automatic finishing device, the automatic finishing product line, numerical controlled finishing equipment or manufacturing system, finishing robot, manufacturing system, the flexible finishing cell, and flexible finishing product line. These automatic finishing equipment and manufacturing systems have been applied in automotive, tractor, tank, aircraft manufacturing, and other industries.

The material conversion processes will be radically advanced as in the widespread application of additive manufacturing and powder metal compaction with advanced heat treatment technology. This revolution may force traditional finishing processes to be replaced by advanced abrasive fine-finishing processes. In the future, precision components with minimum finishing stocks produced by advanced material conversion processes will need to be finished by newly developed abrasive fine-finishing technologies that are capable of both high stock removal and the creation of desired surface integrity.

## References

1. S.C. Yang, *Surface Quality and Finishing Technology* (Machinery Industry Press, Beijing, 2000)
2. Z.W. Qing, *Deburring Technology Handbook* (Navigation Press, Beijing, 1995)

3. Y.S. Hu, *Principle of Mechanical Manufacturing Technology*. (Beijing University of technology press, Beijing, 1992)
4. F.T. Farago, *Abrasive Methods Engineering* (Machinery Industry Press, Beijing, 1991)
5. Y. Li, J.Q. Liu, J.S. Li et al., *Metal Surface Polishing Technology*. Chemical Industry Press, Beijing, 2006
6. S.Q. Yang, S.C. Yang., in *Macroscopical feature of finishing technique and structural thinking of its development*. Academic conference for young scientists and technicians of the national production engineering society, 2007
7. W.H. Li, S.Q. Yang, X.H. Li, W.D. Li, Development status and trends of mass finishing processes. Key Eng. Mater. **621**, 111–120 (2014)
8. Q. Hong, G.C. Wang, Research and development of surface integrity in precision machining. Mod. Manufact. Eng. **8**, 12–15 (2004)
9. I.S. Jawahir, Surface integrity in material removal processes: recent advances. CIRP Ann.—Manufact. Technol. **60**, 603–626 (2011)
10. GB/T3505-2009. *Geometrical Product Specifications-Surface texture: Profile Method-Terms, Definitions and Surface Texture Parameters*. (China Standards Press, Beijing, 2009.11.1)
11. GB/T1031-2009. *Geometrical Product Specifications-Surface Texture: Profile Method-Surface Roughness Parameters and Their Values*. (China Standards Press, Beijing, 2009.11.1)
12. GB/T131-2006. *Geometrical Product Specifications-Indication of Surface Texture in Technical Product Documentation*. China Standards Press, Beijing, 2007.2.1
13. GB/T15757-2002. *Geometrical Product Specifications-Surface imperfections—Terms, Definitions and Parameters*. China Standards Press, Beijing, 2003.1.1
14. F. Hashimoto, H. Yamaguchi, P. Krajnik et al., Abrasive fine-finishing technology. CIRP Ann.—Manufact. Technol. **65**, 597–620 (2016)

# Chapter 2
# Barrel Finishing Technology

## 2.1 Introduction

In the process of the free abrasive tool finishing, the abrasive tool (also known as media) is in a free state, which is based on relative motions and action forces caused by the direct contact with parts to complete the finishing of part surface and to increase the surface qualities of parts. For various methods of free abrasive tool finishing, machining mechanisms, process parameters, finishing effects, and technology scopes have their own uniqueness due to different constituted forms, action modes, machining status of abrasive tools.

Surface barrel finishing, as one of the typical-free abrasive tool finishing methods, has advantages such as low processing cost, good finishing effect, high efficiency, simple device.

### *2.1.1 Connotation*

Surface barrel finishing belongs to free abrasive tool finishing technology. First of all, parts which are in a free or non-free state are placed in the container with machining media (covering abrasive blocks, compounds, as well as water). Based on the driving motion produced by parts, container, or both, the complex relative motions between parts and abrasive blocks are generated in the process of finishing. Under the relative motion, abrasive blocks in a free state perform the micro-grinding action of collision, rolling, sliding, and scratching with a certain amount of force to change the geometrical characteristics of part surface (to reduce the value of surface roughness, to increase material ratio of the profile, to change surface texture, and to remove the defects such as burrs), to improve the physical–

 
S. Yang and W. Li, *Surface Finishing Theory and New Technology*,
https://doi.org/10.1007/978-3-662-54133-3_2

mechanical properties of surface layer (to increase micro-hardness, to form denatured layer on the surface, to improve surface stress state, etc.), to increase the cleanliness of parts, to synthetically improve the surface integrity of parts, to increase the usability and service life of parts and products, and to achieve the goal of surface finishing [1–6].

Surface barrel finishing is a kind of machining methods. According to Einstein's energy equation, the relative motion and interaction between the part and the abrasive blocks can be generated only when the abrasive blocks with certain mass have a certain degree of energy, that is

$$E + m \Rightarrow v \tag{2.1}$$

or

$$E + m \Rightarrow F \tag{2.2}$$

where $E$ represents the energy, $m$ represents the mass, $v$ represents the velocity, and $F$ represents the acting force.

As indicated in Eqs. (2.1) and (2.2), when a certain energy $E$ acts on a certain mass $m$, the velocity and acting force can be generated. The interaction and micro-grinding effect of collision, rolling, sliding, and scratching are produced due to the existence of motion. Then, the finishing of parts is completed.

Results mentioned above also satisfy Poisson equation:

$$\mathrm{MMR} = KFv \tag{2.3}$$

where MMR is the material remove rate, $K$ is the coefficient, $F$ is the acting force, and $v$ is the velocity.

In order to achieve the barrel finishing of part surface, the abrasive blocks and the part must touch each other. The key is how to enable the energy and force to act on the mass, because the same abrasive blocks or parts with different barrel finishing processes can cause different velocities and acting forces, and obtain different finishing effects and efficiencies.

The main effects generated in barrel finishing processes are shown below [3, 4].

(1) Collision effect. When the free abrasive blocks with mass hit the part surface at a certain relative velocity, an impact force $F_N$ is generated. It can cause the part surface to generate elastic and plastic deformation, and to change the microscopic geometry and the physical–mechanical property of part surface layer. According to momentum theorem, the impact force $F_N$ is

$$F_N = \frac{m\left(v_N - v'_N\right)}{\Delta t} \tag{2.4}$$

where $m$ is the mass of abrasive blocks, $v_N$, $v'_N$ are normal components of abrasive blocks velocity on the part surface before and after abrasive block forces on the part surface, respectively, and $\Delta t$ is the action duration.

It can be shown from Eq. (2.4) that the impact force to the part surface can be increased, and the finishing efficiency can be enhanced by increasing the mass of abrasive block $m$ and the change of the velocity $v_N - v'_N$. For example, among various surface barrel finishing processes, larger-sized abrasive blocks (for the same kind of abrasive blocks, larger size means greater mass) can be usually selected to improve the finishing efficiency in the rough machining. When there is a restriction on the size of the abrasive blocks during machining, the denser abrasive blocks should be selected. The forced flow area (means the change of the velocity) is formed to increase the finishing efficiency and improve finishing effects in centrifugal barrel finishing.

(2) Rolling effect. When the free abrasive blocks extrude the part surface with a certain amount of force, the abrasive block must generate any rotation motion due to many other surrounding abrasive blocks. Roll-finishing is formed under the relative movement, micro-peaks on the part surface which yields micro-plastic deformation locally under the effect of rolling, and then it can gain a better surface. Meanwhile, there would be certain degree of surface hardening, and therefore the wearability and the fatigue strength of the part are increased since the residual compressive stress is generated.
For higher normal pressure, the micro-peaks on the part surface are flattened by the plastic deformation on the effect of rolling. The micro-peaks left on the surface are not completely flattened, and the valley is not completely filled due to limit of the normal pressure. The rolling effect of abrasive blocks on part surface includes two forms: surface plastic deformation and elastic deformation. When the contact area is small, the plastic deformation develops in the local contacting area, and the valley is filled by the micro-peaks. When the contact area becomes larger, the finite-sized normal pressure is not enough to generate the plastic deformation of the micro-peaks on the part surface, but only elastic deformation. Under the alternating action of plentiful abrasive blocks, these micro-peaks are fractured in fatigue due to repeating elastic deformations. As a result, it produces chip-breaking.
(3) Micro-grinding effect. Under the action of certain force, the free abrasive blocks slip along the part surface at a certain velocity. The abrasive particle which is bonded to the surface of the abrasive blocks performs the action of sliding, scratching, and cutting, as same as the abrasive particle bonded to the surface of grinding wheel does. When the abrasive particle glides and scrapes across the metal surface of the part one or more times, the plastic deformation limit of the surface metal would be occurred one or more times. If the plastic deformation reaches the yield strength, the microchip might fall off from the metal base, and the micro-grinding is completed.

(4) The effect of liquid medium. Water and specific compound additive as liquid medium are used in many surface barrel finishing processes, and it is very critical to reasonably select the additive and dose for finishing effect and efficiency. On the one hand, they can promote or restrain the grinding and finishing effects. Meanwhile, the compound itself can set off the chemical reaction and involve in finishing to maintain or improve the surface quality of the part (covering luster, color, surface roughness, etc.) and to change finishing time. On the other hand, the compound can regulate and soften water for giving full play to the buffer and flushing action of water (avoiding or reducing the violent collision among parts and abrasive blocks, scratch, deformation, crushing, etc.), changing the pH of the medium, and controlling the formation and the number of the foam. In addition, liquid medium can clean the oil fouling, abrasive dust, microchip and other impurities on the surface of the part, and the abrasive blocks to prevent the corrosion of the part, to keep the abrasive block clean and sharp, and to improve the cutting ability of the abrasive blocks.

To deeply understand the surface barrel finishing in theory, the key is to research the main effect in which the abrasive blocks play on the part during barrel finishing, and even more important is to identify requirements for the main effects. Based on viewpoints of machining process morphology, surface barrel finishing process can be regarded as materials processing, energy conversion process, and the changing of information, and materials machining cannot do it without the energy conversion and the information changing. The material flow in surface barrel finishing processes is the mechanical changing during plastic deforming and chemical changing during corrosive finishing for the solid part to complete the removal of metal on the part surface layer. The energy flow is to use abrasive blocks and definite liquid medium for transferring energy under the action of relative motion and the chemical action, and stressing the part with mechanical force and chemical reaction. The information flow is changed by the mixture medium which makes the part do complex relative motion and completes the free forming of the part with special shape. In surface barrel finishing, the energy conversion includes three aspects, (1) energy types (mechanical energy and chemical energy), (2) energy transfer medium (abrasive blocks with rigid and definite liquid medium), and (3) energy transmission styles (relative motion and chemical action). The change of the information includes two aspects, (1) the forming method (free forming) and (2) the relative motion way (compound motion) [7].

In conclusion, the requirements for producing the main effects in barrel finishing are (1) the complex relative motions between abrasive blocks and part surface, (2) forces driven by relative motions between abrasive blocks and part surface, (3) the abrasive blocks with definite property parameters, and (4) the liquid medium with definite chemical effects.

### 2.1.2 Classification

According to different states of parts during processing, surface barrel finishing includes three major categories of parts in a free state, in a half free state, and in a non-free state. Table 2.1 lists the classification of surface barrel finishing.

### 2.1.3 Functional Characteristics and Application Scope

Surface barrel finishing as one of the typical-free abrasive tolls finishing ways has the following characteristics:

(1) The complex relative motions between parts and abrasive blocks are generated due to one or several kinds of moving ways. Depending on certain relative speed of motion and applied force, the finishing of part surface is implemented [8–10].
(2) Surface barrel finishing works for the deflashing and removing of oxide layer and surface cleaning of middle or small-sized casting-forging stamping parts. It

**Table 2.1** Classification of surface barrel finishing

| Method | Classifications | | |
|---|---|---|---|
| Surface barrel finishing | Parts in free state | Rotary type | Horizontal placement |
| | | | Inclined placement |
| | | Whirling type | Static whirling |
| | | Vibratory type | Vertical vibration |
| | | | Horizontal vibration |
| | | Reciprocating type | |
| | | Centrifugal type | Horizontal placement |
| | | | Inclined placement |
| | | | Vertical placement |
| | | Other types | |
| | Parts in half free state | Spindle type | |
| | | Vibratory type | |
| | Parts in non-free state | Spindle type | Vertical spindle type |
| | | | Intermeshing spindle type |
| | | | Horizontal spindle type |
| | | Non-spindle type | Dynamic whirling |
| | | | Workpiece vibration |
| | | Combined type | Spindle–non-spindle |
| | | | Spindle-vibration |

also works for the deburring, filleting, refining the surface, and decreasing the surface roughness values of parts machined.

(3) Surface barrel finishing can realize the surface finishing of parts with large or middle dimension, such as crankshaft, camshaft, gear, wheel hub, transmission housing [11, 12].
(4) Surface barrel finishing is an all-round finishing method. It can carry out not only the finishing of machined surface, but also the finishing of non-processing surface. It can improve the cleanliness of parts synthetically [13, 14].
(5) Surface barrel finishing has the collision and rolling effects on the surface of parts. It can increase the micro-hardness and improve the stress state, the physical and mechanical performance on the surface layer of parts.
(6) Surface barrel finishing has the characteristics of simple operation, easy management, low processing cost, good finishing effects, efficient production, light pollution, etc [15–17].
(7) For surface barrel finishing, it is constrained by part dimensions, structure shape, finishing area, etc. It could realize the finishing on the apparent surface of middle- and small-sized parts easily. However, for the finishing of large parts, such as inner bores, grooves, and dented surfaces, it becomes much more difficult to implement.
(8) Abrasive blocks are main working media in the surface barrel finishing. The shapes of the abrasive blocks include square, rectangle, rhombus, cylinder, tetrahedron, sphere. The shape, size, and material of the abrasive block make a difference to finishing effects and efficiency, and can be determined by the material, overall dimension, finishing area, and the finishing requirements of workpiece.
(9) The compound used in surface barrel finishing is also the main working medium. The compound and water played a helpful role in grinding, brightness, foaming, lubrication, buffering, cleaning, etc.

Surface barrel finishing can realize deburring, derusting, surface finishing, chamfering, removing various defects left after heat treatment or other processing, pre-treatment before plating and painting, etc. Ferrous metals, nonferrous metals, plastics, composite materials, ceramics, even wood, and other imaginable materials can be finished with high productive efficiency and satisfactory finishing results. Meanwhile, the technology has the advantages of high cleanliness, low equipment and process costs, easy operation, low labor intensity, etc. Just as the other processing methods, surface barrel finishing also has certain scope of application. For example, the barrel finishing of inner bores, grooves, and dented surfaces becomes much more difficult to implement. And, there are also different applicable scopes corresponding to different surface barrel finishing ways. For example, horizontal spindle barrel finishing is mainly applicable to the surface finishing of large- or middle-sized parts, such as crankshaft and camshaft. Table 2.2 shows the characteristics and applicable scopes of most common surface barrel finishing technologies.

**Table 2.2** Characteristics and applicable scopes of surface barrel finishing technology

| Finishing methods | Metal removal capacity | Finishing efficiency | Part surface brightness | Surface roughness decrement | Improvement of physical–mechanical performance | Collision among parts | Degree of automatization | Scope of application |
|---|---|---|---|---|---|---|---|---|
| Rotary type | Very weak | Low | Worse | About 0.5 magnitude | Worse | Great | Average | Small special-shaped parts |
| Vibratory type | Weaker | Average | Better | About 1 magnitude | Worse | Small | Easy | Small and medium special-shaped parts |
| Centrifugal type | Strong | High | Better | 1–2 magnitude | High | Medium | Difficult | Small and medium special-shaped parts |
| Whirling type | Stronger | High | Better | 1–2 magnitude | High | Medium | Easy | Small and medium special-shaped parts (except for sheet parts) |
| Vertical spindle type | Strong | High | Good | 1–2 magnitude | High | None | Easier | Small and medium parts, shaft parts, and special-shaped parts |
| Intermeshing spindle type | Strong | High | Good | 1–2 magnitude | High | None | Easier | Small and medium parts, plate-shaped parts, and special-shaped parts |
| Horizontal spindle type | Strong | High | Good | 1–2 magnitude | High | None | Easy | Parts like crankshaft, camshaft, etc. |

## 2.2 Rotary Barrel Finishing

### *2.2.1 Finishing Principle and Characteristics*

1. Finishing principle

Figure 2.1 shows the schematic of rotary barrel finishing. In the finishing, the workpieces and working media (abrasive blocks, compounds, water included) are put into one drum. When the drum rotates at a certain speed, the workpieces and abrasive blocks in the drum are lifted up with rotation direction in the drum, under the action of gravity, centrifugal force, and friction. In the lifting process, the workpieces and abrasive blocks located at the bottom of the drum are in a dynamic balance, and they remain to be relatively motionless. When the workpieces and abrasive blocks located at the top of the drum are lifted up to a certain height, they will lose their balance and slide down. The workpieces and abrasive blocks on the outer layer subject to least resistance, and their sliding velocities are fast; thus relative movements occur among layers. Even in the same layer, the slipping velocities vary, due to large differences in the shapes, qualities of the workpieces and the abrasive blocks. Therefore, with the rotation of the drum, the workpieces and abrasive blocks in the drum roll over and over, which cause the collision, rolling, rubbing, and scoring among workpieces and abrasive blocks, and the finishing is implemented.

2. Main factors affecting finishing effects

Rotary barrel finishing is one of the barrel finishing technologies at the earliest stage, and it is suitable to the finishing of small special-shaped parts [18]. Its capacity of metal removal is weaker, machining efficiency is lower, certain collision among part surface exists, and the integrity of machined surface is much poor. The

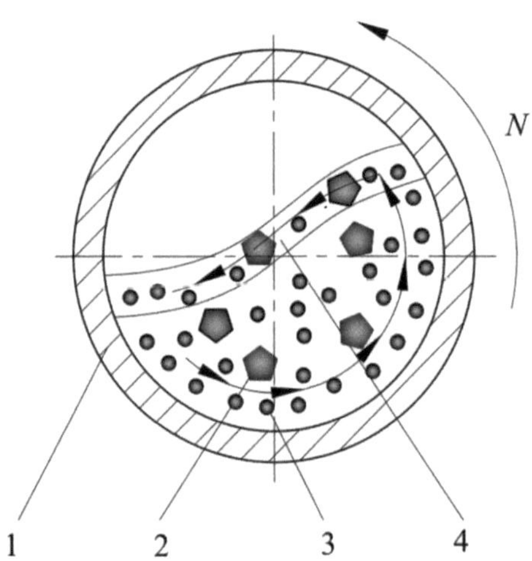


**Fig. 2.1** Schematic of rotary barrel finishing. 1—drum; 2—workpieces; 3—working medium; 4—sliding layer

**Table 2.3** Characteristics of rotary barrel finishing with different drum placements

| Factor | Drum placement | |
|---|---|---|
| | Horizontal placement | Tilted placement |
| Processing condition adjustment | Rotating speed adjustable | Rotating speed and tilt angles adjustable |
| Work-handling | Inconvenient | Convenient |
| Monitoring of machining process | Inconvenient | Convenient |
| Equipment cost | Low | High |
| Machining uniformity | Good | Poor |
| Size range of workpiece | Large | Limited |
| Main purpose | Finishing | Acid cleaning, drying, and small parts finishing |

characteristics of rotary barrel finishing with different drum placement are shown in Table 2.3.

## 2.2.2 *Main Factors Affecting Finishing Effects*

The rotating speed of the drum is an important parameter and has a strong effect, finishing efficiency, and quality. The reasons may be analyzed from the following three points.

(1) As the rotating speed increases, the centrifugal force will increase. And the collision, rolling, rubbing, and scoring among workpieces and abrasive blocks in the drum become serious.
(2) As the centrifugal force increases, the friction forces from the drum wall to the workpieces and the abrasive blocks increase. The workpieces and abrasive blocks can lose their balance and slide down only when they lift up to a turnover. This can increase the impact force of the down-sliding, the sliding distance and the number of the collisions, rolling, rubbing, and scoring suffered by workpieces.
(3) As the rotating speed increases, the rolling frequency of the workpieces and abrasive blocks in the drum quickens, and the sliding number of the workpieces and abrasive blocks per unit of time increases.

Under the influence of three factors mentioned above, the finishing efficiency increases with the increase of the rotating speed. But when the speed increases to more than $\frac{1337}{\sqrt{D}}\,\mathrm{r/min}$ (where $D$ represents rotating radius of the drum wall, mm), the workpieces and abrasive blocks evenly snuggle up on the drum wall along with the rotation of the drum. They are in a dynamic balance and remain relatively motionless. By this time, the rolling forces among the workpieces and abrasive blocks are large, but the workpieces could not be machined due to non-existing relative sliding. Therefore, the rotary barrel finishing is difficult to improve the finishing efficiency and quality due to the rotating speed constraints.

### *2.2.3 Equipment Types and Design*

1. Equipment types

The concrete types of rotary barrel finishing equipment are shown in Fig. 2.2. Where, Fig. 2.2 (a) shows the common tilting type; (b) named bottle type is the improved type of common tilting type; for it combines the advantages of horizontal type and tilting type; (c) designed with material handling port is horizontal type; (d) is multi-compartment type, could apply to machine different shape and sizes of workpiece at the same time; (e) is multiple barrel type, which is one improved type of horizontal type, several drums are fastened together on one connecting plate, the tilt levels of each drum can be adjusted by changing angle of connecting plate; (f) is triple action type, which is designed for improving the phenomena that the easy appearance of workpieces snuggle up on both side of the drum wall in the process of machining flat workpieces with horizontal type rotary barrel finishing equipment; (g) is end loading type, could apply to machine slightness workpieces; (h) designed with one water tank on the bottom is submerged type.

2. Equipment shape

The shape of the drum used for machining small-sized parts is usually 6–8 polygons. When the size of parts increases, the shape can be 10–12 polygons. Figure 2.3 shows several photographs of different rotary barrel finishing equipments.

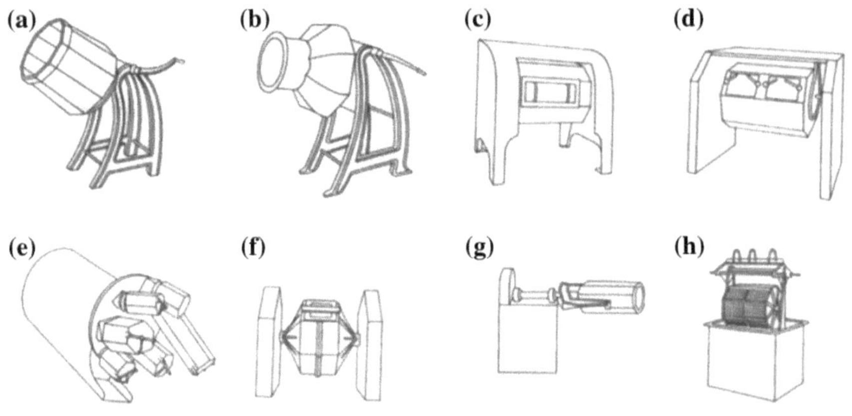


**Fig. 2.2** Types of rotary barrel finishing equipment. **a** Tilting; **b** bottle; **c** horizontal; **d** multi-compartment; **e** multiple barrel; **f** triple action; **g** end loading; **h** submerged

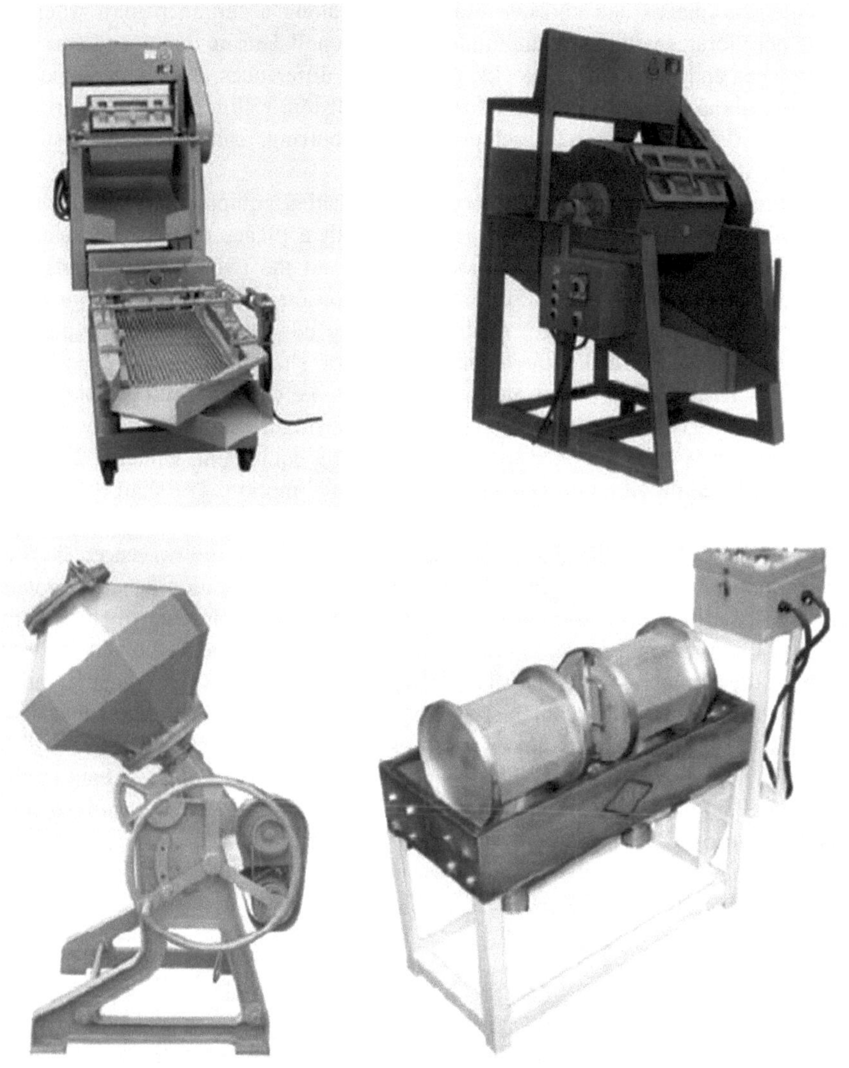

**Fig. 2.3** Photographs of several different rotary barrel finishing equipments

## 2.3 Vibratory Barrel Finishing

### *2.3.1 Finishing Principle and Characteristics*

1. Finishing principle

In the vibratory barrel finishing process, the workpieces, abrasive blocks, and liquid medium are put into a shaped container according to certain proportioning, and the workpieces and abrasive blocks move along a certain pattern when the container vibrates with fixed amplitude and frequency. During the movement of the workpieces and the abrasive blocks, due to their differences in weight, shape, and location, abrasive blocks are forced to make collision, rolling, and micro-grinding to the workpiece surface for achieving the deburring, chamfering, and surface finishing of the workpiece.

Figure 2.4 is a horizontal vibratory barrel finishing equipment, which is driven by a single-shaft inertia vibration generator with a planar motion. The shaft of vibration generator 2 is installed horizontally, and the two eccentric parts with adjustable setting angle and offset are mounted on the shaft. When the horizontal shaft of the vibration generator rotates at high speed, the eccentric parts generate centrifugal force to make the container 6 (slot shaped) generating periodical vibration. The abrasive blocks and workpieces in the container roll oriented along the wall of the container under the action of centrifugal exciting force.

Figure 2.5 is a vertical vibratory barrel finishing equipment, which is driven by single-shaft inertia vibration generator 3 with spatial motion. The shaft of vibration generator is installed vertically, and two eccentric parts are mounted on the upper and the lower ends of the shaft. There is an angle between the two eccentric parts projection in the horizontal plane. When the shaft of the vibration generator rotates at high speed, the eccentric parts not only generate a certain centrifugal exciting force in the horizontal plane, but also a certain exciting torque in the vertical plane to make the container 4 (annular shaped) to generate a complex periodical vibration. Since the bottom of the container is an annular shape, the abrasive blocks and workpieces in the container not only orbit around the center shaft (vertical) of the container, but also roll around the center of the annular. Thus, the resultant motion is an annular spiral motion under the action of horizontal centrifugal exciting force and the vertical exciting torque.

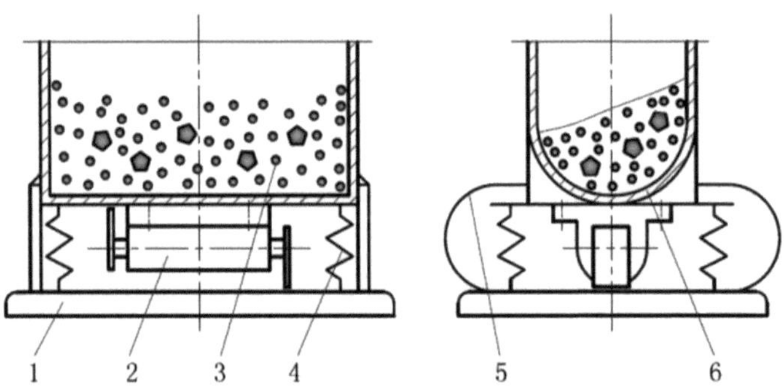


**Fig. 2.4** Horizontal vibratory barrel finishing equipment diagram. 1—Base; 2—vibration generator; 3—workpiece and medium; 4—spiral spring; 5—plate spring; 6—container

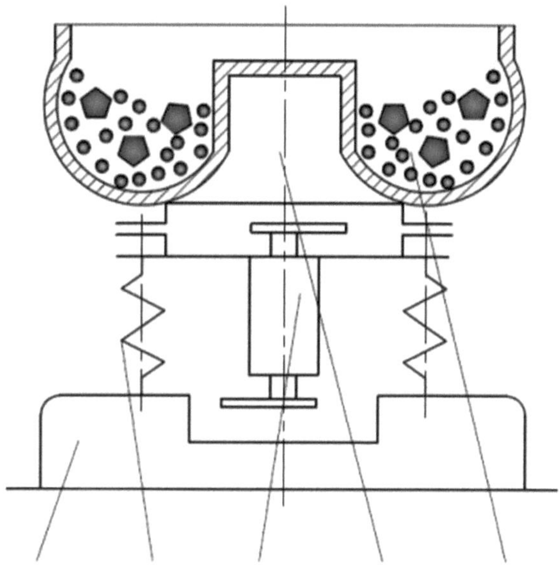


**Fig. 2.5** Vertical vibratory barrel finishing equipment diagram. 1—Base; 2—spiral spring; 3—vibration generator; 4—container; 5—workpiece and medium

2. Force analysis

Here, take the vertical vibratory barrel finishing method as an example to conduct force analysis [19].

Let the weight of the two eccentric parts equal to each other is $m_0$; the angular velocity of the vibration generator shaft is $\omega$; the angle between the two eccentric parts projection in the horizontal plane is $\alpha$; the distance from the gravity center of the eccentric parts to the vertical shaft is $r$; the center-to-center distance between the two eccentric parts is $l_0$; the distance from the upper eccentric part to the gravity center of the container is $l$. The exciting force generated by vibration generator in the horizontal plane is

$$P(t) = 2m_0\,\omega^2 r\cos\frac{\alpha}{2}(\cos\omega t + i\sin\omega t) \tag{2.5}$$

The exciting torque about horizontal shaft generated by vibration generator is

$$M(t) = 2m_0\omega^2 r\left[\left(\frac{l_0}{2} + l\right)\cos\frac{\alpha}{2}(\cos\omega t + i\sin\omega t) + \frac{l_0}{2}\sin\frac{\alpha}{2}(\sin\omega t - i\cos\omega t)\right] \tag{2.6}$$

The exciting force causes abrasive blocks and workpieces in the container to orbit around the center shaft (vertical) of the container, and the exciting torque causes abrasive blocks and workpieces to roll around the center of the annular. The influence of the vibration generator position at an angle between the two eccentric parts and so on can be analyzed by curve plotting.

3. Functional characteristics and application scope

Vibratory barrel finishing is one kind of surface finishing process developed abroad in the 1950s; it has been widely used in the deburring, rounding, chamfering and finishing of small- or medium-sized components made of different materials (like steel, copper, aluminum, bakelite). Horizontal vibratory barrel finishing method can easily make a separation between the workpiece and abrasive blocks in processing, and leave the trace of impulse shock on the workpiece surface, and therefore it is not suitable for the precision finishing. Vertical vibratory barrel finishing method makes the mutual cyclic surrounding among the workpiece and abrasive blocks not leaving any trace of impulse shock on the workpiece surface; therefore, it is suitable for the precision finishing [20–23].

According to the simple analysis of the finishing principle, vibratory barrel finishing processes have the following characteristics.

(1) High finishing efficiency

In the whole machining cycle, the finishing effect of abrasive blocks on the workpiece goes on without a break, so the efficiency is high (its efficiency should be several times even dozens of times higher than that of rotary barrel finishing) [24].

(2) Extensive application range

This kind of finishing method not only machines black metal pieces, nonferrous metal parts, but also machines nonmetal parts, such as plastic parts, ceramic parts. Moreover, it can machine large or medium-sized complicated shaped shell parts with inner holes or hidden surface and other easy-deformed object [20, 21].

(3) Good finishing quality

This kind of finishing method not only removes burrs, but also removes the oxide layer, so as to achieve chamfering and surface polishing [22]. It has a good finishing uniformity, and surface roughness degree can be increased by 1°–2°; the surface micro-hardness can be increased, the residual tensile stress can be relieved, the fatigue resistance can be enhanced (typically about 10%), and also the physical–mechanical properties can be improved.

(4) The equipment is simple in construction, easy to operate, and easy to realize the operation of several sets of equipment by one operator and the automation of manufacture [24].

### 2.3.2 *Main Factors Affecting Finishing Effects*

1. Equipment parameters

(1) The configuration parameters of the eccentric parts
The configuration parameters of the eccentric parts on the vibrating shaft have direct influences on the amplitude, which mainly refer to the quality, the center-of-mass offset of the single eccentric part, the axial distance, and relative angle between the two eccentric parts.
In theory, when the vibrating shaft is rotated, the higher the quality and the center-of-mass offset of the eccentric part are, the higher the centrifugal force developed by the eccentric part is, the more intensive the vibration is, the higher the finishing efficiency is. However, the finishing quality tends to decrease to some extent. For the equipment structure, the quality and the center-of-mass offset of the eccentric part cannot be too large, and otherwise it will not only make the equipment heavier, but also greatly shorten the service life of the support bearing mounted on the vibrating shaft. Conversely, if the quality and the center-of-mass offset of the eccentric part are small, the centrifugal force developed by the vibrating shaft is small, and the vibration becomes more moderate. Although the efficiency is low, a finer surface quality can be obtained, and the deformation of the easy deformability workpiece can be prevented when machining. Therefore, the quality and the center-of-mass offset of the eccentric part need to be decided based on the actual situation, and the counterweight of the eccentric part can be solved. It is convenient to adopt adjustment method using screw fixation for the addition or removal of the counterweight on the eccentric part.
The relative angle between the two eccentric parts not only has an effect on the movement of the workpiece and abrasive blocks in the container, but also affects the finishing efficiency and quality, which shows that it is a very important equipment parameter. The practice shows that the angle between the eccentric parts is related to the effective length of the vibrating shaft, the centrifugal force developed by the eccentric part, and so on. If the setting angle of the eccentric parts is less than 90 °C, the workpiece and abrasive blocks will roll moderately, and the finishing efficiency will be relatively low. Conversely, the movement will be intensive, and resulting in a high efficiency, but the quality will be relatively poor. Generally, it is suitable to select 90 °C as the setting angle of the two eccentric parts [23].

(2) The rigidity coefficient of the spring
The rigidity coefficient of the spring has an impact on the finishing efficiency. Generally, the larger the rigidity coefficient is, the higher the finishing efficiency is, but the louder the noise is. The lower the rigidity coefficient is, the lower the finishing efficiency is, and the lower the noise is.

(3) The rotary speed of the vibrating shaft
The higher the rotary speed of the vibrating shaft is, the higher the vibration frequency is, the better the finishing efficiency is. Generally, the rotary speed of the vibrating shaft is not less than 1500 r/min. The rotary speed of the equipment with small container can be higher and be lower the equipment with large container.

2. Technological parameters

(1) Finishing time
The determination of finishing time relates to the finishing requirements of parts: Deburring processing usually takes 1–2 h. Reduction of the surface roughness value takes 3–5 h. Improvements of surface micro-hardness and residual stress take about 10 h.

(2) Volume ratio and load amount
When finishing small workpieces, the volume ratio of workpiece to abrasive blocks should be controlled at 1:2 $\sim$ 1:6, with the total load amount at 50–80% of the total volume of the container. When finishing large-scale workpieces or parts with special requirements on horizontal vibratory barrel finishing equipment, the clamping device is adopted to fix workpiece. This not only avoids the collision among workpieces, but also improves the finishing efficiency. It is worth noting that when the clamp fixes the workpiece, it must ensure that the workpiece loaded on the clamp can freely rotate for guaranteeing the finishing uniformity of the workpiece; meanwhile, it also allows the workpiece to have enough distance from the wall of the container in order to guarantee the smooth flow of abrasive blocks around the workpiece.

(3) Finishing medium
The abrasive blocks often keep moist condition (as with liquid medium), but they are used as dry state sometimes.
New sensors can be adopted to monitor force status of the workpiece from abrasive blocks during vibratory barrel finishing process. Figure 2.6 shows the curves of normal force with time measured by a sensor A101-1. A miniature color video camera can be used for recording the motion of abrasive blocks, analyzing the motion by replaying the video tape in frames. Figure 2.7 shows the motion photographs of abrasive blocks recorded by a camera IK-SM40A. These advanced test and observation methods can help us to establish the corresponding model of abrasive blocks and analyze the influence of the media on the finishing results. This can contribute to determining practical parameters according to specific finishing requirements. The rest contents will be detailed in Sect. 2.8 of this chapter [25].

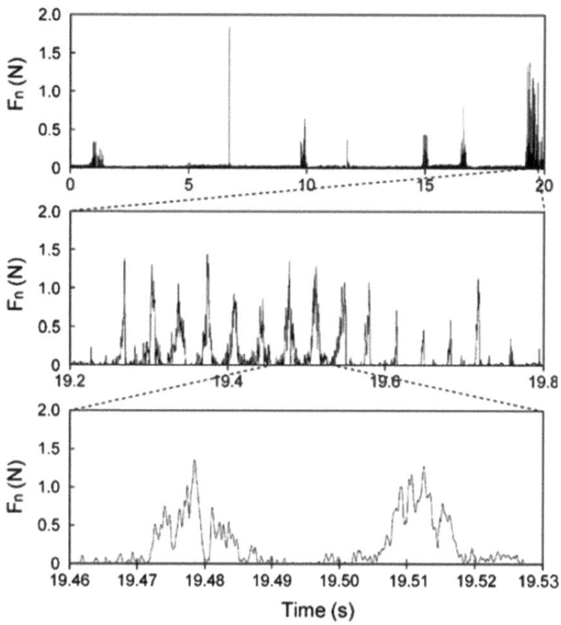


**Fig. 2.6** Curves of measuring normal force with time

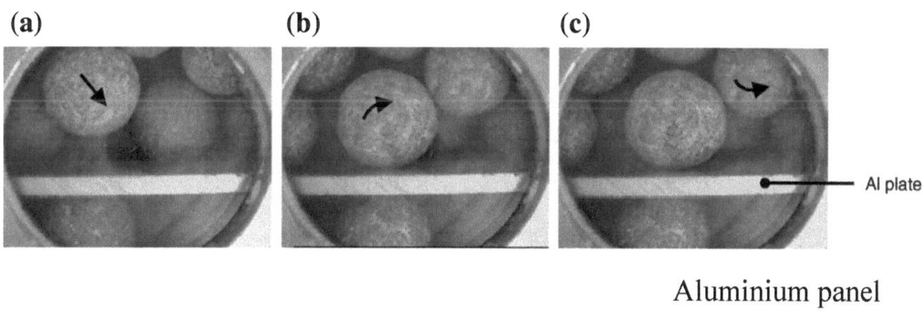


**Fig. 2.7** Motion photographs of abrasive blocks

## 2.3.3 Equipment Types and Design

1. Equipment types
   According to the difference of vibrating object (barrel or workpiece), vibratory barrel finishing equipment can be divided into different types, as listed in Table 2.4.
   Here is a brief description of features of the double vibration source vibratory barrel finishing method, orboresonant resonance barrel finishing method, and cascade barrel finishing method.

**Table 2.4** Types of vibratory barrel finishing equipment

| Type | Vibrating barrel | Vibrating workpiece |
|---|---|---|
| Equipment | Horizontal vibratory barrel finishing equipment | Orboresonant barrel finishing equipment |
| | Vertical vibratory barrel finishing equipment | |
| | Double vibration source vibratory barrel finishing equipment | Cascade barrel finishing equipment |

(1) Double vibration source vibratory barrel finishing method
Figure 2.8 is the structure principle diagram of the double vibration source vibratory barrel finishing method. As shown in the figure, it is basically the same as ordinary horizontal vibratory barrel finishing process and is composed primarily of the container, vibration source, motor, base, spring, and so on. The difference is that there are two vibration sources of this new kind of vibratory barrel finishing method; one installed under the container, which makes the container with workpiece and medium vibrate at certain amplitude and frequency; the other one installed in the middle of the container, which passes the vibration to the mass of workpiece and medium through the driving mechanism of vibration. It uses synthesis vibrations generated by the double vibration source for enabling the vibration frequency and the amplitude to have a wider adjustable range. This method can accelerate or slow down the relative motion of workpiece and medium to suit finishing requirements from various workpieces.

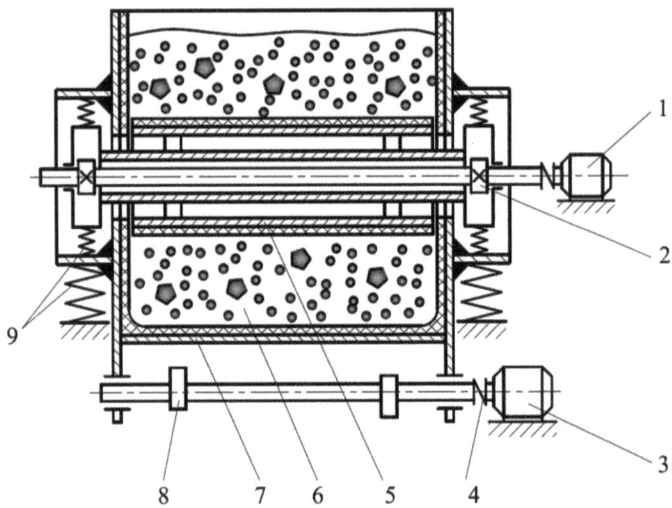


**Fig. 2.8** Structure principle diagram of the double vibration source vibratory barrel finishing method. 1, 3—Motors; 2, 8—vibration generators; 4—coupling; 5—vibration transfer unit; 6—workpiece and medium; 7—container; 9—spiral spring

(2) Orboresonant barrel finishing method

The resonance performance of the steel beam can make the workpiece fixed on the end of the steel beam vibrate at an extremely high frequency. It realizes the finishing of workpiece surface for the workpiece soaked in a motionless container with the medium. The principle originators named it "orboresonant," schematic diagram shown in Fig. 2.9.

As shown in Fig. 2.9, resonant steel beam 2 is supported on the two Support 4, and its position corresponds to the center of vibration device (i.e., the position of the least vibration degree). The maximum vibration amplitude to steel beam will increase at both ends of Clamping 1 fixed with the work-piece. When working, the workpiece soaked in the medium begins to vibrate, and its vibration source is a kind of continuously variable mechanical vibration generator 3. When the speed of rapid repetitive pulse passed from vibration generator approaches the natural frequency of the steel beam, the effectiveness of the input energy will be greatly improved. As the frequency of the vibration generator is 90–100 times/s, the vibration frequency of the workpiece is about 6000 times/s, and its amplitude is about 3 mm. After finishing, container 5 lowers quickly. Meanwhile, the vibrating steel beam continues in the vibrational state under the inertia effect for removing the medium in the workpiece's chamber.

From the foregoing, orboresonant barrel finishing equipment is simple in construction, easy to operate, and there is no impingement between parts. This method can achieve excellent effects for removing burrs and flashes on

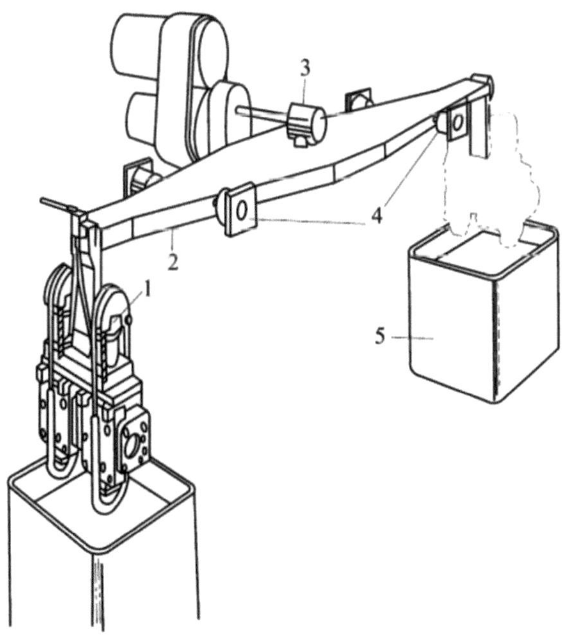


**Fig. 2.9** Principle diagram of orboresonant barrel finishing method. 1—Clamping; 2—resonant steel beam; 3—vibration generator; 4—support for steel beam; 5—containcr

the internal and external surface of more complex castings such as hydraulic components (like pump case, valve body) and gearbox. The finishing time usually lasts 1–5 min.

(3) Cascade barrel finishing method
Cascade barrel finishing method was developed for the Iowa Engineered Processor Corporation in 1983. Its mechanism is a freely cascading stream of the medium flowing into specific passages (e.g., the internal passages of a part). A part or parts are rigidly fixture and are mechanically driven by a vibration unit. A stream of abrasive blocks is fed into the ingress above the part down onto the interior edges of the fixture, while it is vibrated. As the part vibrates, the volume of abrasive blocks is impacted by the internal surfaces. This impact deaccelerates the abrasive blocks and then immediately accelerates abrasive blocks in a rebound direction as the vibration reverses. Because of the rapid vibrations generated within abrasive blocks flow, energy transfer from the forced part to abrasive blocks results in a high finishing efficiency. Figure 2.10 is the structure principle diagram of cascade barrel finishing method.
Let the weight of the two eccentric parts equal to each other is $m$; the angular velocity of the vibration generator shaft is $\omega$; the angle between the two eccentric parts projection in the vertical plane is $\alpha$; the distance from the gravity center of the eccentric parts to the vertical shaft is $r$. The exciting force generated by vibration generator in the horizontal direction is

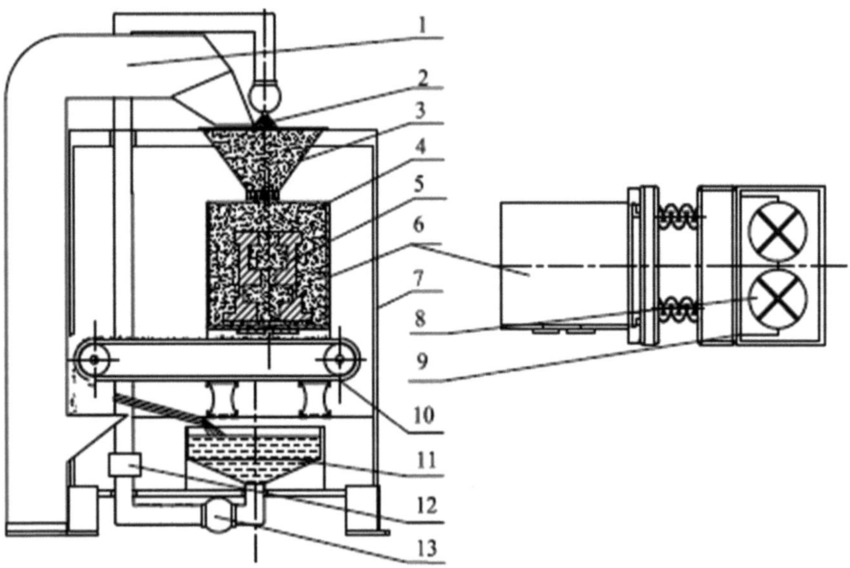


**Fig. 2.10** Structure principle diagram of cascade barrel finishing method. 1—Bucket elevator; 2—liquid medium; 3—hopper; 4—medium; 5—workpiece; 6—container; 7—open type chamber; 8—eccentric part; 9—vibration unit; 10—vibrating screen; 11—filter tank; 12—pump; 13—filter unit

$$P(t) = 4m\omega^2 r \cos \omega t \cos \frac{\alpha}{2} \tag{2.7}$$

Additionally, the vibrating amplitude is one main factor affecting finishing effect and efficiency. Considering only the effect of the horizontal direction springs, the horizontal amplitude of all vibrating parts is

$$A = \frac{4mr}{M}\left|\frac{z^2 \cos\left(\tan^{-1}\left(\frac{2\zeta z}{1-z^2}\right)\right)}{1-z^2}\right| \quad (z \neq 1) \tag{2.8}$$

where $M$ is the quality of all vibrating parts (kg); $\zeta$ is damping ratio; $z = \frac{\omega}{\omega_0}$, and $\omega_0$ is natural frequency of spring (rad/s).

This method is considerably different from the more traditional vibratory processes.

The mass of abrasive blocks causes the part to move, and the flow is primarily dependent upon the driving force transmitted through the mass of abrasive blocks. Cascade finishing process allows finishing a wide range of parts with few restrictions on the physical dimensions of the part.

In this finishing method, the fixture design is very important. The part fixture design must allow that the necessary parts hold, and it must provide the medium ingress and egress features that may not be inherent in the part configuration.

2. Equipment design

(1) Container design

The container capacity is the original parameter, which is determined by the size of the workpiece, production lot, finishing requirements, and so on. In the finishing process, high rigidity is required to meet the requirement that the container is the main part of the vibration. Weld forming and casting forming can be adopted to make the container. When using the steel plates welding formation, one thing to note is that the container must be equipped with ribs to increase its rigidity.

The internal shape of the container is an important condition for forming the continuous turnover flow of workpiece and medium. Given that the container needs to have a good turnover property in the directional excitation of workpiece and medium, the container's inner bottom of the horizontal vibratory barrel finishing equipment often adopts cylindrical structure, and the container's inner chamber of the vertical vibratory barrel finishing equipment adopts ring-shaped curved surface structure. Meanwhile, in order to reduce the noise, extend the service life of the container, reinforce the turnover effect of the container inside wall on the workpiece and medium by the friction, and it is necessary to paste rubber or smear polyurethane liner on the container inside wall.

(2) Recommended values of the main parameters
Tables 2.5 and 2.6 list the main parameters of the common horizontal, vertical vibratory barrel finishing equipment, respectively.

(3) Liquid circulation device
Metal powders and micro-powders of abrasive particles can be removed from the solution in the container through the settler of the circulation system, and the liquid circulation device can ensure the sharpness of the abrasives and the cleanness of liquid medium. The settled liquid can be returned back to the container through the pump of the liquid circulation device and can be reused later. Figure 2.11 is the typical liquid circulation device.

**Table 2.5** Recommended parameters values of horizontal vibratory barrel finishing equipment

| Total capacity of container/L | Centrifugal force exerted by eccentric parts/$\times 10^2$N | Motor power/kW | Coil spring | | Plate spring | |
|---|---|---|---|---|---|---|
| | | | Number | Rigidity coefficient of a single spring/(kg/cm) | Width/mm | Thickness/mm |
| 50 | 50–60 | 1 | 4 | 55–75 | 400 | 40 |
| 100 | 65–85 | 1.7 | 4 | 80–100 | 500 | 40 |
| 150 | 90–120 | 1.7–2.8 | 4 | 100–120 | 600 | 50 |
| 200 | 110–150 | 2.8–4.5 | 4 | 115–140 | 700 | 60 |
| 300 | 160–220 | 4.5–7 | 4 | 140–180 | 800 | 70 |
| 400 | 220–300 | 7–10 | 4 | 180–230 | 900 | 80 |
| 500 | 300–400 | 10–14 | 6 | 150–200 | 1000 | 80 |

**Table 2.6** Recommended parameters values of vertical vibratory barrel finishing equipment

| Total capacity of container/L | Centrifugal force exerted by eccentric parts/$\times 10^2$N | Motor power/kW | Rotating speed of vibration axis/(r/min) | Coil spring | |
|---|---|---|---|---|---|
| | | | | Number | Rigidity coefficient of a single spring/(kg/cm) |
| 10 | 15–18 | 0.25 | 3000 | 6 | 20–30 |
| 25 | 25–30 | 0.6 | 3000 | 12 | 30–40 |
| 50 | 35–45 | 1 | 3000 | 12 | 40–55 |
| 100 | 50–65 | 1.7 | 3000 | 12 | 55–70 |
| 150 | 70–90 | 1.7–2.8 | 3000 | 12 | 70–85 |
| 200 | 90–120 | 2.8–4.5 | 3000 | 18 | 55–70 |

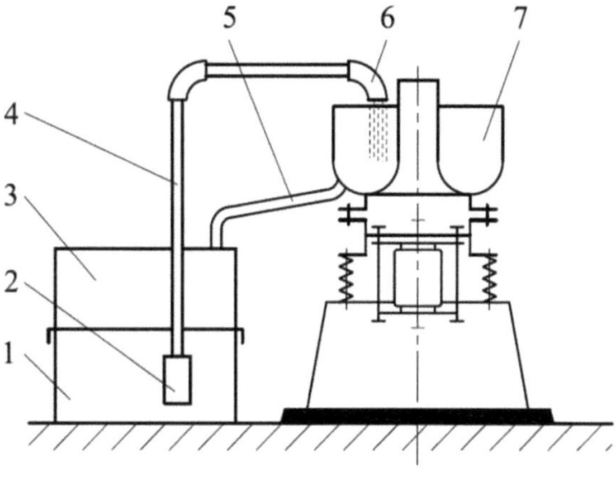


**Fig. 2.11** Liquid circulation schematic diagram. 1—Water tank; 2—pump; 3—settler; 4—pipeline; 5—flexible pipe; 6—valve; 7—vertical vibratory barrel finishing equipment

3. Equipment shapes
   Figure 2.12 shows the photographs of several different vibratory barrel finishing equipments. In practice, most conventional vibratory barrel finishing equipment comes with an amplitude label attached on them from which the amplitude of vibrations can be determined.

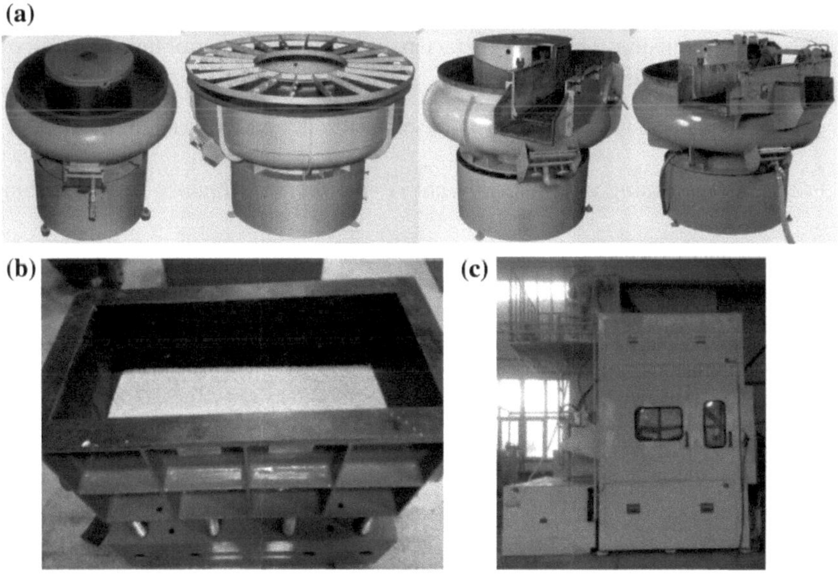

**Fig. 2.12** Photographs of several different vertical vibratory barrel finishing equipment. **a** Vertical vibratory finishing; **b** horizontal vibratory finishing; **c** cascade finishing

## 2.4 Whirling Barrel Finishing

### *2.4.1 Finishing Principle and Characteristics*

1. Finishing principle
   Whirling barrel finishing process, also called centrifugal disk finishing, is a relatively new barrel finishing method developed rapidly in the past 30 years. As shown in Fig. 2.13, the barrel is composed of fixed barrel wall 1 (stationary sidewalls) and rotary disk 2 for a base. As rotary disk 2 rotates at a certain speed, the workpieces and working medium 3 in the barrel are accelerated outwardly under centrifugal force. When the workpieces and working medium contact the stationary sidewalls, the sidewalls act as a brake and the load starts to decelerate, and they are forced upward by the load behind them. When arriving at a certain height, they will fall toward the disk. In the process of continuous rotary motion, the above moving process recurs continually and causes the load of workpieces and working medium to produce a spiral whirling motion. Then, strong finishing action is produced between the workpiece and the abrasive blocks, and the aims of deburring evenly, chamfering, and polishing are achieved. Figure 2.14 is the motion state photograph of whirling barrel finishing process [26–29].
2. Functional characteristics and application scope
   Whirling barrel finishing process is suitable to the finishing of small or medium special-shaped parts (except for sheet-shaped parts). It has a strong finishing capacity and is easy to realize automation control. However, there is a certain degree of collision. The main characteristics of whirling barrel finishing process are shown as follows [27, 28].

   (1) The applied force on the workpiece is strong, and the machining efficiency is high.
   (2) The machining range is wide but is not wide as vibratory barrel finishing process.

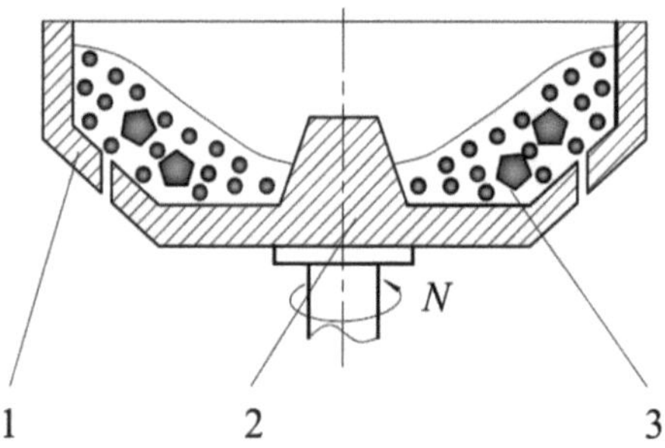


**Fig. 2.13** Whirling barrel finishing principle diagram. 1—Stationary sidewalls; 2—rotary disk; 3—workpieces and working medium

**Fig. 2.14** Motion state photograph of whirling barrel finishing process

(3) The machining process is as follows: feeding of workpieces → water supplying → barrel finishing → water discharge → abrasive blocks/workpieces discharge → separation → rinsing → out. It is easy to realize automation control, relates quite well with the prior process and the next process, and facilitates formation of automatic production line and merging into FMS (flexible manufacturing system).

(4) It has the merits of low noise, low vibration, free of liquid medium splashing. The working environment has been improved.

## 2.4.2 Main Factors Affecting Finishing Effects

1. Speed of the rotary disk

   Figure 2.15 is the status of medium flow in a certain finishing equipment (with a barrel capacity of 290 L, and an inside barrel diameter of 1100 mm) at different

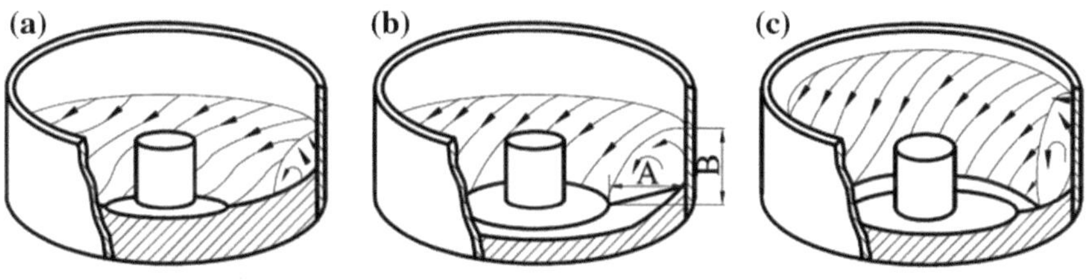


**Fig. 2.15** Status of medium flow at different rotational speed. **a** 60 r/min; **b** 90 r/min; **c** 120 r/min

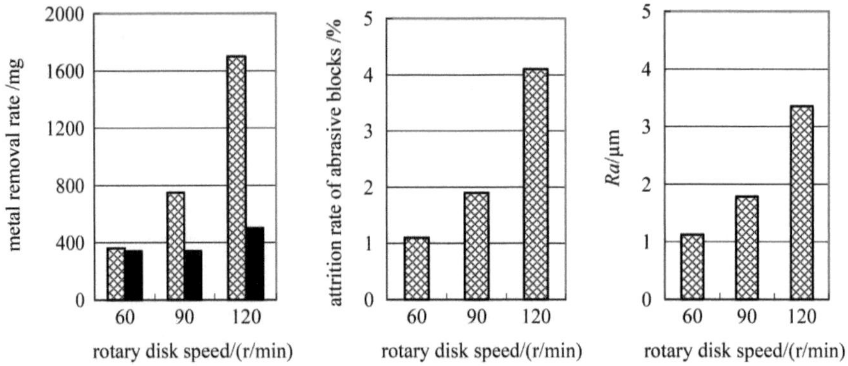


**Fig. 2.16** Impact of the rotary disk speed on metal removal rate, attrition rate of abrasive blocks, and surface roughness value

rotational speed. Under the same condition, the higher the rotational speed of the rotary disk is, the higher the rising altitude of the workpieces and working medium in the barrel along stationary sidewalls under the action of centrifugal force is, and the higher the relative flow speed of the load is. Then, the metal removal rate of workpiece is higher, the surface roughness value after machining is larger, and the attrition rate of the abrasive blocks is proportionally higher. Figure 2.16 shows the relationship among the metal removal rate, the attrition rate of the abrasive blocks, and the surface roughness value after machining under different speed of the rotary disk.

2. Load amount

   In whirling barrel finishing process, the loads of the workpieces, abrasive blocks, water, and compound are put into the barrel for machining. In general, the total volume of the load makes up about 20% of the volume of the barrel. Based on the shape and finishing requirements of the workpiece, the volume ratio of workpieces to abrasive blocks should be controlled 1:3 ~ 1:6. In finishing, the amount of the compound depends on water volume, and 5 g for each liter of water is preferred in general. It is suitable for the amount of the water to just submerge the machining medium, generally about 15% of the barrel volume. If too much water, the action of lubricating and buffering will increase, which reduces the finishing efficiency; if the water is too small, cleaning action of the workpiece and the grinding block may not be well, which will decrease the finishing quality and increase the attrition rate of abrasive blocks [29].

### 2.4.3 Equipment Types and Design

1. Equipment design

(1) Rotating chassis shapes and fixed barrel wall shapes

Figure 2.17 shows the status of medium flow under the condition of different rotary disk shapes and stationary sidewall shapes. When the total volume of the barrel is 290 L, the inside diameter of the barrel is 1100 mm and the rotational speed of the rotary disk is 90 r/min. As shown in Fig. 2.17a, when the stationary sidewall is cylinder and the rotary disk is tilted-line rotating surface, the height *B*, in which the load rises along the stationary sidewall, is less than the width *A* along radius distribution. Coarse-pitch spiraling flow layer with relatively low flow velocity is formed. As shown in Fig. 2.17b, when the stationary sidewall is cylinder and the rotary disk is curved-line rotating surface, the height *B* is greater than the width *A*, and fine-pitch spiraling flow layer with relatively high flow velocity is formed. As shown in Fig. 2.17c, when the stationary sidewall is prismatic surface and the rotary disk is curved-line rotating surface, variable-pitch spiraling flow layer with relatively high flow velocity is formed for that the prismatic surface of the stationary sidewall blocks the balance flow of the medium. Therefore, the motion state becomes relatively complicated.

Figure 2.18 shows the metal removal rate, the attrition rate of the abrasive blocks, and the surface roughness value after machining 2 h under the condition of different rotary disk shapes and stationary sidewall shapes. I, II, III in Fig. 2.18 correspond to the three situations of Fig. 2.18a–c, respectively.

According to the analysis mentioned above, and considering the finishing requirements, the stationary sidewalls weld into a cylindrical and regular polygon with formed plates, and the rotary disk is cast into shape with cast iron material. Their common requirements are lightweight and good rigidity.

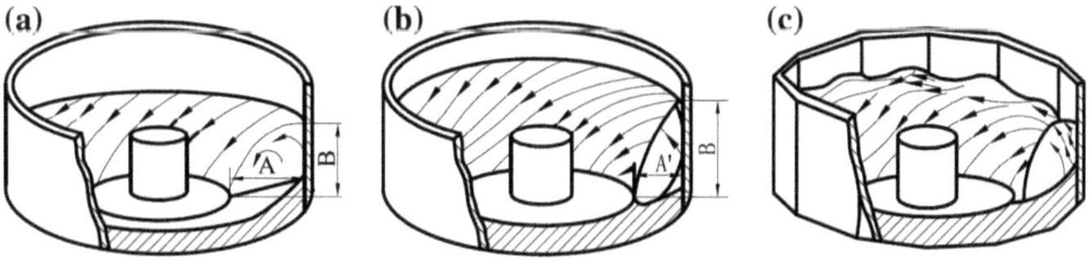


**Fig. 2.17** Status of medium flow under the condition of different rotary disk shapes and stationary sidewall shapes. **a** The stationary sidewall is cylinder, and the rotary disk is tilted-line rotary surface; **b** the stationary sidewall is cylinder, and the rotary disk is curved-line rotary surface; **c** the stationary sidewall is prismatic surface, and the rotary disk is curved-line rotary surface

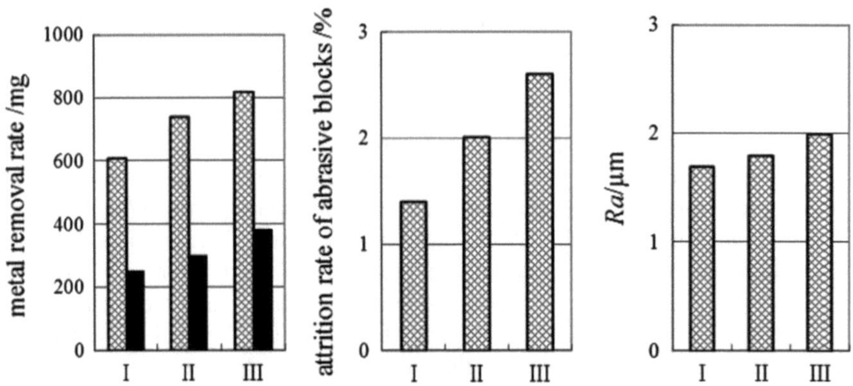


**Fig. 2.18** Impacts of different rotary disk shapes and stationary sidewall shapes on the metal removal rate, the attrition rate of the abrasive blocks, and the surface roughness value. I the stationary sidewall is cylinder, and the rotary disk is tilted-line rotary surface; II the stationary sidewall is cylinder, and the rotary disk is curved-line rotary surface; III the stationary sidewall is prismatic surface, and the rotary disk is curved-line rotary surface

(2) The analysis of structural rationality

The analysis of structural rationality needs to consider the following factors:

(1) Considering the actual processing conditions, the inner wall surface of the barrel, which is assembled by the stationary sidewalls and rotary disk, should adhere or brush a 3–5 mm-thick rubber layer or polyurethane for the sake of preservation and abrasion resistance.
(2) There is a relative movement between the stationary sidewalls and the rotary disk during processing; therefore, there must have a gap between them. In order to prevent the phenomenon of workpieces locking because of the gap, the structural style shown in Fig. 2.19a can be replaced by the style shown in Fig. 2.18b. Meanwhile, considering the inevitable existence of the gap, one container for liquid medium storage should be added below the bottom of the rotary disk in equipment design. In order to improve finishing results, a liquid circulation system can be added by using the container for liquid medium storage.

**Fig. 2.19** Local structure comparison of whirling barrel finishing equipment before and after optimizing. **a** Before optimizing; **b** after optimizing. 1—Stationary sidewalls; 2—gap; 3—rotary disk

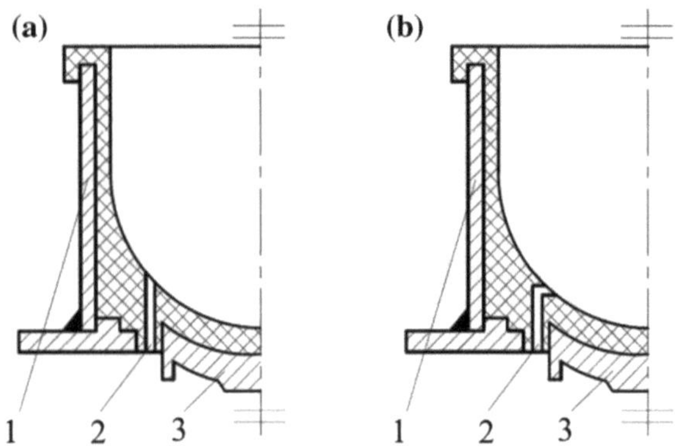

(3) In order to increase the level of automation, the barrel should automatically turn over 180° to discharge the load. Generally, double-hinge support can be adopted, and the operation can be performed by handle.
(4) Considering the suitability of the equipment, the speed of rotary disk must be adjustable, and this function can be realized by using the speed adjusting motor or the frequency conversion control.

2. Equipment shape

Figure 2.20 shows the photographs of ordinary small and full-automatic whirling barrel finishing equipment.

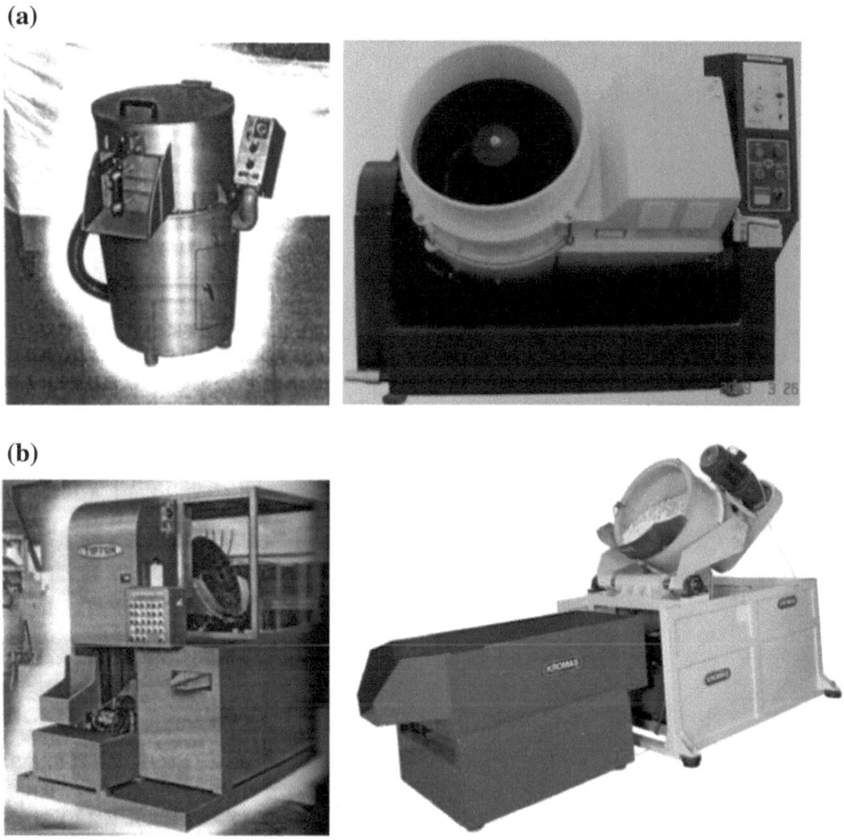

**Fig. 2.20** Photographs of whirling barrel finishing equipment. **a** Small equipment; **b** full-automatic equipment

## 2.5 Centrifugal Barrel Finishing

The centrifugal barrel finishing is the combination of workpieces, abrasive blocks, and liquid medium, which are put into a sealed barrel according to certain proportioning. The barrel undergoes the planetary motion. Under the action of the inertia force, the forced flow of the media in the barrel causes the collision, rolling, and micro-grinding of abrasive blocks onto the workpieces surface to be generated. The surface finishing of the workpieces is achieved [8].

Since the centrifugal barrel finishing process has greatly changed the motion pattern, it overcomes the defects of the rotary barrel finishing process, such as poor finishing quality, low efficiency. This process with centrifugal action results in a very fast, highly controllable operation, and maintains a smooth rubbing action with little or no workpiece impingement, making it possible to produce fine finishes on precision and fragile parts. Another important advantage is its capability to impart high compressive stresses in the surface layer.

### 2.5.1 *Finishing Principle and Characteristics*

1. Finishing principle

Figure 2.21a is the schematic diagram of the centrifugal barrel finishing principle. Several barrels revolve around one fixed shaft (the speed is $N$) and rotate actively round its own shaft (the speed is $n$); therefore, the planetary motion can be formed. When $N \neq 0$ and $n = 0$, the mixture in the barrel can form one curved surface of radius $R_1$ (Fig. 2.21b), and the mutual rolling between the workpieces and abrasive blocks can generate at a given positive pressure, but cannot form the

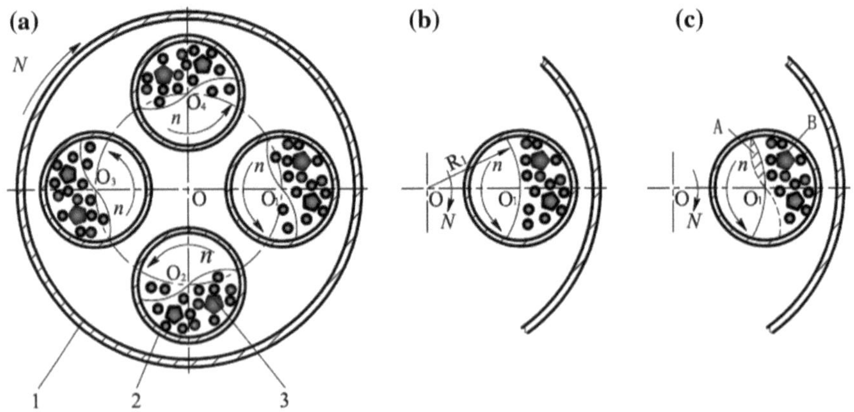


**Fig. 2.21** Schematic diagram of the centrifugal barrel finishing principle. **a** Principle diagram; **b** $N \neq 0$ and $n = 0$; **c** $N \neq 0$ and $n \neq 0$

forced flow. When $N \neq 0$ and $n \neq 0$, the barrel does both the revolution motion and the rotation motion, and the mixture in the barrel can form one new curved surface (shown by the dotted line in Fig. 2.21c). The mixture in the barrel tries to return to its original position under the action of the centrifugal force. The forced sliding flow layer can be formed from *A* to *B* in the figure. The relative slipping motion between the workpieces and abrasive blocks can be generated. By the result of the positive pressure and the forced flow, the mutual collision, rolling, and micro-grinding are forced to form among the workpieces and abrasive blocks [30–32].

2. Dynamics analysis

(1) Magnitude of inertia force and its influence

In centrifugal barrel finishing process, the collision, rolling, and micro-grinding among the workpieces and abrasive blocks result from the joint effects of the inertia force, the counterforce, the gravity, and the friction during their movement. If ignoring the influence of gravity, the forced condition of one particle among the mixture in the barrel is shown in Fig. 2.22. The counterforce $F$ of the barrel wall acting on the mixture can be separated into two component forces: One is the constraining force $F_r$ which has the same size of the inertia force $F_i$ of the particle doing the planetary motion, and has the opposite direction of the inertia force. It forces the particle doing the planetary motion with the barrel; the other is a sliding component force $F_s$, and it is combined with the friction $F_f$ to form the thrusting force $F_t$ which forces the particle doing the relative movement with respect to the barrel. Among the above forces, the sizes of $F$ and $F_r$ depend on $F_i$. Therefore, in centrifugal barrel finishing process, $F_i$ is an important parameter. For one particle in the wall of the barrel, $F_i$ is

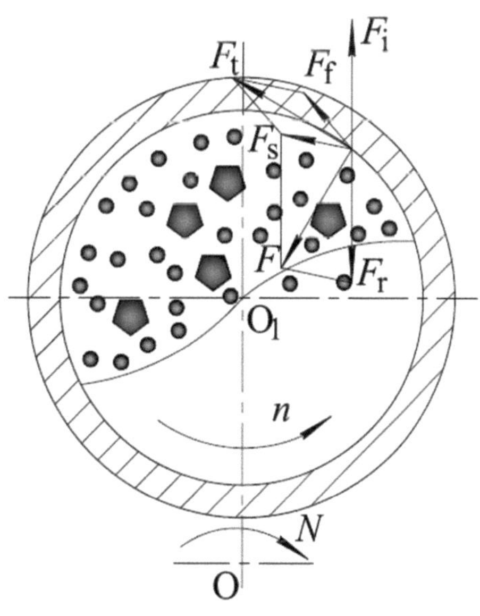


**Fig. 2.22** Force analysis of one particle among the mixture in the barrel

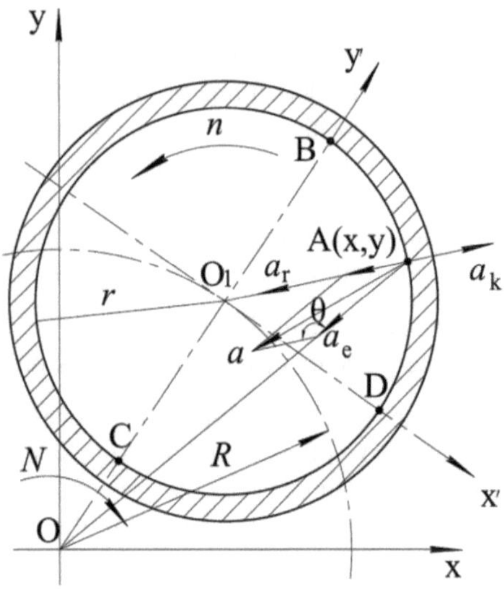


**Fig. 2.23** Acceleration analysis of one particle among the mixture in the barrel

$$F_i = ma \tag{2.9}$$

where $m$ is the mass of the particle and $a$ is the absolute acceleration of the particle.

Figure 2.23 is the acceleration analysis chart of one particle among the mixture in the barrel, and the absolute acceleration $a$ of Point A in the barrel can be computed according to Fig. 2.23. In the figure, $R$ and $r$ are the revolution and rotation radius of the barrel; the convected acceleration $a_e$, the relative acceleration $a_r$, and the Coriolis acceleration $a_k$ of Particle A are

$$a_e = \mathrm{OA} \cdot \omega_N^2 = \sqrt{x^2 + y^2}(2\pi N)^2 \tag{2.10}$$

$$a_r = \mathrm{O_1A} \cdot \omega_n^2 = r(2\pi n)^2 = \left(r\frac{n^2}{N^2}\right)(2\pi N)^2 \tag{2.11}$$

$$a_k = 2\omega_N \cdot V_A = \left(2r\frac{n}{N}\right)(2\pi N)^2 \tag{2.12}$$

$$a^2 = (a_r + a_k)^2 + a_e^2 + 2(a_r + a_k)a_e \cos\theta \tag{2.13}$$

$$\cos\theta = \left(r^2 + x^2 + y^2 - R^2\right)\Big/2r\sqrt{x^2 + y^2} \tag{2.14}$$

where $R$ and $r$ are the revolution and rotation radius of the barrel (mm) and $N$ and $n$ are the revolution and rotation velocity of the barrel (r/min); $\omega_N$ is the revolution angular velocity of Particle A (rad/s); $\omega_n$ is the rotation angular velocity of Particle A (rad/s); $V_A$ is the circumferential velocity of Particle A relative to $O_1$ (m/s).

From Eqs. (2.10)–(2.14), the absolute acceleration of the particle van bet obtained and it is

$$a = (2\pi N)^2 \left[ (x^2 + y^2)\left(1 + \frac{n}{N}\right)^2 + r^2\left(1 + \frac{n}{N}\right)^2\left(2 + \frac{n}{N}\right)\frac{n}{N} - R^2\left(2 + \frac{n}{N}\right)\frac{n}{N} \right]^{1/2} \tag{2.15}$$

When the particle position is Point B which is the furthest point from Center-point O, the absolute acceleration $a$ is

$$a = (2\pi N)^2 \left[ R + r\left(\frac{n}{N} + 1\right)^2 \right] \tag{2.16}$$

When the particle position is Point C which is the nearest point from Center-point O, the absolute acceleration $a$ is

$$a = (2\pi N)^2 \left[ R - r\left(\frac{n}{N} + 1\right)^2 \right] \tag{2.17}$$

When the particle position is Point D which is on the center line of the barrel, the absolute acceleration $a$ is

$$a = (2\pi N)^2 \left[ R^2 + r^2\left(\frac{n}{N} + 1\right)^4 \right]^{1/2} \tag{2.18}$$

Known from Eqs. (2.15)–(2.18), $a$ is related to $N$, $n/N$, $R$, $r$, and the particle position, the maximum appears on Point B, and the minimum appears on Point C. If $n/N = -1$, the value of $a$ always is

$$a = (2\pi N)^2 R \tag{2.19}$$

This shows that the inertia forces of the particle are the same regardless of the positions. Through quantitative graphic methods, the absolute acceleration direction of one particle in the barrel is parallel to the connecting line between the revolution center and the rotation center of the barrel.

(2) Formation of forced flow

The forced flow is relative to the natural flow. In the rotary barrel finishing process, when the barrel rotates at a certain speed, the mixture in the barrel lifts up with a rotation direction along the wall of the barrel, under the action of gravity, centrifugal force, and friction. When the mixture lifts up to a certain height, it will lose the balance, and the mixture of the surface layer will slide down naturally. In the centrifugal barrel finishing process, the barrel does both revolution and rotation; the mixture in the barrel has relatively lifting effect along the wall of the barrel; meanwhile, the mixture of surface layer is forced to slide down. This forced slipping is the forced flow [33].

Known from the foregoing analysis, the barrel must do rotation to form the forced flow into the centrifugal barrel finishing process, that is

$$n \neq 0 \tag{2.20}$$

According to Eq. (2.17), when $\frac{n}{N} > \sqrt{\frac{R}{r}} - 1$ or $\frac{n}{N} < -\left(\sqrt{\frac{R}{r}} + 1\right)$, the acceleration $\alpha$ of the mixture at Point C is negative, and the mixture rotates with the barrel, attaches evenly to the wall of the barrel, and remains relatively motionless. This cannot form the force flow. Therefore, in the process of centrifugal barrel finishing, in order to obtain satisfactory machining effects, $n/N$ should satisfy

$$\sqrt{\frac{R}{r}} - 1 > \frac{n}{N} > -\left(\sqrt{\frac{R}{r}} + 1\right) \tag{2.21}$$

At this point, the mixture in the barrel does both the planetary motion with the barrel and the forced flow relative to the barrel under the action of constraining force $F_r$ and thrusting force $F_t$ [34].

On the basis on Eqs. (2.19) and (2.21), in order to make the workpiece well-proportioned by the force in process, $n/N = -1$ is the best selection. This can guarantee the stability of machining quality and benefit the strength designing of the equipment structure.

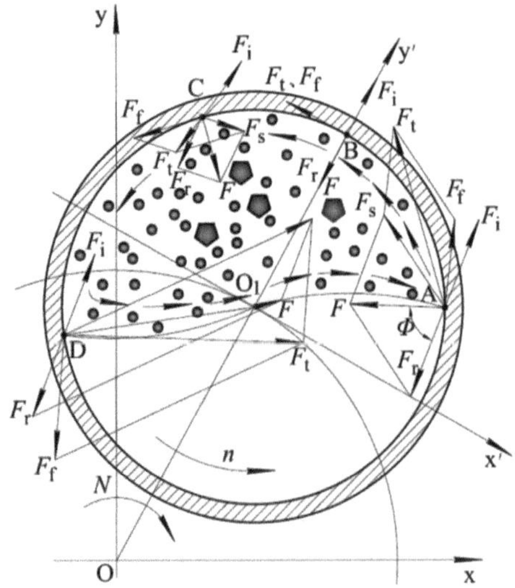


**Fig. 2.24** Force status of the mixture in the barrel at every point

When $n/N = -1$, the force status of the mixture in the barrel at every point is shown in Fig. 2.24. At Point A, for the counterforce $F$ of the barrel wall acting on the particle, its direction is perpendicular to the wall of the barrel. $F$ can be separated into $F_r$ and $F_s$, and the constraining force $F_r$ maintains the planetary motion of the particle; the sliding component force $F_s = \left(F^2 + F_r^2 - 2FF_r \cos \varphi\right)^{1/2}$. $F_s$ and the friction $F_f$ of the barrel wall acting on the particle are combined to the thrusting force $F_t$ which pushes the particle to do the relative movement with respect to the barrel. In Area *AB* along the barrel wall, the thrusting force $F_t$ helps the medium to have a motion along the barrel wall in accordance with the rotation direction of the barrel. But in Area *BCD*, the magnitude and direction of $F_t$ can change with the increase of angle $\Phi$; therefore, the mixture in the barrel tries to do it with its own forced flow relative to the barrel under the action of the thrusting force. The mixture in the inner layer is acted upon the thrusting force from the surrounding mixture, the relative sliding between different mixtures is blocked, the velocity of the relative sliding is slower, and it is in a dynamic balance on the whole and remains to be relatively still. The finishing capacity in this case is very weak. The mixture in the surface layer Area *DA* is acted upon small thrusting force from the surrounding mixture; the fast relative sliding between the mixtures is generated along the action direction of $F_t$, and the forced flow is intense. The finishing capacity in this case is good [1, 35].

The related forced flow process is shown in the high-speed photography of Fig. 2.25. Two hexagon barrels are installed symmetrically on the centrifugal barrel finishing equipment. The white part of the barrel is the mixture of the workpieces and the medium. The factors are as follows: $N = 280$ r/min, $n/N = -1$,

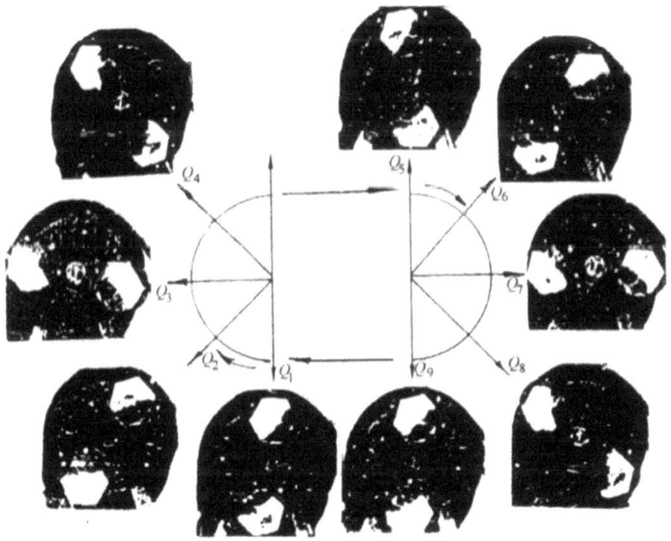

**Fig. 2.25** High-speed photography of related forced flow process

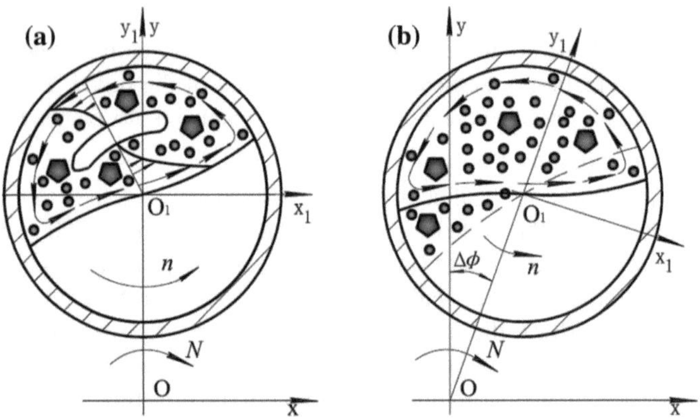


**Fig. 2.26** Analysis of forced flow process of the mixture. **a** Forced flow situation; **b** forced flow process

$R$ = 110 mm, $r$ = 42.5 mm, and the photograph speed of high-speed camera (Model Pentezent-16) is 600 times/s.

The forced flow situation of the mixture in the barrel at the instantaneous moment is shown in Fig. 2.26a. The mixture in the barrel can be divided into three kinds: The mixture in inner layer rotates with the barrel, and it is in a dynamic balance on the whole; the mixture in surface layer does the forced flow in a direction opposite to the rotational direction of the barrel under the action of the thrusting force; the middle area is acted upon the friction torque of the mixture in inner layer and surface layer and can generate core-annulus peristalsis. When the revolution of the barrel reaches $\Delta\Phi$ (Fig. 2.26b), the mixture transforms force fully from the dashed status to the solid-line status. In the process of the forced transition, the relative motion exists not only in the interlayer, but also within the layer due to the differences in the shape, size, and quality of the workpieces and abrasive blocks.

3. Flow field simulation

Based on the motion course of the mixture in the barrel, the corresponding two-phase flow model and its control equations are established. The dynamic process of the mixture motion is simulated and numerically studied using fluid dynamics software Fluent, and the distributing characters of the mixture concentration, velocity, and pressure in the course of the barrel-running are obtained. The impact of various machining conditions (such as barrel shapes, motion parameters, characteristic parameters of abrasive blocks) to the motional behavior of the mixture is analyzed. This supplies theory base for the structure design of this finishing technology and the reasonable selection of the constructional parameters.

In centrifugal barrel finishing process, the barrel contains the mixture of parts and a certain amount of abrasive blocks and liquid medium. When the volume of the abrasive blocks and the parts makes up above 10% of the total volume of the barrel, the interaction among abrasive blocks, parts, and liquid medium can have a non-negligible impact on the movement feature. The liquid medium and abrasive blocks should be taken as a continuous medium, and the motion process of abrasive blocks should be described by Euler method [36].

It's supposed that no physiochemical process, such as dissolution and crystallization, occurs among the abrasive particle phase, the part phase and the liquid phase during establishing and simplifying the model. That is to say, the mass and heat exchangers between any two phases will not be considered, and the pressures among these phases are equal. Based on the mass conservation and the momentum conservation in two-phase flow continuum theory and *k*-ε turbulence law, the dynamical equations of the material and the liquid are established. The boundary conditions and the parameter conditions are unified, and the motion model of the mass material is obtained for analyzing the effect of its movement flow on the processing effect.

Setting the volume fraction of the liquid and the material particle are $\alpha_f$ and $\alpha_k$, respectively, then

$$\alpha_f + \alpha_k = 1 \tag{2.22}$$

And the mass concentration distributions of the particles are

$$n_k = \alpha_k \rho_k \tag{2.23}$$

where $n_k$ is the mass concentration distributions of the particles (kg/m$^3$) and $\rho_k$ is the material density of the particles (kg/m$^3$).

Since there is no mass exchanger between the two phases, the mass conservation equations of the liquid phase and the particle phase are

$$\begin{cases} \frac{\partial}{\partial t}(\alpha_f \rho_f) + \nabla \cdot (\alpha_f \rho_f \vec{V}_f) = 0 \\ \frac{\partial}{\partial t}(\alpha_k \rho_k) + \nabla \cdot (\alpha_k \rho_k \vec{V}_k) - 0 \end{cases} \tag{2.24}$$

where $\rho_f$ is the material density of the liquid (kg/m$^3$) and $v_f$ and $v_k$ are the velocity distributions of the liquid phase and the particle phase (m/s).

The motion of the particles in the fluid medium is a consequence of the resistance of the liquid acting on the particles, and these particles experience centrifugal force and gravity. The fluid resistance is the combined force of the pressure difference resistance and the frictional resistance acting on the particles when the particles move at constant speed in the motionless viscous liquid. Then, the movement of the liquid phase and the particle phase satisfied the following momentum conservation equations

$$\begin{cases} \frac{\partial}{\partial t}(\rho_f \vec{V}_f) + \nabla \cdot (\rho_f \vec{V}_f \vec{V}_f) = -\nabla P + \nabla \cdot \bar{\tau}_f + \vec{R}_{kf} + \vec{F}_{gf} \\ \frac{\partial}{\partial t}(\alpha_k \rho_k \vec{V}_k) + \nabla \cdot (\alpha_k \rho_k \vec{V}_k \vec{V}_k) = -\nabla P + \nabla P_k + \nabla \cdot \bar{\tau}_k + \vec{R}_{fk} + \vec{F}_{gk} \end{cases} \tag{2.25}$$

where $F_{gf}$ and $F_{gk}$ are the combined force of the gravity, the centrifugal force and the Coriolis force acting on the liquid medium and the particles ($N$); $P_k$ is the solid phase pressure of the granular fixing abrasive ($N$); $R_{fk}$ and $R_{kf}$ are the interactive force between the liquid phase and the particle phase, and their sizes are equal; their directions are opposite; $\overline{\tau}_f$ and $\overline{\tau}_k$ are the shear stress tensors of the liquid phase and the particle phase, and they are

$$\begin{cases} \overline{\tau}_k = \mu_k(\nabla\vec{V}_k + \nabla\vec{V}_k^{\mathrm{T}}) + \left(\lambda_k - \frac{2}{3}\mu_k\right)\vec{V}_k\nabla\cdot\overline{I} \\ \overline{\tau}_f = \mu_f\left(\nabla\vec{V}_f + \nabla\vec{V}_f^{\mathrm{T}}\right) + \left(\lambda_f - \frac{2}{3}\mu_f\right)\vec{V}_f\nabla\cdot\overline{I} \end{cases} \tag{2.26}$$

where $\mu_f$, $\mu_k$ and $\lambda_f$, $\lambda_k$ are the first viscosity coefficients and the second viscosity coefficients of the liquid phase and the particle phase and $\overline{I}$ is the unit tensor.

When $N = 158$ r/min and $n = -21$ r/min, the cloud diagram of the volume fraction for particles flow in the barrel over one revolution cycle is shown in Fig. 2.27. As shown in Fig. 2.27, light color represents that the particle density is big enough, and dark color represents that the particle density becomes smaller. They respond to the concentration distribution characteristics of the particles, and this indicates that the interface shapes in different rotation degrees are generally similar. As the centrifugal force moves around periodically, the positions of the interfaces present similar periodic change [37].

4. Discrete element modeling

3D DEM model of centrifugal barrel finishing is constructed on the basis of the Hertz–Mindlin (no slip) contact model. In this model, the normal force component

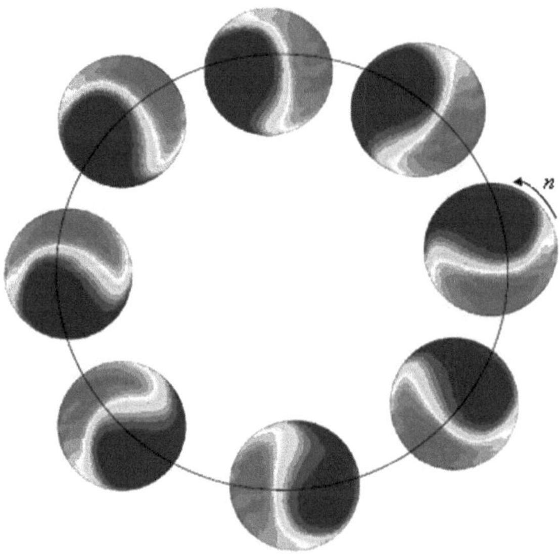


**Fig. 2.27** Cloud diagram of the volume fraction for particles flow in the hexagon barrel

is based on Hertzian contact theory. The tangential force model is based on Mindlin–Deresiewicz work. Both normal and the tangential forces have damping components where the damping coefficient is related to the coefficient of restitution. The tangential friction force follows the Coulomb law of friction model. The rolling friction is implemented as the contact independent directional constant torque model.

The normal force $F_n$ and damping force $F_n^d$ are given by Eqs. (2.27)–(2.28).

$$F_n = \frac{4}{3} E^* \sqrt{R^* \delta_n^3} \tag{2.27}$$

$$F_n^d = -2\sqrt{\frac{5}{6}} \beta \sqrt{S_n m^*} \upsilon_n \overrightarrow{\text{rel}} \tag{2.28}$$

where $E^*$, $R^*$, $\delta_n$, $m^*$, $v_n^{\overline{\text{rel}}}$, $S_n$ are the equivalent Young's modulus, the equivalent radius, the normal overlap, the equivalent mass, the normal component of the relative velocity, the normal stiffness, respectively, which are calculated by Eqs. (2.29)–(2.33).

$$\frac{1}{E^*} = \frac{(1-\upsilon_i^2)}{E_i} + \frac{(1-\upsilon_j^2)}{E_j} \tag{2.29}$$

$$\frac{1}{R^*} = \frac{1}{R_i} + \frac{1}{R_j} \tag{2.30}$$

$$\frac{1}{m^*} - \frac{1}{m_i} + \frac{1}{m_j} \tag{2.31}$$

$$\beta = \frac{\ln e}{\sqrt{\ln^2 e + \pi^2}} \tag{2.32}$$

$$S_n = 2E^* \sqrt{R^* \delta_n} \tag{2.33}$$

The tangential force, $F_t$, the tangential damping $F_t^d$, and the Coulomb friction is given by Eqs. (2.34)–(2.36).

$$F_t = -S_t \delta_t \tag{2.34}$$

$$F_t^d = -2\sqrt{\frac{5}{6}} \beta \sqrt{S_t m^*} \upsilon_t \overrightarrow{\text{rel}} \tag{2.35}$$

$$F_s = \mu_s F_n \tag{2.36}$$

where $\delta_t$ is the tangential overlap and $S_t$ is the tangential stiffness.

$$S_{\mathrm{t}} = 8G^{*}\sqrt{R^{*}\delta_{\mathrm{n}}} \quad (2.37)$$

Here, $G^*$ is the equivalent shear modulus and is calculated as

$$\frac{1}{G^{*}} = \frac{2-\upsilon_i^2}{G_i} + \frac{2-\upsilon_j^2}{G_j} \quad (2.38)$$

For centrifugal barrel finishing in which rolling friction is important, a torque is applying to the contacting surfaces.

$$T_i = -\mu_{\mathrm{r}} F_{\mathrm{n}} R_i \omega_i \quad (2.39)$$

where $\mu_r$ is the coefficient of rolling friction; $R_i$ is the distance of the contact point from the center of mass; $\omega_i$ is the unit angular velocity vector of the abrasive particle at the contact point.

The material properties and the contact parameters given in Tables 2.7 and 2.8, and the geometry parameters and motion parameters provided in Table 2.9 are used in these simulations. About 5000 particles are introduced to the DEM at the speed

**Table 2.7** Material properties used in the DEM simulations

| Material parameters | Material | Poisson's ratio | Shear modulus (Pa) | Density ($kg{\cdot}m^{-3}$) |
|---|---|---|---|---|
| Abrasive blocks | $Al_2O_3$ | 0.21 | 1.24e + 11 | 2675 |
| Barrel | Nylon 66 | 0.28 | 3.2e + 09 | 1150 |

**Table 2.8** Contact parameters of abrasive particles and the barrel

| Interaction parameters | Coefficient of restitution | Coefficient of static friction | Coefficient of rolling friction |
|---|---|---|---|
| Particle–particle | 0.55 | 0.3 | 0.3 |
| Particle–barrel | 0.36 | 0.15 | 0.15 |

**Table 2.9** Geometry parameters and motion parameters

| Parameters | Value/mm |
|---|---|
| Bore diameter of drum/mm | 118 |
| Height of drum/mm | 125 |
| Thickness of drum/mm | 5 |
| Revolution radius/mm | 135 |
| Transmission ratio | −1 |
| Radius of abrasive blocks/mm | 3 |

of 10,000 particles per second when the simulation runs, with the packed weight 50%, and freely falls with the initial velocity 2 m/s. Then, the drum begins to do planetary motion from 1 to 2 s. When simulating, abrasive particles are randomly generated by particle factory from 0 to 1 s (generation rate is 10,000 particles per second) and generate 5000 particles.

The gravity center and the velocity of abrasive blocks group can be obtained from Eqs. (2.40) and (2.41).

$$(x, y, z) = \left(\frac{\sum x_i}{n}, \frac{\sum y_i}{n}, \frac{\sum z_i}{n}\right) \tag{2.40}$$

$$\left(v_x, v_y, v_z\right) = \left(\frac{\sum v_{xi}}{n}, \frac{\sum v_{yi}}{n}, \frac{\sum v_{zi}}{n}\right) \tag{2.41}$$

where $n$ represents the total number of the abrasive blocks and $i$ represents serial number of the abrasive blocks.

Figure 2.28 shows the gravity center change of the abrasive blocks group with $i = -1$. As shown in Fig. 2.28, the $z$-axis of the gravity center of the abrasive blocks group keeps unchanged. In addition, the gravity center trajectory of the abrasive blocks group is a closed circle with the circle center (0, 0). The gravity center trajectory of the abrasive blocks group is same as the revolution center of centrifugal barrel finishing equipment can be obtained.

The velocity variation of gravity center of the abrasive blocks group with time is shown in Fig. 2.29. It can be seen that the gravity center velocity of the abrasive

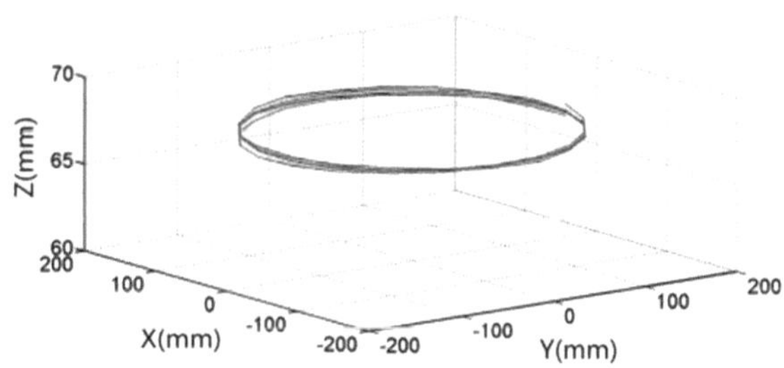


**Fig. 2.28** Position change of the center of gravity of abrasive particles

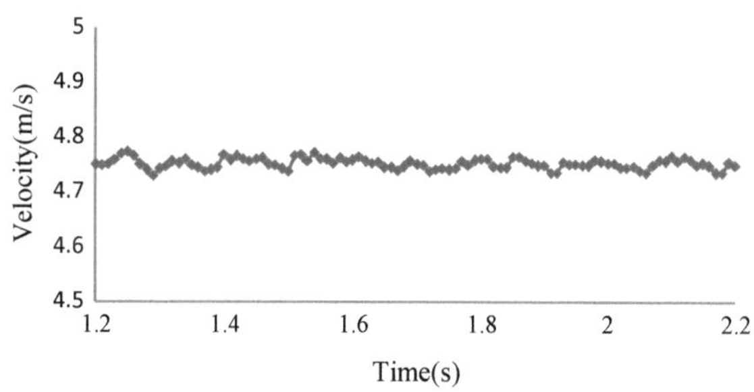


**Fig. 2.29** Gravity center velocity of the abrasive particles group versus time

blocks group ranges from 4.7 to 4.8 m/s, and the average velocity is 4.752 m/s. The deviation of simulation results to theoretical analysis is 0.04%.

The instantaneous distribution of abrasive blocks in the barrel is shown in Fig. 2.30 with simulation time being 2 s. At this time, the instantaneous gravity center coordinates and instantaneous velocity of the abrasive blocks group are (149.577, −23.1086, 65.1121) and (0.730, 4.697, 0.003), respectively. The flow path of abrasive blocks is also shown in Fig. 2.10. When the supportive force and the friction force cannot provide sufficient velocity and acceleration to keep them attached to the wall of the barrel surface, the abrasive blocks move toward the center of gravity and far from the barrel wall, thereby forming a forced flow layer.

When the barrel rotates clockwise, the abrasive blocks rotate together with it because of the friction between the inner wall of the barrel and the adjacent abrasive blocks. When the revolution takes part in, and the centrifugal force formed by revolution is larger than that of formed by rotation, the abrasive blocks separate from the wall of the barrel and move toward the upper right as in Fig. 2.30.

The abrasive blocks in the barrel should be rotary motion relative to the center of gravity under the action of the friction torque. So, the velocity relative to the gravity from the center of the abrasive blocks is proportional to the distance between the gravity center and the abrasive blocks. Therefore, all abrasive blocks velocity is derived. The velocity distribution of abrasive blocks that minus the revolution velocity vector is shown in Fig. 2.31. Specific operation method includes the

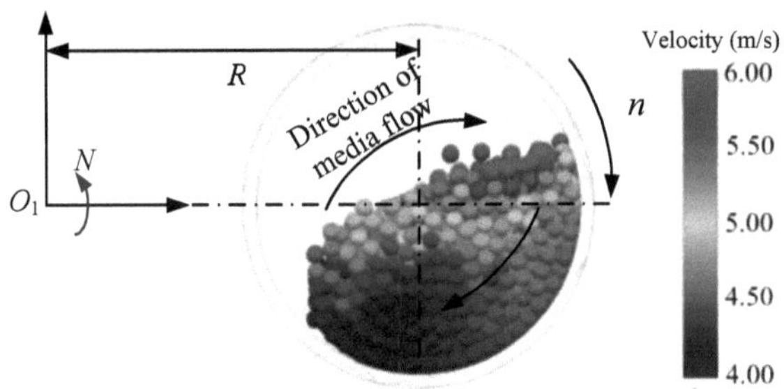


**Fig. 2.30** Instantaneous velocity distribution of abrasive particles within the barrel at 2 s

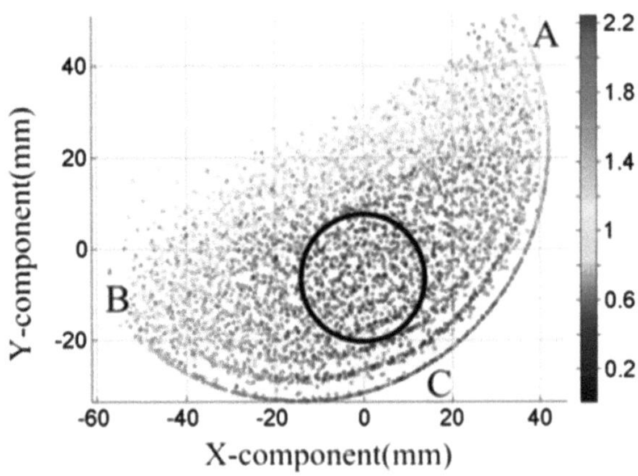


**Fig. 2.31** Velocity distribution of abrasive particles relative to its center of gravity at 2 s

following: Using EDEM ANALYST, export the position and velocity component of all abrasive particles at 2 s, and derive rotation velocity of all abrasive particles, and then the velocity distribution is drawn with the coordinate system of position component in the $x$, $y$ direction and rotation velocity using MATLAB software.

Figure 2.31 shows that the velocity of abrasive blocks is the smallest at the center of gravity, and the peripheral velocities are relatively large. Abrasive blocks in region A is far from the center of the barrel, moving to region C and forming gradually decelerating centripetal motion caused by rotation torques. Then, continue to move to region B and form gradually accelerating eccentric motion. Finally, the velocity reaches the critical value and separates from the wall of the drum surface, close to region A.

Due to the velocities of abrasive blocks in the barrel are different in the macroscale, it is very difficult to describe using conventional methods. Therefore, Maxwell velocity distribution law for large amounts of gas molecular motion state is attempted to the use of describing the abrasive particles motion from the statistics perspective. According to its definition, velocity distributions relative to the center of gravity were classified in the barrel, as shown in Fig. 2.32. It can be seen that the percentage of the abrasive blocks with different velocity range in total amount of abrasive blocks counts is almost at a constant during different moments; this indicates that the velocity distribution follows Maxwell velocity distribution law. Through the use of velocity distribution rules of abrasive blocks in the barrel, it can effectively help designers by revolutionizing the velocity to study velocity distribution of abrasive blocks in the barrel [38, 39].

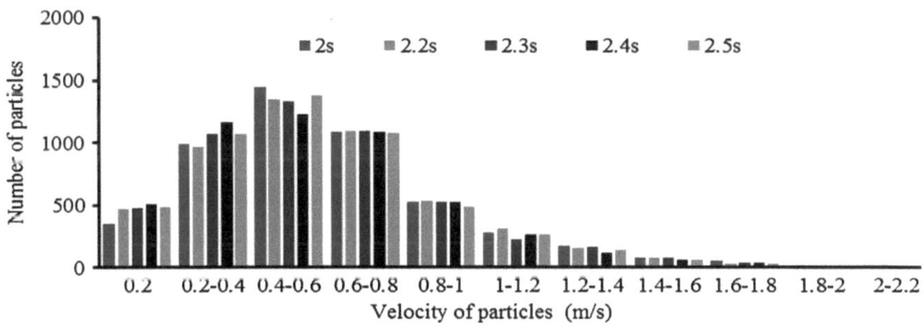


**Fig. 2.32** Number of abrasive particles with different velocity range

5. Function characteristics and applicable scope

The finishing efficiency of the centrifugal barrel finishing process is high, and it is appropriate for the deburring and finishing of small- or medium-sized parts with multiple variety and large number of productions. It can be used for removing the oxide layer and the small flashes on the surface of stamping or casting parts in rough processing, and precision machining before electroplating. It can enhance the surface quality and the surface roughness, with degree of the finished parts that can increase up to 1–2 grades. It can improve the physical–mechanical properties of the part surface layer, increase the surface hardness, and change the surface residual stress. Otherwise, it can also have other advantages such as decreased cost, reduced working intensity [1, 15–17, 20, 40–42].

## 2.5.2 Main Factors Affecting Finishing Effects

In centrifugal barrel finishing process, many factors can influence its finishing quality and efficiency. The factors mainly include the workpiece conditions, finishing equipment, finishing medium, and finishing process [41]. Only through coordinating the matching relationship properly among these four aspects, the barrel finishing process can be completed excellently, efficiently, and beneficially [43]. Figure 2.33 shows the flowchart of the centrifugal barrel finishing process. Once the machined workpiece and the finishing requirements are determined, the factors influencing finishing quality and efficiency become finishing equipment, finishing medium, and various parameters in the course of finishing [44]. The influence of the finishing medium will be expatiated in Sect. 2.8 of this chapter.

1. Kinematic parameters of equipment
   Based on the analysis of finishing principles, the kinematic parameters with the most influence on finishing effects include $n/N$ and $N$.

   (1) $n/N$
   Based on the principle analysis, $n/N$ has to satisfy the following condition:

$$\sqrt{\frac{R}{r}} - 1 > \frac{n}{N} > -\left(\sqrt{\frac{R}{r}} + 1\right), \quad \text{and } n \neq 0$$

   Depending upon the analysis of kinematics research, if $n/N < 0$ (i.e., revolution moves in the opposite direction of rotation), the flow direction of the forced flow layer is basically the same with the orientation of the particle motion, the flow resistance is low, and the sliding speed between abrasive blocks and the workpieces is fast; this is beneficial to improve the finishing efficiency and guarantee the finishing quality. On the contrary, if $n/N > 0$, the sliding speed is slow, and the finishing effects are poor.

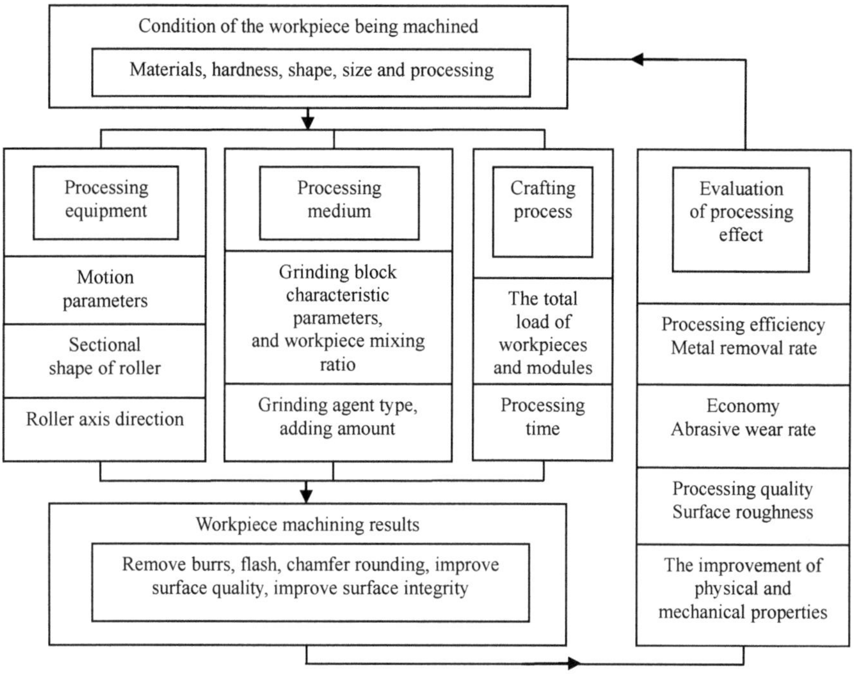


**Fig. 2.33** Flowchart of centrifugal barrel finishing process

For further study of the optimal value of $n/N$, taking $n/N = -1, -1.5$, and $-2$ as samples, the tests are carried out on four hexagonal prism barrels with the total volume of 60 L (15 L × 4), $N = 180$ r/min, $R = 235$ mm, and $r = 150$ mm. The test results are shown in Table 2.10. The test conditions are found as follows: The specimen is cylinder (135 cases) with 45# steel, Φ10 mm × 30 mm, Φ30 mm × 10 mm, Φ20 mm × 5 mm; abrasive blocks used for rough processing are white alumina, oblique triangle prism with 15 mm × 15 mm × 15 mm, and the granularity is 180#; abrasive blocks used for precision processing are aluminum oxide, sphere Φ5 mm, and the granularity is 280#; compounds solution is LC-10, and the load is appropriate; the processing time is 60 min; the temperature in the barrel before finishing is 18 °C.

Known from experimental results, the forced flow of the workpieces and abrasive blocks in the barrel is intensified, and thus the metal removal of the workpiece and the wear of abrasive blocks become much more, and the scratches on the workpieces surface increase. From the economic perspective (the wear of abrasive blocks), the ratio of material removal rate to the wear rate of abrasive blocks is the highest and the economical efficiency is the best during rough processing and precision processing as $n/N$ is $-1$.

**Table 2.10** Records and results of rough machining tests and precision machining tests

| | Roughness processing | | | Precision processing | | |
|---|---|---|---|---|---|---|
| *n*/*N* | −1 | −1.5 | −2 | −1 | −1.5 | −2 |
| Total mass of the workpiece/g | 4067.790 | 4083.925 | 4094.960 | 4130.500 | 4114.190 | 4043.660 |
| Material removal amount/g | 24.130 | 35.520 | 63.050 | 1.745 | 3.040 | 3.238 |
| Material removal rate/(g/kg·h) | 5.932 | 8.698 | 15.397 | 0.4225 | 0.7389 | 0.8008 |
| Total mass of abrasive blocks/g | 7982.925 | 7665.570 | 7982.480 | 7958.320 | 7935.430 | 7951.760 |
| Wear amount of abrasive blocks/g | 317.355 | 670.345 | 1176.215 | 6.560 | 13.930 | 43.930 |
| Wear rate of abrasive blocks/(g/kg·h) | 39.754 | 87.449 | 147.350 | 0.824 | 1.755 | 5.525 |
| Material removal rate/Wear rate of abrasive blocks | 1:6.70 | 1:10.05 | 1:9.57 | 1:1.95 | 1:2.38 | 1:6.89 |
| The temperature in the barrel after finishing/°C | 44 | 49 | 73 | 47 | 63 | 69 |

(2) *N*

When $n/N$ is −1, the absolute acceleration $\alpha$ of any particle in the barrel is $(2\pi N)^2R$. This is the inertia force of the particle, which performs the planet motion in the barrel, and it is determined by the two parameters $N$ and $R$. Since the value of $R$ is determined by structure design, the value of $R$ directly determines the magnitude of the inertia force.

If the value of $N$ is too large, the inertia force of the mixture in the barrel will be very large, and the forced flow will be vigorous, thus causing the colliding and scoring among abrasive blocks and the workpieces to be serious. This can cause the scratching and bumping on the workpiece surface, an increase in the surface roughness value, a breakage of the abrasive blocks, and so on. If the value of $N$ is too small, the inertia force of the mixture in the barrel will become very small, and the forced flow will be slow, thus causing the normal action of rolling and scoring between the workpieces and abrasive blocks to be weakened, which can ultimately decrease the finishing efficiency.

Based on a large number of experiments, the recommended revolution speed $N$ is

$$N = K/\sqrt{D}(\mathrm{r/min}) \tag{2.42}$$

**Table 2.11** Values range of *K* under different conditions

| Conditions | The workpiece of medium or large hardness | | | The workpiece of low hardness | | |
|---|---|---|---|---|---|---|
| | Rough processing | Self-finishing process | Finishing processing | Rough processing | Self-finishing process | Finishing processing |
| *K* | 5200–6000 | 4500–5200 | 3800–4500 | 4700–5500 | 4000–4700 | 3200–4000 |

where *D* represents the revolution radius of the barrel (mm) and *K* represents the coefficient, and $K = 3000$–$6000$. The value range of *K* under different conditions is shown in Table 2.11.

*N* is restricted by the workpiece hardness and finishing requirements, and it can also affect the surface quality of the workpiece and the finishing efficiency.

2. Geometrical parameters of equipment

The revolution radius *R* and the rotation radius *r* of the barrel are determined by the structure design. In general case, the value of $n/N$ is determined by the value of $R/r$. Under the condition of $n/N = -1$, the magnitude of *R* directly affects the inertia force of the mixture in the barrel, and the magnitude of *r* responds directly to the size of the barrel volume. At a constant *R* and *r*, the geometric parameters influencing the finishing effects include the shape of the barrel section and the axis direction of the barrel.

(1) Shape of barrel section

The shapes of the barrel section used in the centrifugal barrel finishing process have two general types: circle and regular polygon. Different section shapes can cause variety of the force and forced flow of the mixture in the barrel, and further affect the finishing effects.

On the testing finishing equipment with the same parameters, the movements of the mixture in the barrels with circle and hexagon section are observed by means of high-speed camera. Results show that the forced flow of the workpieces and abrasive blocks in the barrel with circle section is relatively slow, and there is an obvious core-annulus peristaltic area between the workpieces and abrasive blocks. The forced flow of the mixture in the barrel with hexagon section is vigorous, and the core-annulus peristaltic area disappears; the acting force between the workpieces and abrasive blocks increases at that moment, the contacting times between each other increase, and the collision, rolling, and micro-grinding effects are evident. These benefit the improvement of the finishing efficiency, but will also bring scratches to the workpieces surface. If the edge number of the barrel section shape reduces to three or four sides, the forced flow will be intensified, but the finishing efficiency will decrease as the length of the slipping zone becomes shorter. The experimental results are shown in Fig. 2.34.

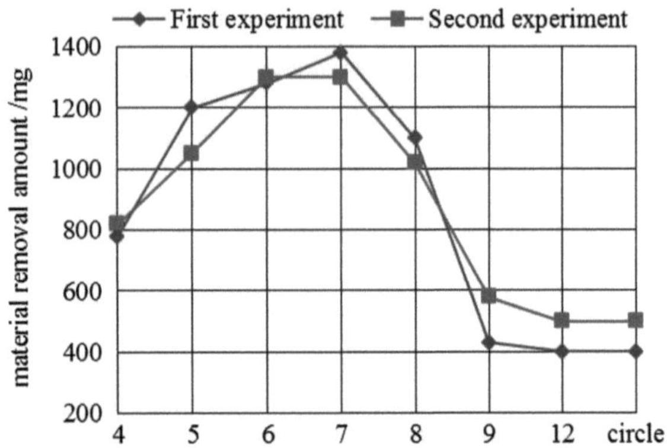


**Fig. 2.34** Impact of barrels section shape on finishing efficiency

Therefore, the finishing requirements of the workpieces should be considered when choosing the edge number of the barrel section shape. When the workpiece surface demands higher quality and no scratches are allowed, the circular section is a better choice; when the workpiece needs higher finishing efficiency and is less likely to show scratches, regular hexagon or octagon section are better choices.

(2) The axis direction of the barrel

On the premise of the axis direction of the barrel is horizontal, if the geometric axis is consistent to the rotary axis, the mixture in the barrel only shows the forced flow in the section perpendicular to the rotary axis (i.e., two-dimensional forced flow). As shown in Fig. 2.35, if the axis direction of the barrel sets at an angle $\alpha$ to the rotary axis, the counterforce $F$ of the mixture in the barrel from barrel wall has two components when the barrel does a planetary motion: One component is $F$r, which forces the workpiece to do a planetary motion following the barrel. Meanwhile, to execute a circumferential forced flow relative to the barrel, the other component $F$s forced the mixture to do an axial motion relative to the barrel, and the flow proceeds to the opposite direction when the barrel rotates over 180°. In the whole process of processing, the mixture in the barrel executes ∞-shaped

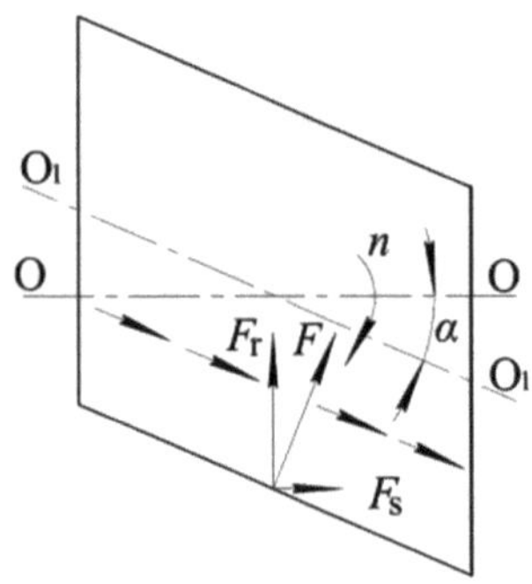


**Fig. 2.35** Impact analysis of the axis direction of the barrel on finishing effects

comprehensive flow (i.e., three-dimensional forced flow), and this can increase the relative motion between the workpieces and abrasive blocks, which improve the finishing efficiency.

3. Process parameters

(1) Load of the mixture

The load of the mixture is the percentage of the total volume of the mixture loaded in the barrel to the total volume of the barrel. In barrel finishing processes, the loading capacity has a greater influence on the flowing layer length of the mixture in the barrel than the relative sliding velocity and the number of interactions between abrasive blocks and the workpieces. So, the loading capacity would affect the material removal rate, the surface roughness value, deformation degree of the workpiece, and so on.

For the appropriate loading capacity (Fig. 2.36a), the local relative stationary status will not be formed in the mixture center. A large relative sliding velocity and a longer flowing layer will be obtained in the sliding flowing layer, and then the times of the relative sliding and scoring between abrasive blocks and workpieces will increase, which will lead to high finishing efficiency.

If the loading capacity is too large (Fig. 2.36c), the core-annulus peristaltic area in the mixture center will increase, abrasive blocks and workpieces in the area will be in the moderate flow state, and the finishing action will be small. Meanwhile, the length of flowing layer is very short, which leads to lower finishing efficiency.

Based on a lot of experimental results, the influence and the adaptation situation of different loading capacity are summarized, shown in Table 2.12.

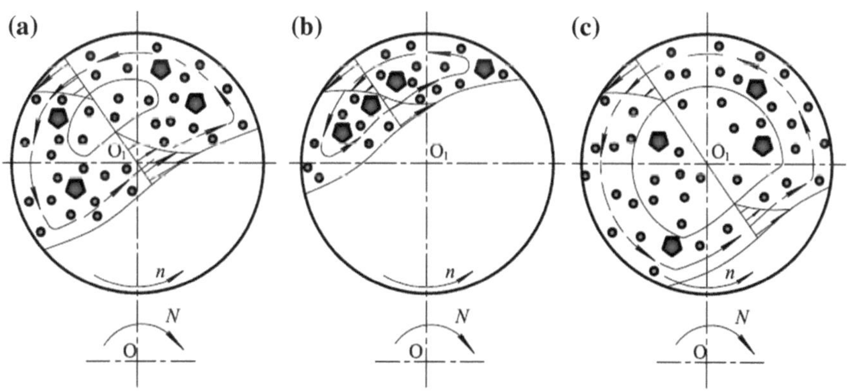


**Fig. 2.36** Flow status analysis of the mixture in the barrel with different loading capacity. **a** 60% loading capacity; **b** 30% loading capacity; **c** 90% loading capacity

**Table 2.12** Influence and adaptation situation of different loading capacities

| Loading capacity (volume ratio)/% | Material removal rate | Decreasing of surface roughness value | Deformation degree of the workpiece | Adaptation situation |
|---|---|---|---|---|
| <50 | Low | Low | Large | No using generally |
| 50–65 | Large | Moderate | Moderate | Heavy cutting, difficult deformation parts |
| 65–90 | Low | High | Low | Easy-deformed flat parts, thin bars, and low stiffness parts |

(2) Mixing ratio

The mixing ratio of abrasive blocks and the workpieces is the mass or volume ratio of abrasive blocks to workpieces loaded in the barrel. If mixing ratio is low, it indicates that there are more workpieces in the barrel, the abrasive blocks are reduced, and the barrel finishing efficiency is low. As more workpieces are added, the possibility of mutually collision and pounds among the workpieces is enhanced, the deformation often happens in finishing the easy-deformed flat parts, thin bars, or soft material workpieces, and this can affect surface quality. If the mixing ratio is large, it indicates that there are fewer workpieces in the barrel, and the abrasives are more. This can not only impact the finishing efficiency, but also aggravate the actions of abrading and collision among abrasive blocks, and the finishing economy will be poor.

Experimental results show that the mass mixing ratio of abrasive blocks to workpieces between 1:1 and 2:1 is suitable. The mixing ratio can be chosen according to the workpiece material, shape, size, and finishing requirements.

(3) Finishing time

The length of the barrel finishing time directly affects the finishing quality, finishing efficiency, and economy. The barrel finishing process can be divided into three stages (Fig. 2.37):

Intense finishing stage: This is the preliminary stage of the process; the burrs, sharp corners, and profile peaks of the workpieces surface are removed effectively by the actions of forced collision, rolling, and grinding of abrasive blocks to the workpieces, and the surface roughness value is reduced significantly. In this stage, the material removal rate of the workpiece is large, and accordingly the wear rate of abrasive blocks is also large.

Uniform finishing stage: In this stage, the action of abrasive blocks to the workpiece is mainly sliding. The profile peaks on the workpieces surface are continually smoothed and grinded, and the concave–convex surface is not found. The relationship of material removal rate and the surface roughness value versus the finishing time appears to be linear. This is a stable and uniform finishing process.

**Fig. 2.37** Impact rules on finishing effects of different finishing time

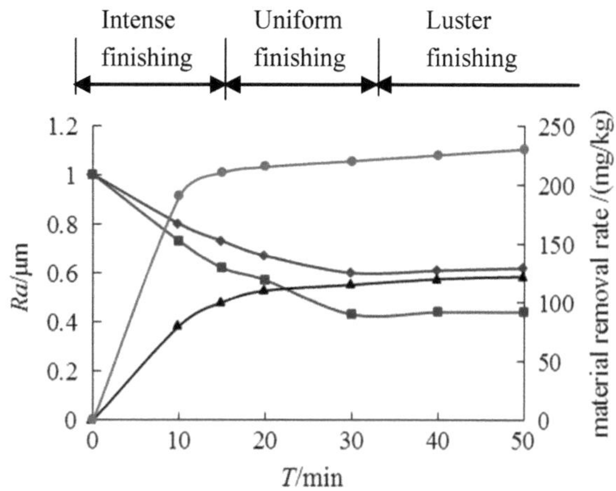


Luster finishing stage: In this stage, the original profile peaks completely disappear, the contact area between abrasive blocks and the workpieces surface increases, the pressure is lower (that means the cutting force per unit area decreases), the material removal amount of the workpiece decreases significantly, and the decrease of the surface roughness value is no longer significant. Therefore, this stage only exists with the luster finishing effect. In the luster finishing stage, the surface stress distribution would be improved and the surface hardness would be enhanced (the workpieces surface would gain cold hardening effect) as the abrasive blocks have this intense rolling effect on the workpieces surface. The surface roughness value no longer decreases and remains a relatively stable value during this finishing stage. The workpieces surface is considered to have reached the ultimate roughness value ($Ra_j$), that is to say, the ultimate roughness value always exists with certain finishing conditions.

For different materials, shapes, finishing requirements of workpieces, and abrasive blocks, the finishing time of these three stages mentioned above is different. Large amount of experimental results demonstrates that the removing burrs, flashes, oxide skins, and other similar rough processing needs 40–60 min; removing edges and corners, chamfering, and other similar rough processing needs 30–40 min; semi-finishing processing of various workpieces needs about 30 min; finish processing of various workpieces needs about 15 min [45].

### *2.5.3 Equipment Types and Design*

1. Equipment types

Figure 2.38 shows the basic structure diagram of the centrifugal barrel finishing equipment ($n/N = -1$). When finishing, abrasive blocks, workpieces, and liquid medium (water and compounds) are put into sealed barrel 8 with a certain proportion, and tie-rod belt wheel 3 generates rotation by means of the belt transmission and the motor. The barrel not only rotates around central shaft 4 (speed $N$), but also makes retrograde motion around its shaft (speed $n$) through planetary driving wheel system comprised of tie-rod belt wheel 3, center belt wheel 5, and planetary belt wheel 6. Figure 2.39 is the three-dimensional model generated by Pro/E software.

The classification of centrifugal barrel finishing equipment is shown as follows:

(1) They can be classified into horizontal type and vertical type, based on the installed direction of the barrel shaft. Generally, horizontal type equipment is the most applied, and vertical barrel is often used on the small equipment in which the volume of single barrel is less than 2L (vertical type has advantage of high automation, such as the mixture loading or unloading).
(2) They can be classified into coaxial type and tilting type, based on the relationship between the geometric axis of the barrel and its rotational axes.
(3) They can be classified into circular barrel type and regular polygon barrel type, based on the cross-sectional shape of the barrel. The regular polygon type is most commonly chosen to be regular hexagon and regular octagon barrel.
(4) They can be classified into double-barrel type, triple-barrel type, quadruple-barrel type, and hexa-barrel type, based on the number of barrels. The number of the barrels is usually an even number, thereby ensuring dynamic balance in processing.
(5) They can be classified into belt-driven type, chain-driven type, and gear-driven type, based on the mode of transmission. Gear-driven is compact in structure, but its lubrication condition is poor, and it has a great noise. Belt-driven and chain-driven have to set tensioning mechanism.
(6) They can be classified into constant-speed type and change-speed type, based on the change of revolution speed. Change-speed type is achieved by adopting gearbox, adjustable speed motor, and stepless speed regulation device. During the time of the start-up and the finishing termination, the revolution speed will have a sudden increase or decrease. This can cause violent collision and affect the finishing effect. In the design of finishing equipment, it is very necessary to implement measures for achieving slow-start and slow-stop.
(7) They can be classified into stationary barrel type and detachable barrel type, based on the operating mode of the barrel. Considering the discharging requirements, stationary barrel type must add discharging mechanism, so that the barrel can achieve rotations, but cannot obtain revolutions, which pouring the workpieces and abrasives out by its rotation.

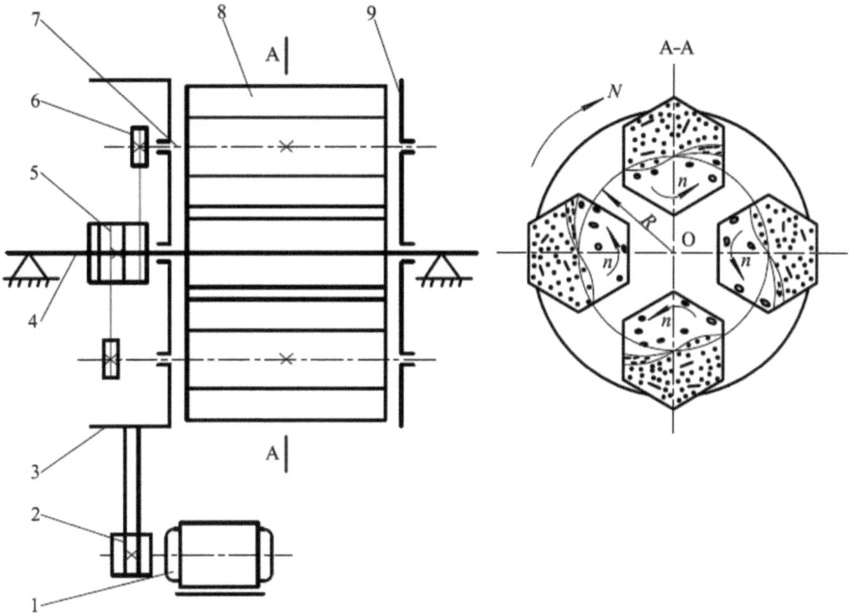


**Fig. 2.38** Basic structure diagram of centrifugal barrel finishing equipment. 1—Motor; 2—motor belt wheel; 3—tie-rod belt wheel; 4—central shaft; 5—center belt wheel; 6—planetary belt wheel; 7—rotary shaft of the barrel; 8—barrel; 9—tie-rod

**Fig. 2.39** Three-dimensional model of centrifugal barrel finishing equipment

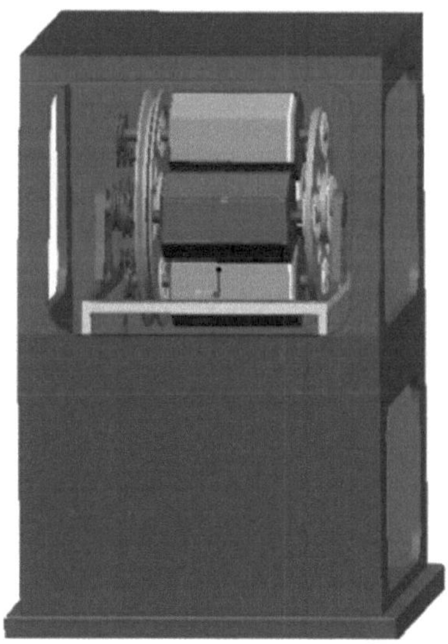

(8) They can be classified into single support type (cantilever structure) and double support type, based on the supporting form of the central shaft. Single support type is only suitable for low-volume barrels.
(9) They can be classified into stationary central shaft type and detachable central shaft type, based on the fixation of central shaft or not. Stationary central shaft type (Fig. 2.38) needs a large-scale tie-rod belt wheel, and this can consume less speed reducers and other speed reducing mechanisms, but becomes difficult when changing the driving belt. Detachable central shaft type usually adopts a small-scale tie-rod belt wheel which is placed besides the support, and the driving belt is convenient to exchange, but the speed reducing mechanism must be added.

As planetary mechanism has great inertial motion after finishing, the brake device should be also be set. Based on the difference in the structural form, the brake device can be mechanical type and electromagnetic type. For the barrel operating (barrels or mixtures loading/unloading) reliability, it is necessary to set the position with a locking mechanism.

Otherwise, the revolution-driven and rotation-driven of the equipment will be separated. The double driving structure diagram is shown in Fig. 2.40.

2. Equipment design

(1) Structure of critical component

Among centrifugal barrel finishing equipment, the barrel not only is the loading container of the mixture, but also is the carrier which realizes the finishing process.

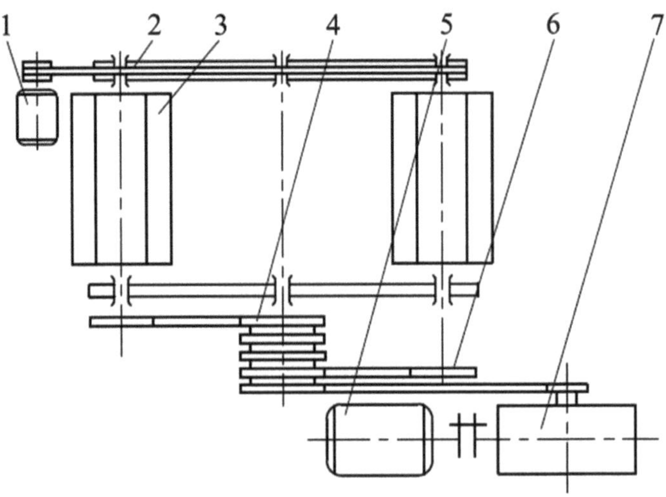


**Fig. 2.40** Double driving structure diagram. 1—Revolution control motor; 2—tie-rod belt wheel; 3—barrel; 4—primary belt wheel of rotation; 5—rotation control motor; 6—barrel belt wheel; 7—speed reducing mechanism

During the finishing process, the barrel carries the mixture (workpieces, abrasive blocks, and liquid medium) to make the planetary motion at the high speed, and the following is the requirements for the barrel.

(1) It must have good sealing property.
(2) The lining on the inner wall should resist corrosion and wear, and has a big coefficient of friction acting on the mixture.
(3) A single barrel and a set of barrels of the same equipment should have a good rotary balance.

The barrel component is mainly composed of barrel body, barrel cover, and pressing mechanism. The barrel body mainly includes the shell and the lining. The shell of the barrel body can employ cast iron materials or welded structure, and the method of the rotating counterweight should be considered in the structure. The best way to process the shell of the barrel body is with a welded structure, which is corrected after the aging treatment. The lining of the barrel body can adopt complete sulfide rubber or combined rubber. The effect of sulfide barrel body is good, but the cost is high; therefore, the combining way can be used generally. Besides, the lining of the barrel body can also adopt polyurethane coating. The pressing mechanism should be safety, reliability, and easy operating.

(2) Parameters determination

When determining parameters, the following factors should be considered.

(1) Total volume of the barrels reflects the maximum working capacity of the centrifugal barrel finishing equipment.
(2) Number of the barrels and the volume of single barrel determine the structure of the finishing equipment.
(3) Structure size of single barrel (diameter $d$ and length $L$): If no special requirements are present, $d/L$ can be about 0.618; this makes a more harmonious shape. In case of the special requirements, the specialization should be satisfied.
(4) Determination of barrel orbital diameter $D$ can guarantee the structure compactness.
(5) Design of planet transmission mechanism: It should guarantee $\frac{n}{N} = -1$ or $\sqrt{\frac{R}{r}} - 1 > \frac{n}{N} > -\left(\sqrt{\frac{R}{r}} + 1\right)$.
(6) Determination of revolution speed $N$: Coefficient $K$ should be reasonably selected based on $N = K/\sqrt{D}$.

3. Equipment shapes

Figure 2.41 shows the photographs of two centrifugal barrel finishing equipment [46, 47]. Figure 2.42 shows the photographs of different centrifugal barrel finishing equipments produced by Langfang Beifang Tianyu Mechanical and Electrical Technology Co., Ltd.

**Fig. 2.41** Photographs of centrifugal barrel finishing equipment

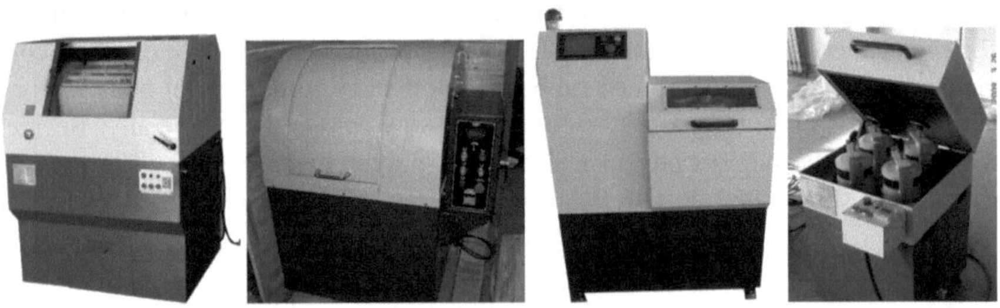

**Fig. 2.42** Photographs of centrifugal barrel finishing equipment produced by Beifang Tianyu

# 2.6 Vertical Spindle Barrel Finishing

The vertical spindle barrel finishing is that the workpieces installed on the spindle are inserted into the barrel with media (abrasive blocks, compounds, and water) vertically or at a tilting angle to a vertical direction. The relative motion and interactions between the abrasive blocks and the workpiece surface can be generated by the rotational motion of the spindle and the barrel or the revolution and the rotation of the spindle, and then the deburring process or the surface finishing of the workpieces is achieved.

## *2.6.1 Finishing Principle and Characteristics*

1. Finishing principle

Figure 2.43 shows the schematic diagrams of several typical spindle barrel finishing ways. In Fig. 2.43a, the barrel revolves at *N*, the spindle revolves at *n*, and the rotating axis of the spindle is parallel to the rotating axis of the barrel. In Fig. 2.43b, the barrel is fixed, the spindle revolves along *N* and rotates along *n*, and

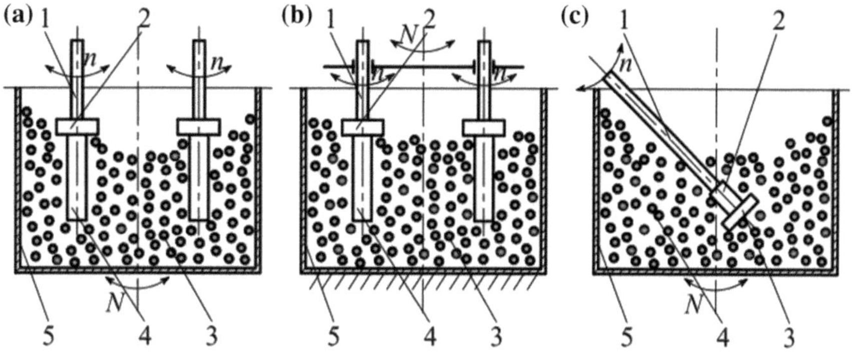


**Fig. 2.43** Schematic diagrams of several typical spindle barrel finishing ways. **a** Parallel spindle type I; **b** parallel spindle type II; **c** intermeshing spindle type. 1—Spindle; 2—holding device; 3—medium; 4—parts; 5—barrel

the revolution axis of the spindle is parallel to the rotation axis. As shown in Fig. 2.43a, b, both types are collectively called parallel spindle barrel finishing. In Fig. 2.43c, the barrel revolves at $N$, the spindle revolves at $n$, and the rotating axis of the spindle is laid crosswise to the rotating axis of the barrel, and this situation is called intermeshing spindle barrel finishing.

When $N \neq 0$, as shown in Fig. 2.43a, b, a certain amount of the medium in the barrel produces centrifugal force under the action of the revolution, due to the mass exists. The abrasives form a rigid staging area in a certain range of the barrel wall due to the centrifugal force. In Fig. 2.43b, the medium in the barrel is free-depositional incompact status.

When $N \neq 0$ and $n = 0$, the workpiece surface facing at the direction of the abrasive blocks flow is subjected to the impacting action (or barrier action) of the abrasive blocks, and the rolling force and the friction of the abrasive blocks to the workpiece surface are produced during the impacting (barrier) process. Under the action of these forces, the phenomenon of eccentrically worn occurs on the surface of the workpiece. Comparing between Fig. 2.43a, b, the machining range in Fig. 2.43a is in the rigid staging area, and the machining range in Fig. 2.43b is in incompact status, so the action force on the workpiece in Fig. 2.43a is larger than that of in Fig. 2.43b when the revolution speed $N$ and the revolution radius of the workpiece are the same. This can cause more severe eccentrically worn phenomenon.

When $N \neq 0$ and $n \neq 0$, various places over the workpiece surface are under the action of the above several forces. Given the machining medium, the finishing effects depend heavily on the complexity of the relative movement between the abrasive blocks and the workpiece surface.

2. Kinematic analysis

(1) Kinematic analysis of parallel spindle barrel finishing

Kinematic analysis of parallel spindle barrel finishing is shown in Fig. 2.44. The rotating speed of the barrel is set to $N$; the rotating speed of the workpiece is $n$; $v_N$ is the velocity of the abrasive blocks at Moving-point A because of the rotation ($N$) of the barrel; $v_n$ is the relative velocity of the abrasive blocks at Moving-point A because of the rotation ($N$) of the workpiece; $v$ is the resultant velocity (i.e., cutting velocity) of the abrasive blocks relative to the workpiece at Moving-point A; $q$ is the distance from Moving-point A to the rotating axis of the workpiece; $Q$ is the distance between the rotating axis of the barrel and the rotating axis of the workpiece; $\theta$ is the angle between the connecting line from Moving-point A to the center of the workpiece and the connecting line between the two center points; $\theta = 0°$ is at the intersection of the extension of line OO′ and the surface of the workpiece; the size of angle $\theta$ is in the counterclockwise direction; $\varphi$ is the acute angle (cutting angle) between the cutting velocity direction of the arbitrary points on the outside circle surface of the workpiece and the tangent line of that point, when $\theta$ is in the range of 0°–180°, $\varphi$ is positive, when $\theta$ is in the range of 180°–360°, $\varphi$ is negative.

Based on the equation of velocity synthesis,

$$v = v_N + v_n \tag{2.43}$$

Through the analysis, calculation, and sorting of the geometric relationship, it can obtained that

$$v = 2\pi N\left[q^2\left(1-\frac{n}{N}\right)^2 + Q^2 + 2Qq\left(1-\frac{n}{N}\right)\cos\theta\right]^{1/2} \tag{2.44}$$

$$\varphi = \arcsin\frac{-2\pi NQ\sin\theta}{v} \tag{2.45}$$

The cutting velocity $v$ and the cutting angle $\varphi$ of certain point on the workpiece surface can be calculated by substituting $N$, $n$, $q$, $Q$, and $\theta$ into Eqs. (2.44) and (2.45).

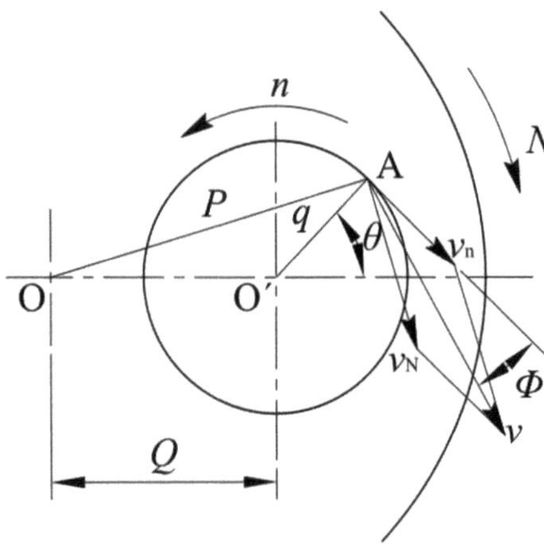


**Fig. 2.44** Kinematic analysis of parallel spindle barrel finishing

The higher the cutting velocity $v$ is, the longer the relative friction of the cutting route per unit time is. The removal thickness of the workpiece is related not only to the cutting velocity $v$ but also to the size of the cutting angle. It is for this reason that the workpiece cannot achieve excellent effects [26, 27].

(2) Kinematic analysis of intermeshing spindle barrel finishing

Kinematic analysis of intermeshing spindle barrel finishing is shown in Fig. 2.45. In the figure, $OO'$ ($O_1O'$) is the rotating axis of the barrel; $O_2O_3$ is the rotating axis of the workpiece; $N$ is the rotating speed of the barrel; $n$ is the rotating speed of the workpiece; $O_2O_3$ is in Plane $P_1$, $OO'$ ($O_1O'$) is in Plane $P_2$, and $P_1$ // $P_2$; $OO'$ ($O_1O'$) is in Plane $P_3$, and $P_3 \perp P_2$; Plane $P_3$ and $O_2O_3$ meet in Point $O_2$; $O_3$ is the end plane center of the workpiece; $q$ is the rotating radius of the workpiece about $O_2O_3$; $Q$ is the distance between Plane $P_1$ and Plane $P_2$, that is the shortest distance ($O_1O_2$) between Line $O_2O_3$ and Line $OO'$ ($O_1O'$); $\beta$ is the angle (spindle swinging angle) between Line $O_2O_3$ and the horizontal plane; $L$ is the distance between $O_2$ and $O_3$.

(1) Establishment of moving point trajectory equations

As shown in Fig. 2.45, considering the characteristic and requirements of keeping invariable relative motion and postulating to the barrel stops, the workpiece rotates on axis $O_2O_3$ at rotational speed $n$. Meanwhile, the spindle axis $O_2O_3$ used for clamping the workpiece rotates on axis $OO'$ opposite to the original rotation direction of the barrel, and its rotating speed is $N$. The kinematic analysis model of the abrasives relative to one point on the workpiece surface is shown in Fig. 2.46.

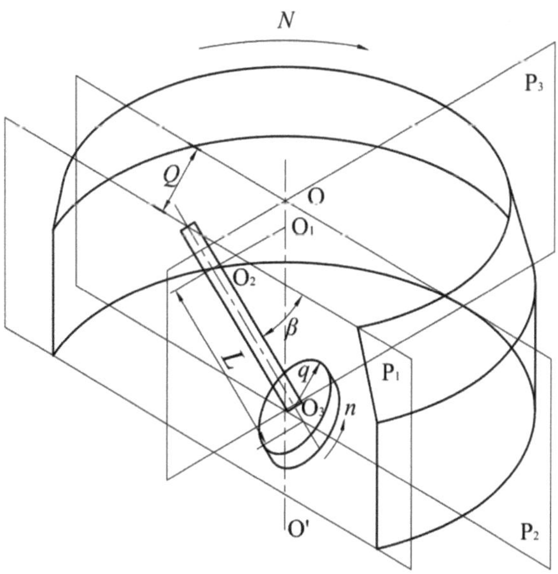


**Fig. 2.45** Kinematic analysis of intermeshing spindle barrel finishing

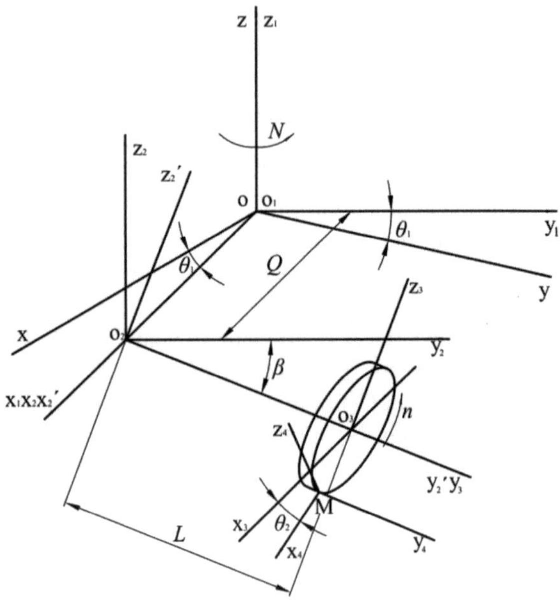


**Fig. 2.46** Kinematic analysis model of intermeshing spindle barrel finishing

For Moving-point M on the outer circumference of the workpiece end face, geometric transformation coordinates shown in Fig. 2.46 are built, and its transformation processes are as follows:

$$Oxyz \xrightarrow{\text{Going } \theta_1 \text{ around the } z \text{ axis}} O_1\,x_1\,y_1\,z_1 \xrightarrow{\text{Translating } Q \text{ along } x_1} O_2\,x_2\,y_2\,z_2$$

$$\xrightarrow{\text{Going } \beta \text{ around the } x_2 \text{ axis}} O_2\,x_2'\,y_2'\,z_2' \xrightarrow{\text{Translating } L \text{ along } y_2'} O_3\,x_3\,y_3\,z_3$$

The corresponding geometric transform relation is

$$\begin{vmatrix} x \\ y \\ z \end{vmatrix} = \begin{vmatrix} \cos\theta_1 & -\sin\theta_1 & 0 \\ \sin\theta_1 & \cos\theta_1 & 0 \\ 0 & 0 & 1 \end{vmatrix} \begin{vmatrix} x_1 \\ y_1 \\ z_1 \end{vmatrix} \tag{2.46}$$

$$\begin{vmatrix} x_1 \\ y_1 \\ z_1 \end{vmatrix} = \begin{vmatrix} 1 & 0 & 0 \\ 0 & 1 & 0 \\ 0 & 0 & 1 \end{vmatrix} \begin{vmatrix} x_2 \\ y_2 \\ z_2 \end{vmatrix} + \begin{vmatrix} Q \\ 0 \\ 0 \end{vmatrix} = \begin{vmatrix} Q + x_2 \\ y_2 \\ z_2 \end{vmatrix} \tag{2.47}$$

$$\begin{vmatrix} x_2 \\ y_2 \\ z_2 \end{vmatrix} = \begin{vmatrix} 1 & 0 & 0 \\ 0 & \cos\beta & \sin\beta \\ 0 & -\sin\beta & \cos\beta \end{vmatrix} \begin{vmatrix} x_2' \\ y_2' \\ z_2' \end{vmatrix} \tag{2.48}$$

$$\begin{vmatrix} x_2' \\ y_2' \\ z_2' \end{vmatrix} = \begin{vmatrix} 1 & 0 & 0 \\ 0 & 1 & 0 \\ 0 & 0 & 1 \end{vmatrix} \begin{vmatrix} x_3 \\ y_3 \\ z_3 \end{vmatrix} + \begin{vmatrix} 0 \\ L \\ 0 \end{vmatrix} = \begin{vmatrix} x_3 \\ L+y_3 \\ z_3 \end{vmatrix} \tag{2.49}$$

In the $O_3x_3y_3z_3$ coordinate system, the position of Moving-point M changes with $\theta_2$. $\theta_2$ is defined as the angle between the line from the outside point on the workpiece to $O_3$ and the line from M to $O_3$, then

$$\begin{vmatrix} x_3 \\ y_3 \\ z_3 \end{vmatrix} = \begin{vmatrix} q \cos \theta_2 \\ 0 \\ -q \sin \theta_2 \end{vmatrix} \tag{2.50}$$

Through the deduction of inversion equation, it can be obtained that

$$\begin{vmatrix} x \\ y \\ z \end{vmatrix} = \begin{vmatrix} Q \cos \theta_1 + q \cos \theta_1 \cos \theta_2 - L \sin \theta_1 \cos \beta + q \sin \beta \sin \theta_1 \sin \theta_2 \\ Q \sin \theta_1 + q \sin \theta_1 \cos \theta_2 + L \cos \theta_1 \cos \beta - q \sin \beta \cos \theta_1 \sin \theta_2 \\ -L \sin \beta - q \cos \beta \sin \theta_2 \end{vmatrix} \tag{2.51}$$

Since $\theta_1$ and $\theta_2$ vary simultaneously with time $t$, and $\theta_1 = 2\pi Nt$, $\theta_2 = 2\pi nt$ are substituted to Eq. (2.51), there is

$$\begin{vmatrix} x \\ y \\ z \end{vmatrix} = \begin{vmatrix} Q \cos(2\pi Nt) + q \cos(2\pi Nt) \cos(2\pi nt) - L \sin(2\pi Nt) \cos \beta + q \sin \beta \sin(2\pi Nt) \sin(2\pi nt) \\ Q \sin(2\pi Nt) + q \sin(2\pi Nt) \cos(2\pi nt) + L \cos(2\pi Nt) \cos \beta - q \sin \beta \cos(2\pi Nt) \sin(2\pi nt) \\ -L \sin \beta - q \cos \beta \sin(2\pi nt) \end{vmatrix} \tag{2.52}$$

Equation (2.52) is the movement trajectory equation of Moving-point M.

(2) Velocity component of Moving-point M and the resultant cutting velocity in cylindrical coordinates

Taking the trajectory Eq. (2.52) of Moving-point M in the O$xyz$ coordinate system as a research object, the derivative of both sides is taken, and the velocity component of M in the O$xyz$ coordinate system is obtained, that is

$$\begin{vmatrix} \dot{x} \\ \dot{y} \\ \dot{z} \end{vmatrix} = \begin{vmatrix} -2\pi[NQ \sin(2\pi Nt) + Nq \sin(2\pi Nt) \cos(2\pi nt) + nq \cos(2\pi Nt) \sin(2\pi nt) + NL \cos(2\pi Nt) \cos \beta \\ \quad + Nq \sin \beta \cos(2\pi Nt) \sin(2\pi nt) - nq \sin \beta \sin(2\pi Nt) \cos(2\pi nt)] \\ 2\pi[NQ \cos(2\pi Nt) + Nq \cos(2\pi Nt) \cos(2\pi nt) - nq \sin(2\pi Nt) \sin(2\pi nt) - NL \sin(2\pi Nt) \cos \beta \\ \quad + Nq \sin \beta \sin(2\pi Nt) \sin(2\pi nt) - nq \sin \beta \cos(2\pi Nt) \cos(2\pi nt)] \\ -2\pi nq \cos \beta \cos(2\pi nt) \end{vmatrix} \tag{2.53}$$

The velocity component of M in the $O_1x_1y_1z_1$ coordinate system is making the above-mentioned component do counterclockwise rotation $2\pi Nt$ about the $Z$-axis, and there is

$$\begin{vmatrix} \dot{x}_1 \\ \dot{y}_1 \\ \dot{z}_1 \end{vmatrix} = 2\pi \begin{vmatrix} q(N\sin\beta - n)\sin(2\pi nt) - NL\cos\beta \\ q(N - n\sin\beta)\cos(2\pi nt) + NQ \\ -nq\cos\beta\cos(2\pi nt) \end{vmatrix} \tag{2.54}$$

Since the transformation of $O_1x_1y_1z_1 \rightarrow O_2x_2y_2z_2$ is the translation along $x_1$-axis, so

$$\begin{vmatrix} \dot{x}_2 \\ \dot{y}_2 \\ \dot{z}_2 \end{vmatrix} = \begin{vmatrix} \dot{x}_1 \\ \dot{y}_1 \\ \dot{z}_1 \end{vmatrix} = 2\pi \begin{vmatrix} q(N\sin\beta - n)\sin(2\pi nt) - NL\cos\beta \\ q(N - n\sin\beta)\cos(2\pi nt) + NQ \\ -nq\cos\beta\cos(2\pi nt) \end{vmatrix} \tag{2.55}$$

The velocity component of M in the $O_2x_2'y_2'z_2'$ coordinate system is making the velocity component in $O_2x_2y_2z_2$ coordinate system to do clockwise rotation $\beta$ about the $x_2$-axis, and there is

$$\begin{vmatrix} \dot{x}_2' \\ \dot{y}_2' \\ \dot{z}_2' \end{vmatrix} = 2\pi \begin{vmatrix} q(N\sin\beta - n)\sin(2\pi nt) - NL\cos\beta \\ N\cos\beta[q\cos(2\pi nt) + Q] \\ N\sin\beta[q\cos(2\pi nt) + Q] - nq\cos(2\pi nt) \end{vmatrix} \tag{2.56}$$

Since the transformation of $O_2x_2'y_2'z_2' \rightarrow O_3x_3y_3z_3$ is the translation along $y_2'$-axis, so

$$\begin{vmatrix} \dot{x}_3 \\ \dot{y}_3 \\ \dot{z}_3 \end{vmatrix} = \begin{vmatrix} \dot{x}_2' \\ \dot{y}_2' \\ \dot{z}_2' \end{vmatrix} = 2\pi \begin{vmatrix} q(N\sin\beta - n)\sin(2\pi nt) - NL\cos\beta \\ N\cos\beta[q\cos(2\pi nt) + Q] \\ N\sin\beta[q\cos(2\pi nt) + Q] - nq\cos(2\pi nt) \end{vmatrix} \tag{2.57}$$

With this, if M is set as the origin, the direction of $O_3M$ is $x_4$ axle, $y_4$ axle is parallel to $y_3$, and rectangular coordinate system $Mx_4y_4z_4$ is set up. Under this coordinate system, the velocity component of moving point is making the velocity component in the $O_3x_3y_3z_3$ coordinate system do counterclockwise rotation $2\pi nt$ about the $y_3$-axis, and there is

$$\begin{vmatrix} \dot{x}_4 \\ \dot{y}_4 \\ \dot{z}_4 \end{vmatrix} = 2\pi \begin{vmatrix} -NL\cos\beta\cos(2\pi nt) - NQ\sin\beta\sin(2\pi nt) \\ N\cos\beta[q\cos(2\pi nt) + Q] \\ q(N\sin\beta - n) - NL\cos\beta\sin(2\pi nt) + NQ\sin\beta\cos(2\pi nt) \end{vmatrix} \tag{2.58}$$

According to the composition equation of velocity, if the cutting speed is $v$, there is

$$\begin{aligned} v^2 &= \dot{x}^2 + \dot{y}^2 + \dot{z}^2 = \dot{x}_1^2 + \dot{y}_1^2 + \dot{z}_1^2 = \dot{x}_2^2 + \dot{y}_2^2 + \dot{z}_2^2 = \dot{x}_2'^2 + \dot{y}_2'^2 + \dot{z}_2'^2 \\ &= \dot{x}_3^2 + \dot{y}_3^2 + \dot{z}_3^2 = \dot{x}_4^2 + \dot{y}_4^2 + \dot{z}_4^2 \end{aligned} \tag{2.59}$$

The results calculated, derived, and arranged based on different components are the same, and $2\pi nt$ is recorded as $\theta_2$; then,

$$v = 2\pi N\left\{q^2\left(\sin\beta - \frac{n}{N}\right)^2 + Q^2 + 2Qq\left(\sin\beta - \frac{n}{N}\right)\sin\beta\cos\theta_2 + \cos\beta\left[q^2\cos\beta\cos^2\theta_2 + 2Qq\cos\beta\cos\theta_2 - 2qL\sin\beta\sin\theta_2 + 2qL\frac{n}{N}\sin\theta_2 + L^2\cos\beta\right]\right\}^{1/2} \quad (2.60)$$

(3) Cutting angle at certain point on the workpiece cylindrical surface, and the axial end surface

To better describe machinability at a certain point on the workpiece, it is not enough to only consider the value of its cutting speed, and the direction of the cutting speed $v$ relative to special machined surface should also be considered. During crossed spindle barrel finishing process, the surface of the workpiece finished by the abrasive blocks includes the workpiece cylindrical surface and the axial end surface.

For cylindrical surface, the acute angle between the defined direction of the cutting speed $v$ at Moving-point M and the tangent plane across the workpiece is the cutting angle (symbolized by $\varphi_r$) of the cylindrical surface; for the axial end surface, the acute angle between the defined direction of the cutting speed $v$ at M and the axial end surface of the workpiece is the cutting angle (symbolized by $\varphi_0$) of the axial end surface.

In the $Mx_4y_4z_4$ coordinate system, the cutting angle $\varphi_0$ of the cylindrical surface is the angle between the direction of the cutting speed $v$ and coordinate plane $Mx_4y_4$; the cutting angle $\varphi_0$ of the axial end surface is the angle between the direction of the cutting speed v and coordinate plane $Mx_4z_4$. If these information is presented in the form of cosine, there are

$$\cos\varphi_r = \sqrt{\dot{y}_4^2 + \dot{z}_4^2}/v \quad (2.61)$$

$$\cos\varphi_o = \sqrt{\dot{x}_4^2 + \dot{z}_4^2}/v \quad (2.62)$$

If this information is presented in the form of sine, there are

$$\sin\varphi_r = \dot{x}_4/v \quad (2.63)$$

$$\sin\varphi_o = \dot{y}_4/v \quad (2.64)$$

It can be derived from Eqs. (2.61)–(2.64) that

$$-90^\circ \le \varphi_r \le 90^\circ \quad -90^\circ \le \varphi_0 \le 90^\circ$$

$\dot{x}_4$, $\dot{y}_4$ are plugged separately into Eqs. (2.63) and (2.64), it can be obtained that

$$\varphi_r = \arcsin\frac{-2\pi N(L\cos\beta\cos\theta_2 + Q\sin\beta\sin\theta_2)}{v} \quad (2.65)$$

$$\varphi_o = \arcsin\frac{2\pi N\cos\beta(q\cos\theta_2 + Q)}{v} \quad (2.66)$$

For special cases, if $\beta = 90°$, then the workpiece spindle is parallel to the revolving axis of the barrel, and Eqs. (2.60) and (2.65) are equivalent to Eqs. (2.44) and (2.45). It can be seen that the axial end surface cannot be processed properly from Eq. (2.66), when $\varphi_0 = 90°$.

It can be seen from Eqs. (2.60), (2.65), and (2.66) that, when $\beta \neq 90°$, and $N$, $n$, $Q$, $q$, $L$, $\beta$, and $\theta_2$ are given, the cutting speed $v$, the cutting angle $\varphi_r$ of the cylindrical surface, and the cutting angle $\varphi_0$ of the axial end surface at arbitrary point on the workpiece surface can be obtained. This illustrates that the workpiece can carry out surface finishing for both the cylindrical surface and the axial end surface.

(4) Impacts of every parameter on the cutting speed and the cutting angle

It can be seen from Eqs. (2.60), (2.65), and (2.66) that the cutting speed $v$ and the cutting angle $\varphi_r$, $\varphi_0$ are affected by parameters such as $n/N$, $N$, $Q$, $q$, $L$, $\beta$, and $\theta_2$. When $\theta_2$ is in the range of 0°–360°, the effect condition of every parameter on $v$, $\varphi_r$, and $\varphi_0$ can be analyzed.

① Impacts of $n/N$ on $v$, $\varphi_r$, and $\varphi_0$

The preparing conditions of Fig. 2.47 are $N = 50\,\mathrm{r/min}$, $\beta = 60°$, $Q = 0.3\,\mathrm{m}$, $L = 0.3\,\mathrm{m}$, and $q = 0.05\,\mathrm{m}$, as shown in the figure.

Within the range of 360° on the cylindrical surface of the workpiece, the cutting angles $\varphi_r$ in the 180° range are negative. That is, there are no finishing effects; within the range of 360° on the axial end surface of the workpiece, all of the cutting angles $\varphi_r$ are positive value, and that is there are finishing effects for every degree.

In all of those areas, there are finishing effects, and the change of the average value of $v$ is smaller with $n/N$, that is $n/N$ has less effect upon finishing. It could also be said that the value of $n$ has less effect upon machining at a constant $N$.

When $n/N = 0$, the phenomenon of non-uniformity finishing will occur on the workpiece.

② Impacts of $N$ on $v$, $\varphi_r$, and $\varphi_0$

The preparing conditions of Fig. 2.48 are $n/N = -2$, $\beta = 60°$, $Q = 0.3\,\mathrm{m}$, $L = 0.3\,\mathrm{m}$, and $q = 0.05\,\mathrm{m}$, as shown in the figure.

The change of $N$ does not affect the cutting angles $\varphi_r$ and $\varphi_0$ in terms of other conditions unchanged. The key to representing the strength of process ability is the cutting speed.

With the increase of the value of $N$, the cutting speed $v$ increases linearly. So, to improve finishing efficiency, the value of $N$ can be increased.

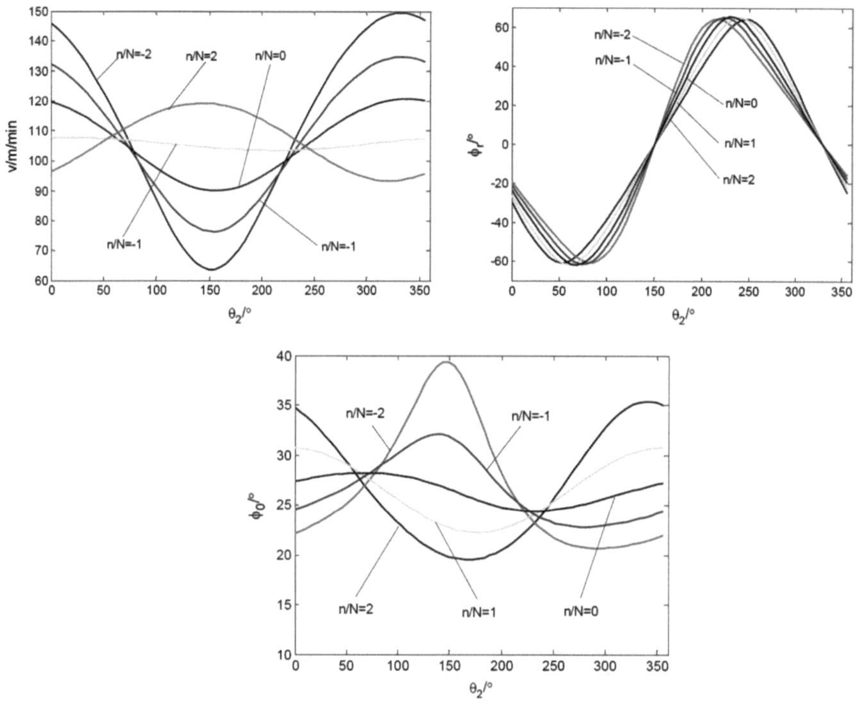


**Fig. 2.47** Impacts of $n/N$ on $v$, $\varphi_r$, and $\varphi_0$

③ Impacts of $\beta$ on $v$, $\varphi_r$, and $\varphi_0$

As shown in the figure, which is drawn by the condition of $n/N = -2$, $N = 50\,\text{r/min}$, $Q = \sqrt{0.16 - L^2\cos^2\beta}$, $L = 0.3\,\text{m}$, and $q = 0.05\,\text{m}$.

Comparing $\beta \neq 90°$ with $\beta = 90°$, the change of the cutting speed is small, but the cutting angle changes greatly.

When $\beta = 90°$, the maximum value of $\varphi_r$ is 90° and $\varphi_0 = 0°$. This shows that there are obvious collision points on the cylindrical surface, and the axial end surface does not obtain any finishing effects during parallel spindle barrel finishing.

④ Impacts of Q on $v$, $\varphi_r$, and $\varphi_0$

As shown in the figure which is drawn by the condition of $n/N = -2$, $N = 50\,\text{r/min}$, $\beta = 60°$, $q = 0.05\,\text{m}$, and $L = \sqrt{0.16 - Q^2}/\cos\beta$.

The average value of $v$ increases along with increasing $Q$, and the maximum value of $\varphi_r$ decreases along with increasing $Q$. Overall, a large $Q$ is good for the finishing effect of the cylindrical surface.

When $Q = 0$, the value of $\varphi_0$ has negative territory, and the value of $\varphi_0$ in positive territory is no remarkable. This shows that the finishing effect of the axial end surface is poor when $Q = 0$.

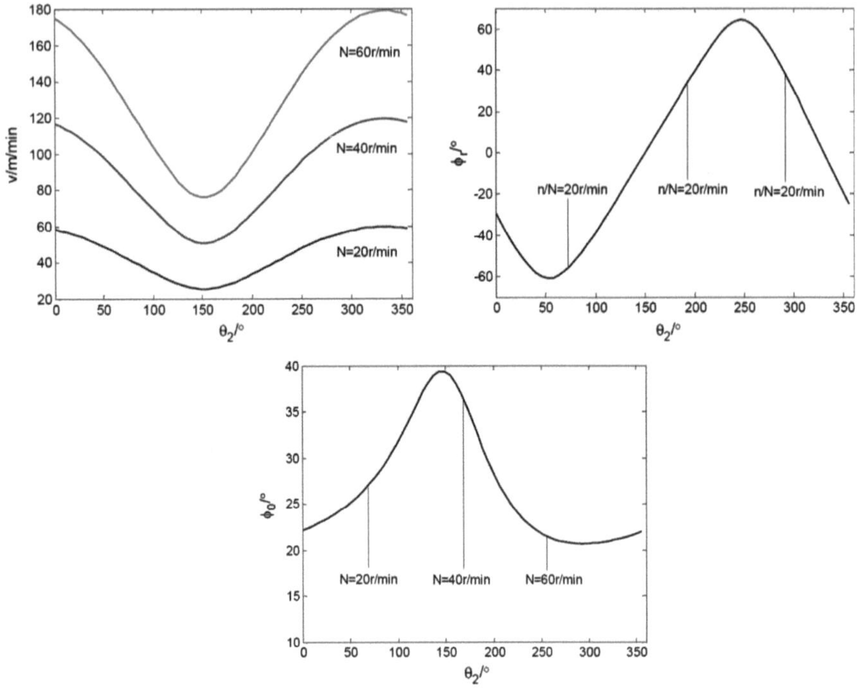


**Fig. 2.48** Impacts of $N$ on $v$, $\varphi_{\mathrm{r}}$, and $\varphi_0$

⑤ Impacts of $L$ on $v$, $\varphi_{\mathrm{r}}$, and $\varphi_0$

As shown in the figure which is drawn by the condition of $n/N = -2$, $N = 50\,\mathrm{r/min}$, $\beta = 60°$, $q = 0.05\,\mathrm{m}$, and $Q = \sqrt{0.16 - L^2\cos^2\beta}$.

The processing effect of $L$ on the effective finishing area is not very obvious. Note that the value of $L$ has a smaller effect, then the possibility of the cylindrical surface can be plumped, and it may occur to the cylindrical workpiece during barrel finishing.

⑥ Impacts of $q$ on $v$, $\varphi_{\mathrm{r}}$, and $\varphi_0$

As shown in the figure which is drawn by the condition of $n/N = -2$, $N = 50\,\mathrm{r/min}$, $\beta = 60°$, $Q = 0.3\,\mathrm{m}$, and $L = 0.3\,\mathrm{m}$.

Increasing with $q$, the machinability of the cylindrical surface is basically stable. Increasing with $q$, the machinability of the axial end surface has the change, and therefore the adjustment of other parameters should be noticed when finishing disk parts with high requirements on the axial end surface [48].

3. Functional characteristics and application scope

Vertical spindle barrel finishing belongs to the surface barrel finishing that the workpiece is in a non-free state. Compared with the surface barrel finishing technology, the workpiece is in a free state, and it can effectively avoid the knocking and scratch between workpieces.

Parallel spindle barrel finishing, including vertical spindle type and planetary vertical spindle type, mainly applies to finishing small-sized axis parts or the peripheral surface of disk parts. Because its finishing depth along the axial direction is inconsistent, the axial machining uniformity of the parts is not very ideal when finishing large medium-sized parts. Intermeshing spindle barrel finishing is mainly used to the finishing of disk parts end surface and can also implement the finishing of the peripheral surface. It applies to finishing the surface of small- and medium-sized axis/disk parts.

## 2.6.2 Main Factors Affecting Finishing Effects

According to the analysis of finishing principle, the influence factors of the intermeshing vertical spindle barrel finishing are more than the parallel vertical spindle barrel finishing, including kinematic parameters, geometric parameters, and finishing time. In order to illustrate the factors of intermeshing vertical spindle barrel finishing, the schematic diagram of its variable basic parameters is described in Fig. 2.49. In the figure, the anticlockwise of the roller in vertical view is positive

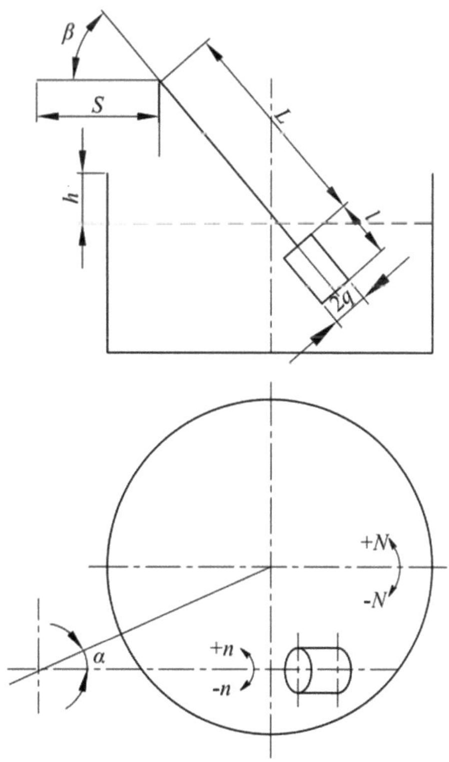


**Fig. 2.49** Schematic diagram of variable basic parameters

(+$N$), and the clockwise of that is negative (−$N$); the anticlockwise of the projection direction from the truncation spindle to workpiece is positive (+$n$), and the clockwise of that is negative (−$n$). In addition, $h$ is the distance between the horizontal plane of solid medium in static roller and the upper edge of roller mouth, $L$ can be adjusted using a flexible spindle jig, $\alpha$ is spindle deflection angle, corresponding to the angle between the roller axis and rotation axis in the previous machining principle analysis, and $\beta$ is spindle oscillating angle.

1. Kinematic parameters of equipment

In the finishing process of the intermeshing vertical spindle barrel, the equipment kinematic parameters consist of the revolving speed and direction of a barrel and that of a spindle. Theoretical results demonstrate that the kinematic parameters affecting cutting speed and angle are the revolving speed of a barrel $N$, and the ratio between the spindle speed $n$ and the revolving speed of a barrel $n/N$. The variation curves under different test conditions are shown in Figs. 2.50, 2.51, 2.52, 2.53, and 2.54.

Figures 2.50 and 2.51 show, respectively, the different thickness curves of surface material removals, as the spindle is fixed under conditions of $\alpha = 25°$, $\beta = 60°$, $N = -50$ r/min, $S = 490$ mm, $L = 485$ mm, $h = 230$ mm, and the processing time is 15 min. $\theta$ is arranged from 0 to 360. At $\theta = 0$, the farthest distance of the processing plate far away from the central position of the roller can reach to the maximum. All angles are subject to the rule that the direction of a spindle −$n$ rotates to the rotational direction.

Further, from Figs. 2.50 and 2.51, the finishing inhomogeneous can be observed as the spindle is fixed, which fits theoretical calculations well.

Figure 2.52 shows variation curves of surface material removal in thickness versus $n$ and $|n/N|$ under conditions of $\alpha = 25°$, $\beta = 60°$, $N = -50$ r/min, $S = 490$ mm, $L = 485$ mm, $h = 230$ mm, and the time is 20 min. As shown in Fig. 2.52, when $N$ is fixed, the less effect of $n$ and $|n/N|$ on surface material removal in the thickness and the more the power variation can be found. What is more, the thickness of axial end surface material removal is much higher than that of cylindrical surface material removal. It is caused by different cutting angles of different surface materials. For a cylindrical surface within 360°, only half of its surface can be regarded as a processing zone, but for an axial end surface, the whole surface is a processing zone. Thus, with the view of improving thickness of surface material removal, the spindle speed is not too high.

Figure 2.53 shows variation curves of surface material removal in thickness versus $N$, as the spindle speed is fixed under conditions of $\alpha = 25°$, $\beta = 60°$, $n = 88$ r/min, $S = 490$ mm, $L = 485$ mm, $h = 230$ mm, and the time is 20 min. From Fig. 2.53, it can be seen that with the increase of roller rotation speed, whatever for axial end surface or cylindrical surface, the thickness of surface material removal can be increased, and the non-load power and test load power are also increased. Moreover, the thickness variation of axial end surface material removal is more obvious than that of the cylindrical surface. Thus, to obtain a good variation

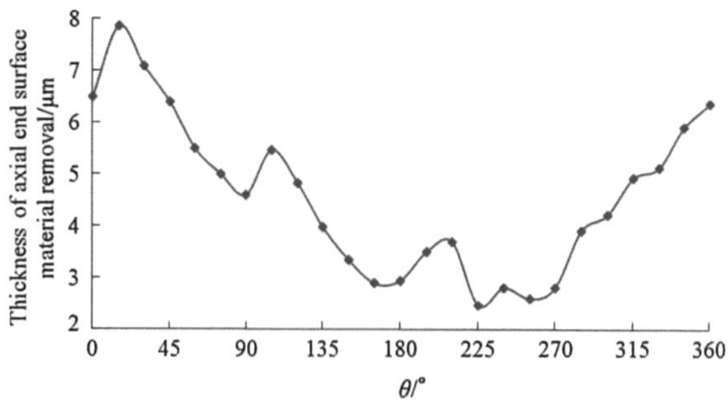


**Fig. 2.50** Thickness distribution curves of axial end surface material removal, as the spindle is fixed

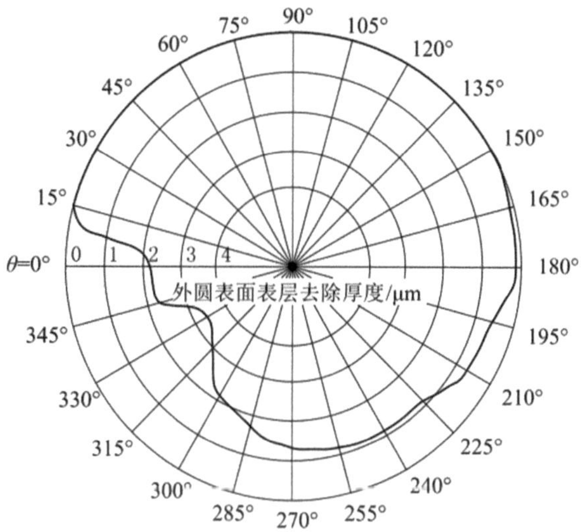


**Fig. 2.51** Thickness distribution curves of cylindrical surface material removal, as the spindle is fixed

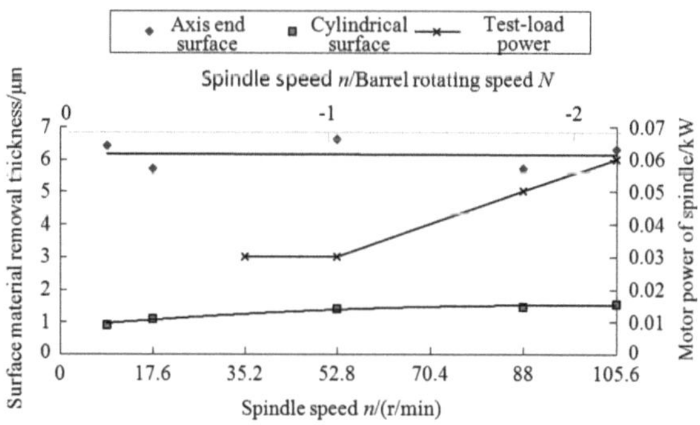


**Fig. 2.52** Variation curves of surface material removal in thickness versus $n$ and $|n/N|$, as the revolving speed of a barrel is fixed

consistency, the rotation direction of the roller needs to be adjusted periodically during the whole processing. Furthermore, the revolving speed of the roller must be selected as high as possible in design under the consideration of the equipment ultimate reliability.

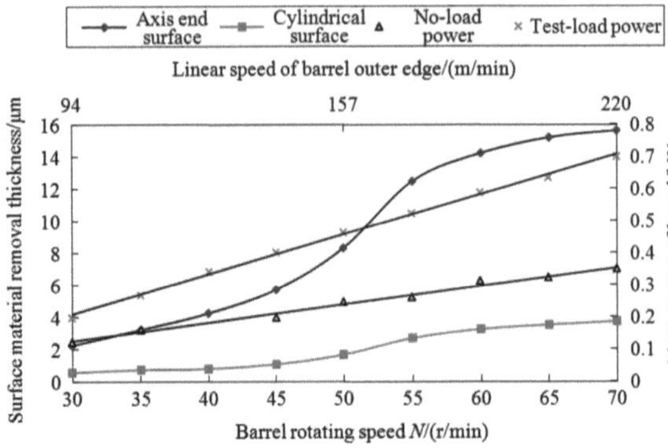


Fig. 2.53 Variation curves of surface material removal in thickness versus $N$, as the spindle speed is fixed

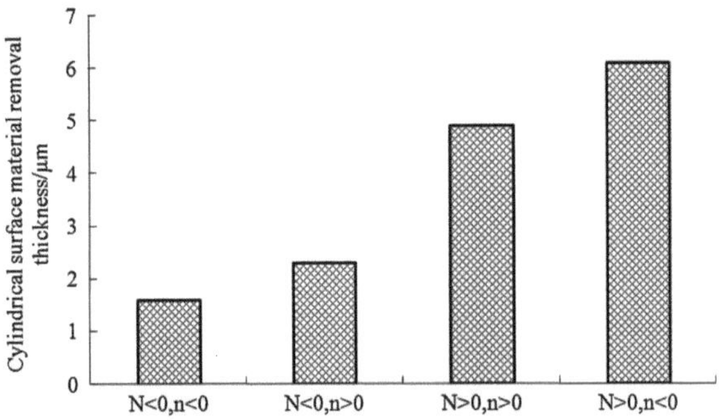


Fig. 2.54 Effect of the roller and spindle rotation direction on the thickness of cylindrical surface material removal

Figure 2.54 shows effect of the roller and the spindle rotation direction on the thickness of cylindrical surface material removal under conditions of $\alpha = 25°$, $\beta = 45°$, $|N| = 50$ r/min, $|n| = 88$ r/min, $S = 490$ mm, $L = 485$ mm, $h = 230$ mm, and the time is 50 min. As shown in Fig. 2.54, the thickness of surface material removal of counterclockwise rotary of the barrel ($N > 0$) is significantly greater than that of clockwise rotary ($N < 0$). This is mainly stems from spindle inclination. When $N > 0$, the flow direction of the solid medium impacting the sample cylindrical surface moves toward the bottom of the barrel. Owing to the fact that the bottom of the barrel is filled with solid medium, the momentum of effective solid medium mainly consumed in removal of metal sample surface. When $N < 0$, the flow direction of the solid medium impacting the sample cylindrical surface moves toward the top of the barrel. On account to the top of the barrel being a blank space, some of the momentum of the effective solid medium may have been released, and then the remaining momentum is used to remove the metal material. When the rotation direction of the barrel is fixed, contrary to that of the spindle, the thickness of surface material removal is slightly greater than both of the same rotation directions. It is the result of the different cutting speed of a solid medium on sample. For example, when the rotation direction of the barrel is contrary to the spindle, the entire linear velocity of the solid medium on the outside of a sample is much larger and vice versa.

Therefore, from Figs. 2.53 and 2.54, it is entirely possible that the average thickness of cylindrical and an axial end surface material removal can keep a good variation consistency, by adjusting the rotation direction of the barrel and spindle periodically.

2. Geometrical parameters of equipment

Figure 2.55 shows the effect of the spindle deflection angle $\alpha$ on the thickness of sample surface material removal under conditions of $\beta = 60°$, $N = -50$ r/min, $n = 88$ r/min, $S = 490$ mm, $L = 485$ mm, $h = 230$ mm, and the time is 30 min.

In Fig. 2.55, in view of the thickness of sample axial end surface material removal, compared with the calculation results of $\alpha = 0$, the result of $\alpha \neq 0$ is much better. This is mainly influenced by the cutting angle. However, for a cylindrical surface, the results of the thickness of material removal of $\alpha \neq 0$ are slightly inferior to that of $\alpha = 0$. This situation is also caused by the cutting angle.

Figure 2.56 shows the effect of the spindle oscillating angle $\beta$ on the thickness of sample surface material removal under conditions of $\alpha = 25°$, $N = -50$ r/min, $n = 88$ r/min, $S = 490$ mm, $L = 485$ mm, $h = 230$ mm, and the time is 30 min.

From Fig. 2.56, it can be seen that for an axial end surface, the results of the thickness of material removal of $\beta \neq 90°$ are obviously less than that of $\beta = 90°$. This is to say, due to the existing cutting angle, the effect of the intersection vertical spindle barrel finishing is superior to the parallel vertical spindle barrel finishing.

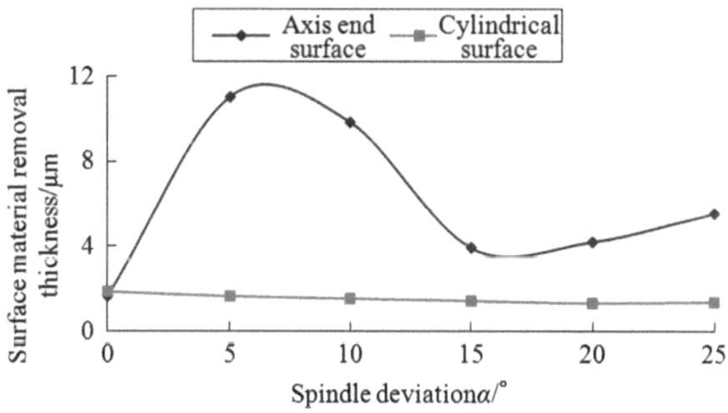


**Fig. 2.55** Effect of the spindle deflection angle $\alpha$ on the thickness of sample surface material removal

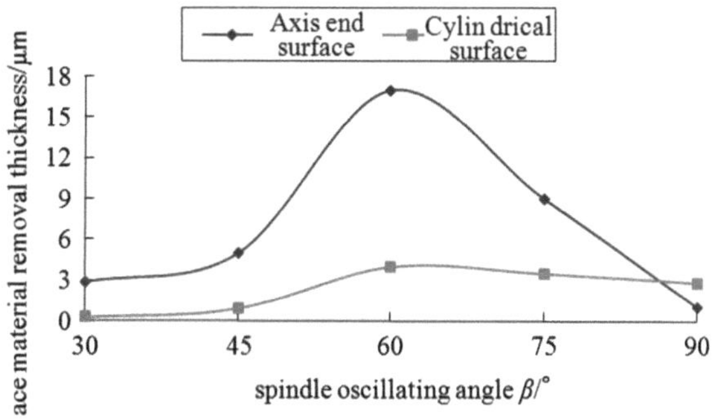


**Fig. 2.56** Effect of the spindle oscillating angle $\beta$ on the thickness of sample surface material removal

For a cylindrical surface, the difference of the results of the thickness of material removal of $\beta \neq 90°$ and $\beta = 90°$ is not clear, but the situation of $\beta = 60°$ is much better than $\beta = 90°$.

The results of theoretical and experiment results demonstrate that by increasing the effective radius of the roller during processing, the thickness of material removal of the axial end surface of the parallel vertical spindle and the axial end and cylindrical surface of the intersection vertical spindle can increase. Thus, when the structure size of a barrel is fixed during finishing, for improving the cutting velocity effectively, the distance between the sample position and the center of the barrel needs to be as far as possible. Furthermore, in the process of special equipment design, the diameter of the barrel should be taken into account big enough.

In view to the improvement of the applied force of the barrel finishing or enhancing the momentum of the effective solid medium, the results show that as the effective thickness of the barrel is increased, the cylindrical surface of the parallel vertical spindle and the axial end surface of the intermeshing vertical spindle can increase. Thus, in the equipment design, the effective thickness of the barrel, which is the height of the loaded solid medium, should also be considered. This can cause the thickness non-uniformity of workpiece material removal along the surface of the principal axis. As a result, this method is just fit for shaft parts or short axis. For processing precision long axis, the method of the horizontal spindle barrel finishing or adjusting turning processing of the vertical spindle barrel finishing again can be used.

3. Technological parameters

Figure 2.57 shows the effect of time on the thickness of sample surface material removal and surface roughness under conditions of $\alpha = 25°$, $\beta = 45°$, $N = -50$ r/min, $n = 88$ r/min, $S = 390$ mm, $L = 630$ mm, and $h = 230$ mm.

Experimental results show the continuation of processing after working for 15 min, and the value of surface roughness *Ra* has changed a little, whatever for the cylindrical or the axial end surface material of the sample. That means the value of

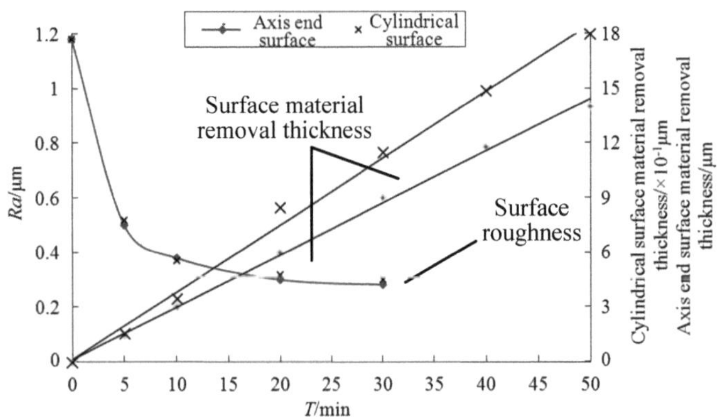


**Fig. 2.57** Effect of time on the thickness of sample surface material removal and surface roughness

surface roughness *Ra* can tend to be stable after processing at a certain time $T_j$, when employing specific barrel medium. At this point, the value *Ra* can be referred to as the limit surface roughness under certain conditions. As processing time increases, thickness of the sample surface material removal is increased steadily. In addition, as shown in Fig. 2.57, compared to the cylindrical surface, the thickness of sample surface material removal of the axial end surface has greater differences. In order to solve this problem, some solutions such as adjusting the values of $\alpha$, $\beta$, or transforming the rotating direction of the barrel and spindle by the automatic cycle can be used [49].

Moreover, the time has large influence on the surface topography, physical, and mechanical properties of the workpiece. The surface topography of the sample before and after the barrel finishing is presented using scanning electron microscope KYKY1000B, as shown in Fig. 2.58. From Fig. 2.58, before barrel finishing, the ups and downs change of the material surface can be found and material surface texture can be identified clearly. However, after the barrel finishing, the material surface topography can be refined, which illustrates that the barrel finishing can refine the surface topography and improve the surface integrity of materials.

The side of the sample (i.e., the sample section) was firstly polished metallographically and then corroded with nitric acid of 4%. The surface thickness of the denatured layer of the sample was recorded using scanning electron microscope (see Table 2.13), and further the denatured layer was observed which is described

**Fig. 2.58** Surface topography of the sample HT before and after the barrel finishing (200×). **a** Before finishing (traditional grinding); **b** barrel finishing for 15 min (cylindrical surface); **c** barrel finishing for 15 min (axial end surface)

**Table 2.13** Observation data of denatured surface layer thickness

| Specimen material | Finishing time/min | Denatured surface layer thickness of specimen with axis end surface/μm | Denatured surface layer thickness of specimen with cylindrical surface/μm |
|---|---|---|---|
| Q235 | 0 | <5 | <5 |
| | 15 | 42 | 24 |
| | 45 | 56 | 66 |
| | Over 5 h | 26 | 30 |
| HT | 0 | <5 | <5 |
| | 15 | 34 | 26 |

in Fig. 2.59. Experimental results show the barrel applied force of the flow medium on the workpiece during the barrel finishing can make the material surface enact plastic deformation and then form the structural skin layer (denatured layer). In the process of formation of the denatured layer, it must be accompanied by lattice movement, dislocation, and compression of metal crystal, which has an important value for improving the abrasive resistance, the corrosion resistance, and the fatigue strength.

The surface micro-hardness of the sample before and after barrel finishing is observed using a micro-hardness tester HVS-1000. The results indicate that the micro-hardness of the sample is increased by 8–15 HV on average.

The values of surface residual stress of the sample before and after barrel finishing can be obtained using X-ray stress measurement method (see Table 2.14). The results show improvement of surface residual stress of the sample, corresponding to the time when other conditions are invariable. At time of 15 min in finishing, the surface residual stress of the sample is obviously increased, and the increment value can reach −300 MPa. At time of finishing 45 min, the surface

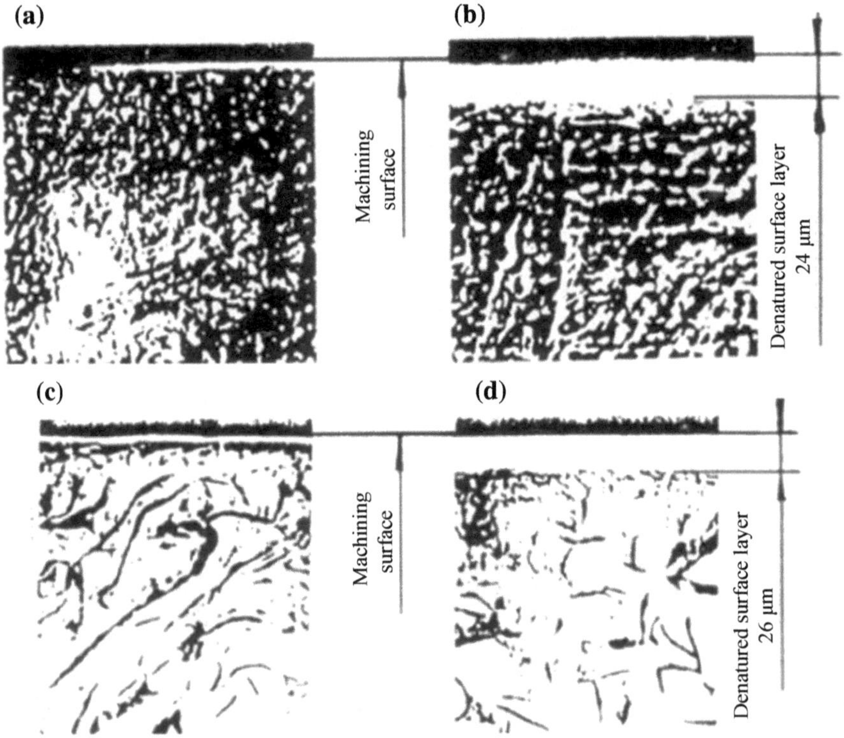


**Fig. 2.59** Denatured surface layer of workpiece before and after barrel finishing (200×). **a** Before finishing (Q235 common grinding); **b** finishing 15 min (Q235 excircle surface); **c** before finishing (HT common grinding); **d** finishing 15 min (HT excircle surface)

**Table 2.14** Observation data of the surface residual stress under different conditions

| Processing condition of specimen | Surface stress measurements of specimen cylindrical surface/MPa | Surface stress measurements of specimen axis end surface/MPa |
|---|---|---|
| Before finishing (common grinding) | – | −52 |
| Finishing 15 min | −304 | −349 |
| Finishing 45 min | −386 | −424 |
| Finishing over 5 h | −326 | −376 |

residual stress of the sample can still increase, but the increment value is smaller. When the time is longer than 5 h, the surface residual stress of the sample begins to decrease. This is caused by serious wear of the sample surface in barrel finishing with long life. The degree of wear of the sample can be confirmed by the measurement date of the surface thickness of the denatured layer [50].

### 2.6.3 Equipment Types and Design

1. Equipment types and structure

According to whether the barrel is fixed or not, the vertical spindle barrel finishing equipment can be divided into rotary barrel type and fixed barrel type; according to the position of the spindle axis, it can be divided into parallel spindle type, crossed spindle type, and intermeshing spindle type.

Figure 2.60 shows the general structural schematic diagram of the parallel spindle barrel finishing equipment, which consists primarily of the spindle reducer-rotary assembly, the slider, the vertical base, the base, the barrel reducer-rotary assembly, the barrel, the work-holding assembly, and et al. There are multiple spindles which can do positive or counter revolution movement to spindle box 1. The spindle box is fixed on slider 2, and the slider is installed on framework 3 along the guide rail for reciprocating motion. Turntable 7 is installed on base 5. Barrel 9 is fixed on the turntable and does the positive or counterrevolution movement with the turntable. Holding device 4 connects into the spindle.

Figure 2.61 shows the general structural schematic diagram of the intermeshing spindle barrel finishing equipment, which consists primarily of the base, the vertical base, the spindle deflection angle adjusting assembly, the spindle shifting assembly, the spindle reducer-rotary assembly, the spindle swing angle adjusting assembly, the work-holding assembly, the barrel, the barrel reducer-rotary assembly. Its most distinctive feature is that the spindle is three-dimensional adjustable structure. That is to say, the rotation axis of the spindle and the rotation axis of the vertical barrel can be adjusted to three states: parallel, crossing, and intersection for suiting different workpieces and different processing requirements.

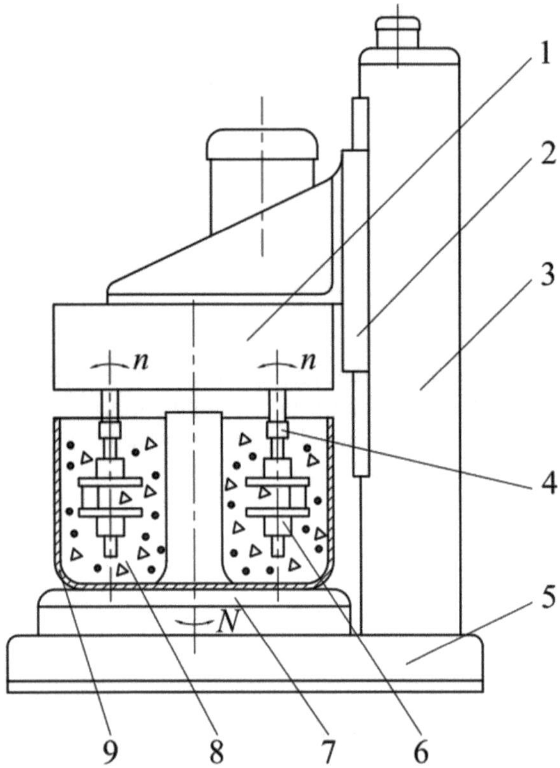


**Fig. 2.60** General structural schematic diagram of parallel spindle barrel finishing equipment. 1—Spindle reducer-rotary assembly; 2—slider; 3—vertical base; 4—work-holding assembly; 5—base; 6—workpiece; 7—barrel reducer-rotary assembly; 8—processing medium; 9—barrel

In addition to the essential functions (main functions) of realizing workpiece processing, the above two equipments should have the following functions:

(1) Processing safety and reliability considered, the barrel does the high-speed rotating in processing, and the safety protection devices are needed. The structure of the safety protection device can be fixed and remain active.
(2) Improving machining efficiency considered, pneumatic or hydraulic control work-holding assembly, increasing the number of the spindle, and other measures can be adopted.
(3) Ensuring the processing quality should be considered, and liquid medium (grinding compound and water) as processing medium can automatically recycle throughout the progressive processing. The filter is set in the re-circulating device, and this can efficiently reduce the impact of the processing impurities in the processing quality. When setting the liquid medium to re-circulate the device, the feature of the liquid medium movement with the rotating barrel should be considered fully for good match between the liquid medium re-circulating device and the barrel.

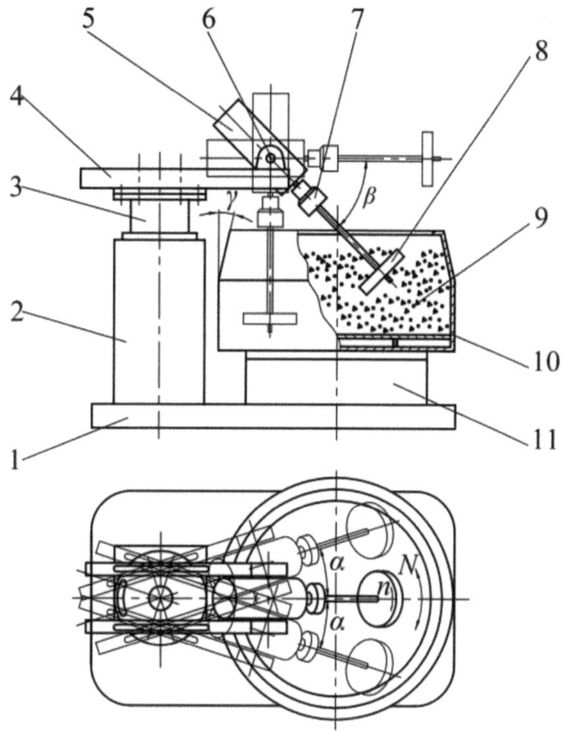


**Fig. 2.61** General structural schematic diagram of intermeshing spindle barrel finishing equipment. 1—Base; 2—vertical base; 3—spindle deflection angle α adjusting assembly; 4—spindle shifting assembly; 5—spindle reducer-rotary assembly; 6—spindle swing angle $\beta$ adjusting assembly; 7—work-holding assembly; 8—workpiece; 9—processing medium; 10—barrel; 11—barrel reducer-rotary assembly

(4) Improving processing adaptability should be considered; the rotating speed adjustments of the spindle and the barrel are quite necessary. This can be realized through method that sets frequency conversion speed regulator.

2. Equipment design

(1) Design of barrel

The design of the barrel bases is the simplest of mechanical principle. When the abrasives are rotating in the barrel, certain centrifugal force can generate due to the presence of the abrasives own quality. The abrasives group can generate discrete flow away from the rotating axis under the action of the centrifugal force. The abrasives group can self-climb along the barrel wall during the flow as a result of the restrictions imposed by the limited barrel wall. The number of the abrasives at the center position of the barrel reduces accordingly, and then one revolution paraboloid that exists a minimum value forms on the abrasives-boundary. In case of the packed weight of the abrasives limited, as the rotating speed of the barrel is improving, the minimum value of the revolution paraboloid on the abrasives-boundary becomes smaller. Limited by the height dimensions of the barrel, the degree of the abrasives group self-climbing along the barrel wall will become more intense, which even results the abnormal circumstances that the abrasives are thrown out. Even if the abnormal phenomenon does not occur, the circumstance that the abrasive splashes can occur during the processing. Given the above analysis, the vertical structural barrel that is used during parallel spindle barrel finishing must have a certain height. During intermeshing spindle barrel finishing, the whole structure of the barrel can be designed into two parts: The lower part is designed into regular twelve-prime structure; the upper part is designed into upper reentrant regular twelve-prismatic table structure. As a result, the lower part has a larger rotating radius in favor of workpiece processing for ensuring productivity; the upper reentrant form of the upper part can efficiently reduce the climbing and splash of the abrasives group during processing.

The whole structure of the barrel often adopts welded steel plate. There should be a certain thickness of polyurethane or vulcanized rubber anti-corrosive coating on the barrel inner wall. This can not only avoid the corrosion of the bottom and the wall of the barrel, but also increase the friction property between the barrel and the abrasives as appropriate for the purpose of the effective rotating of the abrasives in the barrel.

The result that the abrasives process of the workpiece must make a part of metal particle and burrs on the workpiece surface comes off during barrel finishing. Meanwhile, there are also some tiny abrasive falling off the surface of the abrasives. These metal micro-powders, burrs, and abrasive particles are mixed in the barreling medium during sustaining processing, and even some can adhere to the surface of the abrasives. This can affect not only the processing efficiency, but also the processing quality. In order to reduce or eliminate the influence of these micro-contaminant, a partition with one groove about 1 mm wide is added on the bottom of the barrel. One depositing chamber is formed between the partition and the bottom of the barrel. There are two symmetrical sewage outlets on the wall of the depositing chamber for easy periodic cleaning. Otherwise, dewatering equipment can be designed on the wall or at the bottom of the depositing chamber because of the cleaning of the abrasives and the replacement problem of the liquid medium during sustaining processing.

(2) Design of barrel reducer-rotary assembly

The barrel reducer-rotary assembly serves two functions: One is to ensure the implementation of the barrel rotating speed; the other is to guarantee the stationary of the barrel motion process with large diameter.

The effective working linear velocity of the barrel is typically in 80–280 m/min. under the condition to maintain the rotating speed of the barrel. The adoptive transmission route is as follows: electromotor—belt driving—cylindrical worm-driven reducer—rotary table (barrel). In order to guarantee the stationarity of the barrel rotating motion, the major diameter rotary table should be connected to the output end of the cylindrical worm-driven reducer, and the lower end surface of the rotary table should be supported on the hoop-type guide rails of the rotary table base.

(3) Design of finishing automatic cycle process

The technological process of the vertical spindle barrel finishing is shown in Fig. 2.62. Whereby, special attention should be paid to the determination and the transformation of the spindle rotating direction ($n$) and the barrel rotating direction ($N$) during barrel finishing. For the ordinary uniform workpiece with cylindrical surface, no matter what directions $n$ and $N$ take, the full surface finishing of the workpiece can achieve. Based on the theoretical analysis and the experimental study, the situation of $n/N < 0$ is better than the situation of $n/N > 0$; for the workpiece with non-uniform cylindrical surface such as gear and cam, the direction of the spindle rotating ($n$) must be changed during finishing in order to ensure that the full surface of the workpiece can be processed; furthermore, the continuous uniform processability of the finished surface considered, the automatic circulation process of the spindle rotating ($n$) and the barrel rotating ($N$) is adopted. And

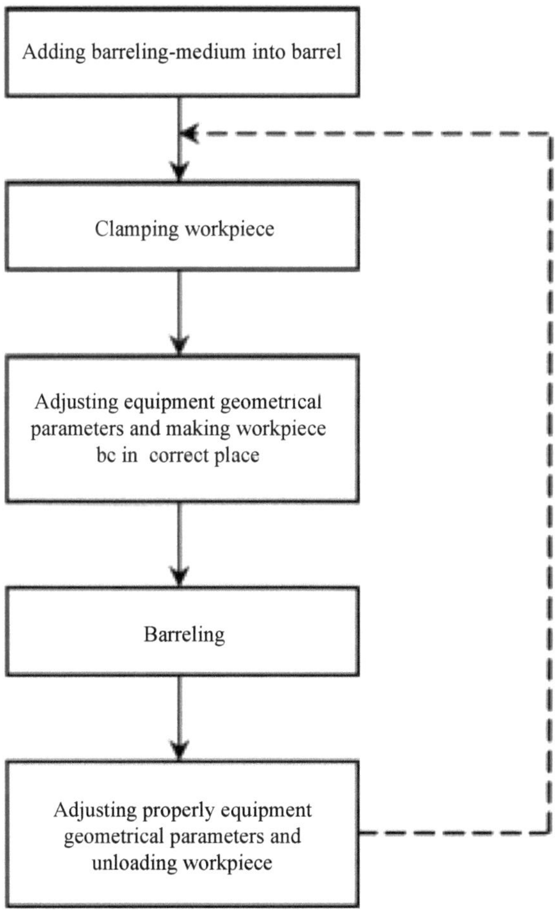


**Fig. 2.62** Technological process of vertical spindle barrel finishing

$$(N, -n, t) \Leftrightarrow (0, 0, \Delta t) \Leftrightarrow (-N, n, t)$$
$$\text{or } (N, n, t) \Leftrightarrow (0, 0, \Delta t) \Leftrightarrow (-N, -n, t)$$

In the above circulation process, t is the circulating half period, usually 1–2 min, and $\Delta t$ is controlled by electrical appliance, usually 5–7 s.

(4) Equipment shape

Figures 2.63 and 2.64 show the photographs of the parallel spindle and the intermeshing spindle barrel finishing equipment which are produced by Langfang Beifang Tianyu Mechanical and Electrical Technology Co., Ltd.

# 2.7 Horizontal Spindle Barrel Finishing

## *2.7.1 Finishing Principle and Characteristics*

**Fig. 2.63** Photograph of parallel spindle barrel finishing equipment

**Fig. 2.64** Photograph of intermeshing spindle barrel finishing equipment

1. Finishing principle

Based on design request, crankshafts should have good machining precision, and its surface quality and integrality are also important. At present, crankshafts are finished mostly by abrasive cloth and manual steel brush, which is not ideal on finishing quality, efficiency, and cost, and especially cleanness degree does not meet design and using request. Therefore, a new process, horizontal spindle barrel finishing, is developed.

Figure 2.65 shows the principle diagram of horizontal spindle barrel finishing. Parts like crankshaft rotate round their axes of main journal in opposite direction. Reciprocating motion of the container (slot barrel), which is the loaded machining medium (abrasive blocks, compounds, water included), is driven by crank and slider mechanism. With rotary motion and reciprocating motion, the abrasive blocks have the action of colliding, extruding, and the micro-grinding (scraping and scoring) against the crankshaft surface. Consequently, it has effects of finishing crankshafts, increasing surface quality, and improving the physical–mechanical properties [51, 52].

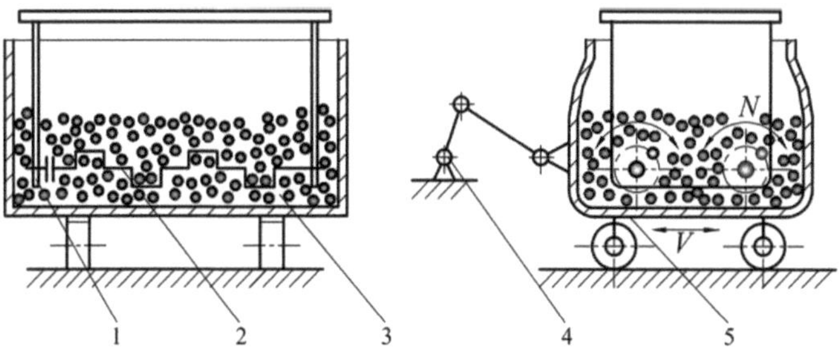


**Fig. 2.65** Principle diagram of horizontal spindle barrel finishing. 1—Gearing; 2—crankshaft; 3—machining medium; 4—driven mechanism; 5—slot barrel

The acting force is necessary to achieve horizontal barrel finishing. But, under certain circumstances of the machining medium, the impact of the acting force mainly depends on the complex motion condition between the abrasive blocks and the workpiece surface.

2. Motion analysis

According to the characteristic and request of relative motion invariability, model diagram of motion analysis for arbitrary point M of connecting rod journal surface is set up, as shown in Fig. 2.66, where O is the rotary center of crankshaft (main journal), $O_1$ is the center of connecting rod journal, $l$ is the distance between the center of main journal and connecting rod journal, $r$ is radius of connecting rod journal, $n$ is rotational speed of crankshaft, $S$ and $f$ are the reciprocating amplitude and frequency.

In the course of finishing ($\theta = 2\pi nt$), the velocity component of the abrasive block to the arbitrary point M ($\theta_1 \in [0, 2\pi]$) on the connecting rod journal is

$$\begin{bmatrix} \dot{x} \\ \dot{y} \end{bmatrix} = \begin{bmatrix} 2\pi n(l \sin\theta + r \sin(\theta + \theta_1)) \pm v_1 \\ 2\pi n(l \cos\theta + r \cos(\theta + \theta_1)) \end{bmatrix} \tag{2.67}$$

where

$$v_1 = 2Sf \tag{2.68}$$

According to the equation of velocity composition, composite velocity $v$ can be obtained [53].

Under certain circumstances of the machining medium, finishing effects of horizontal spindle barrel finishing have relations with composite velocity $v$. From Eq. 2.67, influencing factors on composite velocity include $n$, $S$, $f$, $r$, $l$, $\theta_1$. Their specific impacts are analyzed through the MATLAB software as follows.

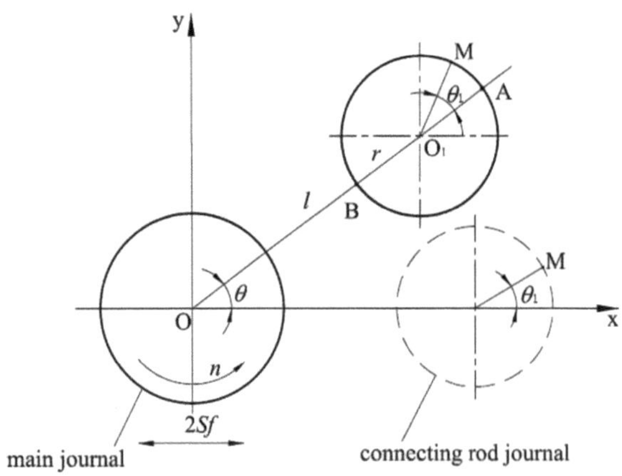


**Fig. 2.66** Motion analysis diagram of connecting rod journal surface

**Fig. 2.67** Effect curve of composite velocity with different rotational speeds

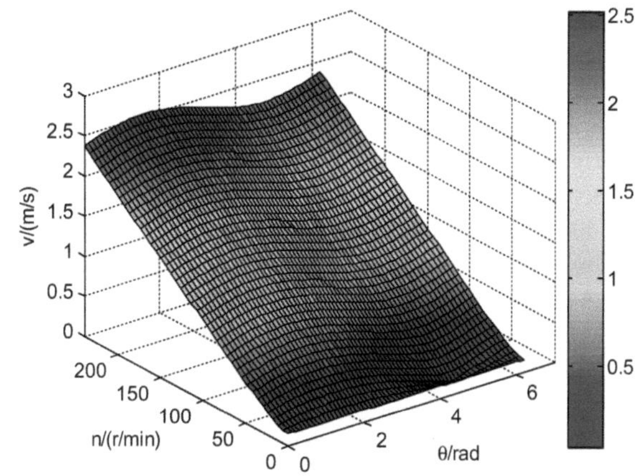


(1) Effects of different rotational speed $n$. Under the conditions of $r$ = 0.0345 m, $l$ = 0.06 m, $\theta_1 = 0°$, $S$ = 0.08 m, $f$ = 60 times/min, $n$ = 108 r/min, Fig. 2.67 shows effect curve of composite velocity with different rotational speeds. With the increase of $n$, $v$ increases, so they are helpful to improve efficiency.
(2) Effect of different reciprocating amplitudes $S$ and frequencies $f$. Under $n$ = 108 r/min and other same conditions, Fig. 2.68 shows effect curves of composite velocity with different reciprocating amplitudes and frequencies. With the increase of $S$ and $f$, $v$ increases, so they are helpful to improve efficiency. $S$ and $f$ have similar rule to $v$. In view to improve efficiency, it should increase $S$ or $f$. But large $S$ leads to large equipment dimension, and the lateral size of the container increases at least by $2(S_2 - S_1)$.

Additionally, too large $f$ would form instantaneous empty field between the crankshaft and particles according to the barrel finishing mechanism, so finishing effect is not ideal. This phenomenon is simulated by Fluent software. Boundary conditions are the velocity inlet of particles with 0.2 m/s and outflow outlet.

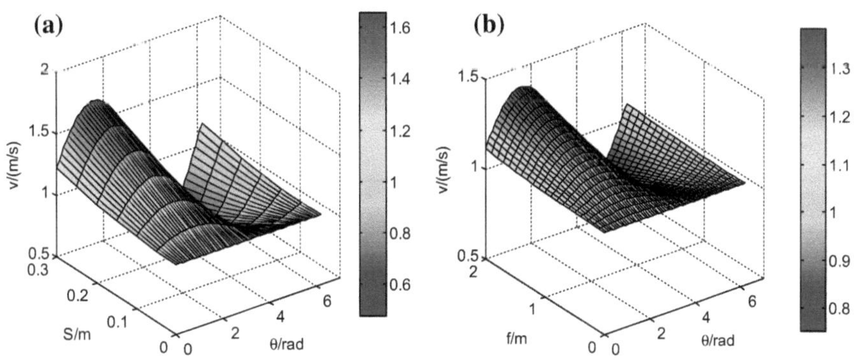


**Fig. 2.68** Effect curves of composite velocity with different reciprocating amplitude and frequency. **a** $S$; **b** $f$

Crankshaft surface is wall surface condition, and its angular velocity is 11.3 rad/s. The $k$-$\varepsilon$ model is adopted. SIMPLE arithmetic is selected, which can directly compute velocity and pressure field. Simulation results with different reciprocating frequency $f$ are shown in Fig. 2.69. Under constant rotational speed, the crankshaft surface is basically surrounded by abrasive blocks as low reciprocating frequency (such as 50 times/min), and abrasive blocks are driven to rotate a certain scope by crankshaft rotation, which is main proportion to reciprocating motion. With the increase of reciprocating frequency, the velocity of abrasive blocks increases, which leads to untimely surrounded crankshaft, and forms instantaneous empty field. At the same time, the degree and abrasive blocks are driven to rotate, which are relatively weakened, thus affecting finishing effects and efficiency. Therefore, it is proper for the main journal to improve the finishing effect and efficiency by selecting large rotational speed.

(3) Effect of different distance between the center of main journal and connecting rod journal $l$ and radius of connecting rod journal $r$. Under the other same conditions, Fig. 2.70 shows effect curves of composite velocity with different $l$ and $r$. With the increase of $l$ or $r$, $v$ increases. For the specific crankshaft finishing, the structure size is fixed. But there is different productivity for the crankshaft with different size (i.e., different crankshaft should use different processing time).

3. Finishing uniformity analysis

Finishing uniformity is the consistency of processing results on different positions of the part surface. Under conditions of $r = 0.0345$ m, $l = 0.06$ m, $\theta_1 = 0°$, $S = 0.08$ m, $f = 60$ times/min, $n = 108$ r/min, composite velocity curves of all positions on connecting rod journal surface with one rotation cycle are plotted, as

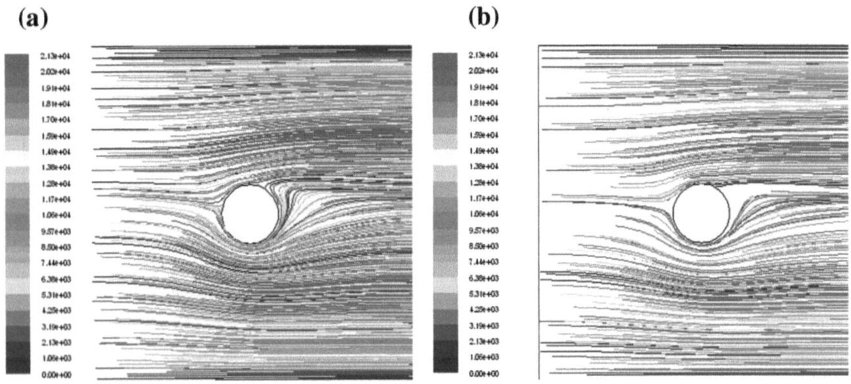

**Fig. 2.69** Simulation results with different reciprocating frequency. **a** 50 times/min; **b** 62.6 times/min

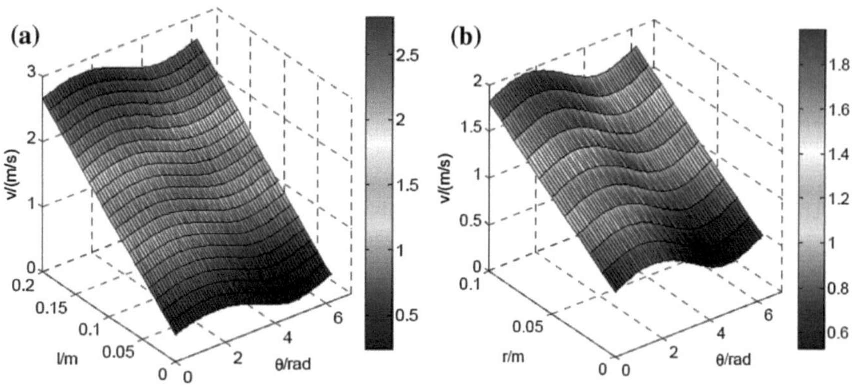


**Fig. 2.70** Effect curves of composite velocity with different distance between the center of main journal and connecting rod journal and radius of connecting rod journal. **a** $l$; **b** $r$

shown in Fig. 2.71. The velocity of different points on connecting rod journal surface is changeable. When $\theta_1$ varies from 0 to $\pi$, $v$ decreases from the maximum value to minimum value. Similarly, $v$ increases from the minimum value to the maximum value as $\theta_1$ varies from $\pi$ to $2\pi$. Moreover, finishing depth is also different which leads to different acting force, so that the surface of connecting rod journal is not uniform.

When the distance between the center of main journal and connecting rod journal $l$ is zero, the part is not an eccentric shaft. According to Eq. (2.67), the composite velocity of the abrasive block to the arbitrary point M ($\theta_1 \in [0, 2\pi]$) is

$$v = ((2\pi nr)^2 \pm 2\pi nrv_1 \sin\theta_1 + v_1^2)^{1/2} \tag{2.69}$$

**Fig. 2.71** Effect curve of finishing uniformity with different point of connecting rod journal

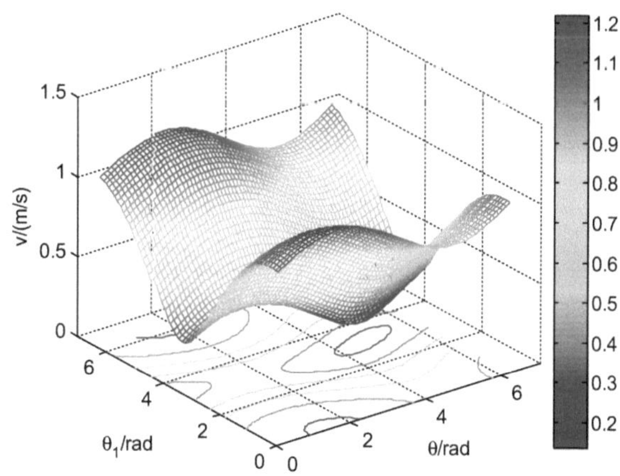

In the course of finishing ($\theta_1 = 2\pi nt$), the velocity and the processing state are consistent to the different positions of the surface, and only the phase changes, leading to the uniform of the finishing effect. Therefore, the finishing uniformity is for eccentric shaft parts (such as crankshaft, camshaft). For the crankshaft, finishing effect of main journal is basically uniform, and finishing uniformity is mainly aimed at the connecting rod journal.

The composite velocity of different points on the part surface is analyzed as below. According to Fig. 2.66, two most representative points are selected for simplifying analysis that is outside Point A ($\theta_1 = 0°$) and inside Point B ($\theta_1 = 180°$) on connecting rod journal surface.

(1) Finishing uniformity effect with influencing factors

Effect rules of composite velocity *v* on Point A and Point B with different *n* are shown in Fig. 2.72. When *n* is zero (crankshaft does not rotate), *v* is equivalent of all points to the connecting rod journal surface. With the increase of *n*, their difference is gradually increasing. Therefore, *n* should be smaller to improve the uniformity.

Effect rules of composite velocity *v* on Point A and Point B with different *S* and *f* are shown in Fig. 2.73. It is shown that *v* has same rules of all points to the connecting rod journal surface with different *S* and *f*, which shows that large *S* and *f* would have good uniformity and efficiency. As shown in Fig. 2.73, velocity difference between Point A and Point B is mainly caused by certain rotational speed.

Effect rules of composite velocity *v* on Point A and Point B with different *l* and *r* are shown in Fig. 2.74. Figure 2.75 shows the effect relationship with different *l* and *r* as $\theta$ is 0°. Because parameters setting of curve drawing is $l = 0.06$ m and $r = 0.0345$ m, *v* deceases as *l* and *r* increase to 0.06 m and 0.0345 m. Generally, the difference of *v*, of all points to connecting rod journal surface, is gradually increasing as the *l* and *r increase* and has the same changing trend. But when *l* and

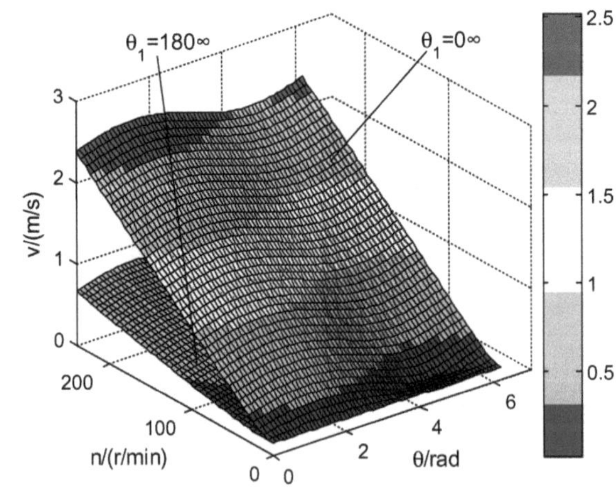


**Fig. 2.72** Effect curves of finishing uniformity with different *n*

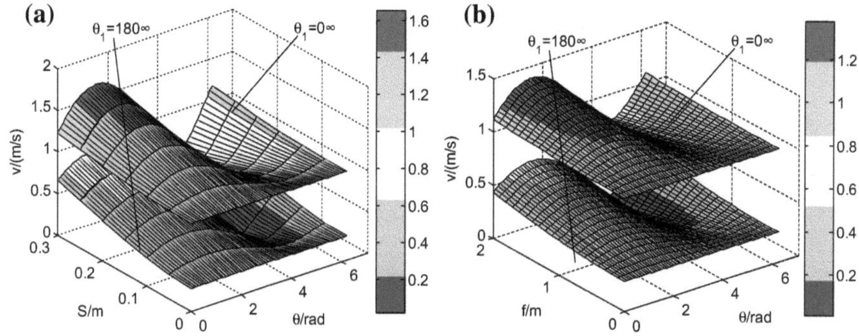


**Fig. 2.73** Effect curves of finishing uniformity with different $S$ and $f$. **a** $S$; **b** $f$

$r$ are relatively large, the proportion of the rotary motion is increased in the same reciprocating motion, that is,

$$\frac{v - v_1}{v} = 1 - \frac{v_1}{v} \tag{2.70}$$

Therefore, the velocity difference on different points is more significant, and this indicates that finishing uniformity is more important to crankshafts with larger size.

(2) Solving methods

According to the above analysis of the finishing uniformity of the connecting rod journal surface, the different rotating speed, the reciprocating amplitude and frequency, and the crankshaft size will affect finishing uniformity.

For improving uniformity, the composite velocity $v$ of arbitrary point of the connecting rod surface is expected to equality. If the velocity is equal, $v$ can be derived from Eq. (2.67).

$$v = 2Sf \tag{2.71}$$

Equation (2.71) means that $n$ is zero (the workpiece does not rotate), or $r$ is zero (changing into a point), or $l$ is zero (the workpiece is non-eccentric).

For crankshaft, both $r$ and $l$ cannot be zero. From Figs. 2.74 and 2.75, it is known that the uniformity is good as $r$ and $l$ are small. When $r$ and $l$ are relatively larger, the velocity difference of different positions on the connecting rod journal is caused, which leads to the poor finishing uniformity. That is to say, for the small-sized crankshaft, the uniformity of points on the connecting rod journal surface is relatively good, and the key is how to improve the finishing uniformity of large- or medium-sized crankshaft.

If the velocity is equal, $n$ must be zero. But the processing position of connecting rod journal surface is not changeable, and the cutting angle of different points is different, so the finishing effect and uniformity are obviously bad. At this time, we can use the reciprocating motion of the barrel as the main movement, and the

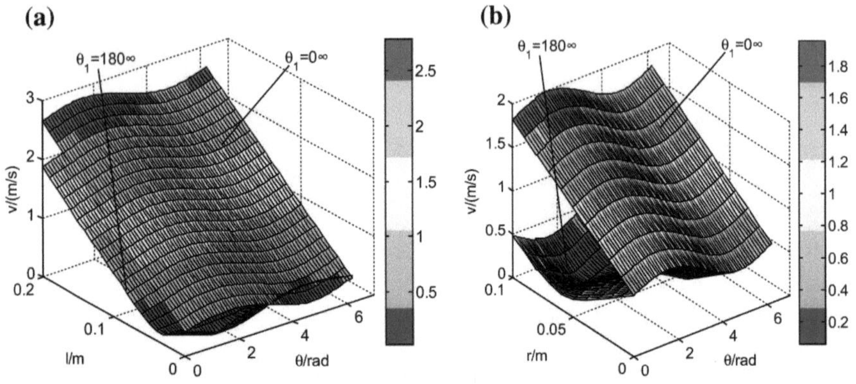


**Fig. 2.74** Effect curves of finishing uniformity with different $l$ and $r$. **a** $l$ ($r$ = 0.0345 m); **b** $r$ ($l$ = 0.06 m)

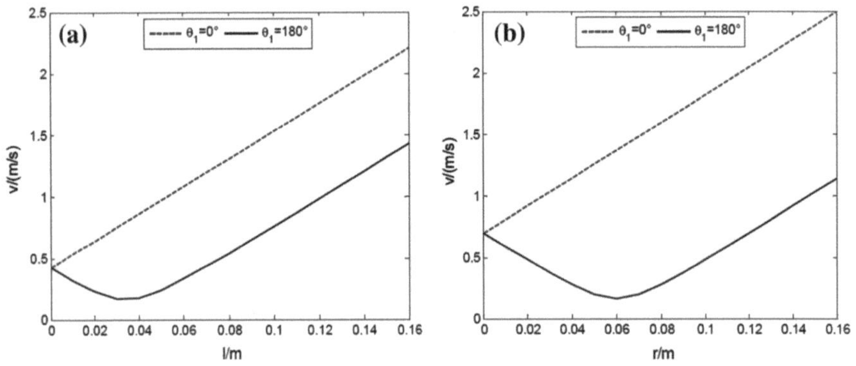


**Fig. 2.75** Effect curves of finishing uniformity with different $l$ and $r$ as $\theta$ is 0°. **a** $l$ ($r$ = 0.0345 m); **b** $r$ ($l$ = 0.06 m)

periodic intermittent rotary motion of the crankshaft as the feed movement, to reach the goal of finishing uniformity.

According to above analysis, finishing effect of the eccentric shaft cannot be absolutely consistent by this process. But differences can be minimized to meet used requirements and customer needs. The concrete analysis is done as follows.

When the connecting rod journal is located at underside, their equations of composite velocity on Point A and Point B are

$$v_A = -2\pi n(l + r) \pm 2Sf, \; v_B = -2\pi n(l - r) \pm 2Sf \tag{2.72}$$

For definite crankshaft, its size is unchangeable (e.g., $l$ = 105 mm, $r$ = 65 mm). From Eq. (2.72), under different motion parameters, composite velocities and the sum of their absolute values on Point A and Point B can be computed in a period of reciprocating motion with the same direction of rotation motion, as shown in Table 2.15.

From Table 2.15, the absolute value of the sum of the composite velocity may be equal by choosing suitable parameters. From Eq. (2.72), the condition of velocity equality is

$$-2\pi n(l+r)+2Sf\geq 0 \tag{2.73}$$

That is,

$$Sf\geq \pi n(l+r) \tag{2.74}$$

So, relations of $S$, $f$, and $n$ can be obtained.

Equation (2.73) or (2.74) is the analysis result when the connecting rod journal is located at underside. In the scope of $\theta$ in [0, 2π], $Sf$ should be π/2 times with the rule of sine or cosine. In practice, $S$ should be π/2 times, for high frequency is not reasonable. In addition, the value of $f/n$ may as well be decimal fraction, and this is conducive to be similar for the processing state of different points in the statistical sense, which can improve finishing uniformity of connecting rod journal.

The above analysis shows that large rotational speed and certain reciprocating motion speed can be desirable for small-sized crankshaft to achieve the required processing effectiveness and processing efficiency. For medium or large-sized crankshaft, larger reciprocating motion amplitude, certain reciprocating motion frequency, and smaller rotary speed should be chosen to improve finishing uniformity. In a certain finishing depth, through the reasonable selection of motion parameters, surface finishing of the eccentric shafts can be achieved to meet used requirements and customer needs [11, 12].

**Table 2.15** Composite velocities and their absolute value sum on point A and point B under the different motion parameters

| Motion parameters | | | | V/(mm/min) | | Absolute value sum/(mm/min) |
|---|---|---|---|---|---|---|
| n/(r/min) | f/(times/min) | S/mm | | + | − | |
| 60 | 60 | 80 | B | −5460 | −24,660 | 30,120 |
| | | | A | −54,480 | −73,680 | 12,8160 |
| 5 | 60 | 80 | B | 8345 | −10,855 | 19,200 |
| | | | A | 4260 | −14,940 | 19,200 |
| 10 | 60 | 150 | B | 15,490 | −20,510 | 36,000 |
| | | | A | 7320 | −28,680 | 36,000 |
| 10 | 45 | 200 | B | 15,490 | −20,510 | 36,000 |
| | | | A | 7320 | −28,680 | 36,000 |
| 11 | 40 | 200 | B | 13,239 | −18,761 | 32,000 |
| | | | A | 4252 | −27,748 | 32,000 |

4. Functional characteristics and application scope

Horizontal spindle barrel finishing process is a new technology for the surface finishing problem of the eccentric shafts (including crankshaft, camshaft). By this process, surface roughness value decreases, burrs is removed, surface texture is more isotropic, micro-hardness increases, material ratio of the profile increases, and denatured layer forms. So, the process can improve the geometric, physical and mechanical features of the surface layer, increase the cleanliness of parts (meet requirements of the design and the usage), resulting in improved surface integrity of parts. The specific situations include:

(1) Due to decrease of the surface roughness value, increase of the micro-hardness, and formation of the compressive stress, the fatigue strength increases and the fracture rate reduces, which improves the service life of the parts.
(2) Because of burr removing, non-working surface processing, and the casting or forging residual sand removing, the cleanliness of the whole machine improves, and it is convenient for assembly.
(3) The initial running-in of parts is done while finishing, and this improves the used performance of the whole machine.
(4) After finishing, surface texture tends to be isotropic, and independent micro-pit in part surface is formed, which is conducive to the improvement of storage capacity. Thereby, when the machine starts, the surface wear reduces.
(5) The application of this technology can replace the original four steps for belt polishing, that is aging after shot peening, felt polishing, manual cleaning, and second manual cleaning, so as to reduce the process cost and improve the processing efficiency.

In addition, the labor intensity is reduced, and the production efficiency is improved, which is convenient for realizing mechanization and automation.

Horizontal spindle barrel finishing process is an effective method for eccentric shaft surface finishing and will have a positive impact on the internal combustion engine industry and related parts processing industry, with a wide range of social and economic benefits [54].

## 2.7.2 *Main Factors Affecting Finishing Effects*

In experiments, three measuring points are chosen on main journal surface, two points in the direction of 180° and another point with 90°; measuring positions on the connecting rod journal surface are the outside point, the inside point, and the point clockwise 90°. When measuring, the average value of each position, measuring three times, is taken as the surface roughness value to reduce the man-made error in the measurement.

Because the surface roughness of the workpiece is not consistent at different points, the influence of each parameter can be determined by the decrease rate of the surface roughness value.

$$\Delta Ra = \frac{Ra^1 - Ra^i}{Ra^1} \qquad (2.75)$$

where $\Delta Ra$ is the decrease rate of the surface roughness $Ra$, $Ra^1$ is the surface roughness $Ra$ before finishing, and $Ra^i$ is the surface roughness $Ra$ as $i$th measurements.

1. Motion parameters

(1) Rotary motion parameter

Rotary motion parameter is a linear velocity of the part surface. For certain part, it is rotational speed. Under conditions of $S$ = 0.08 m, $f$ = 75 times/min, finishing depth is 240 mm, the shape of abrasive blocks is sphere Φ3 mm, the material is aluminum oxide, compound is LC-10, and decrease rate curves of the surface roughness value $Ra$ on main journal surface, to different rotational speeds, are shown in Fig. 2.76. It shows that $\Delta Ra$ reached a stable value in 16 min as the rotary speed is 108 r/min; but when the rotary speed is 54 r/min, $\Delta Ra$ is still in a state of decline in 24 min; while the rotary speed is 0 r/min, the decline is slower and processing time is longer. This indicates that the decrease of the surface roughness value is more rapid as rotational speed increases, which indicates that the rotational speed can improve the finishing efficiency.

Figure 2.77 shows $\Delta Ra$ curves on connecting rod journal surface with finishing time to different rotational speeds. When $n$ is 54 r/min, $\Delta Ra$ on the middle point is 16.7% in 4 min, while 9.4% when $n$ is 108 r/min. This indicates that the velocity difference between outside point and inside point at 54 r/min is smaller than 108 r/min, so it is reasonable for improving finishing uniformity. Therefore, in order to improve the uniformity, $n$ should be low; but too small rotational speed will decrease the finishing efficiency, which can be solved by selecting reciprocating motion parameters. In addition, the original surface roughness of the workpiece is

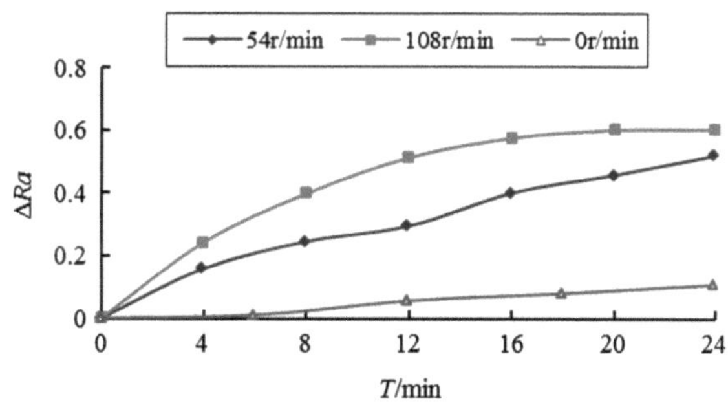


**Fig. 2.76** $\Delta Ra$ curves on main journal surface with finishing time to different rotational speeds

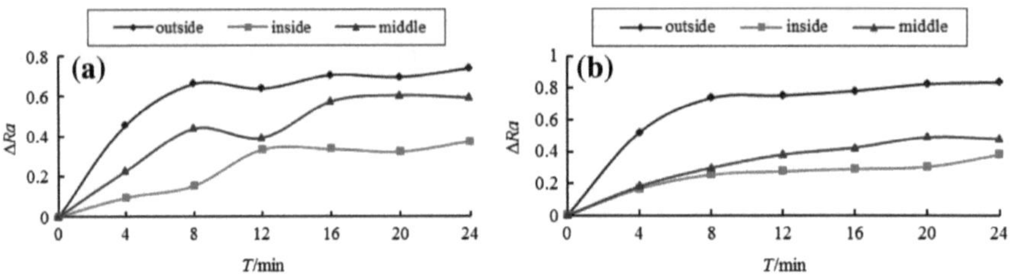


**Fig. 2.77** Δ*Ra* curves on connecting rod journal surface with finishing time to different rotational speeds. **a** $n$ = 108 r/min; **b** $n$ = 54 r/min

larger, better uniformity than the experimental results can be obtained in a certain finishing time when the original surface roughness value is small.

(2) Reciprocating frequency

Under $n$ = 54 r/min and other same conditions, Δ*Ra* curves on main journal surface with finishing time to different reciprocating frequency are plotted, as shown in Fig. 2.78. With the increase of reciprocating frequency *f*, opportunities and amounts of collision, rolling, micro-grinding (sliding and scratching) are increased between abrasive blocks and the workpiece surface, the surface roughness value decreases faster, and the finishing efficiency is high at the same processing time. When the reciprocating frequency increased to a certain extent, mutual contact between abrasive blocks and the workpiece surface is not full due to the increment of empty area, and micro-grinding (scratching and scratching) function is not sufficient. Therefore, the decline of the surface roughness value slows down. In a word, there should be a better match between rotational speed and reciprocating frequency for improving the surface roughness.

When $n$ is 108 r/min, Δ*Ra* curves on connecting rod journal surface with finishing time to different reciprocating frequency are shown in Fig. 2.79. Different frequency leads to different uniformity, and *Ra* with reciprocating frequency of 75 times/min is consistent than 62.6 times/min and 87.6 times/min, so the uniformity is good. The main reason is that different reciprocating frequencies lead to different medium flows, different function amounts of collision, rolling, grinding (slip and scratching), and different empty areas.

**Fig. 2.78** Δ*Ra* curves on main journal surface with finishing time to different reciprocating frequency

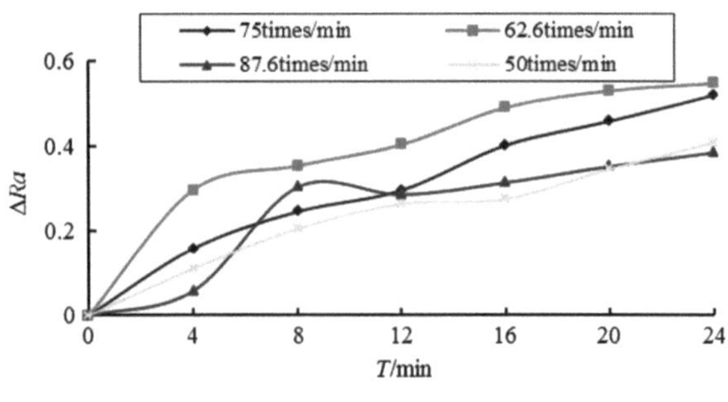

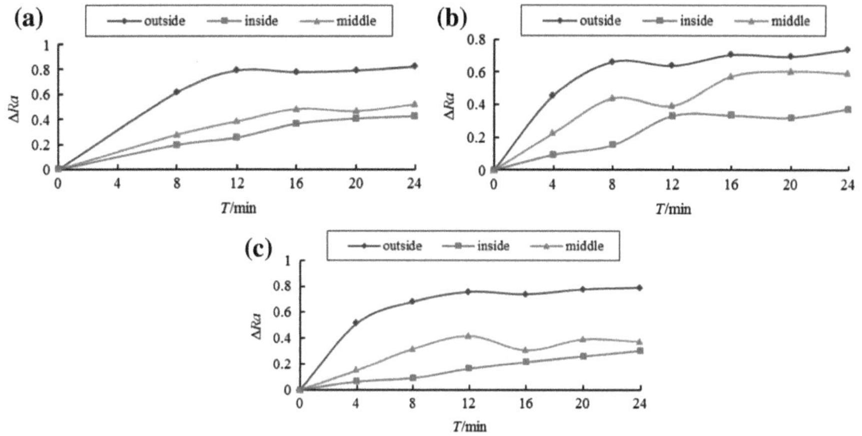


**Fig. 2.79** $\Delta Ra$ curves on connecting rod journal surface with finishing time to different reciprocating frequency. **a** $f = 62.6$ times/min; **b** $f = 75$ times/min; **c** $f = 87.6$ times/min

(3) Reciprocating amplitude

Under conditions of W1600 equipment, $n$ = 75.6 r/min, and $f$ = 62.6 times/min, $\Delta Ra$ curves on main journal surface with finishing time to different reciprocating amplitude are plotted, as shown in Fig. 2.80. Experimental results show that $\Delta Ra$ is larger as reciprocating amplitude is 240 mm than as 200 mm, or that stable value of the surface roughness can be achieved in a relatively short period of time. Due to the increase of the reciprocating amplitude, the amount and length of sliding friction and scratching between abrasive blocks and the workpiece surface will increase, and this is conducive to realize surface micro-machining, to improve the finishing efficiency.

Under conditions of $n$ = 75.6 r/min and $f$ = 43.8 times/min, $\Delta Ra$ curves on connecting rod journal surface with finishing time to different reciprocating amplitude are shown in Fig. 2.81. Compared to the reciprocating amplitude of 80 mm, the uniformity of the connecting rod journal surface can be obviously improved as reciprocating amplitude of 200 and 240 mm. When the reciprocating

**Fig. 2.80** $\Delta Ra$ curves on main journal surface with finishing time to different reciprocating amplitude

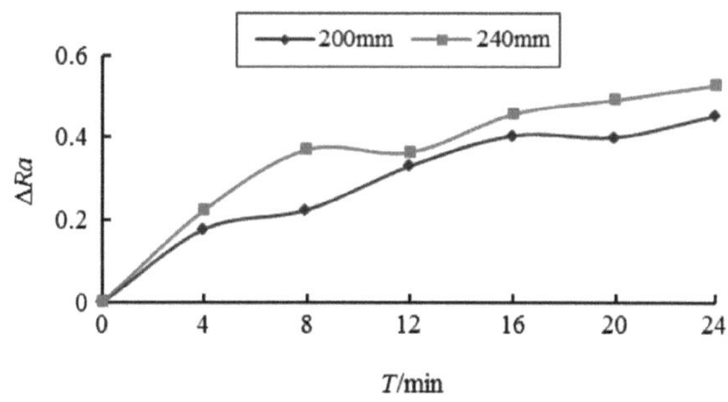

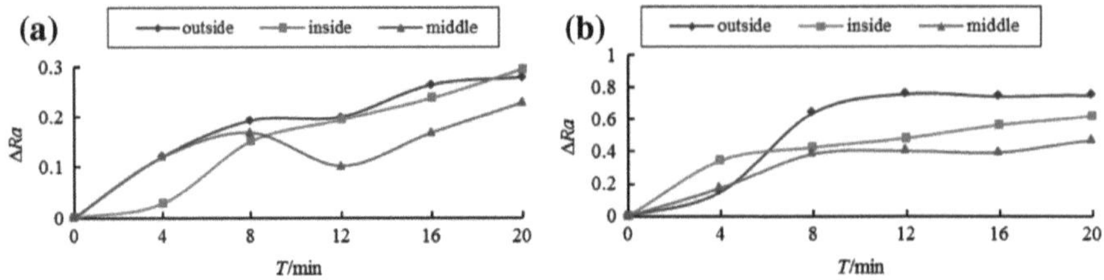


**Fig. 2.81** $\Delta Ra$ curves on connecting rod journal surface with finishing time to different reciprocating amplitude. **a** $S$ = 240 mm; **b** $S$ = 200 mm

amplitude increases, the amount and the length of sliding friction and scratching on different points of connecting rod journal surface are relatively consistent, and empty area between the workpiece and abrasive blocks is weakened, which is helpful to improve the finishing uniformity. Additionally, $\Delta Ra$ is more obvious at $S$ = 200 mm than $S$ = 200 mm, which indicates that too large reciprocating amplitude may reduce the effective impact of the collision, the rolling, the sliding, and the scratching.

In a word, of processing, the reciprocating amplitude ratio will have a greater effect on improving the finishing uniformity.

2. Process parameters

(1) Finishing depth

Finishing depth is the distance between the rotary axis of the crankshaft and the upper surface of the working medium.

In experiments, ring workpieces are adopted with 40 Cr material, quenching and tempering 28–32 HRC, outer diameter 55 mm, inner diameter 30 mm, length 20 mm. Through measuring weight of ring workpiece before and after finishing, the removal thickness is converted by

$$m_1 = \rho(\pi R_1^2 - \pi(R_1 - h_1)^2)w \tag{2.76}$$

where $m_1$ is removal mass, $R_1$ is outer diameter, $w$ is workpiece length, and $h_1$ is removal thickness.

From Eq. (2.76), the removal thickness can be expressed as

$$h_1 = R_1 - \sqrt{R_1^2 - \frac{m_1}{\rho \pi w}} \tag{2.77}$$

Under conditions of W1300 equipment, $n$ = 108 r/min, $f$ = 75 times/min, and $S$ = 80 mm, removal thickness curves with finishing time to different finishing depth are shown in Fig. 2.82. From Fig. 2.82a, as finishing depth increases, the removal thickness increases in the same time. Figure 2.82b shows the trend of removal thickness variation with different finishing depth as finishing time is

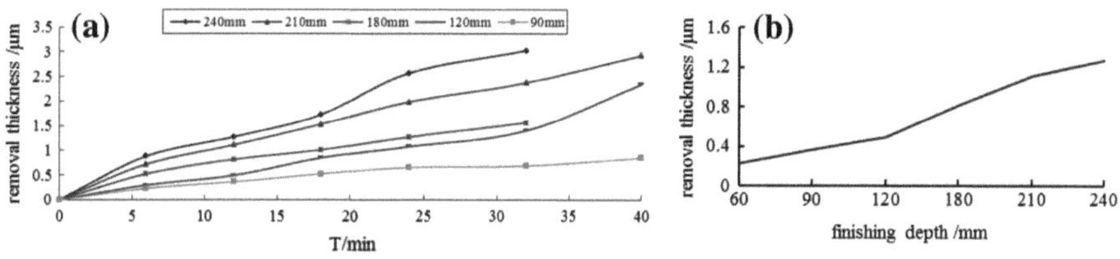


**Fig. 2.82** Removal thickness curves with finishing time to different finishing depth. **a** Curves with finishing time; **b** trend curves

12 min, which indicates finishing depth has significant influence to removal thickness. When the depth is up to a certain extent, the increment of the removal thickness is weak. In order to achieve finishing uniformity of connecting rod journal surface, the part should be at a reasonable depth. Additionally, the depth should be determined by considering the equipment structure.

Figure 2.82b shows that the removal thickness is very small when finishing depth is small and is in a slow growth stage; when finishing depth increases to 120 mm, the removal thickness increases significantly with finishing and is in the rapid growth stage; when finishing depth increases to a certain extent (e.g., 210 mm), it is in stable growth stage. Figure 2.83 shows schematic diagram of the influence analysis on the finishing uniformity of connecting rod journal with finishing depth. In Fig. 2.83, three stages of the change of the removal thickness of the surface layer of the workpiece are as follows: the slow growth stage ($<H_1$), the rapid growth stage (between $H_1$ and $H_2$), and the stable growth stage ($>H_2$). In the rotary process of the connecting rod journal with rotary center O, finishing depth of each

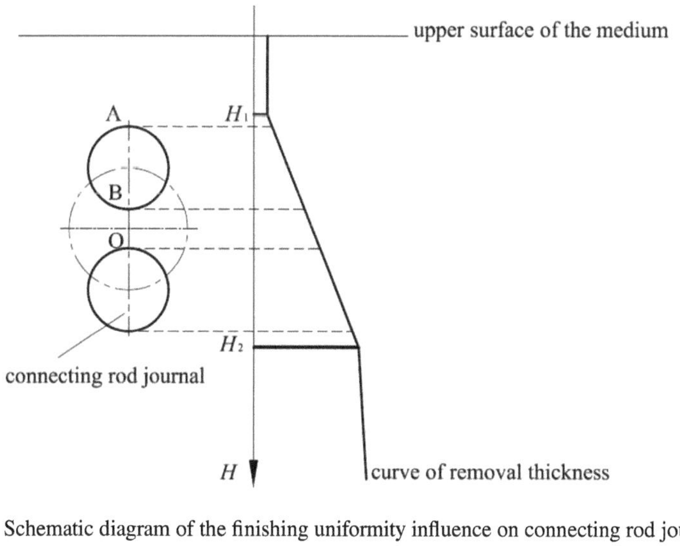


**Fig. 2.83** Schematic diagram of the finishing uniformity influence on connecting rod journal with finishing depth

point experiences periodic changes. Two most representative points, Point A and Point B on the connecting rod journal surface are selected for simplifying analysis. When the connecting rod journal is located at the top position, it is more reasonable that Point A is near $H_1$ or Point B is in the rapid growth stage, so the average value of finishing depth on each point is closer in a rotary cycle, so as to improve the finishing uniformity. If Point B is above $H_1$, the range of finishing depth and removal thickness is small, therefore, influencing the uniformity. If Point A is under $H_2$, finishing depth of each point is all in the stable growth stage, the uniformity is relatively good, but the size of finishing equipment also increases significantly.

The above analysis shows that the finishing uniformity of the connecting rod journal is related to the diameter of the connecting rod journal, the distance between the center of the connecting rod journal and the center of the main journal. In the case of certain crankshaft, the finishing uniformity can be improved by selecting the reasonable finishing depth.

(2) Finishing time

Under conditions of W1300 equipment, $n$ = 108 r/min, $f$ = 62.6 times/min, $S$ = 80 mm, and finishing depth 80 mm, the rule of surface roughness value *Ra* with finishing time is shown in Fig. 2.84. The surface roughness value decreases rapidly in initial time, and *Ra* is 0.183 μm when finishing time is 24 min. Then, *Ra* declines lowly in 24–40 min and is essentially unchangeable to 0.15 μm in the last. It indicates that the surface roughness value tends to stabilize in a certain time under certain condition, and the stable *Ra* is also known as the limit surface roughness value.

The surface roughness parameters are tested by Mahr surface roughness instrument, and curves are shown in Figs. 2.85 and 2.86.

From Figs. 2.85 and 2.86, surface roughness value *Ra* can be changed from 0.55 to 0.09 μm. When finishing 8 min, *Ra* decreases obviously and material ratio of the profile also increases significantly; variation trend of surface roughness parameters with finishing time is gradually slow, which also indicates that there exists a limit surface roughness value under certain condition.

By JSM-6700F scanning electron microscope, under the 500 magnification, the surface micro-topography with different finishing time is shown in Fig. 2.87. Before finishing, the workpiece has parallel surface texture, and anisotropy is particularly evident; after finishing 8 min, parallel surface texture disappeared, and

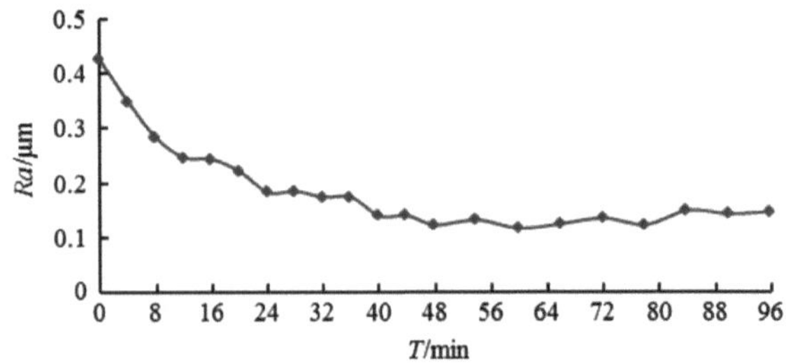


**Fig. 2.84** Curve of surface roughness value *Ra* with finishing time

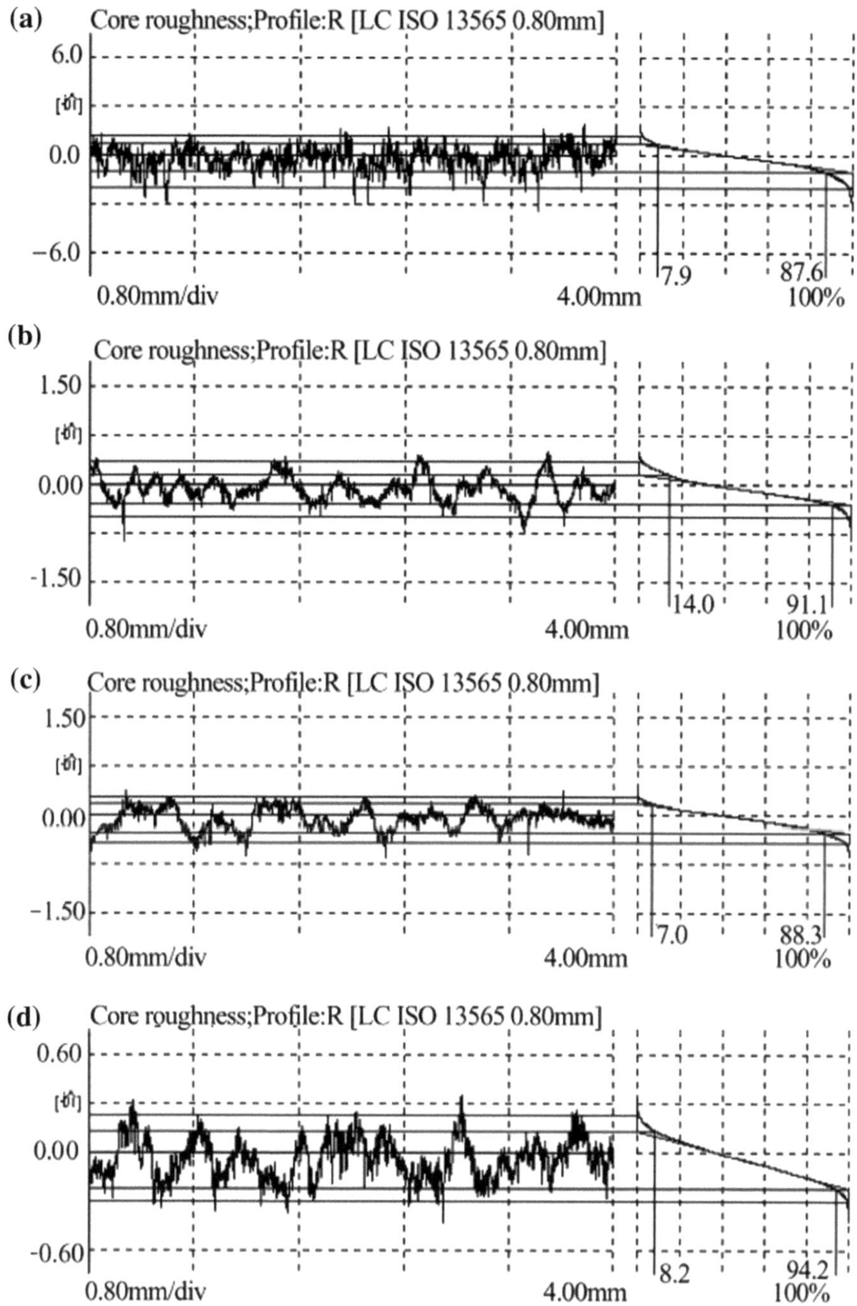


**Fig. 2.85** Curves of surface roughness value with finishing time. **a** Before finishing; **b** 16 min; **c** 32 min; **d** 40 min

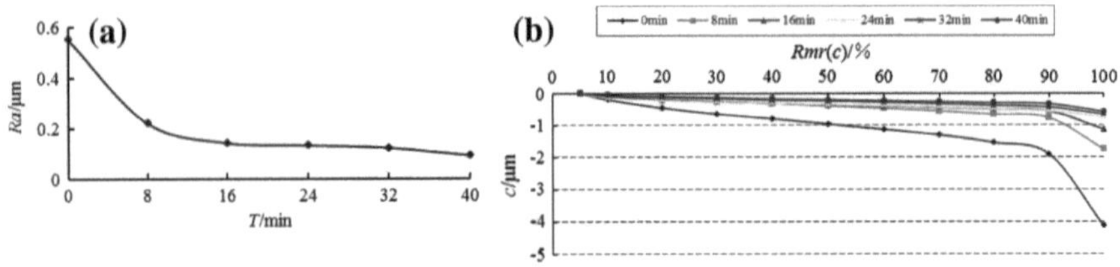


**Fig. 2.86** Curves of surface roughness value with finishing time. **a** *Ra*; **b** *Rmr*

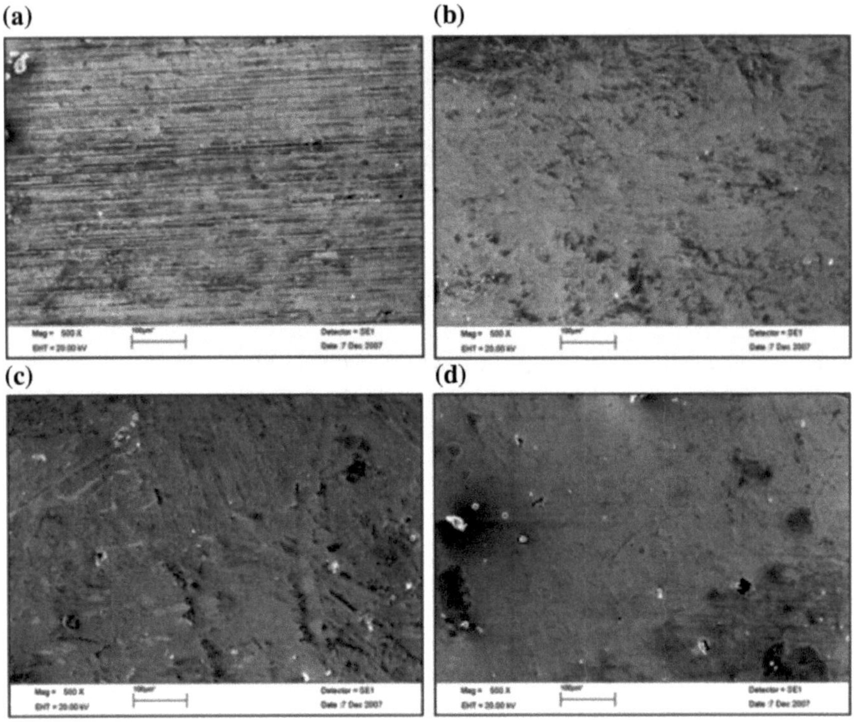


**Fig. 2.87** The surface micro-topography with different finishing time under the 500 magnification

the surface morphology refines and has isotropic trend; as the process continues, the surface of isotropic trend is more obvious. In addition, independent micro-pits appear in the workpiece surface, which is beneficial to the storage and reduces the surface abrasion. It is explained that the oil storage capacity of the workpiece can be improved by surface finishing.

Under conditions of $n$ = 108 r/min, $f$ = 50 times/min, $S$ = 80 mm, and finishing depth 80 mm, the crankshaft is finished in long time with one kind of abrasive blocks, and finishing effect is shown in Fig. 2.88. Due to blocking effect of the crank arm, abrasive blocks are arranged in a more orderly manner between two arms and their random motion reduces. Therefore, main journal forms a ring-shaped surface texture after long finishing time. When there are grooves in the surface, it is easier to process the groove and deepen its influence. For the non-eccentric axis, there will be no ring-shaped surface texture because the medium is in a free state. This also shows that the better the flow of the medium is, the better the finishing effect. On the one hand, the solution is to reduce the influence by selecting the appropriate finishing time; on the other hand, it can reduce the orderly manner by selecting a variety of abrasive blocks [55].

3. Structure parameters

Horizontal spindle barrel finishing process mainly depends on the rotary motion of the workpiece and the reciprocating motion of the container to achieve processing. Therefore, structure parameters include realizing rotary motion and reciprocating motion, and structure parameters of the container.

The driven mechanism of rotary motion mainly comprises a spindle frame, a live center frame, and a beam mechanism, and also a motor, a reducer, and a hydraulic system. The spindle frame is fixed on the left end of the beam; the live center frame, clamping the workpiece, is on the axis of the beam on the right end and can do reciprocating motion along the beam by hydraulic system. The motor and the reducer, driving the spindle to rotate through a chain drive, are laid on the beam which supports the spindle frame and the live center frame.

The design of the spindle frame mainly refers to the width and height. The width is equal to the sum of the distance of the two live centers and reserved distance of two sides, and the reserved distance should be greater than the maximum diameter. The height mainly depends on the container height and is approximately equal to the container height plus 200–300 mm, to meet motion range of the live center frame. The live center frame is corresponding to position with the spindle frame, so their width and height are same. But their top structure is slightly different, that is, the spindle frame is bolted to an end of the beam, and the live center frame is sleeve-jointed on the axis of the beam on the right end. The beam mechanism connects the spindle frame with the live center frame, its width is approximately equal to the width of the barrel, and the length is approximately the length of the equipment, so that the whole equipment coordinates each other.

**Fig. 2.88** Finishing effect of the crankshaft in long time with one kind of abrasive blocks

The mechanism realizing the reciprocating motion of the container is driven by an electric motor and a reducer, and forms a crank connecting rod mechanism by an eccentric wheel and a connecting rod connected to the container. The length of the crank determines the reciprocating amplitude of the container (the amplitude is equal to 2 $l$).

According to different types of crankshafts, structure parameters of the container can be determined according to the following equation.

$$L_c = L_1 + 2L_2 + 2s_1 \tag{2.78}$$

$$W_c = 2(l + r + S + s_2) + (q - 1)s_3 \tag{2.79}$$

$$H_c = H + l + r + s_2 + s_4 \tag{2.80}$$

where $L_c$, $W_c$, and $H_c$ are longitudinal dimension, transverse dimension, and vertical dimension of the container, respectively (mm); $L_1$ is the length of the crankshaft (mm); $L_2$ is the width of the spindle frame clamping the crankshaft (mm); $s_1$ is the minimum distance between the inside wall of the container and the outside wall of the spindle frame (the principle is to facilitate the lifting of the clamping part, mm); $l$ is the distance between the center of main journal and connecting rod journal (mm); $r$ is radius of connecting rod journal (mm); $S$ is reciprocating amplitude of the container (mm); $s_2$ is the minimum distance between the inside wall of the container and the workpiece (to determine by the experience or the experiment results, mm); $q$ is the number of parts processed simultaneously; $s_3$ is the distance of the axis between two workpieces (mm); $H$ is the finishing depth (mm); $s_4$ is the minimum distance of the top surface between the container and the medium (mm).

According to the experience or experiment results, $s_1$ and $s_2$ should be 100–150 mm, to reduce the crushing of abrasive blocks; $s_3$ should be the sum of $l$ and $r$, also plus 100–150 mm, to reduce the crushing of abrasive blocks; $s_4$ should be 100–150 mm, to reduce splashing of the medium. When small-sized parts are finished, $q$ can be selected to 2 or 4; but $q$ can be selected to 1 when using finishing large-sized parts.

### *2.7.3 Equipment Types and Its Design*

1. Equipment design

According to the decomposition principle of the system engineering, the function decomposition is established, which is the function structure diagram (the function tree). Through understanding functions relationship and then combining them, the system plan can be obtained. Figure 2.89 is the function tree of the horizontal spindle barrel finishing equipment.

As a kind of barrel finishing equipment, horizontal spindle finishing equipment has no strict requirement of the stiffness, the accuracy as other barrel finishing equipment, and the design goal is to have good function and comprehensive performance.

Rotary motion and reciprocating motion function are main functions of horizontal spindle finishing equipment. Figure 2.90 shows the assembly diagram of rotary spindle component and reciprocating motion component.

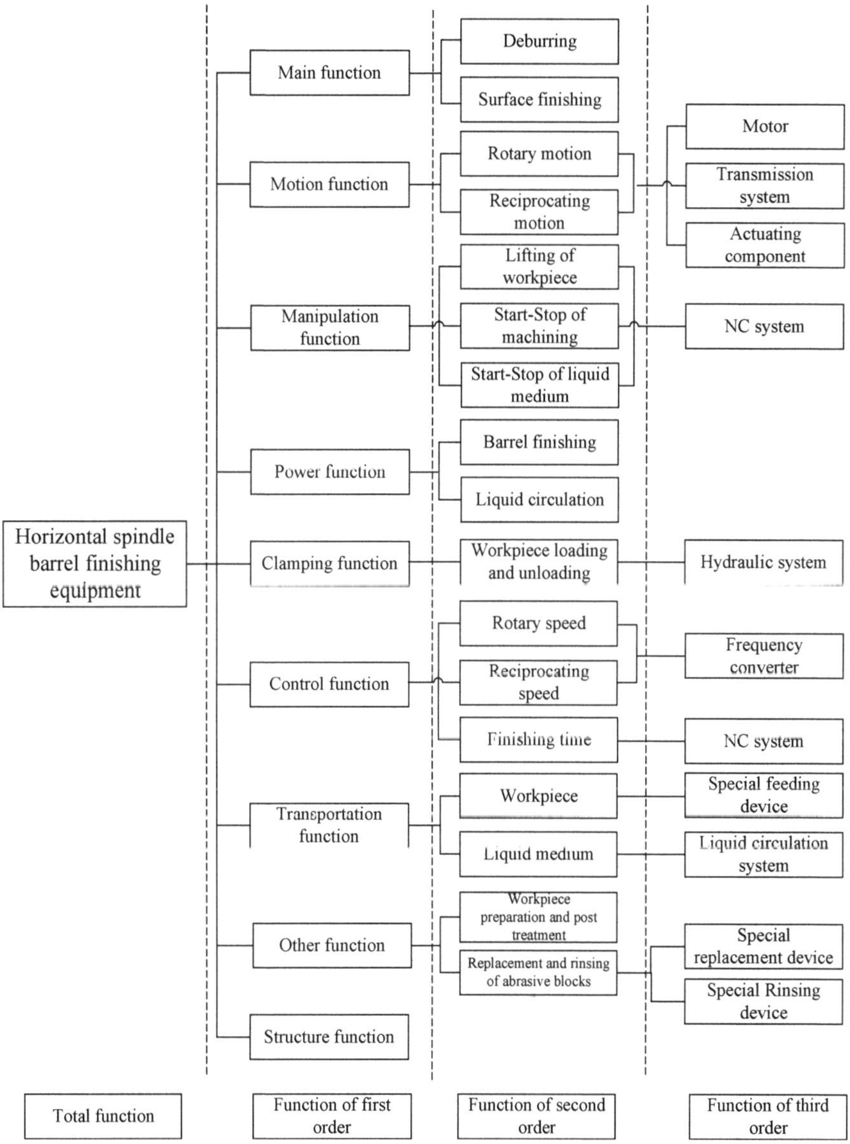


**Fig. 2.89** Function tree of the horizontal spindle barrel finishing equipment

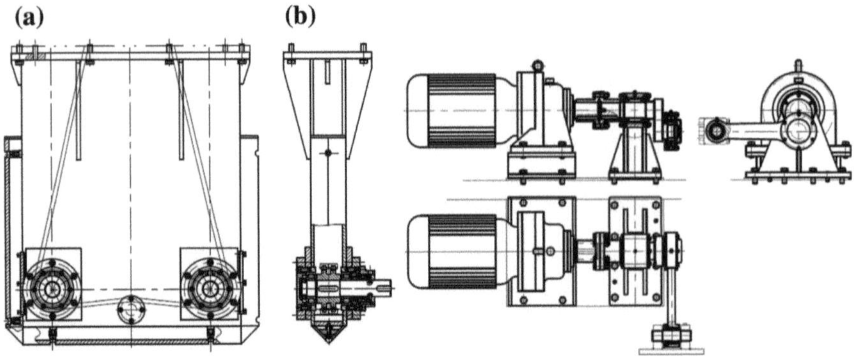


**Fig. 2.90** Assembly diagram of rotary spindle component and reciprocating motion component. **a** Rotary spindle component; **b** reciprocating motion component

For a certain processing parts, the range of rotary speed, reciprocating frequency, and amplitude can be determined based on the theoretical analysis and the experimental study, and then overall scheme is selected by analyzing various schemes. Finally, the specific design is done. In addition, in order to meet specific functional requirements, system reliability, compact in structure, stable operation, technological property of the structure, manufacturing economy, and other comprehensive performances are also considered in the schemes election. Finally, the rotary motion is realized through the motor, reducer, and chain transmission mechanism; and the reciprocating motion through the motor, the reducer, and the crank link mechanism.

Through Pro/E software, three-dimensional solid modeling of horizontal spindle finishing equipment is done using the parameterized modeling method, and expected assembly effect can be inspected; assembly process can be reasonably planned. Figure 2.91 shows an assembly drawing of horizontal spindle finishing equipment with W1200 type.

2. Auxiliary device design

The pumping device is mainly used for loading and unloading all kinds of abrasive blocks, which mainly comprises storage tank, cyclone separator, bag filter, centrifugal fan, motor, ash storage box, and electric control components, as shown in Fig. 2.92. Its working principle is as follows: The fan is driven to rapidly rotating after the motor begins to work, so that negative pressure forms in the cyclone separator, and abrasive blocks are inhaled from a feed inlet. When the discharge valve is opened, abrasive blocks are unloaded by the discharge outlet.

The rinsing device is mainly used for rinsing all kinds of abrasive blocks and can be used either alone or in combination with a pumping device. The rinsing device mainly consists of a vibrating screen, water circulation system (main water tank, filter box, paper filter box, water pump, and pipeline), frame body, cover body, and electric control system, as shown in Fig. 2.93. Its working principle is as follows: abrasive blocks vibrate in the vibrating screen according to the predetermined

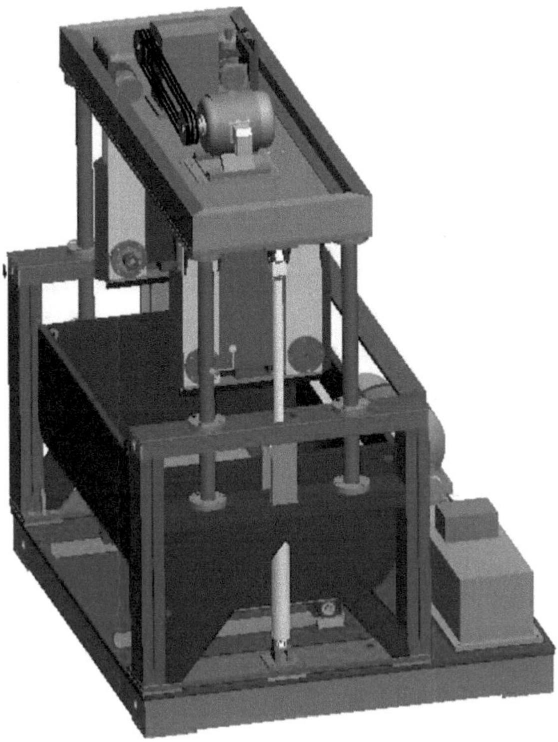

**Fig. 2.91** Assembly drawing of horizontal spindle finishing equipment

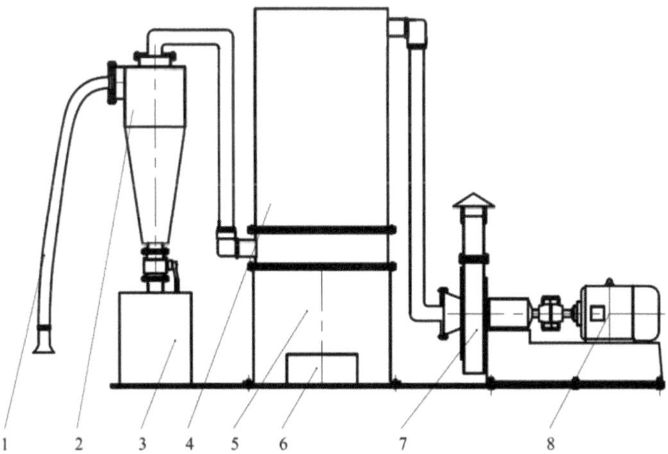


**Fig. 2.92** Overall diagram of abrasive blocks pumping device. 1—Inlet; 2—cyclone separator; 3—storage tank; 4—bag filter; 5—ash bucket; 6—ash storage box; 7—centrifugal fan; 8—motor

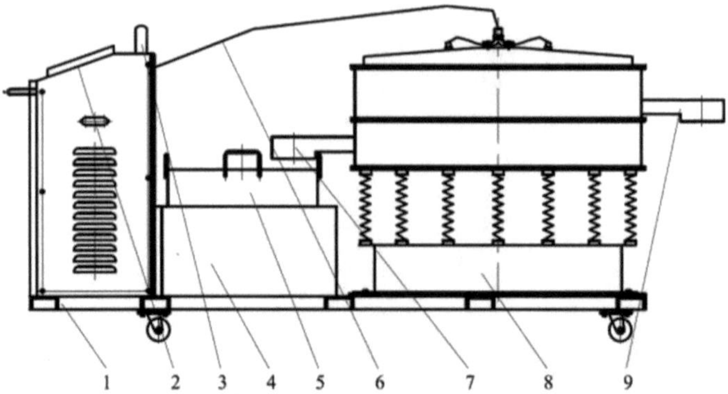


**Fig. 2.93** Overall diagram of abrasive blocks rinsing device. 1—Frame body; 2—control panel; 3—alarm lamp; 4—main water tank; 5—filter box; 6—pipeline; 7—water outlet; 8—vibrating screen; 9—discharge outlet

vibration trajectory, broken abrasives in abrasive blocks are filtered and separated, and friction is produced between abrasive blocks. At the same time, rinsing liquid with a certain pressure sprays to the surface of abrasive blocks under the action of the pump, and impurities such as oil are washed away. According to different conditions of abrasive blocks, the purpose of rinsing can be achieved by properly setting washing and vibration time.

The pumping device and rinsing device are mainly suitable for horizontal spindle barrel finishing equipment and have advantages such as compact in structure, strong power, high efficiency, convenient movement, low noise, simple operation, and safe, reliable, and long service life, which can effectively reduce labor intensity and improve production efficiency.

3. Equipment shape

Figure 2.94 shows photographs of horizontal spindle finishing equipment for different types, produced by Langfang Beifang Tianyu Mechanical and Electrical Technology Co., Ltd. Figure 2.95 shows the photograph of abrasive blocks pumping device, and Fig. 2.96 shows the photograph of abrasive blocks rinsing device produced by Langfang Beifang Tianyu Mechanical and Electrical Technology Co., Ltd.

## 2.8 Finishing Medium

In general barrel finishing processes, the abrasive blocks, compounds, and water, which are used in the barrel finishing, are usually known as the finishing medium. Their features have a great influence on finishing effects of the workpiece.

Fig. 2.94 Photographs of horizontal spindle finishing equipment for different types. **a** W900; **b** W2000; **c** W3000

Fig. 2.95 Photograph of abrasive blocks pumping device

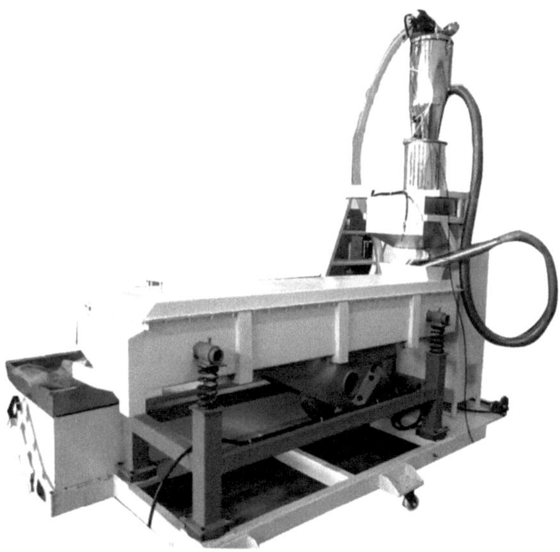

Fig. 2.96 Photograph of abrasive blocks rinsing device

## 2.8.1 Abrasive Blocks

The abrasive tool used for barrel finishing is in various shapes of blocks generally made of abrasives and binders, and referred to as abrasive blocks.

Selection of the proper abrasive blocks for a specific application is critical for optimum deburring and surface finishing results. Major selection factors to consider are composition (type), size, shape, and weight of the abrasive blocks to be used. Other important considerations include the capability, the economy, and the versatility of abrasive blocks. Workpiece material, size and geometry, burr sizes and locations, finishing requirements, finishing equipment used, subsequent operations, and the functions of the parts are also critical factors.

1. Functions and basic requirements of abrasive blocks

(1) Functions of abrasive blocks

As the main finishing medium, abrasive blocks are the main factors for realizing finishing effects.

(2) Condition characteristics and basic requirements of abrasive blocks

Under the action of mechanical transmission, abrasive blocks in the barrel produced a variety of forces, and therefore they (among abrasive blocks, workpieces, and abrasive blocks) present some colliding, rolling effects. This requires that abrasive blocks must be compact, hard, and tough. At the same time, abrasive blocks should be capable of cutting the workpiece and abrasion resistance.

During processing, abrasive blocks and the workpiece have been soaked in a mixture of water and compounds, and the mixture can wash off small chips which plugging in the surface of abrasive blocks. Meanwhile, the chemical reaction between the surface and the mixture is helpful to improve finishing effects. This requires abrasive blocks having characteristics with good compactness, low water absorption, water resistance, oil resistance, and acid alkali resistance.

2. Specifications and performance of abrasive blocks

(1) Types of abrasive blocks

In the development and application of barrel finishing technology, many natural materials often substitute for abrasive blocks in the early period. Table 2.16 lists types and applicability of main substituted abrasive blocks. With the technology development and the application maturity, these abrasive blocks have been unable to meet demand. Therefore, preformed abrasive blocks are produced which can be artificially controlled and have stable processing performance. Table 2.17 lists types, features, and applicability of preformed abrasive blocks.

(2) Manufacture of abrasive blocks

Figure 2.97 shows the manufacture process of sintered abrasive blocks with high temperature. For a variety of abrasive blocks with column shaped, the hole design of extrusion die for extrusion molding is the key factor, and the schematic diagram is shown in Fig. 2.96. From Fig. 2.98, *d* is determined by the shape of green bodies.

**Table 2.16** Types and applicability of substituted abrasive blocks

| Types | | Applicability |
|---|---|---|
| Natural stone products | Cobble stone | For the removal of large flash and burr of casting, forging, and stamping parts |
| | Granite sand | For rounding or polishing of a variety of machined metal parts and die casting parts |
| | Dolomite | For polishing of softer metal |
| Agricultural and wood products | Hardwood sawdust | Their cutting capacity is weak, but they not only have a cutting action, but also have good processing effect with the adhesion layer of fine abrasive. These materials can produce a good luster on some workpieces, particularly in the jewelry industry |
| | Nutshells, ground corncobs, walnuts | |
| | Leather or blankets flakes | |
| | Rice nutshells | |
| Industrial and agricultural used products | Used ceramic | For the removal of small burr of black or nonferrous metal parts; rounding and chamfering of nonferrous metals |
| | Used wheel | For ferrous metal, casting and forging parts can remove oxide layer, flash and burr efficiently; but the surface roughness is not well |
| | Used file | |
| | Glass ball | Cutting capacity is weak, commonly used in metal polishing |
| | Steel ball | For surface hardening and burnishing of a variety of parts |
| | Copper ball (powder) | For the removal of small burr and polishing of a variety of parts |

**Table 2.17** Types, characteristics, and applicability of preformed abrasive blocks

| Types | | Features | Applicability |
|---|---|---|---|
| Sintered abrasive blocks with high temperature | Ceramic abrasive blocks | Dense microstructure, fine-grain size, high hardness (about 7 degrees), low abrasion, water resistance, oil resistance, acid alkali resistance | For the removal of small burr of black or nonferrous metal parts; polishing of surface roughness requirement |
| | Fused (crystalline) alumina grinding blocks (pure clay high-temperature alumina crystalline polymer) | Strong cutting ability, high hardness (about 9.3 degrees) | Can remove oxide layer, deburr, round of black metal parts |
| | Alumina grinding blocks (clay and mixed sintered silicon carbide) | Relatively strong cutting ability, high hardness, good toughness | Commonly used in the removal of a small piece of black metal burrs and polish |
| | A variety of corundum grinding blocks (corundum, corundum, etc.) | Strong cutting ability | Can be used for a variety of metal parts deburring and polishing |
| | Silicon carbide abrasive (clay and silicon carbide powder mixed sintering) | High hardness and brittle texture, wear big | For nonferrous metal parts deburring and polishing |
| | High-hardness abrasive grinding block-based (synthetic diamond, cubic boron nitride, etc.) | Cutting ability, long service life | Because the price is expensive, rarely used |
| Resin abrasive blocks (bonding abrasives into resin binder by low-temperature mixture) | | Lower density, flexible and softer, strong cushioning effect | Well suited for finishing softer metals and fragile parts, achieving pre-plating requirements, and producing smooth finishes; it is also quieter than sintered abrasive blocks |
| Metallic abrasive blocks | | Larger density, good surface conditions, a variety of shapes and sizes | Used primarily for burnishing to achieve maximum luster for decorative purposes, for light deburring applications, and for heavy duty cleaning |

When $d$ is less than or equal to 10 mm, take $\alpha$ 12° ~ 13°, and when $d$ is more than 10 mm, take $\alpha$ 17° ~ 20°. In general, $d/D$ can be selected from 1/1.6 to 1/1.3. In order to ensure a smooth surface and uniform quality after green bodies forming, $L$, known as shaping area, should be 2–2.5 $d$. If $L$ is small, green bodies will produce elastic expansion, cause transverse cracks, and be easy to swing in the extruding process. Otherwise, green bodies tend to increase the internal stress and produce longitudinal cracks. After determining the structure of the extrusion die hole, the

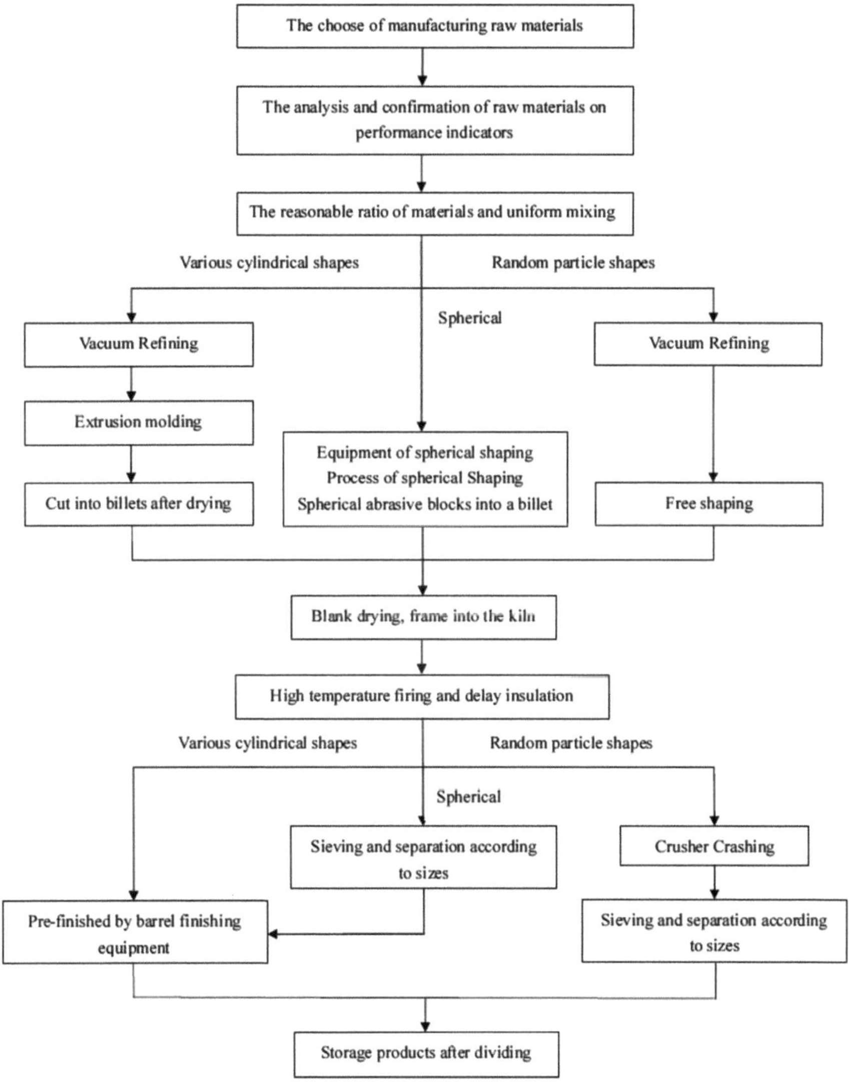


**Fig. 2.97** Manufacture process of sintered abrasive blocks with high temperature

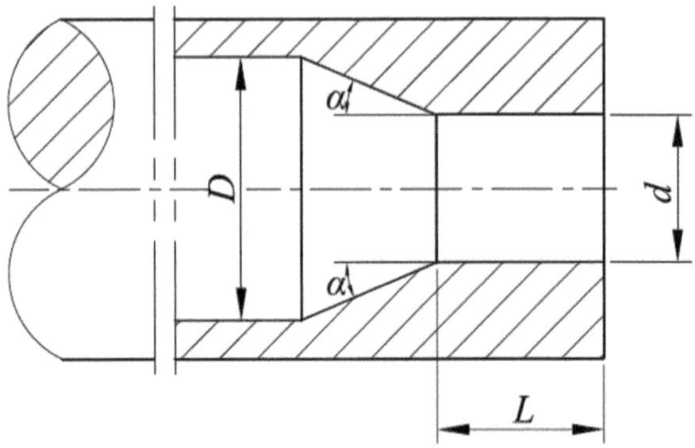


Fig. 2.98 Schematic diagram of the hole design of extrusion die

rate of extrusion should be considered. If this rate is quick, spring back may appear which can lead to deform. When lean materials are used, it needs to add the right amount of binders (polyvinyl alcohol, carboxymethyl cellulose, tung oil or dextrin, and so on). Figure 2.99 shows partial photographs of manufacture process of sintered abrasive blocks with high temperature.

For spherical abrasive blocks, the key is the forming process after the equipment and basic parameters determined, and the control of the raw material and the spray is laid special stress. In addition, the continuous process will go on without adding materials and water after the spherical formation, to increase the density and improve the surface smooth degree. Figure 2.100 shows partial photographs of manufacture process for spherical abrasive blocks.

Various resin abrasive blocks are made of compression metal molding method, and it has special advantages such as apparent angular, accurate edge, small

**Fig. 2.99** Photographs of manufacture process of sintered abrasive blocks with high temperature. **a** Extrusion forming materials; **b** sintered product

**Fig. 2.100** Photographs of manufacture process for spherical abrasive blocks. **a** Spherical forming; **b** screening

shrinkage, and large strength. Figure 2.101 shows photographs of the mold and molded resin abrasive blocks. Casting process is another method for resin abrasive blocks. Figure 2.102 shows partial photographs of casting process for resin abrasive blocks.

(3) Shapes and sizes of abrasive blocks

Preformed abrasive blocks are directly made by using a variety of raw materials, or sintered together by adding binders. Different shapes and sizes of abrasive blocks have advantages for specific applications, primarily depending upon configurations of workpieces. Table 2.18 lists shapes and sizes of sintered abrasive blocks with high temperature. Table 2.19 lists shapes and sizes of resin and metallic abrasive blocks. Figure 2.103 lists a variety of abrasive blocks.

(4) Performance of abrasive blocks

At present, there is JB/T10153-2013 including partial performance and test methods of abrasive blocks in China. Based on many years of theoretical study and experimental analysis, the performance is tested referring similar products, and this plays a positive role for production guiding and the usage of abrasive blocks. Table 2.20 lists performance indexes and test standards of abrasive blocks. According to lots of testing results, the hardness is more than 115HRA, the

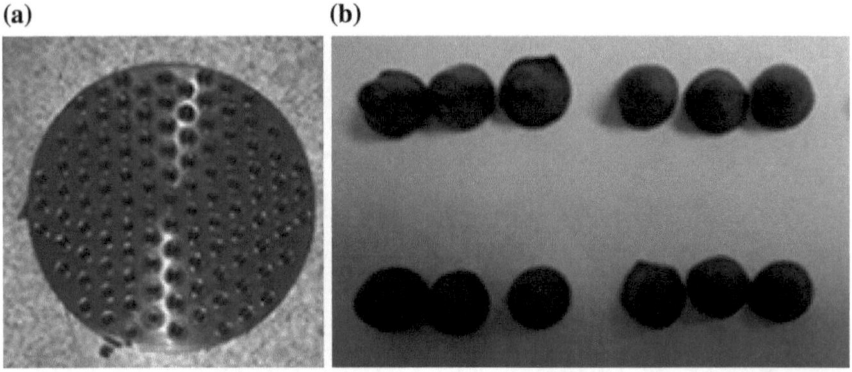

**Fig. 2.101** Photographs of the mold and molded resin abrasive blocks. **a** Mold; **b** molded resin abrasive blocks

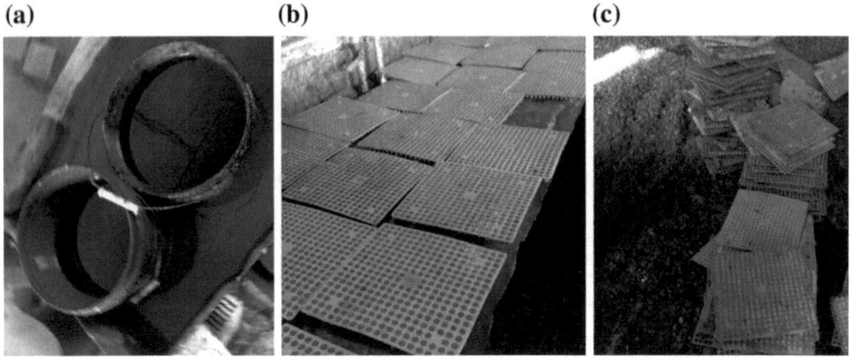

**Fig. 2.102** Photograph of casting process for resin abrasive blocks. **a** Raw materials; **b** curing molding; **c** demolding

compressive strength is not less than $9 \times 10^7$ N/m$^2$, the toughness is more than 1 kN m/m$^2$, and the water absorption rate is less than 0.7%.

Quality evaluation of abrasive blocks with the same material, the hardness, the size, and the shape by different manufacturers or different batches of the same manufacturer is difficult through single performance index. A new index $G$ can be adopted to a comprehensive comparison evaluation of finishing quality and using results of abrasive blocks.

$$G = \frac{C}{C'} \tag{2.81}$$

$$C = \frac{M - m}{M} \times 100 \tag{2.82}$$

**Table 2.18** Shapes and sizes of sintered abrasive blocks with high temperature

| Shape (code) | Shape sketch | Sizes/mm | Shape (code) | Shape sketch | Sizes/mm |
|---|---|---|---|---|---|
| Equilateral triangle prism (T) |  | 3 × 3, 4 × 4, 6 × 4, 10 × 6, 15 × 8, 18 × 10, 25 × 12, 30 × 15, 35 × 20 | Arrowhead prism (V) |  | 13 × 13, 20 × 20, 25 × 25, 30 × 30 |
| Oblique triangle prism (TP) |  | 3 × 5, 4 × 6, 6 × 10, 10 × 15, 15 × 20, 18 × 25, 25 × 30, 30 × 20 | Tri-stars prism (ST) |  | 10 × 10 × 10, 20 × 20 × 20, 25 × 25 × 25, 30 × 30 × 30 |
| Equilateral cylinder (R) | 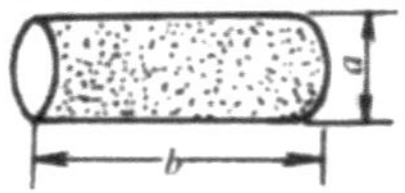 | 3 × 6, 4 × 8, 6 × 12, 8 × 16, 10 × 20, 15 × 30, 18 × 35, 25 × 40, 30 × 50 | Four-pointed stars prism (FST) | 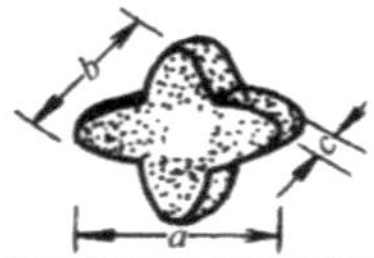 | 45 × 22 × 15, 35 × 7×11, 18 × 10 × 9 |
| Oblique cylinder(RP) | 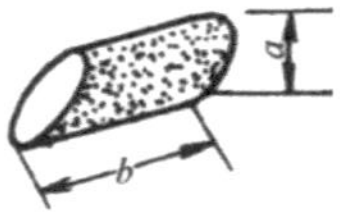 | 3 × 6, 4 × 8, 6 × 12, 8 × 16, 10 × 20, 15 × 30, 18 × 35, 25 × 40, 30 × 50 | Elliptical cylinder (E) | 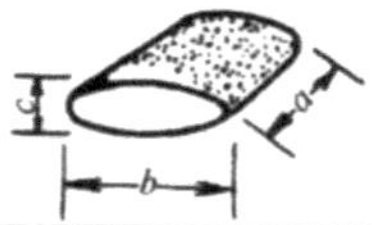 | 15 × 4×15, 15 × 6×15, 20 × 8×20 |
| Quadrants prism (F) | 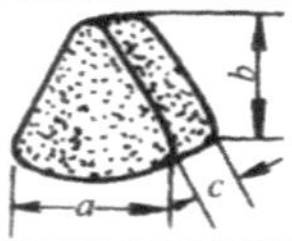 | 7 × 7×5, 10 × 10 × 7, 15 × 15 × 10, 20 × 20 × 12, 30 × 30 × 14 | Equilateral tetrahedron (FT) |  | 10,12, 15, 20 |

(continued)

**Table 2.18** (continued)

| Shape (code) | Shape sketch | Sizes/mm | Shape (code) | Shape sketch | Sizes/mm |
|---|---|---|---|---|---|
| Diamond prism (D) | 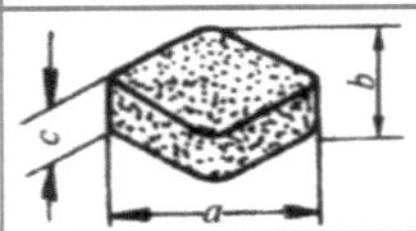 | 23 × 16 × 10, 36 × 25 × 18, 46 × 30 × 20, 55 × 35 × 24 | Sphere (S) | 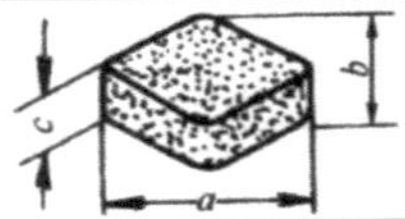 | 1.5, 2, 3, 4, 5, 6, 8, 10, 12, 15 |
| Cube (C) |  | 10, 15, 20 | Random shape (G) |  | 1.0–1.5, 1.5–2.0, 2.0–2.5, 2.5–3.0, 3.0–4.0, 4.0–5.0 |

Table 2.19 Shapes and sizes of resin and metallic abrasive blocks

| Shape (code) | | Shape sketch | Sizes/mm | Shape (code) | | Shape sketch | Sizes/mm |
|---|---|---|---|---|---|---|---|
| Resin abrasive blocks | Cone (PLCO) | 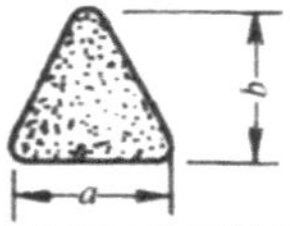 | 10 × 10, 15 × 15, 20 × 20, 25 × 25, 30 × 30, 35 × 35, 40 × 40, 45 × 45, 50 × 50, 60 × 60 | Metallic abrasive blocks | Sphere (SS) |  | 1.588, 2.381, 3.175, 3.969, 4.762, 5.556, 6.350 |
| | Equilateral tetrahedron (PLFT) |  | 10, 12, 15, 20 | | Double pins (STS) | 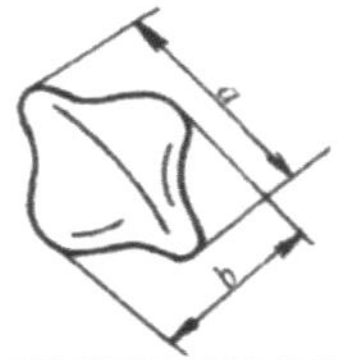 | 4.5 × 3.2, 5.1 × 3.7, 7.0 × 4.7, 8.9 × 6.3 |
| | Equilateral triangle prism (PLT) |  | 6 × 4, 10 × 6, 15 × 8, 18 × 10, 25 × 12, 30 × 15, 35 × 20 | | Pin (SP) | 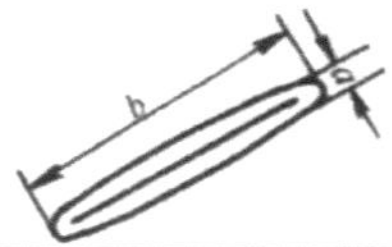 | $\varphi 2 \times 8.5$, $\varphi 2 \times 10$ |
| | | | | | Oval (SE) | 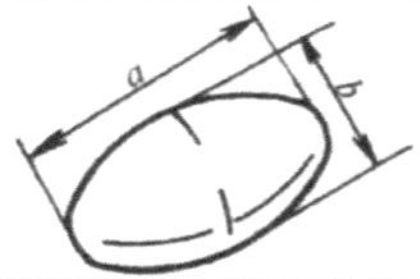 | 7.0 × 4.7, 8.9 × 6.3 |

**Fig. 2.103** Variety of abrasive blocks

**Table 2.20** Performance indexes and test standards of abrasive blocks

| Performance indexes | Test standards |
|---|---|
| Hardness | GB/T 2490 |
| Compressive strength | GB/T 4740 |
| Toughness | GB/T 4742, GB/T 3810 |
| Water absorption rate | GB/T 10153 |
| Wear rate of abrasive blocks | GB/T 10153 |
| Material removal rate of the workpiece | GB/T 10153 |
| Geometry precision | GB/T 10153 |

$$C' = \frac{M' - m'}{M'} \times 100 \tag{2.83}$$

where $C$ and $C'$ are material removal rate of the workpiece and wear rate of abrasive blocks; $M$ and $m$ are initial weight and final weight of the workpiece; $M'$ and $m'$ are initial weight and final weight of abrasive blocks.

Under the condition of ensuring processing quality, the larger the $G$ is, the better the performance of abrasive blocks is.

(5) Failure mode of abrasive blocks

Abrasive blocks have three failure modes:

(1) The initial failure

If the naked eye can discern uneven hole in the surface of abrasive blocks newly purchased (through pre-finished), the surface of abrasive blocks forms piebald holes after normal use and has obvious influence on the workpiece quality. So, it should not be used. If this kind of abrasive blocks is broken, small holes are recognizable at the break position, known as porosity. The reason may be not enough vacuum refining time or vacuum in the manufacturing of abrasive blocks. Such abrasive blocks quickly wear in normal use. Performance indexes, having obvious difference mentioned above, should not be used.

(2) Crushing failure

Due to manufacturing defects or improper use, the edges of abrasive blocks are seriously off or abrasive blocks are broken. The broken fragments or edges have a great impact on workpiece surface quality. Therefore, abrasive blocks should be cleaned and sieved, or else it should not be used.

(3) Wear failure

Under conditions of normal use, edges of abrasive blocks will be blunted, and their sizes will be smaller. When the volume of a single abrasive block is reduced to about 0.5 of the original size, it has great influence on the finishing quality and productivity, and it should be considered to replace at this moment.

3. Impact on finishing results of abrasive blocks parameters

Characteristic parameters of abrasive blocks include composition, abrasive size, hardness, microstructure, the shape and size, the binder similar to the methods of grinding wheel. However, because of different working features between abrasive blocks and the wheel, the abrasive size, and the hardness, the shape and the size are main factors [56]. Under definite experimental condition (machine type is 2M3110, the workpiece is 45 steel $\phi 10 \times 10$, and finishing time is 30 min), the experimental data comparisons of finishing effects of abrasive blocks parameters are shown in Table 2.21.

(1) Abrasive size

The size refers to the abrasive size of abrasive blocks.

(1) Effects on the workpiece surface roughness value. In general grinding processes, the smaller the abrasive size is, the smaller the surface roughness value after processing is. In the barrel finishing processes, sharp corners on the abrasive are removed after pre-finished or a period of use, which can improve conditions of the coequal highness. When the mutual force between abrasive blocks and the workpiece is low, smaller surface roughness of the workpiece can obtained no matter the abrasive size is large or small. Therefore, the influence on surface roughness of abrasive size is not obvious, and variety of surface roughness is obvious only when the abrasive size is very large or small. Under definite experimental condition (machine type is 2M3110, the workpiece is 45 steel, finishing time is 30 min, and *Ra* is 1.8 μm before finishing), influence rules of abrasive size on the workpiece surface roughness value are shown in Fig. 2.104a, and the results show that surface roughness is basically same after finishing with sphere shape abrasive blocks.
(2) Effects on the material removal rate. Generally speaking, if the abrasive size of abrasive blocks is smaller, the metal removal rate and finishing efficiency increases owing to the decrease of abrasive number per unit area and increase of intensity of pressure. But the relationship between the material removal rate and

**Table 2.21** Comparisons of finishing effects of abrasive blocks parameters

| Abrasive blocks | | | *Ra*/μm | | Material removal rate *C*/(g/kg) | Wear rate *C*′/(g/kg) |
|---|---|---|---|---|---|---|
| Materials | Abrasive size | Sizes/mm | Before finishing | After finishing | | |
| Corundum | 80[#] | Oblique triangle prism 15 × 15 × 15 | 1.8 | 1.6 | 13.88 | 88.29 |
| | 150[#] | | 1.8 | 1.3 | 11.22 | 66.72 |
| | 240[#] | | 1.8 | 1.3 | 12.35 | 65.83 |
| | W20 | | 1.8 | 1.2 | 10.39 | 53.34 |
| | 80[#] | Oblique cylinder Φ10 × 20 | 1.8 | 1.4 | 15.48 | 106.75 |
| | 240[#] | | 1.8 | 1.0 | 11.83 | 77.67 |
| | W20 | | 1.8 | 1.0 | 10.33 | 54.16 |
| | 80[#] | Sphere Φ5 | 1.8 | 0.6 | 3.67 | 37.61 |
| | 240[#] | | 1.8 | 0.5 | 2.96 | 24.08 |
| | W40 | | 1.8 | 0.5 | 2.05 | 14.14 |
| Alumina | 240[#] | Sphere Φ3 | 1.8 | 0.9 | 0.38 | 2.17 |
| | | Sphere Φ5 | 1.8 | 0.6 | 0.54 | 1.25 |
| | | Sphere Φ8 | 1.8 | 0.4 | 3.21 | 4.32 |

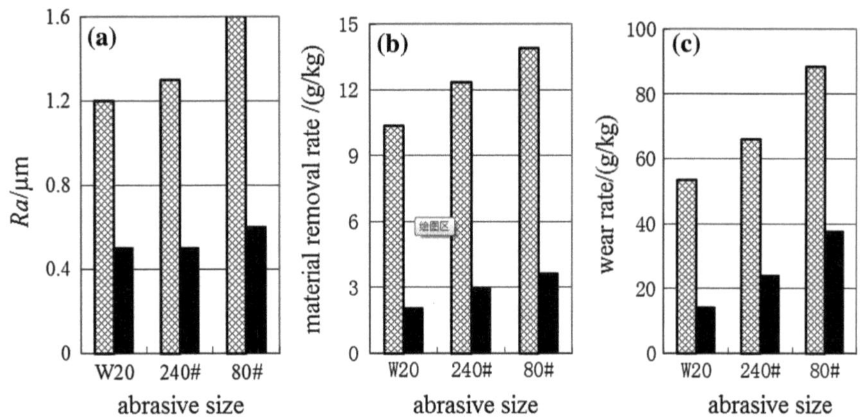


**Fig. 2.104** Influence rules of abrasive size. In the figure, filling with black solid represents abrasive blocks with sphere, another with oblique triangle prism

the abrasive size is not a linear function. The relationship is that when the abrasive size is small, material removal rate increases rapidly with the increment of abrasive size; when reaching a limit size, material removal rate will increase lowly, and abrasive particle size is about 80 μm at the same time. Under the same experimental conditions in Fig. 2.104a, the influence rules of abrasive size on material removal rate are shown in Fig. 2.104b.

(3) Effects on wear rate of abrasive blocks. The larger the abrasive size of abrasive blocks is, the larger the bonding force between the abrasives is, the lower

breaking possibility of abrasives from block body is. Therefore, abrasive blocks more easily maintain its original shape, and the wear rate is small. Under certain experimental conditions, influence rules of abrasive size on wear rate are shown in Fig. 2.104c. It can be seen that the wear rate is smaller with larger abrasive size. From the point of lower wear rate of abrasive blocks, abrasive blocks with larger abrasive size are more suitable to deburring and finishing.

(4) Selection principle of abrasive size. Above all, the selection should meet deburring and finishing requirements of the workpiece. Then, abrasive size with high material removal rate (high finishing efficiency) and low wear rate (good economy) should be considered. According to this principle, abrasive size is generally about 80 μm or more for the less obvious influence on surface roughness.

(2) Hardness

The hardness of abrasive blocks generally refers to difficulty degree of surface abrasives off from the block body with the external force. If abrasives wear off easily, the hardness of abrasive blocks is soft, and otherwise is hard. Although this concept is different with metal hardness, it has relationships with blocks hardness. Therefore, abrasive blocks with smaller abrasive size (80 μm or more is provided in GB 2491) can use Rockwell hardness to express its hardness.

In the barrel finishing processes, main functions of abrasive blocks on the workpiece surface are collision, extrusion, and micro-grinding, and therefore extremely sharp cutting edges of abrasives and automatic peeling-off capacity of the dull abrasives may not be required. Since the cutting force during processing is small, and water plays a role of cleaning and cooling effect on abrasive blocks, workpiece overheating, and chip clogging the surface of abrasive blocks, and other issues will not appear. Therefore, the use of hard abrasive blocks also did not lead to workpiece burning and larger thermal deformation due to overheating.

Using hard abrasive blocks has following advantages:

(1) Rockwell hardness value of abrasive blocks must be greater than the workpiece hardness, so hard abrasive blocks can be used in a variety of workpiece hardness, which may reduce the hardness type of abrasive blocks and be easy for production, storage, and usage of abrasive blocks.
(2) Harder the abrasive blocks are, the smaller wear rate of abrasive blocks is, and this can increase the life span of abrasive blocks (good economy).
(3) Hard abrasive blocks are easy to maintain the original shape and less prone to breaking in the finishing process. Especially for finishing requirements with low surface roughness of the workpiece, the surface roughness can be affected greatly by peeled chips and edges.

The hardness of abrasive blocks has relations with the material removal rate. In general, the hardness of abrasive blocks must be greater than the workpiece hardness in the barrel finishing processes (but harder abrasive blocks are not related to greater material removal rate). According to the relevant literature, when the ratio of abrasive blocks hardness and workpiece hardness exceeds $K$ ($K$ is 1.3–1.7 based on many experimental results), the material removal rate does not increase any more. So, the hardness of abrasive blocks selected should be greater 1.3–1.7 times than the workpiece hardness to obtain large material removal rate. If the hardness is not enough ($K$ is less than 1.3), material removal rate declines sharply. Generally, the Rockwell hardness of hardened steel is about 60–70HRC, so the Rockwell hardness of abrasive blocks should be above 100HRC.

(3) Shape and size

Different with ordinary grinding processes, abrasive shape and size influence on finishing effects are very obvious, including surface roughness, surface damage and deformation of the workpiece, finishing efficiency, and economy. Therefore, they are key factors of realizing finishing request.

(1) Effects on the workpiece surface roughness value. With the same material and same abrasive size, different shape of abrasive blocks will lead to great difference of the surface roughness. The experimental results show that the more the angular and sharp edges and corners of abrasive blocks are, the greater the surface roughness values after finishing are. Under definite experimental condition (the workpiece is 45 steel, abrasive blocks are $80^{\#}$Corundum, finishing time is 30 min, and $Ra$ is 1.8 μm before finishing), test results are shown in Table 2.22. Among them, sphere is the best, followed by oblique cylinder, and oblique triangle prism. Therefore, if low surface roughness or good surface brightness is required, sphere shape should be chosen.

Under other same conditions, the high pressure, large contact area between the abrasive block and the workpiece, and large scratching traces of the workpiece surface may produce with large abrasive blocks. Therefore, after finishing surface roughness and brightness are not well. Under definite experimental conditions (the workpiece is 45 steel, abrasive blocks are $80^{\#}$ Corundum, finishing time is 30 min, and $Ra$ is 1.8 μm before finishing), test results are shown in Table 2.23.

The surface roughness value above mentioned should be the ultimate value of surface roughness. Due to the material removal rate with small size abrasive blocks is low, the surface roughness of the workpiece reached the limit value for a long time. When the ultimate value of surface roughness with large size abrasive blocks obtains, the ultimate value cannot obtain with small size abrasive blocks at the same time. Therefore, the surface roughness value is larger with sphere Φ3 in Table 2.24.

(2) Effects on the material removal rate. With the same material and the same abrasive size, different shapes of abrasive blocks will also lead to great

**Table 2.22** Test results of *Ra* with different shapes

| Shapes | Oblique triangle prism 15 × 15 × 15 | Oblique cylinder Φ10 × 20 | Sphere Φ5 mm |
|---|---|---|---|
| *Ra* value after finishing/μm | 1.6 | 1.4 | 0.6 |

**Table 2.23** Test results of *Ra* with different sizes

| Sizes/mm | Sphere Φ3 | Sphere Φ5 | Sphere Φ8 | Sphere Φ10 |
|---|---|---|---|---|
| *Ra* value after finishing/μm | 0.50 | 0.45 | 0.50 | 0.70 |

difference of the material removal rate. Experimental results shown in Table 2.25 indicate that the more the angular and sharp edges and corners of abrasive blocks are, the greater the material removal rate is. Among them, equilateral tetrahedron is the largest, followed by oblique triangle prism, oblique cylinder, and sphere. Therefore, if only for rough machining for deburring and descaling, it has maximum efficiency with equilateral tetrahedron or oblique triangle prism shape.
The material removal rate of the workpiece increases with abrasive blocks size increasing for greater pressure on the workpiece. The relationship between the diameter of abrasive blocks (spherical alumina) and material removal rate is shown in Table 2.26.

(3) The relationship between the shape and the size of abrasive blocks and the workpiece. The shape and the size of abrasive blocks should provide access to all workpiece surfaces requiring deburring or finishing, not lodge in holes or recesses, and permit easy separation from parts at the end of the cycle in mass production.

**Table 2.24** Test results of the material removal rate with different shapes

| Shapes and sizes | Sphere Φ10 mm | Oblique cylinder Φ10 mm × 20 mm | Oblique triangle prism 15 mm × 15 mm | Equilateral triangle prism 10 mm × 10 mm |
|---|---|---|---|---|
| Material removal rate/(g/kg) | 6.89 | 10.33 | 10.49 | 28.83 |

**Table 2.25** Test results of the material removal rate with different sizes

| Sizes | Sphere Φ3 | Sphere Φ5 | Sphere Φ8 | Sphere Φ10 |
|---|---|---|---|---|
| Material removal rate/(g/kg) | 0.38 | 0.54 | 3.21 | 6.89 |

**Table 2.26** pH value of various compounds and main compositions

| Workpiece materials | Black metal | Copper, aluminum | Zinc alloy | Stainless steel |
|---|---|---|---|---|
| pH value | >8 | About 7 | <7 | >10 |
| Main compositions | Triethanolamine, sodium dodecylbenzene sulfonate, oleic acid, sodium nitrate, sodium phosphate, and the like | | | |

When using large size abrasive blocks, it is easy for low-rigidity workpiece to deform due to large impact force on the workpiece. Abrasive blocks also help to keep workpieces from impinging on each other, but this effect is not well if abrasive blocks are small size or low edges [57].

For the workpiece with holes or recesses, the size of the abrasive block should be 1/3 or less of the hole or recess size, to ensure holes or recesses finishing and not lodging in.

When finishing more sophisticated workpieces such as thread or gear, abrasive blocks with less angular and small size should be chosen to avoid bumping. The radius of spherical abrasive blocks for finishing machining should be less than fillet radius of the workpiece. Therefore, a mixture of abrasive blocks shapes is sometimes used.

## 2.8.2 *Liquid Medium*

Compounds, known as liquid medium shown in Fig. 2.105, are added to parts and abrasive blocks to increase deburring action, remove scale and heat-treat tints, scour surfaces, and clean parts, abrasive blocks, and the container. In addition, compounds modify luster or color of workpieces. They can add corrosion-inhibiting coatings for metal workpiece such as steels and aluminum. In some cases, abrasive blocks in the compounds may be used to keep flat parts from sticking together.

Most compounds are combinations of chemicals that dissolve in water and form solutions to maintain consistency or to modify the action of abrasive blocks on the workpiece. The use of too little, too much, or an improper compound can adversely affect the cycle time and/or ability of abrasive blocks to perform as intended.

Generally, good compounds include water, water conditioners, cleaners, wetting agents, lubricants, foam control agents, pH buffers (pH control chemicals), rinse agents, suspension agents, and emollients. Some also contain metal brighteners and corrosion inhibitors. The list of chemicals is extremely long, and the amount of

**Fig. 2.105** Photographs of compounds produced by Langfang Beifang Tianyu Mechanical and Electrical Technology Co., Ltd

various chemicals used is almost as much as the variety of chemicals, depending upon finishing results desired and economics of the manufacturer, and the user.

There are some finishing applications where no compound solutions are used. For example, loads of workpieces are sometimes finished dry, perhaps with saw-dust, ground corncobs, or other abrasive blocks.

1. Cleaning functions

All mass finishing operations require clean machine compartments, abrasive blocks, and workpieces for effective and efficient results. Compound solutions are often required to clean workpieces by removing tarnish or scale, emulsifying oils, and suspended solids. Compound solutions are also expected to maintain the cleanliness of workpieces and abrasive blocks during the finishing cycle. Buildup of dirt on the abrasive block causes glazing and reduces cutting capability. Strongly alkaline formulations are often used for cleaning iron and steel parts, while milder alkaline to mildly acidic compositions is generally best for nonferrous metals and alloys. Some workpieces require cleaning and/or descaling prior to mass finishing.

Therefore, the replacement of compounds in mass finishing processes should pay attention to realize timing cleaning of abrasive blocks. Generally, replacement time should be 3–4 h in rough machining, and it can be 6–8 h in finish machining.

2. pH control

pH control chemicals hold the pH at the required level of alkalinity, neutrality, or basicity. The pH is controlled to make the chemical system of pH sensitive work properly.

Selecting compounds for different workpiece materials, the pH value and main compositions are shown in Table 2.27.

3. Loads

Compounds loads directly affect the deburring and finishing effect. If the load is too less, lubrication and cushioning action are not well, causing roughness and scratch of workpiece surface, even damaging workpiece surface. If the load is too much, the processing efficiency will decrease. Practice shows that compounds loads in a sealed drum are appropriate for just submerging the workpiece and abrasive blocks.

Experimental conditions: The type centrifugal barrel finishing equipment is 2M3160 ($N = -n = 180$ r/min), workpieces are 45 steel of $\varphi 10$ mm × 15 mm by finishing turning, sphere abrasive blocks are alumina material of $\varphi 5$ mm, and the processing time is 20 min. Finishing effects comparison is shown in Table 2.28. Experimental results show that adding proper amount of compounds is beneficial for improving finishing quality and efficiency.

**Table 2.27** Finishing effects comparison of compounds

| Finishing status | The surface roughness *Ra* value/μm | | Brightness after finishing | Material removal rate |
|---|---|---|---|---|
| | Before finishing | After finishing | | |
| Compounds LC-10 | 1.6 | 1.25 | Good | Larger |
| Without compounds | 1.6 | 1.45 | Bad | Smaller |

**Table 2.28** Equipment selection and notes

| Shapes and sizes of the workpieces | Equipment type | Notes and explanations |
|---|---|---|
| Miscellaneous parts with small sizes | Centrifugal type | The flat parts are easy to stick, thin-walled or bar parts are easy to deform in batch production. Larger loads of the mixture or more less amounts of parts may appropriately reduce these phenomena |
| | Vibratory type | Solid parts mixed with the medium are difficult to vibrate, and finishing results are poor. Except for selecting abrasive blocks with large size, other barrel finishing processes may be used |
| | Whirling type | It is possible for flat parts to insert the connection position between stationary sidewalls and rotary disk. In general, the thickness of the workpiece should excess 2 mm. |
| | Rotary type | This process has low material removal capacity and poor finishing efficiency and also has little applications at present |
| Shaft disk parts with small or medium sizes | Vertical spindle type | Shaft parts should not be too long for the workpiece inserts into the barrel through cantilever action. If necessary, parts may turn 180° and be fixed to deburr or finish. For short shaft parts, several parts can be deburred or finished simultaneously on a spindle. In addition, finishing uniformity of end surface on disk parts is not well |
| Shaft parts with large sizes | Horizontal spindle type | Practical finishing problems of the crankshaft can be solved by this process. The surface quality can be improved synthetically, especially cleanliness degree of crankshaft parts |
| Disk parts with large or medium sizes | Intermeshing spindle type | All surfaces of disk parts can be achieved to finish, but the adjustment of equipment parameters is complicated |
| Casing parts with medium sizes | Vibratory type | Enough space of the equipment size must be to ensure enough space of parts vibration in the container |

# 2.9 Mass Finishing Applications

## *2.9.1 Introduction*

1. Three stages of mass finishing application

Practical applications of the mass finishing technology involve three stages: pre-treatment before mass finishing, mass finishing, and post-treatment after mass finishing.

(1) Pre-treatment

The pre-treatment of the workpieces mainly get rid of oil dirt residual by last procedure, in order to avoid reducing the cleaning function of the liquid medium, darkening the workpiece surface, weakening the cutting force of abrasive blocks, and decreasing the finishing effect.

For some systems to work properly, heavy oil must be removed before it covers barrel liners and fouls drain systems. Special cleaners of some design will be needed before finishing of parts. This is different from the cleaning process that is needed after finishing.

Different kinds of metal cleaning solutions are mainly used for the pre-treatment, and the most common is water-based cleaning solutions (mass proportion is often 5%) in which workpieces are soaked or washed. This kind of solutions can be continuously used in two weeks. It is necessary adding appropriate cleaning solution (5–10% of the load) every 3 days.

(2) Mass finishing

Equipment type, finishing parameters, finishing media, and so on are main considerations for selecting or determining mass finishing processes, and mainly based on the workpiece material, shape and size, finishing requirements, etc. The selection of equipment type is shown in Table 2.29. Finishing parameters and finishing media of different equipment are described as the above chapter.

(3) Post-treatment

The post-treatment of workpieces involves the sieving between abrasive blocks and workpieces, cleaning of abrasive blocks and workpieces, drying and preventing rust of workpieces.

The common sieving device for abrasive blocks and workpieces includes mechanical sieving and magnetic sieving. Mechanical sieving device may be either a part of the mass finishing equipment, or an auxiliary device, and its principle is shown in Fig. 2.106. Separation workpieces 5 sieving from abrasive blocks 4 is realized by vibrating screen 3 driven mechanically. In general, the efficiency of vibrating sieving method is high, and the way of swing or rotating the screen may also be used.

Magnetic sieving is to separate workpieces from abrasive blocks through sticking workpieces by magnetic forces, so this method is only suitable for the

**Table 2.29** Methods and applicable conditions of drying and preventing rust methods

| The methods of drying and preventing rust | The applicable conditions |
|---|---|
| Soaking in sodium nitrite solution, the mixture of sodium nitrite and sodium carbonate, sodium silicate solution | Better effects, but harmful to humans health, so applications should be minimized |
| Oven drying and anti-rust oil soaking | Large investment and cost; not suitable to the small flat workpiece; the fit face of the workpiece taking from the oven may be rusted because there is water between workpiece and dryer frame |
| Blow-drying with a hairdryer and drying with fresh cloth | Low efficiency, so only for experiments |
| Alcohol drying and anti-rust oil soaking | The processing effects are remarkable, but the cost is high and the service cycle of the alcohol is short |
| Dryers drying and anti-rust oil soaking | Economical and practical, but inapplicable to heavy thin parts(less than 0.1 mm) |

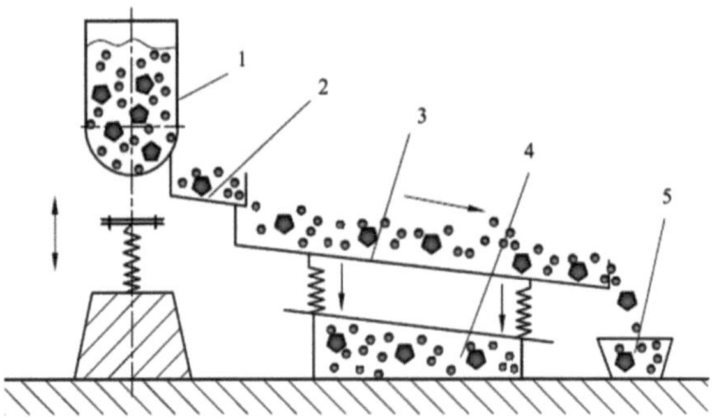


**Fig. 2.106** Principle diagram of mechanical sieving device. 1—Mass finishing equipment; 2—discharging exit; 3—vibrating screen; 4—abrasive blocks; 5workpieces

separation of ferromagnetic material parts. In general, the common magnetic sieving devices have both conveyor belt and magnetic disk. The principle diagram of conveyor belt sieving with magnetic forces is shown in Fig. 2.107. Workpieces and abrasive blocks climb to the feeder export through slight vibrating of spiral feeder 1, then fall on magnetized driving wheel 2, and sent to magnetized-driven wheel 5 with conveyor belt 8. Abrasive blocks fall into abrasive blocks in Box 6 through conveyor belt end because the media is not magnetized. But workpieces continue to move with the conveyor belt in suction of magnetized-driven wheel until conveyor belt disjointed from magnetized-driven wheel, and magnetic forces become weaken. At that time, workpieces fall into workpieces inbox 7 under their weights and separate with abrasive blocks. The principle diagram of the magnetic disk sieving is shown in Fig. 2.108. Its fundamental principle is the same with conveyor belt sieving. Magnetic disk 4, constituted non-ferromagnetic material such as Al alloys dish and some magnet, is installed over barrel finishing equipment 7. After barrel finishing, disk Bracket 2 is rotated to put the magnetic disk into the barrel.

**Fig. 2.107** Principle diagram of conveyor belt sieving with magnetic forces. 1—Spiral feeder; 2—driving wheel; 3—workpieces; 4—abrasive blocks; 5—driven wheel; 6—abrasive blocks inbox; 7—workpieces inbox; 8conveyor belt; 9—barrel finishing equipment

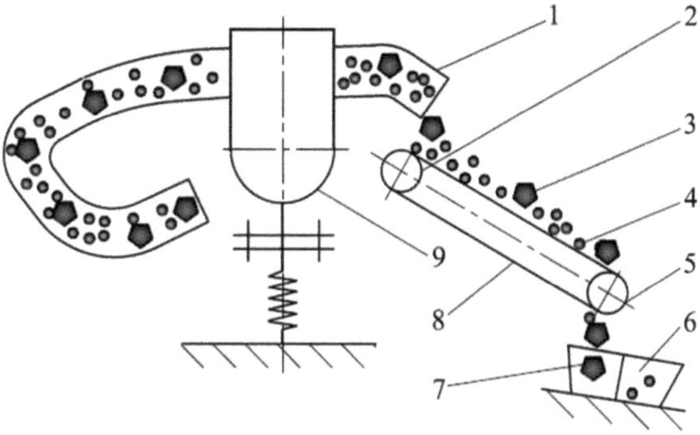


**Fig. 2.108** Principle diagram of the magnetic disk sieving. 1—Hand wheel; 2—bracket; 3—scraper; 4—magnetic disk; 5—workpieces; 6—abrasive blocks; 7—barrel finishing equipment

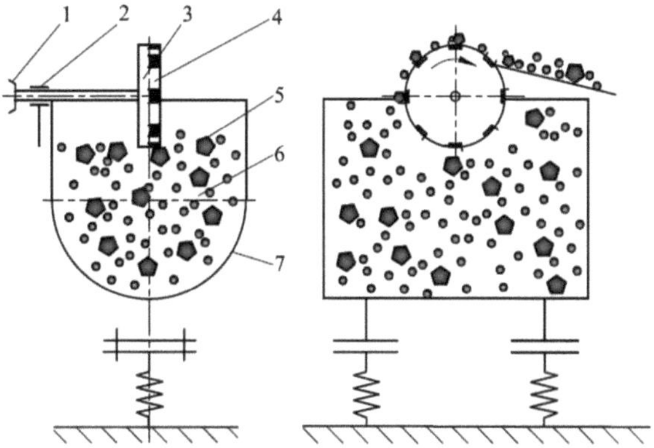


Workpicccs arc magnetized and hold by magnets when the disk rotates in the mixture. Workpieces rotate with the disk together and fall into workpieces inbox when passing through Scraper 3.

Barrel finishing operations have the built-in ability to clean parts as part of the process. In some situations, however, it is necessary to add an ultrasonic cleaning step, which requires an ultrasonic cleaner in the work area. Spray cleaners may also be appropriate. Water softeners may be necessary for finishing and cleaning operations. If workpieces cleaning is not dry, drying and preventing rust should be taken into account.

Experiments show that workpiece surface is bright after barrel finishing, and the surface is oxidized and becomes black soon after the surface active metal molecular is exposed in the air. The surface will become rusty after cleaning and dipping into the corrosion resistant fluid in half an hour. The dirt, oxide layer, rust spots, and other defects of workpieces after barrel finishing are removed, and this leads to expose fresh metal material into air. In addition, the surface of the workpiece has a water layer after cleaning with water solutions, which is enough to form a electrolyte solution that is necessary for electrochemical corrosion. Water ionization degree is small, but can be still ionized $H^+$ and $[OH]^-$. This ionization process increases with the raising temperature, that is,

$$H_2O \rightarrow H^+ + [OH]^-$$

There are $CO_2$ and $SO_2$ dissolved in water and form the following reaction:

$$CO_2 + H_2O \rightarrow H_2CO_3 \rightarrow H^+ + [HCO_3]^-$$

Corrosion cell is formed with Fe and the impurity soaked in the ionic solution with various kinds of $H^+$, $[OH]^-$, $[HCO_3]^-$, etc., and Fe is the anode and the impurity is the cathode. In general, the water layer contains oxygen, anodic Fe is oxidized to $Fe^{2+}$ ions and forms $[OH]^-$ions in cathode by obtaining oxygen. Corrosion reaction is

$$2Fe + O_2 + 2H_2O = 2Fe(OH)_2$$

There are many methods to drying and preventing rust, and different methods have their advantages and disadvantages. Specific methods and using cases are shown in Table 2.29.

As stated above, practical process of barrel finishing technology is as follows: degreasing → cleaning → deburring and finishing → sieving → cleaning → drying → anti-rust, as shown in Fig. 2.109. The workpiece that switches to later processes soon or is not allowed to dig oils will adopt dehydration methods of head drying, which can get the ideal effect.

2. The method of experimental study

(1) Research methods of finishing procedure and mechanism

Common methods about finishing procedure and mechanism are shown in Fig. 2.110. These methods can model, deepen finishing theory, and optimize parameters through recording the mixture motion (by high-speed camera), testing the force situation, simulating (by finite element analysis and discrete element analysis), etc., finally to gain finishing results as needed.

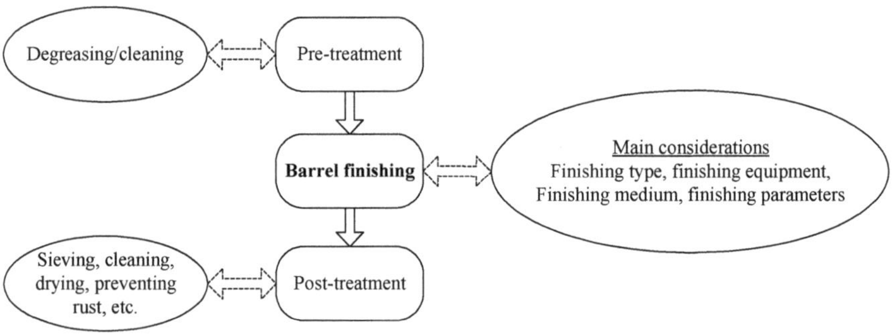


**Fig. 2.109** Technical process of barrel finishing

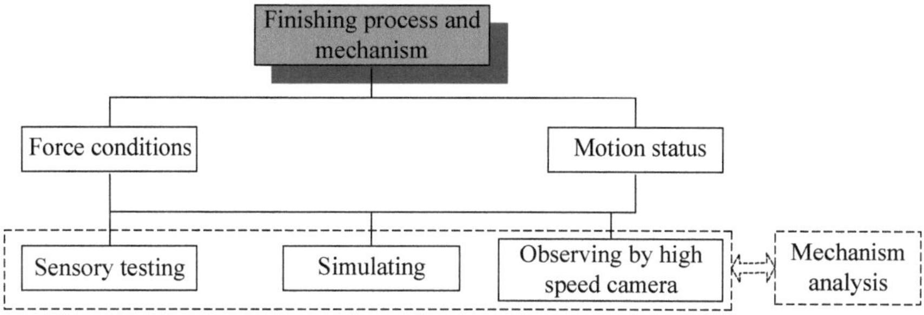


**Fig. 2.110** Common study methods of finishing process and mechanism

(2) The method of finishing effects

Common study methods about finishing effects are shown in Fig. 2.111. These methods mainly study the material removal ability of the workpiece and the improvement of surface geometric features (surface roughness, surface texture, surface defect, etc.) and the physical and mechanical properties of surface layer (surface micro-hardness, surface metamorphic layer, surface stress state,etc.) under various actions of influence factors. Additionally, the processing capacity of certain barrel finishing technique, the influence rule of surface quality, and surface integrity can be studied, to provide reasonable parameters for industrial application.

## 2.9.2 Applications

1. Barrel finishing processes of miscellaneous parts with small sizes

Miscellaneous parts have a great variety and can divide into castings, punching parts, powder metallurgical parts, machining parts, spring parts, and special parts according to final forming method.

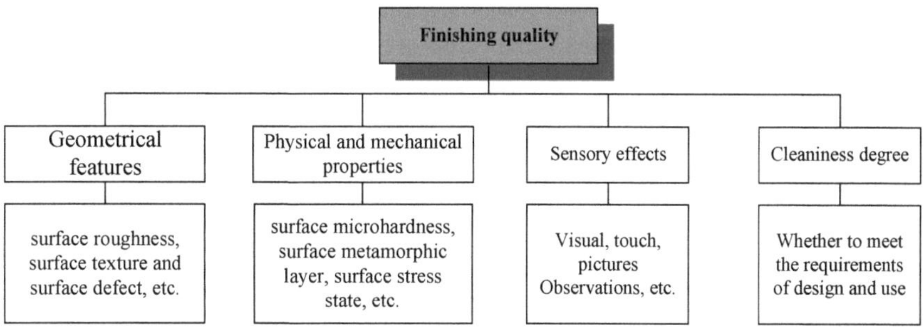


**Fig. 2.111** Common study methods of finishing results

Case 1. Safes-handle

(1) Workpiece conditions and requirements of barrel finishing

Safes-handle is steel material, its blank is casting, and large flash is removed with grinding wheel after casting. Surface requirements of safes-handle need to be smooth and meet electroplating requirements.

(2) Finishing equipment

The horizontal inclined centrifugal mass finishing equipment is adopted. The drum total volume is 4 × 15 L. The revolution speed of the drum is 180 r/min, and the rotational speed of the drum is −180 r/min.

(3) Process arrangement and finishing medium

Rough processing:
Abrasive blocks are brown alumina, oblique triangle prism with 15 mm × 15 mm × 15 mm, the granularity is $180^{\#}$, and the total volume is approximately 40% of the drum volume. Compounds' solution is LC-10 (suitable to ferrous metal), and the load is about 50 g for each drum. The total volume of parts added is approximately 40% of the drum volume. Enough water is added to cover the mixture of abrasive blocks and parts, above 10 mm than the mixture. The processing time is 40 min.
Finishing processing:
Abrasive blocks are white alumina, sphere with Φ5 mm, and the granularity is $240^{\#}$; the total volume is approximately 55% of the drum volume. The total volume of parts added is approximately 20% of the drum volume. Compounds' solution is LC-10, and the load is about 80 g for each drum. Enough water above 15 mm of the mixture is added. The processing time is 20 min.

(4) Processing results

After rough processing, the surface of the workpiece is smooth and even. By finishing processing, workpiece surface is smooth and light, and meets electroplating requirements.
Using similar barrel finishing method, all kinds of miscellaneous casting or forging parts with small sizes can be deburred or finished, such as various wrench by forging forming, pipe joints of malleable steel casting, hose connections of cast steel, engine connection rod, valve pendulum pole, cipher lock case.
Case 2. Lock body plate

(1) Workpiece conditions and requirements of barrel finishing

The shape of lock body plate is shown in Fig. 2.112. The material of the lock body plate is Q235. All edges of lock body plate have burrs after stamping process. The size of the workpiece is small and thin. Finishing requirements are removing all burrs and rust spots.

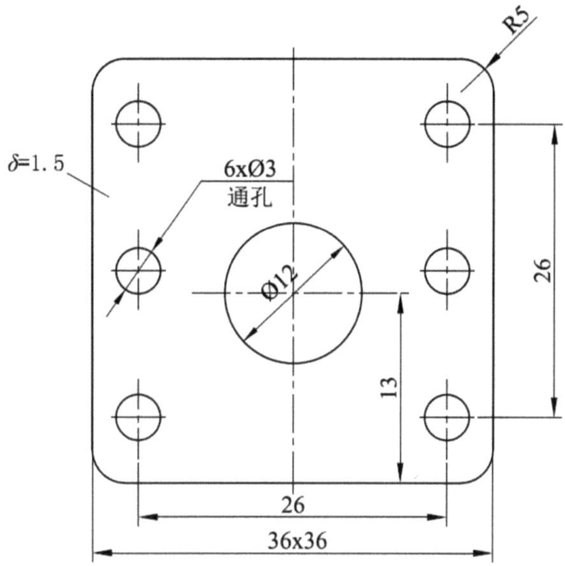


**Fig. 2.112** Diagram of lock body plate

(2) Finishing equipment

Vertical vibratory barrel finishing equipment is adopted. The total volume of the barrel is 150 L, vibratory frequency is 1800 times/min, and amplitude is 3–4 mm.

(3) Finishing medium and finishing parameters

The abrasive blocks are brown alumina, oblique triangle prism with 12 mm × 12 mm × 12 mm, the granularity is $240^{\#}$, and the total volume is approximately 60% of the barrel volume. The total volume of parts added is approximately 1/4 volume of abrasive blocks used. Dry finishing is applied to avoid workpieces sticking together. The processing time is about 30 min.

(4) Processing results

After finishing, workpiece surface is gray and is not so bright, but all burrs are removed and the workpiece has no distortion.

Using similar barrel finishing method, deburring ability is good for all kinds of flat parts, such as motor stator plates, chain plates of stepless speed regulator.

Case 3. Universal wheel frame

(1) Workpiece conditions and requirements of barrel finishing

Universal wheel frame is formed with stamping and bending using Q235 steel plates. There are burrs at the stamping circumference and bending marks. Finishing requirements are used to remove all burrs, smooth and bright surfaces, and to meet the electroplating requirements.

(2) Finishing equipment

The horizontal centrifugal barrel finishing equipment is adopted. The drum total volume is 4 × 20 L. The revolution speed of the drum is 180 r/min, and the rotational speed of the drum is −180 r/min.

(3) Process arrangement and finishing medium

Rough processing:
The abrasive blocks are brown alumina, oblique triangle prism with 15 mm 15 mm × 15 mm, the granularity is 200#, and the total volume is approximately 45% of the drum volume. Compounds solution is LC-10, and the load is about 40 g for each drum. The total volume of parts added is approximately 15% of the drum volume. Enough water is added to cover the mixture of abrasive blocks and parts, and it is above 10 mm than the mixture. The processing time is 30 min.
Finishing processing:
Abrasive blocks are white alumina, sphere with Φ5 mm, the granularity is 240#, and the total volume is approximately 55% of the drum volume. The total volume of parts added is approximately 15% of the drum volume. Compounds solution is LC-10, and the load is about 80 g for each drum. Enough water above 15 mm of the mixture is added. The processing time is 15 min.

(4) Processing results

After rough processing, workpiece surfaces have no rust or scratch, and all burrs are removed. By finishing processing, workpiece surface is smooth and light, and meets electroplating requirements.
Using similar barrel finishing methods, all kinds of stamping parts can be finished, such as stamping parts on textile machine/camera/printer/bicycle.

Case 4. Metal glasses frame

(1) Workpiece conditions and requirements of barrel finishing

Metal glasses frame is mainly welded using thin rod blank with alloy material. Its surfaces are rough and have welding burrs. The surface requirements need to be smooth and bright, and meet the electroplating requirements.

(2) Finishing equipment

Vertical vibratory barrel finishing equipment is adopted. The total volume of the barrel is 150 L, vibratory frequency is 1800 times/min, and amplitude is 3–4 mm.

(3) Process arrangement and finishing medium

Rough processing:
Abrasive blocks are ceramic, and oblique triangle prism with 10 mm × 10 mm × 10 mm; the total volume is approximately 60% of the barrel volume. The total volume of parts added is approximately 1/6 ~ 1/5 the volume of abrasive blocks used. Compounds solution is LC-13 (suitable for nonferrous alloy materials). The water in the container, using water circulation device, is not less than 10 kg, and the mass portion of compounds in water is about 5‰.

Finishing processing:

The abrasive blocks are high alumina ceramics, it has sphere with Φ5 mm, and the total volume is approximately 70% of the barrel volume. The total volume of parts added is approximately 1/6 the volume of used abrasive blocks. The water in the container, using water circulation device, is not less than 15 kg, and mass portion of compounds in water is about 1%.

(4) Processing results

After rough processing, workpiece surfaces are smooth, and all burrs are removed. By finishing processing, workpiece surface is smooth and light, and it meets electroplating requirements.

Using similar barrel finishing methods, all kinds of thin rod parts and stamping flat parts can be finished, such as camera's optical eye, bearing cage.

Case 5. Magnetic sheet for neodymium iron boron (NdFeB) material

(1) Workpiece conditions and requirements of barrel finishing

The hardness of magnetic sheet is high, and the size is generally 10 mm × 10 mm × 2 mm or Φ15 mm × 1.5 mm. Edges of magnetic sheet after forming are sharp. Finishing requirements are the uniform edge rounding and edge radius of 0.1–0.2 mm.

(2) Finishing equipment

The horizontal inclined centrifugal mass finishing equipment is adopted. The drum total volume is 4 × 8 L, and each drum is detachable. The revolution speed of the drum is 200 r/min, and the rotational speed of the drum is −200 r/min.

(3) Finishing medium and finishing parameters

Abrasive blocks are high alumina ceramics, oblique triangle prism with 15 mm × 15 mm × 15 mm, and the total volume is approximately 45% of the barrel volume. The total volume of parts added is approximately 15% of the drum volume. Compounds solution is LC-10, and mass portion of compounds in water is about 5‰. Enough water above 15 mm of the mixture is added. The processing time is 30 min.

(4) Processing results

After the barrel finishing, edges of magnetic sheet are uniform and the edge radius is about 0.1–0.2 mm. Electroplating requirements can be improved, and electroplating quality is increased.

Case 6. Push rod

(1) Workpiece conditions and requirements of barrel finishing

A kind of push rods for safe device is shown in Fig. 2.113. Its material is 45 steel. Requirements are as follows: Cylindrical dimension precision is high, and the right cone mating with bearing seat hole is smooth and it cannot have burrs.

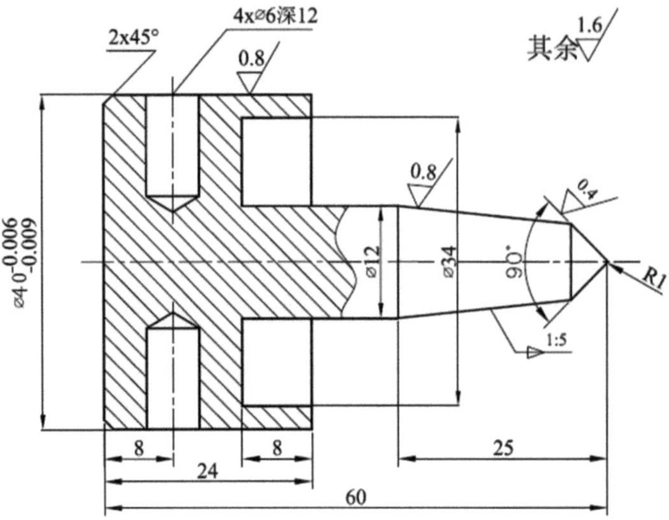


**Fig. 2.113** Diagram of the push rod

(2) Finishing equipment

The horizontal centrifugal barrel finishing equipment is adopted. The drum total volume is 4 × 15 L. The revolution speed of the drum is 180 r/min, and the rotational speed of the drum is −180 r/min.

(3) Finishing medium and finishing parameters

Abrasive blocks are white alumina, sphere with Φ5 mm, and the granularity is $240^{\#}$; the total volume is approximately 65% of the drum volume. The total volume of parts added is approximately 8–10% of the drum volume. Compounds solution is LC-10, and the load is about 30 g for each drum. Enough water above 10 mm of the mixture is added. The processing time is 15 min.

(4) Processing results

The surface roughness value *Ra* decreases from 0.8–1.2 μm to 0.6 μm. Edge radius can be assured. The hardness of surface layer increases in an average of 14 HV. All burrs are removed. The cone surface after assembly can be flexible to rotate and can ensure the service performance.

Using similar barrel finishing method, black metal machined parts with high dimensional accuracy can be finished, such as valve rod, valve lifter, fuel injection nozzle, bearing inner ring and outer ring, $CO_2$ welding nozzle, insulating bushing, conductive nozzle.

Case 7. The base of sight device

(1) Workpiece conditions and requirements of barrel finishing

The base of sight device for certain shotgun is shown in Fig. 2.114. Its material is aluminum alloy, and the shape is complex and is made of spherical surface, cylindrical surface, oblique plane, etc. Precisions of form and position by turning and milling processes are high. Finishing requirements are removing all burrs, and surface roughness value *Ra* is 0.8 μm.

(2) Finishing equipment

The horizontal centrifugal barrel finishing equipment is adopted. The drum total volume is 4 × 15 L. The revolution speed of the drum is 180 r/min, and the rotational speed of the drum is −180 r/min.

(3) Finishing medium and finishing parameters

Abrasive blocks are high alumina ceramics, oblique cylinder with $\Phi$3 mm 5 mm, the granularity is 240#, and the total volume is approximately 70% of the drum volume. The total volume of parts added is approximately 8–10% of the drum volume. Compounds solution is LC-12 (suitable for zinc aluminum alloy), and the load is about 40 g for each drum. Enough water above 10 mm of the mixture is added. The processing time is 12 min.

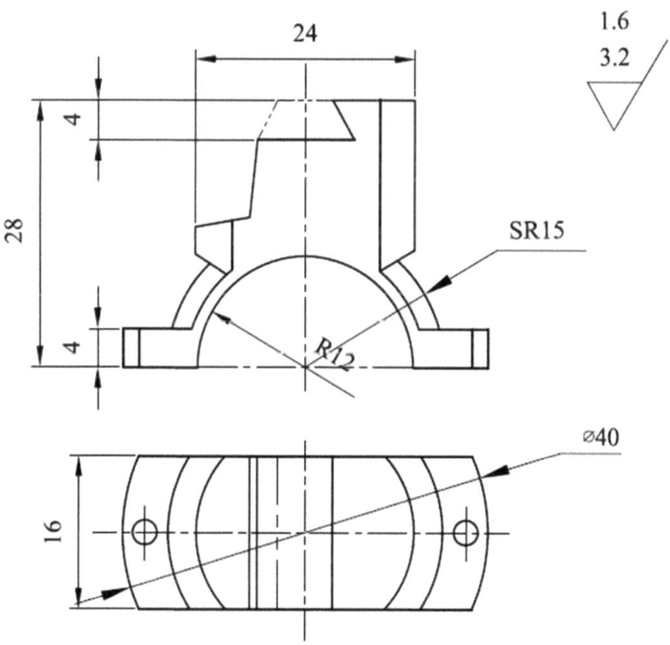


**Fig. 2.114** Diagram of the base of sight device

(4) Processing results

After finishing, all burrs are removed and the edges are filleted. The surface roughness value *Ra* decreases from 1.6 to 0.8 μm, and the surface is smooth and the surface color is uniformly bright.

Using similar barrel finishing method, nonferrous metal complex parts with high dimensional precision can be finished, such as copper torch with argon arc welding, electrode, shunt, fittings, chuck, chuck base.

The process methods and equipment used for above cases are not unique. Other equipment with corresponding finishing parameters also can be chosen. For example, whirling barrel finishing equipment can also be selected to finish case 3, 6, and 7.

Except for parts mentioned above, some parts with special requirements can be finished with barrel finishing processes and obtain ideal effects. Here are some examples.

(1) Gears of the gear pump in one factory (gear material is 40Cr) were finished in 20 min by centrifugal barrel finishing process. After finishing, the gear machining precision degree did not change, the surface hardness increased to 15 HV, the surface roughness degree of teeth surface and end surface increased by one degree, the surface tensile stress (33.82 MPa) is converted to compressive stress (−589.4 MPa), and edges radius is 0.1–0.2 mm. After assembly with finished gears, the noise of the pump decreased by 2 dB, the scratches of end surface and pump body decreased, and using performance of the gear pump is improved.
(2) Springs in one factory, finished by barrel finishing, have good finishing quality: The burrs were removed, the physical and mechanical properties were improved, and service life was extended. Through fatigue tests, cycles of the spring are varied from 36,000 times to 52,000 times by barrel finishing.
(3) For riveting parts (such as new handcuffs) and hinging casting parts (such as universal ball hinge), comprehensive effects of deburring and surface finishing are well after barrel finishing.

Figure 2.115 shows comparison photographs before and after finishing of miscellaneous parts with small sizes.

2. Barrel finishing processes of shaft parts

Case 1. 195 crankshaft

(1) Workpiece conditions and requirements of barrel finishing

The material of 195 crankshaft is ductile cast iron. The break rate of the machined product is about 3‰ during normal use. Finishing requirements are surface comprehensive effects are good, and physical and mechanical properties are obviously improved after barrel finishing [58].

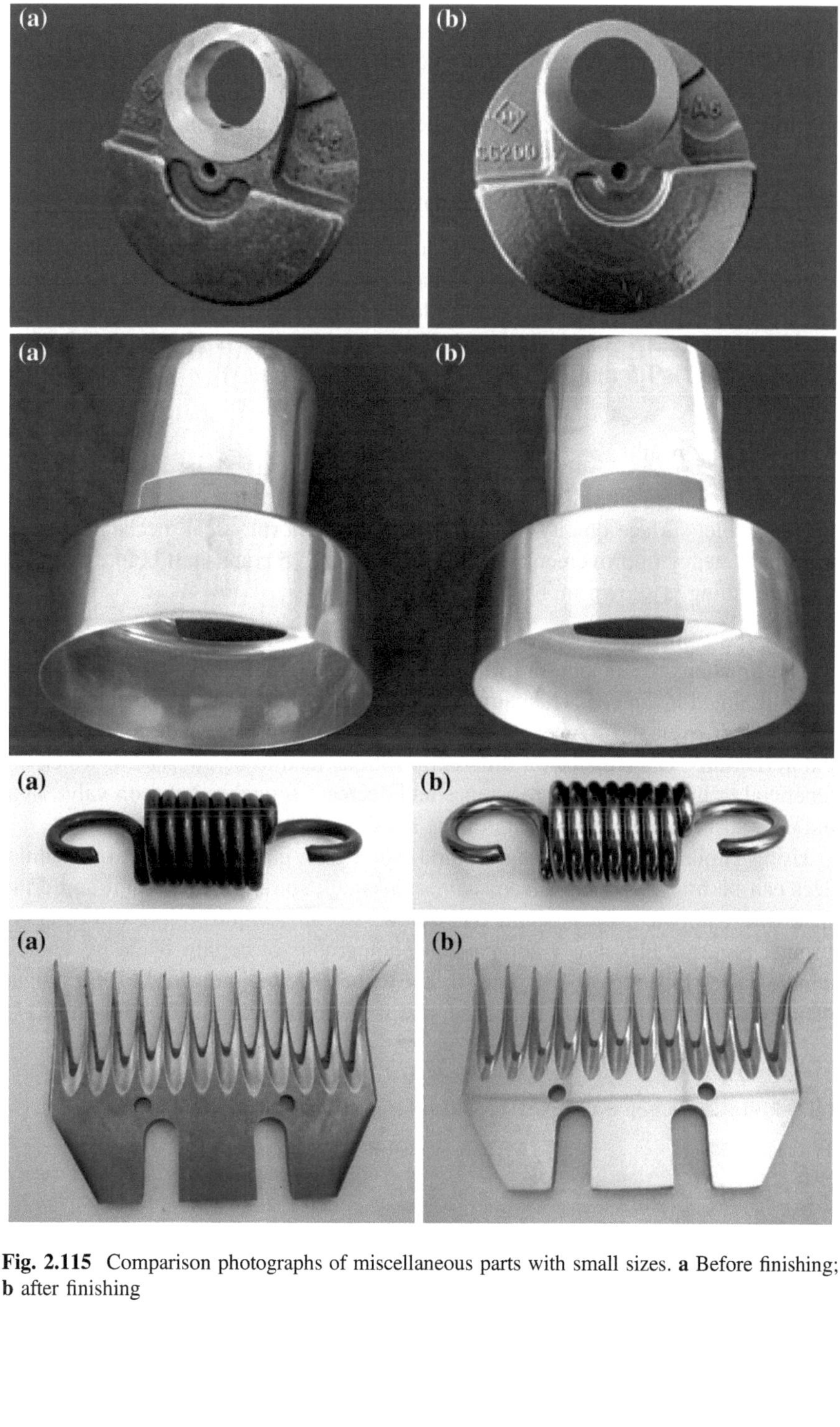


**Fig. 2.115** Comparison photographs of miscellaneous parts with small sizes. **a** Before finishing; **b** after finishing

(2) Finishing equipment

The vertical spindle barrel finishing equipment is adopted. The diameter of the barrel is 900 mm, and the height is 650 mm. The rotational speed of the barrel is 35 r/min or 150 r/min for the spindle. There are four spindles, and four workpieces are loaded at one time.

(3) Finishing medium and finishing parameters

The abrasive blocks are white alumina, sphere with Φ5 mm, and the granularity is 240#. The height of abrasive blocks added is approximately 500 mm. Compounds solution is LC-10, and the load is about 500 g. Enough water above 10–20 mm of abrasive blocks is added. The processing time is 15 min, and the time for half period $t$ is 1.5 min. The cycling process is done as $(N, -n, t) \Leftrightarrow (0, 0, \Delta t) \Leftrightarrow (-N, n, t)$.

(4) Processing results

Experimental results indicate that machining precision of the journals is unchangeable, surface quality obviously increases, and physical mechanical properties have heavy improvement. The testing data of 195 crankshaft before and after barrel finishing is shown in Table 2.30.

Case 2. 488 camshaft

The finishing equipment and finishing process are similar with 195 crankshaft finishing noted in Case 1. The camshaft is loaded into the spindle twice by turning the camshaft over 180°. After finishing, all burrs can be removed and the surface texture is finer. The stresses on different positions tend to be consistent, which are beneficial to improve fatigue resistance and decrease scratches between valve stem and cam surface. The testing data of the stress is shown in Table 2.31.

Using similar barrel finishing method, shaft disk parts with small or medium sizes can be finished, such as camshaft, crankshaft, spiral grooved shaft and drum on textile equipment, gear shaft, ball screw, worm, damping pipe, turbocharger turbine. Figure 2.116 shows comparison photographs of certain blade by vertical spindle barrel finishing process, and surface texture shown in Fig. 2.117. By fatigue testing, fatigue life is greatly improved, and it is shown in Fig. 2.118. Figure 2.119

**Table 2.30** Comparison data of 195 crankshaft before and after barrel finishing

| Finishing state | Ra/μm | | Micro-hardness/HRC | | Connecting rod journals stress/MPa | | Metallographic observed under the electron microscopy |
|---|---|---|---|---|---|---|---|
| | Main journal | Connecting rod journal | Main journal | Connecting rod journal | tangential | normal | |
| Before finishing | 0.63 | 0.67 | 40 | 39 | −149 | −423 | Loose |
| After finishing | 0.35 | 0.38 | 45.6 | 46 | −703 | −655 | Dense layer with about 24 μm |

**Table 2.31** Comparison data of camshaft stress before and after finishing

| Finishing state | Tangential stress/MPa | | | Axial stress/MPa | | |
|---|---|---|---|---|---|---|
| | Top | Side | Bottom | Top | Side | Bottom |
| Before finishing | −357 | −102 | −281 | −324 | +167 | −350 |
| After finishing | −572 | −543 | −487 | −494 | −563 | −519 |

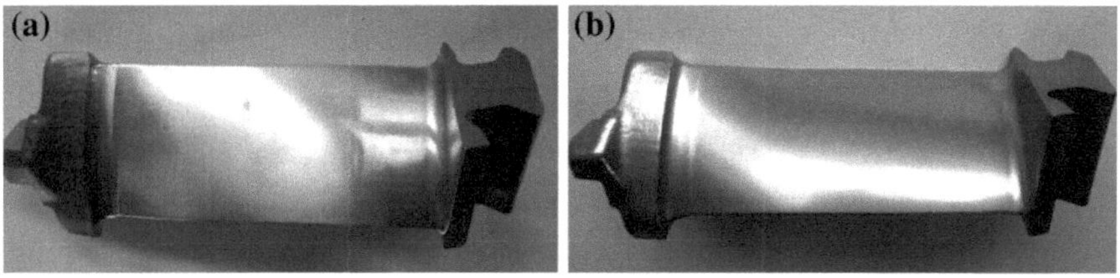


**Fig. 2.116** Comparison photographs of certain blade by vertical spindle barrel finishing process. **a** Before finishing; **b** after finishing

(a) (b)

**Fig. 2.117** Comparison of surface texture of certain blade by vertical spindle barrel finishing process. **a** Before finishing; **b** after finishing

shows comparison photographs of certain labyrinth disk by vertical spindle barrel finishing process.

For shaft parts with large or medium sizes (especially for camshaft, crankshaft), better finishing effects can be received by horizontal spindle barrel finishing process [59]. Comparisons of surface roughness value *Ra* for certain camshaft and crankshaft are shown in Tables 2.32 and 2.33.

Comparison curves of the material ratio of the profile for certain four cylinder diesel engine camshafts are shown in Fig. 2.120.

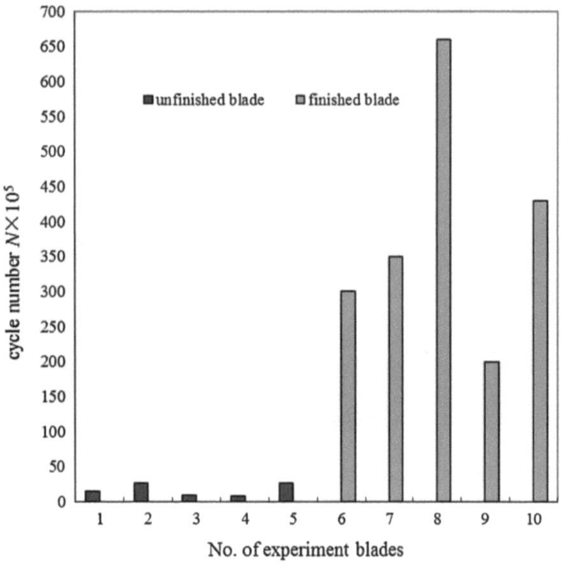


**Fig. 2.118** Comparison of fatigue life of certain blade by vertical spindle barrel finishing process

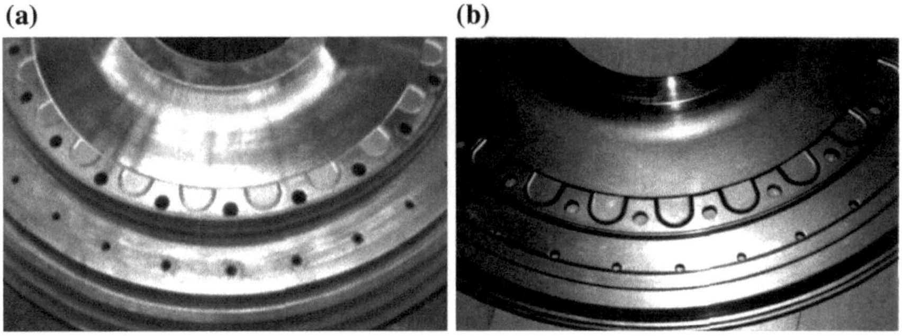


**Fig. 2.119** Comparison photographs of certain labyrinth disk by vertical spindle barrel finishing process. **a** Before finishing; **b** after finishing

**Table 2.32** Comparison of surface roughness value *Ra* for certain crankshaft

| Finishing state | Connecting rod journals | | Main journals |
|---|---|---|---|
| | The outside position | The inside position | |
| Before finishing | 0.605 | 0.475 | 0.149 |
| After finishing | 0.170 | 0.276 | 0.06 |

Figure 2.121 shows comparison photographs of surface textures (500 times magnification) for M11 camshaft. Figure 2.122 shows comparison photographs of certain camshaft and crankshaft.

**Table 2.33** Comparison of surface roughness value *Ra* for certain camshaft

| Finishing state | The surface of bearing journal | The surface of small camshaft | The surface of large camshaft |
|---|---|---|---|
| Before finishing | 0.361 | 0.442 | 0.486 |
| After finishing | 0.181 | 0.131 | 0.148 |

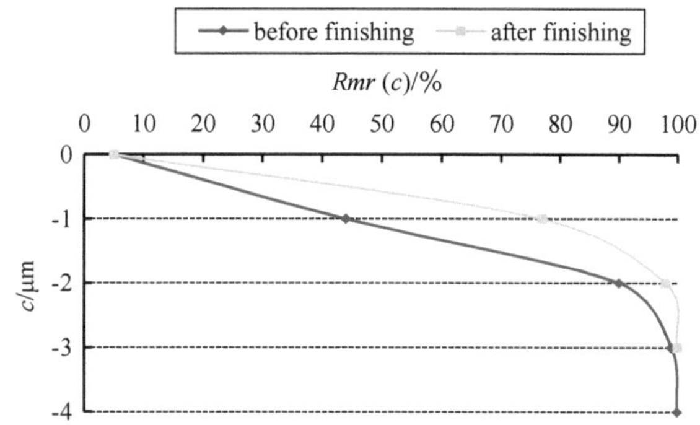


**Fig. 2.120** Comparison curves of the material ratio of the profile for certain four cylinder diesel engine camshafts

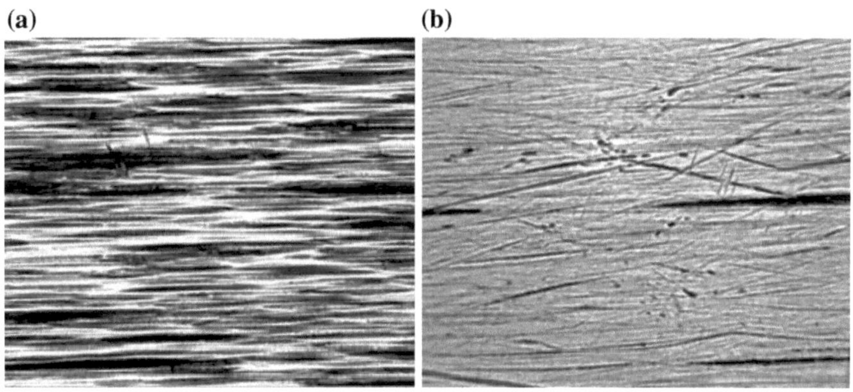


**Fig. 2.121** Comparison photographs of surface textures (500 times magnification) for M11 camshaft. **a** Before finishing; **b** after finishing

Comparisons of some military camshaft after using 350 h are shown in Fig. 2.123. It can be shown that there are obvious scratches on the unfinished camshaft surface and not on finished surface. This phenomenon also illustrates that the changes of the surface roughness, surface textures, and physical–mechanical properties of the surface layer (surface micro-hardness, surface metamorphic layer, the surface stress state, etc.), can also lead to the wear resistance increase and scratches decrease after barrel finishing.

(a) (b)

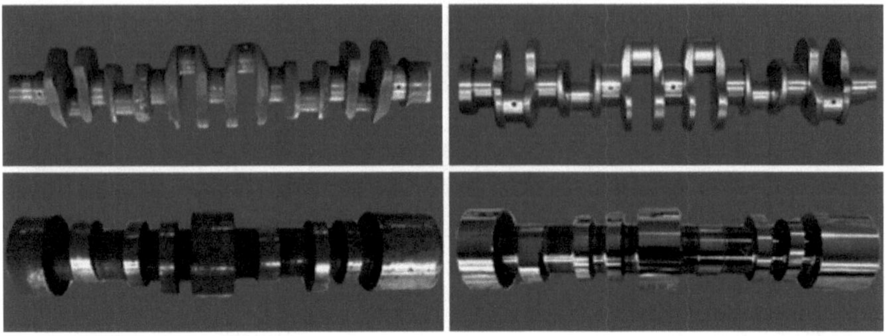

**Fig. 2.122** Comparison photographs of certain camshaft and crankshaft. **a** Before finishing; **b** after finishing

(a) (b)

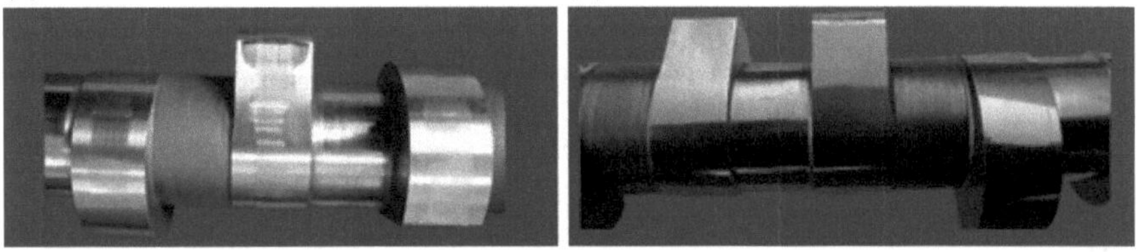

**Fig. 2.123** Comparison photographs of some military camshaft after finishing 350 h. **a** Before finishing; **b** after finishing

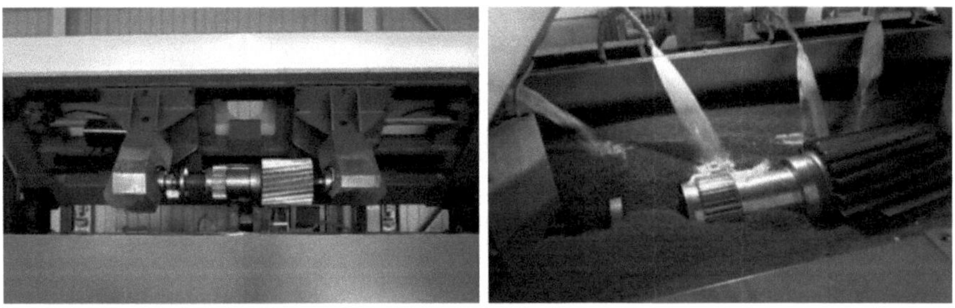

**Fig. 2.124** Finishing status of gear shaft with large gear module

By horizontal spindle barrel finishing process, gear shaft with large gear module can also be finished, and the finishing status is shown in Fig. 2.124.

3. Barrel finishing processes of disk parts

Case 1. Steel wheel hub for 130 automobile

(1) Workpiece conditions and requirements of barrel finishing

The blank of wheel hub for 130 automobile is welded after being rolled using steel plate and steel strip. The diameter of wheel hub is 440 mm, and the thickness is 200 mm. The rim of hub surface has burrs. The surface of steel plate and steel strip is untreated and has rust. One of the welding joints of cylinder surface has been razed. There are welding joints at the same circle of the axial end surface. Finishing requirements are as follows: All burrs are removed, all rust spots of outer and inner surface are cleaned, surfaces of welding joints are smooth, and electroplating requirements are met.

(2) Finishing equipment

Intermeshing spindle barrel finishing equipment is adopted. The diameter of the barrel is 1200 mm. The rotational speed of the barrel is 50 r/min or 88 r/min for the spindle. Spindle deviation angle $\alpha$ is 15° and $\beta$ is 30°.

(3) Finishing medium and finishing parameters

Abrasive blocks are high alumina ceramics, sphere with Φ3 mm, and the granularity is $200^{\#}$. The height of abrasive blocks added is approximately 2/3 of the barrel height (about 250 kg). Compounds solution is LC-10, and the load is about 500 g. Enough water just covering abrasive blocks is added. The processing time is 20 min, and the time for half period $t$ is 1.5 min. The cycling process is done as $(-N, -n, t) \Leftrightarrow (0, 0, \Delta t) \Leftrightarrow (N, n, t)$.

(4) Processing results

The wheel hub surfaces after finishing are smooth, light, and shining. The edges burrs are removed. Both surfaces of welding joints are smooth and meet electroplating requirements. Comparison photographs before and after finishing are shown in Fig. 2.125.

(a) (b)

**Fig. 2.125** Comparison photographs of wheel hub for 130 automobile. **a** Before finishing; **b** after finishing

Case 2. High-end aluminum alloy wheel hub

(1) Workpiece conditions and requirements of barrel finishing

Finishing requirements are as follows: All burrs and flash around the decorating holes are removed, all edges filleted are uniform and smooth, and each adjacent surface cannot be scratched.

For now, the used technological process on these issues at workshops is manually polishing using rotary scraper blade, but polishing quality is bad (main problems are remaining burrs, non-uniform edge radius, matte surface, easy to scratch adjacent surface, and low production efficiency).

(2) Finishing equipment

Intermeshing spindle barrel finishing equipment is adopted. The diameter of the barrel is 1200 mm. The rotational speed of the barrel is 50 r/min and 88 r/min for the spindle. Spindle deviation angle $\alpha$ is 15° and $\beta$ is 30°.

(3) Finishing medium and finishing parameters

Rough processing:

The abrasive blocks mixed with different shapes and sizes are used to finish. The processing time is 10 min. Other conditions are the same with Case 1.

Semi-finishing processing:

Conditions are the same with Case 1.

Finishing processing:

The abrasive blocks are stainless steel, sphere with Φ5 mm, and some kerosene is added. The processing time is 5 min.

(4) Processing results

By rough processing, burrs are removed, edge filleting is uniform, and there are no scratches on the surface. After semi-finishing processing, the surface is smoother, and the surface texture is more isotropic. After finishing processing, all finished surfaces are smooth and bright. The comparison photographs of aluminum alloy wheel hub are shown in Fig. 2.126.

Based on the process principle of intermeshing spindle barrel finishing, two kinds of automatic production line for finishing high-end aluminum alloy wheel hub are developed independently, as shown in Fig. 2.127. Compared with other finishing processes for wheel hub, this process has good quality, high efficiency, and low cost. Finishing effects are shown in Fig. 2.128.

Case 3. Gear (gear module $m$ = 3, teeth number $z$ = 78)

In the standard of cylindrical gears accuracy (GB 10095), edge filleting has specific requirements [60]. Three edges along different directions from the teeth tip should be filleted (shown in Fig. 2.129), in order to improve bearing capacity, running stability, and reliability.

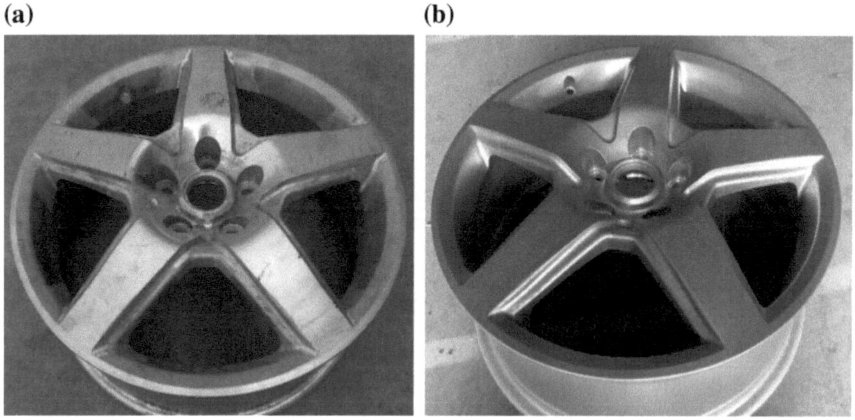

**Fig. 2.126** Comparison photographs of aluminum alloy wheel hub. **a** Before finishing; **b** after finishing

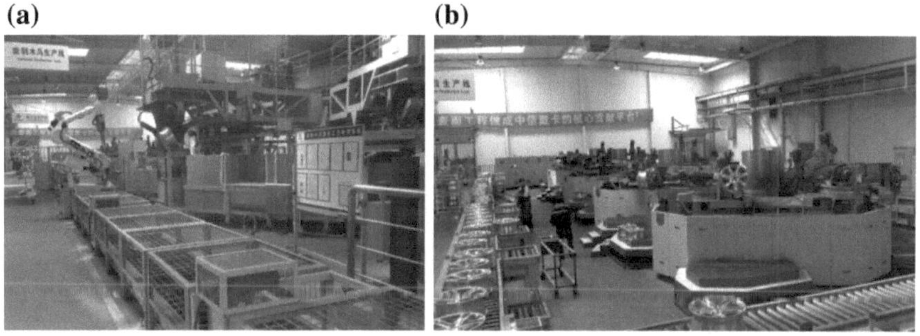

**Fig. 2.127** Automatic production line for finishing high-end aluminum alloy wheel hub. **a** C410; **b** DQ500

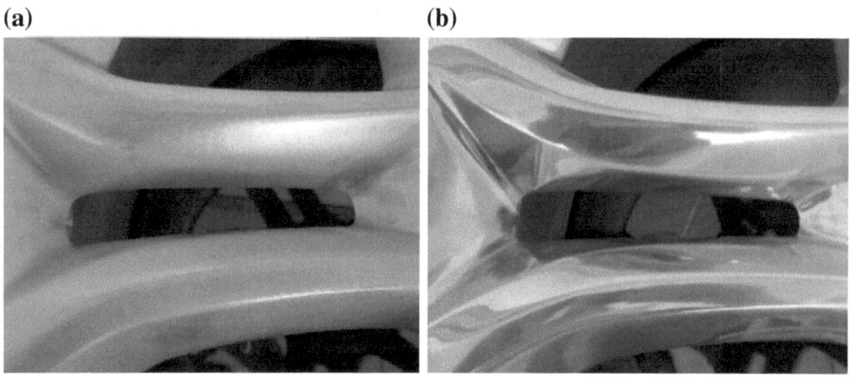

**Fig. 2.128** Comparison photographs of high-end aluminum alloy wheel hub. **a** Before finishing; **b** after finishing

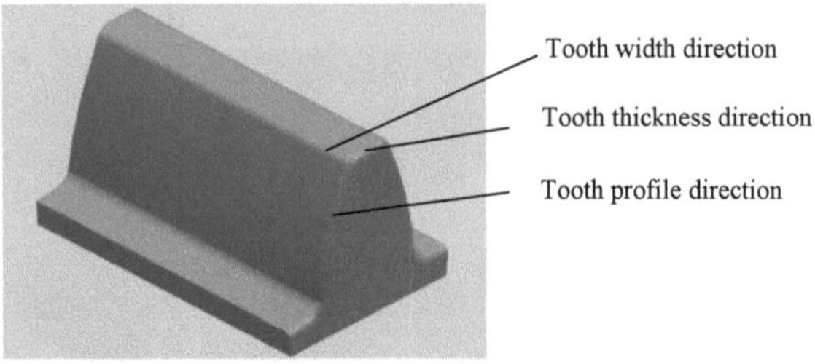


**Fig. 2.129** Diagram of edge filleting for gears

For certain gear of textile equipment accessories (its material is 40Cr, and edge filleting along different directions need to be done in around 0.3 mm according to the drawing designs) produced by international joint ventures in China, the intermeshing spindle barrel finishing process is used to finish in about 15 min. Experimental results show that edges along different directions have small radius, all burrs are removed, and teeth surface roughness, surface hardness, and surface stress are improved. The comparison photographs before and after mass finishing are shown in Fig. 2.130.

Using similar barrel finishing methods, all kinds of gears, sprockets, cycloid gears, synchronous belt wheels, cams, automobile wheel hubs, motorcycle wheel hubs and rims, turbocharger turbines and cases, high-end nutrition pots with large or medium sizes can be finished. Figure 2.131 shows the comparison photograph of certain gear.

4. Parts cleanliness improvement

Crankshaft is one of the most critical components of the engine, and its cleanliness degree has important influence on engine reliability. Therefore, engine manufacturing enterprises not only control its material, machined sizes, and tolerances, but also control its cleanliness degree, which is one of the most important factors.

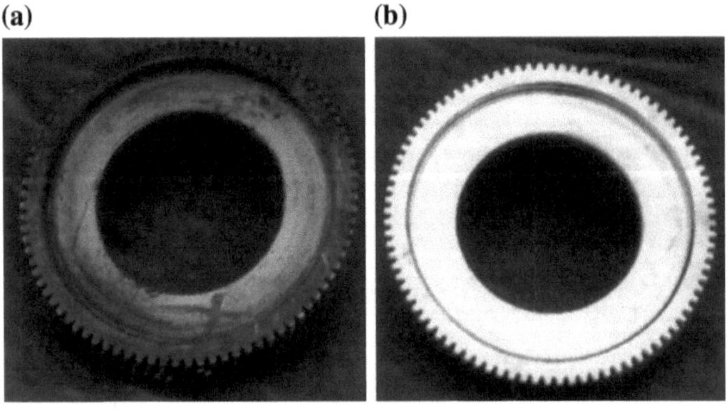


**Fig. 2.130** Comparison photographs of certain gear of textile equipment accessories. **a** Before finishing; **b** after finishing

**Fig. 2.131** Comparison photographs of certain gear. **a** Before finishing; **b** after finishing

The cleanliness is machinery residual impurity degree of products system and the channels. Effective factors on the cleanliness are residual impurities during processing, so improving the engine cleanliness is to reduce the abrasive wear caused by these impurities. Main source of these impurities is residual impurities which are not removed, such as the burrs, metal debris. These impurities could produce scratching or scoring to the cylinder, damage crankshaft, crankshaft bearing, hinder cooling water flow,etc., which affects the product life and performance. Therefore, it is important to improve the cleanliness degree of crankshafts.

Table 2.34 shows contrast data of cleanliness degree by crankshaft finishing. Before finishing, its cleanness degree cannot meet the design and use request. By barrel finishing, its cleanness degree improves (the reason includes the following: Burrs and residue on the casting are removed, and non-machined surfaces are finished). Therefore, it can further improve its usage performance and meet requirements.

**Table 2.34** Contrast data of crankshaft cleanliness before and after barrel finishing

| Company name | Crankshaft type | Cleanliness degree [mg] | | |
|---|---|---|---|---|
| | | Requirement | Before finishing | After finishing |
| Tianrun Crankshaft Ltd. | YC4108 | $\leq 66$ | >100 | $40.8 \pm 10$ |
| Jiangsu Songling Crankshaft Ltd. | 6 cylinder | $\leq 200$ | >400 | $180 \pm 20$ |
| Xiangtan Xinghuang Crankshaft Ltd. | G6135 | $\leq 200$ | >400 | $160 \pm 30$ |
| Hangzhou Xianglong Crankshaft Ltd. | MR479Q | $\leq 15$ | >20 | $10 \pm 2$ |

5. Improvement of the engine performance

The engine is a main power of automobile, engineering machinery, agricultural machinery, etc. It is an eternal subject developing the engine performance. Among factors influencing engine performance, surface quality of parts with frication pairs is one main factor. For a long time, the traditional manufacture level is relatively steady, and some technical requests of mechanical parts with frication pairs for mass products, such as deburring, edge filleting, are often be ignored. There are absences of understanding and realizing method about the surface texture, surface layer properties (physics, mechanics, chemistry, and optics), and the surface layer structure. Researches on improving the engine performance are often focused on modifying the engine's local structure, changing working process,etc. Owing to the difference of surface quality, the domestic running-in process is long in time and low in quality. There is more consumption in human and material resource.

With the development of special manufacturing process, surface finishing technique has been approbated through a long period application. It has functions on deburring, decreasing surface roughness, improving surface quality and physical–mechanical properties,etc. Aimed at the specific structure and the surface characteristic of engine's parts with frication pairs, it is imperative to develop the engine performance, decrease the fuel consumption, and improve the cleanness of the machine oil through using various finishing process.

Taking LL480QB diesel engines as the general testing subject, 17 kinds of parts are selected considering the overall structure of the engine (the friction parts mainly concentrated in the crank mechanism and the distribution mechanism) and the impact degree of the engine performance for surface quality and surface integrity of specific parts. Specific finishing technology is chosen according to the parts structure, the parts material, surface conditions, and finishing requirements (as shown in Table 2.35).

(1) Contrast analysis of finishing results

From the optical observation and the photograph contrast, parts have obvious burrs and sharp edges before finishing, and some parts look black (oxidized), dim, and rusty. After finishing, burrs and rusts are removed, sharp edges are filleted, and surfaces are smooth and glitter.

Testing comparisons of some typical parts is shown in Tables 2.36, 2.37, 2.38, 2.39, and Fig. 2.132. After finishing, the dimension precision can be insured because the dimension changes are within 0.001–0.005 μm. The surface roughness value *Ra* of all kinds of parts is basically reduced by around 0.5, and the material ratio of the profile Rmr(*c*) increases apparently, which can effectively improve wearing performances of frication pairs. The piston grooves and sharp edges of piston rings are filleted to around *R* 0.1 mm which has great significance on running-in quality of the cylinder liner and the decrement of fuel consumption.

Micro-hardness of the surface layer clearly increases, and the stress may vary from tensile stress to compressive stress (or the original compressive stress obviously increases), so the fatigue strength is developed and the service life is

**Table 2.35** Specific finishing processes of main friction pairs

| No. | Parts name | Parts material | Finishing processes | Finishing time/min |
|---|---|---|---|---|
| 1 | Crankshaft | QT800-2 | Horizontal spindle barrel finishing | 15 |
| 2 | Connecting rod assembly (outside surface) | | Vertical spindle barrel finishing | 10 |
| | Connecting rod assembly (two holes) | | Magnetic abrasive finishing | 8 |
| 3 | Piston pin | 20Cr10Si2Mo | Centrifugal barrel finishing | 8 |
| 4 | Piston | 66-1eutectic silicon alloy | Vertical spindle barrel finishing | 2 |
| 5 | Piston ring components | Alloy casting | Centrifugal barrel finishing | 8 |
| 6 | Cylinder liner | Boron Cast Iron | Gas-particle two-phase flows finishing | 10 |
| 7 | Camshaft | 45 steel | Horizontal spindle barrel finishing | 15 |
| 8 | Rocker shaft | 20 steel | Vertical spindle barrel finishing | 10 |
| 9 | Intake valve | 4Cr9Si2 | Vertical spindle barrel finishing | 10 |
| 10 | Exhaust valve | 4Cr10Si2Mo | Vertical spindle barrel finishing | 10 |
| 11 | Inner spring of valve | steel wire 4-10/50CrVA-2Y | Centrifugal barrel finishing | 10 |
| 12 | Outer spring of valve | steel wire 4-10/50CrVA-2Y | Centrifugal barrel finishing | 10 |
| 13 | Bump spur gear | 45 steel | Intermeshing spindle barrel finishing | 12 |
| 14 | Crankshaft timing gear | 45 steel | Intermeshing spindle barrel finishing | 12 |
| 15 | Camshaft gear | 45 steel | Intermeshing spindle barrel finishing | 12 |
| 16 | Timing idler gear | 45 steel | Intermeshing spindle barrel finishing | 12 |
| 17 | Idler gear | 40Cr | Intermeshing spindle barrel finishing | 12 |

prolonged. The surface texture of cylinder sleeve, crankshaft, and camshaft is P-shaped, and it is more suitable for oil storing than ring texture. The service life of valve springs increases by an average of 40% [58].

Deburring and improving edge quality can improve the suitability of frication pairs' parts and the engine cleanliness degree. Improving surface quality can enhance matching, sealing, and force transferring between parts with frication pairs;

**Table 2.36** Comparison surface integrity indexes of crankshaft before and after finishing

| Finishing state | *Ra*/μm | | Micro-hardness/HRC | | Tangential stress/MPa | | Axial stress/MPa | | Measured diameter basic size/mm | |
|---|---|---|---|---|---|---|---|---|---|---|
| | Main journal | Connecting rod journal | Main journal | Connecting rod journal | Main journal | Connecting rod journal | Main journal | Connecting rod journal | Main journal | Connecting rod journal |
| Before finishing | 0.63 | 0.67 | 377 | 367 | −293 | −149 | −171 | −423 | −0.015 | −0.008 |
| After finishing 15 min | 0.35 | 0.38 | 449 | 444 | −622 | −703 | −507 | −655 | −0.017 | −0.012 |

**Table 2.37** Comparison surface integrity indexes of camshaft before and after finishing

| Finishing state | *Ra*/μm | Micro-hardness/HV | Tangential stress/MPa | | | Axial stress/MPa | | |
|---|---|---|---|---|---|---|---|---|
| | | | Top | Side | Bottom | Top | Side | Bottom |
| Before finishing | 1.82 | 357 | −289 | +51 | −20 | −454 | −365 | −238 |
| After finishing 15 min | 0.95 | 393 | −611 | −524 | −714 | −899 | −898 | −775 |

**Table 2.38** Varieties of main dimensions and *Ra* of piston before and after finishing

| Finishing state | Ra/μm | Actual average size of tapered portion/mm | Actual average size of cylinder portion/mm |
|---|---|---|---|
| Before finishing | 1.6 | 79.890 | 79.325 |
| After finishing 2 min | 0.4 | 79.886 | 79.320 |

**Table 2.39** Surface integrity comparisons of idler gear before and after finishing

| Finishing state | Ra/μm | | Micro-hardness of the end face/HV | Stress of the teeth surface/MPa |
|---|---|---|---|---|
| | Tooth surface | End face | | |
| Before finishing | 0.71 | 0.42 | 525 | 33.82 |
| After finishing 12 min | 0.38 | 0.25 | 561 | −589.40 |

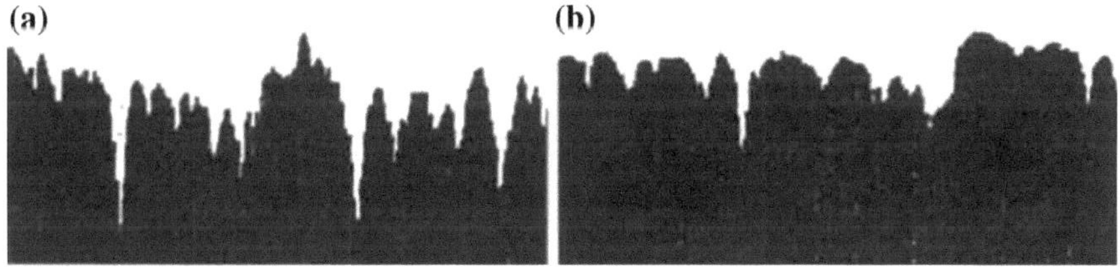


**Fig. 2.132** Comparison photographs of surface texture of cylinder liner. **a** Before finishing; **b** after finishing

therefore, its initial abrasion can be decreased. Service life of parts and engines can be prolonged by improving physical–mechanical properties.

(2) Contrast testing of the engine performance

The mainly contrast of the engine performance is the contrast of running-in indexes in the factory. The normal standard is obtained by experience and analogy method based on the machining process, the condition of part quality, etc., which is shown in Table 2.40. According to the normal standard, oil samples of the engine

**Table 2.40** Normal running-in standards

| No. | Rotational speed/(r/min) | Power/kW | Dynamometer reading/N | Running time/min | Accumulated running time/min |
|---|---|---|---|---|---|
| I | 700–2000 | - | - | 20 | 20 |
| II | 2200 | - | 32.5 | 35 | 55 |
| III | 2400 | 15 | 57.5 | 40 | 95 |
| IV | 2600 | 21 | 80.8 | 30 | 125 |
| V | 3000 | 26 | 92.3 | 15 | 140 |
| VI | 850 | - | - | 10 | 150 |

(No. 1 is the engine with unfinished parts, and No. 2 is the engine with finished parts) are extracted to take spectral analysis, and the fuel consumption rate is tested at regular intervals under various conditions. The principle of oil sample spectral analysis is that the wearing of frication pairs' parts becomes stable and normal when they are well-grounded, so the content of metal elements suspended in the engine oil becomes stable. The FAS-2C spectral analysis instrument is used in the experiment, and 20 ml oil samples are extracted to take spectral analysis at regular intervals when the engine runs in every operating mode, and the running-in quality is analyzed based on rules of elements' percentage in oil samples. The contents of 12 elements are separately tested, and contents of Fe, Al, and Cu are main factors that reflect the engine running-in performance (The content of Fe element is the most obvious, and other elements are essential elements of the fuel and its additive.). Table 2.41 is main experimental data on the normal standard [61, 62].

The contrast curves of Fe element content by spectral analysis are shown in Fig. 2.133.

Besides, the surface roughness *Ra* of friction pairs' unfinished/finished parts is tested before and after running-in, and contrast data are shown in Table 2.42.

Experimental results mentioned above show that ① in no-load and low-load running-in periods, Fe element content of oil sample in No. 2 is fewer by 18–35% than that in No. 1. The reason is that burrs are entirely removed, the surface roughness decreases, and the micro-peak wearing is light. These can indicate that the finishing process improves the engine cleanliness. ② In high-load period, Fe element content in No. 2 is basically stable, and this shows that the running-in time can be properly shorten. ③ The factory running-in fuel consumption in No. 2 reduces by around 30% more than that in No. 1, and the fuel consumption rate decreases by 3–4 g/kW h under the condition of 26 kW and 3000 r/min. ④ Fe element content in No. 2 decreases during continuous sampling under VI conditions, and its main reason is the obvious vibration behavior of the engine [63].

Through above analysis of testing results, there is an "extra running-in time" in the original running-in standards. New running-in standards after optimization are shown in Table 2.43. The main experimental data of the engine with unfinished parts under new running-in standards are shown in Table 2.44. The main experimental data of the engine with finished parts under new running-in standards are

**Table 2.41** Main experimental data on the normal standard

| Running-in conditions | | | | Accumulated running time/min | Fe element content/ppm | | Fuel consumption rate/(g/kW h) | |
|---|---|---|---|---|---|---|---|---|
| No. | *N*/(r/min) | *P*/kW | *Me*/N·m | | №1 | №2 | №1 | №2 |
| I | 0 | - | 0 | 0 | 2 | 4 | - | - |
| | 700 | - | 0 | 5 | 16 | 12 | - | - |
| | 1000 | - | 0 | 10 | 20 | 15 | - | - |
| | 1500 | - | 0 | 15 | 25 | 19 | - | - |
| | 2000 | 8 | 0 | 20 | 25 | 24 | - | - |
| II | 2200 | - | 32.5 | 25 | 31 | 31 | - | - |
| | | | | 30 | 33 | 34 | - | - |
| | | | | 35 | 36 | 34 | - | - |
| | | | | 40 | 36 | 35 | - | - |
| III | 2400 | 15 | 57.7 | 48 | 39 | 36 | - | - |
| | | | | 56 | 39 | 37 | - | - |
| | | | | 64 | 42 | 37 | - | - |
| | | | | 72 | 41 | 37 | - | - |
| | | | | 80 | 40 | 36 | - | - |
| IV | 2600 | 21 | 80.8 | 88 | 40 | 35 | 276.23 | 255.07 |
| | | | | 96 | 39 | 38 | 276.35 | 254.88 |
| | | | | 104 | 42 | 40 | 274.88 | 253.89 |
| | | | | 112 | 42 | 40 | 274.88 | 252.17 |
| | | | | 120 | 40 | 40 | 274.14 | 251.28 |
| V | 3000 | 26 | 92.3 | 125 | 45 | 43 | 272.42 | 273.69 |
| | | | | 130 | 49 | 44 | 320.86 | 273.48 |
| | | | | 135 | 47 | 43 | 317.19 | 270.45 |
| VI | 600(667) | - | 0 | 140 | 47 | 43 | - | - |
| | | | | 145 | 47 | 38 | - | - |
| | | | | 150 | 46 | 38 | - | - |

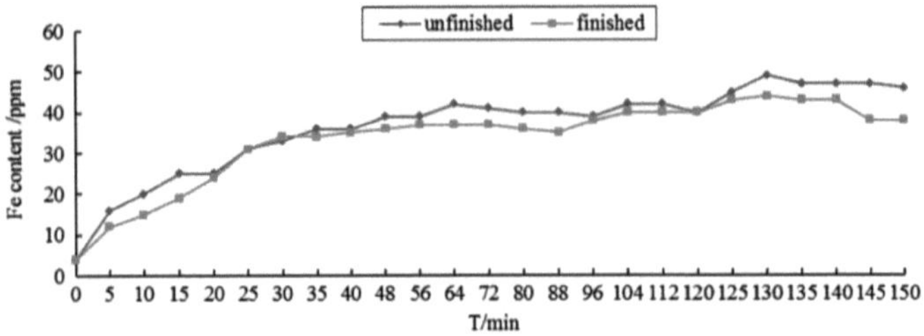


**Fig. 2.133** Contrast curves of Fe element content

**Table 2.42** Surface roughness comparisons of typical parts before and after running-in

| Parts | | Cylinder liner | Rocker shaft | Piston pin | Camshaft | Crankshaft | | Intake valve | |
|---|---|---|---|---|---|---|---|---|---|
| | | | | | | Main journal | Connecting rod journal | Journal | End surface |
| Before running-in | Unfinished | 0.34 | 0.6 | 0.2 | 0.261 | 0.826 | 1.235 | 0.595 | 8.125 |
| | finished | 0.345 | 0.3 | 0.1025 | 0.126 | 0.66 | 0.875 | 0.25 | 7.5 |
| After running-in | Unfinished | 0.33 | 0.4 | 0.145 | 0.235 | 0.72 | 1.05 | 0.57 | 8.125 |
| | finished | 0.305 | 0.225 | 0.094 | 0.115 | 0.63 | 0.6525 | 0.245 | 7.45 |

**Table 2.43** New running-in standards

| No. | Rotational speed/(r/min) | Power/kW | Dynamometer reading/N | Running time/min | Accumulated running time/min |
|---|---|---|---|---|---|
| I | 700–2000 | - | - | 20 | 20 |
| II | 2200 | - | - | 35 | 55 |
| III | 2400 | 15 | 62.5 | 40 | 95 |
| IV | 2600 | 21 | 80.7 | 30 | 125 |
| V | 3000 | 26 | 86.7 | 15 | 140 |

**Table 2.44** Main experimental data of the engine with unfinished parts under new running-in standards

| Running-in conditions | | | | Accumulated running time/min | Fe element content /ppm | Fuel consumption rate/(g/kW h) | Air leakage/ (L/h) |
|---|---|---|---|---|---|---|---|
| No. | *N*/ (r/min) | *P*/ kW | *Me*/ N·m | | | | |
| I | 0 | - | 0 | 0 | 3 | 0 | - |
| | 700 | - | 0 | 3 | 15 | - | 11.1–11.8 |
| | 1500 | - | 0 | 6 | 20 | - | 13.5–13.6 |
| | 2000 | 8 | 40 | 10 | 25 | - | 18.1–19.2 |
| II | 2400 | 15 | 62.5 | 15 | 33 | 382.03 | 14.6–15.0 |
| | | | | 20 | 36 | 377.50 | 11.8–12.4 |
| | | | | 25 | 37 | 369.99 | 13.5–14.0 |

(continued)

**Table 2.44** (continued)

| Running-in conditions | | | | Accumulated running time/min | Fe element content /ppm | Fuel consumption rate/(g/kW h) | Air leakage/ (L/h) |
|---|---|---|---|---|---|---|---|
| No. | *N*/ (r/min) | *P*/ kW | *Me*/ N·m | | | | |
| III | 2600 | 21 | 80.7 | 30 | 40 | 254.05 | 18.3–18.7 |
| | | | | 35 | 40 | 253.22 | 16.1–16.5 |
| | | | | 40 | 42 | 252.04 | 16.2–19.7 |
| IV | 3000 | 26 | 86.7 | 45 | 45 | 255.62 | 19.7–20.2 |
| | | | | 50 | 47 | 252.84 | 19.4–20.1 |
| | | | | 55 | 49 | 251.84 | 19.0–19.7 |
| V | 600 | - | - | 58 | 48 | - | 11.2–12 |
| | | | | 60 | 44 | - | 11.8–12.5 |

shown in Tables 2.45 and 2.46. The contrast curves of Fe element content by spectral analysis under new running-in standards are shown in Fig. 2.134.

The data of air leakage in Tables 2.44 and 2.46 is measured by leakage monitor produced by Changsha test instrument factory. The basic principle of leakage

**Table 2.45** New running-in standards

| No. | Rotational speed/(r/min) | Power/kW | Dynamometer reading/N | Running time/min | Accumulated running time/min |
|---|---|---|---|---|---|
| I | 1000–2000 | 0 | 0 | 10 | 10 |
| II | 2400 | 15 | 62.5 | 15 | 25 |
| III | 3000 | 26 | 86.7 | 15 | 40 |
| IV | 850 | 0 | 0 | 5 | 45 |

**Table 2.46** Main experimental data of the engine with finished parts under new running in standards

| Running-in conditions | | | | Accumulated running time/min | Fe element content /ppm | Fuel consumption rate /(g/kW·h) | Air leakage/ (L/h) |
|---|---|---|---|---|---|---|---|
| No. | *N*/ (r/min) | *P*/ kW | *Me*/ N m | | | | |
| I | 0 | – | 0 | 0 | 3 | 0 | – |
| | 700 | – | 0 | 3 | 18 | – | 21.6–22.3 |
| | 1500 | – | 0 | 6 | 19 | – | 18.4–19.0 |
| | 2000 | 8 | 40 | 10 | 20 | – | 14.8–15.3 |
| II | 2400 | 15 | 62.5 | 15 | 29 | 279.12 | 12.6–13.3 |
| | | | | 20 | 31 | 278.08 | 12.0–12.5 |
| | | | | 25 | 34 | 275.27 | 12.4–12.6 |
| III | 3000 | 26 | 86.7 | 30 | 38 | 249.58 | – |
| | | | | 35 | 38 | 247.18 | – |
| | | | | 40 | 38 | 246.43 | – |
| IV | 600 | – | – | 45 | 38 | – | – |

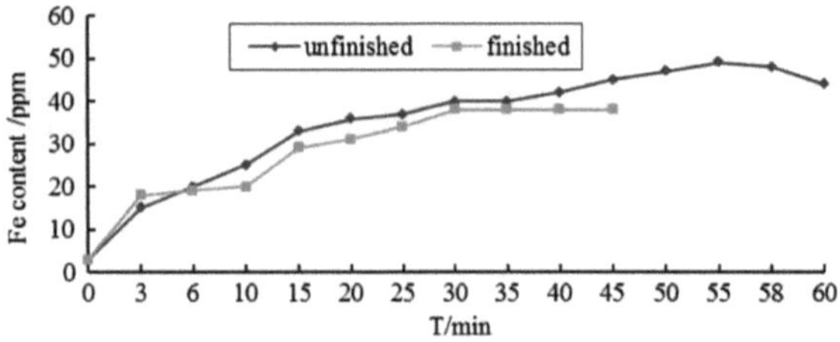


**Fig. 2.134** Contrast curves of Fe element content under new running-in standards

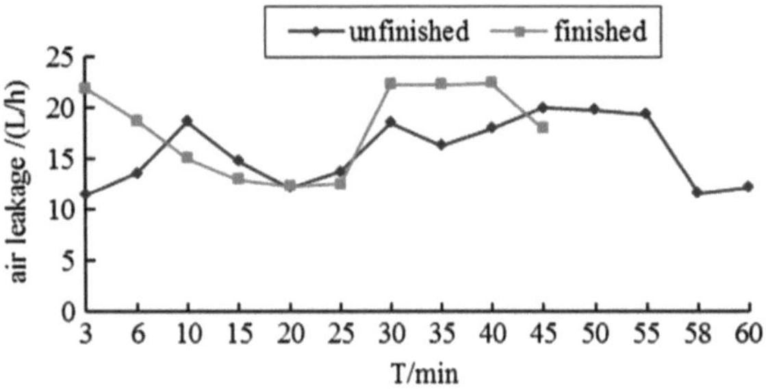


**Fig. 2.135** Contrast curves of air leakage under the new running-in standards

analysis is as follows: The crankshaft case leakage and the cylinder compression pressure tend to be stable after well running-in of friction pairs parts and good seal of cylinder sleeves, piston, and piston rings. The contrast curves of air leakage under new running-in standards are shown in Fig. 2.135. Compared with the engine with unfinished parts, the leakage of the engine with finished parts can achieve a relatively stable value within a very short time and the value is relatively small. When the engine starts, the leakage with finished parts is high because of edge filleting (lead to the weak seal between the piston edge and cylinder liner).

The synthetic analysis indicates that ① statistics data shows that the fuel consumption rate of the engine with finished parts decreases 3–4 g/kW h more than those with unfinished parts under the conditions of 26 kW and 3000 r/min. ② There is 15 min "extra running-in time." This means that running-in time of the engine with unfinished parts can decrease from 60 to 45 min, and running-in time of the engine with finished parts can decrease from 45 to 30 min. ③ Running-in curve is getting closer to that ideal. Besides, it is possible to decrease running-in time, and it provides experimental basis to achieve international running-in standards.

The comparisons of external characteristics for two 4D30YB engines are shown in Fig. 2.136. Solid lines $1^{\#}$ in the figure represents the performance testing results of the engine with unfinished parts running 20 h under the usual experience over 2.5–7 h running-in standards. Dotted lines $2^{\#}$ in the figure represents the performance testing results of the engine with finished parts over 1–4 h running-in

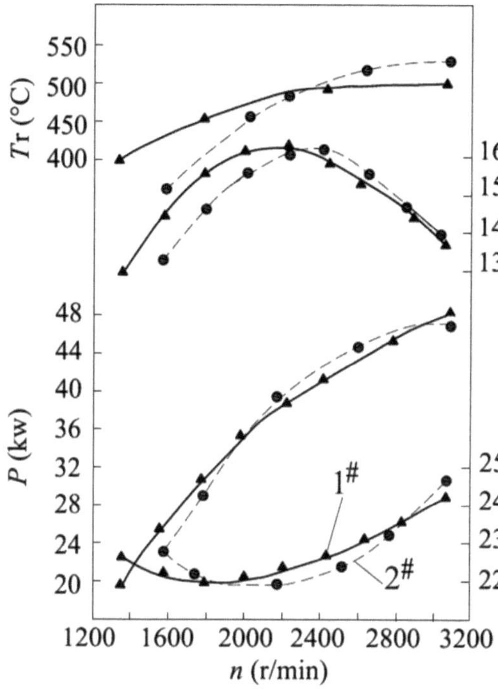


**Fig. 2.136** Comparisons of external characteristics for 4D30YB engines

standards. It can be shown that the power and the torque of the engine with finished parts increase, the heavy oil consumption rate decreases, and the exhaust temperature rises when rotational speed is greater than 2000 r/min, which indicates that the power and economy performance of the engine with finished parts have all raised.

Experimental analysis comprehensively shows that running-in time of the engine with finished parts can be reduced by around 50%, and this can save a large number of diesel oil, man-hour, water and electricity fee, etc., and produce huge economy and society benefits; the abnormal cases, such as scuffing of cylinder bore, do not occur after disassemble testing; the noise of the engine with finished parts decreases more than 1 dB.

Finishing processes lead to improving surface quality and surface integrity for frication pairs' parts with the action of colliding, extruding, and the minim grinding (scraping and scoring). Therefore, the fatigue strength, the wear resistance, and the corrosion resistance are improved. Finally, the engine cleanness is increased, the valid power is developed, the fuel consumption is decreased, the running-in time is shortened, and the service life is prolonged. Barrel finishing for friction pairs' parts is one of the effective ways to improve the engine performance.

# References


1. S.C. Yang, M.Z. Wang, Y.X. Zhang, *Surface Quality and Finishing* (Machinery Industry Press, Beijing, 2000)
2. W.H. Li, *Surface Barrel Finishing Technology Theory and Practical Study* (Taiyuan University of Technology, Taiyuan, 2009)
3. W.H. Li, S.C. Yang, S.Q. Yang. Experimental research of affecting main performance of the engine by surface finishing technology[G]. Proceedings of the International Conference on Surface Finishing Technology and Surface Engineering, Frontiers of Design and Manufacturing, 2006
4. S.Q. Yang, W.H. Li, S.C. Yang, Experimental research on improving performance of the engine by surface finishing technology. Trans. Chin. Soc. Agric. Mach. **37**(2), 17–20 (2006)
5. A.F. Kenton, Back to the basis of surface finishing. Met. Finish. **105**(10), 84–85 (2007)
6. L. Li, Barrel finishing technology and process. For. Mach. Woodwork. Equip. **32**(2), 46–48 (2004)
7. S.C. Yang, H.F. Li, The application and development of our precision surface finishing and deburring technology. New Technol. New Process **3**, 35–36 (1995)
8. S.C. Yang, M.Z. Wang, S.Q. Yang, Surface quality and barrel finishing technology. Shan Dong Intern. Combust. Engine **1**, 14–16 (2001)
9. S.C. Yang, S.Q. Yang, The development direction of the free abrasive finishing technology. Machinist Met. Cutting **1**, 11–12 (2002)
10. S.Q. Yang, S.C. Yang, G.L. Liu, Application examples of barrel finishing technology. Mach. Manufact. Eng. **11**, 21–23 (1999)
11. W.H. Li, S.Q. Yang, S.C. Yang, Theoretic analysis and experimental research on barrel finishing uniformity of crank shafts with larger size. Key Eng. Mater. **359–360**, 394–398 (2008)
12. W.H. Li, S.Q. Yang, S.C. Yang, Theoretic analysis and simulation on horizontal spindle barrel finishing. Key Eng. Mater. **416**, 332–336 (2009)
13. W.H. Li, S.Q. Yang, S.C. Yang, Technique and theoretical analysis for the finishing of large discal parts. Comput. Appl. Technol. **29**(2–4), 168–172 (2007)
14. W.H. Li, S.Q. Yang, S.C. Yang, Technique project and theoretic research on surface finishing to the wheel hub. Key Eng. Mater. **375–376**, 385–389 (2008)
15. 研磨布紙加工技術研究会. 新しい研磨技術[M]. ォーム社, 1992
16. 田中義信, 津和秀夫, 井川直哉. 精密工作法[M]. 共立出版, 1995
17. 木下直治, 高哲孝哉. 精度設計と部品仕上システム技術[M]. 日經技術圖書株式會社, 1990
18. A. Boschetto, A. Ruggiero, F. Veniali, Deburring of sheet metal by barrel finishing. Key Eng. Mater. **344**, 193–200 (2007)
19. C.X. Long, An acting force analysis on vibrating tumbling machine. Bearing **1**, 27–32 (1989)
20. S. Wang, R.S. Timsit, J.K. Spelt, Experimental investigation of vibratory finishing of aluminum. Wear **243**, 147–156 (2000)
21. M.R. Baghbanan, A. Yabuki, R.S. Timsit et al., Tribological behavior of aluminum alloys in a vibratory finishing process. Wear **255**(7–12), 1369–1379 (2003)
22. J. Domblesky, R. Evans, V. Cariapa, Material removal model for vibratory finishing. Int. J. Prod. Res. **42**(5), 1029–1041 (2004)
23. D. Ciampini, M. Papini, J.K. Spelt, Impact velocity measurement of media in a vibratory finisher. J. Mater. Process. Technol. **183**, 347–357 (2007)
24. David A. Davidson, Mass finishing processes. Met. Finish. **105**(10), 72–83 (2007)
25. A. Yabuki, M.R. Baghbanan, J.K. Spelt, Contact forces and mechanisms in a vibratory finisher. Wear **252**, 635–643 (2002)
26. Y.Y. Guo, Y.X. Zhang, Principle analysis of gyro-finishing. Instrum. Tech. Sens. 1996(5): 17–19, 21
27. S.L. Shi, M.Z. Wang, Y.X. Zhang, Discussion cyclone barrel polishing polished crankshaft. Intern. Combust. Engines **6**, 14–17 (1994)

28. S.L. Shi, Y.X. Zhang, S.C. Yang, Experimental research of whirling barrel finishing. Instrum. Tech. Sens. **4**, 14–16 (1994)
29. S.Q. Yang, H.F. Li, D.G. Hou. Whirling barrel finishing technology and application[J].New Technology & New Process, 1995(6): 19–20
30. M.Z. Wang, Y.W. Wang, J.J. Wang, in *Analysis of Centrifugal Barrel Finishing Principle*. International Conference Proceedings of Second-precision Surface Finishing and Deburring Technology (Machinery Industry Press, Dalian, 1992)
31. H.C. Wang, Q.M. Cai, The rotary power analysis of planetary roller polishing machine. Tianjin Text. Inst. Technol. **16**(5), 35–39 (1997)
32. H.C. Wang, T.W. Chen, D.L. Zhang, et al., in *Experimental and Theoretical Study on the Factor of Productivity in the Centrifugal Polishing Process*. Proceedings of the International Conference on Surface Finishing Technology and Surface Engineering, Frontiers of Design and Manufacturing (2006)
33. M.Z. Wang, Barrel finishing process. New Technol. New Process, 1995(5): 18–19
34. P.K. Zhang, H.L. Wang, New technology of complex shaft outer surface finishing. Machinery **39**(446), 13–14 (2001)
35. H.C. Wang, H.T. Zhang, Q. Wang et al., Experiment study on the dynamic characteristics of the abrasive flow in centrifugal grinding. Key Eng. Mater. **304–305**, 369–373 (2006)
36. T.W. Chen, *Theoretical and Experimental Research of Abrasive Flow Regime Characterized by Centrifugal Grinding Process*. Tianjin Polytechnic University, Tianjin (2007)
37. D.L. Zhang, X. Fang, Motion characteristics of abrasive blocks in the barrel with FLUENT software. CFD China.com, 2004(3): 10–47
38. Z. Li, The Optimization and Application of DEM Simulation in Centrifugal Barrel Finishing Process. Taiyuan University of Technology, Taiyuan (2016)
39. W.H. Li, L. Zhang, X.H. Li, S.Q. Yang, Discrete element simulation of particle behavior in mass finishing process: a parametric study. J. Balkan Tribol. Assoc. **22**(3), 2731–2750 (2016)
40. A. Sofronas, S. Taraman, Model development and optimization of vibratory finishing process. Int. J. Prod. Res. **17**, 23–31 (1979)
41. David A. Davidson, Precision finishing processes in centrifugal barrel equipment. Met. Finish. **104**(7–8), 65–67 (2006)
42. K. Kitajima, A. Yamamoto, M. Sasai, Finishing characteristics and residual stress on workpiece surface in dry barrel finishing. Key Eng. Mater. **257–258**, 189–194 (2004)
43. S.L. Shi, M.Z. Wang, Y.X. Zhang, Centrifugal barrel finishing principle and examples. Acta Armamentarii **17**(3), 230–233 (1996)
44. S.Q. Yang, Centrifugal barrel finishing technology and selection of practical process. Grinder and Grinding, 1997(3): 62–63
45. Y.X. Zhang, Y.Y. Guo, Analysis of principle and experiment of centrifugal barrel finishing. J. Shanxi Agric. Univ. **14**(3), 320–322 (1994)
46. Mass Finishing Inc. Centrifugal barrel finishing machines. Met. Finish. **101**(11), 49 (2003)
47. David A. Davidson, Surface finishing reaches new heights. Met. Finish. **103**(3), 25–28 (2005)
48. S.Q. Yang, S.C. Yang, M.Z. Wang, The technological plan and theory analysis of skew spindle high pressure finishing. China Mech. Eng. **10**(4), 397–400 (1999)
49. S.Q. Yang, M.Z. Wang, Y.W. Wang, Experimental equipment and experimental research of skew spindle high pressure finishing. China Mech. Eng. **10**(5), 502–505 (1999)
50. S.Q. Yang, Y.W. Wang, S.C. Yang, The comprehensive improvement of skew spindle high pressure finishing to parts surface integrity. China Mech. Eng. **10**(6), 626–629 (1999)
51. W.H. Li, S.C. Yang, S.Q. Yang, Study of shaft planar barrel finishing. J. Taiyuan Univ. Technol. **33**(1), 51–53 (2002)
52. W.H. Li, S.C. Yang, S.Q. Yang, Experiment of finishing crank parts with spindle planetary barrel finishing. J. Taiyuan Univ. Technol. **33**(1), 166–168 (2002)
53. S.Q. Yang, M.Z. Wang, S.C. Yang, Crankshafts' parts of barrel finishing technology development and research. China Mech. Eng. **12**(3), 300–302 (2001)
54. W.H. Li, S.C. Yang, S.Q. Yang, Process characteristics research on horizontal spindle barrel finishing. Adv. Mater. Res. **53–54**, 15–19 (2008)

55. W.H. Li, H.L. Chen, S.Q. Yang, in *Process Characteristics and Finishing Effects Research on Horizontal Spindle Barrel Finishing*. The 4th IEEE Conference on Industrial Electronics and Applications (2009)
56. W.H. Li, S.Q. Yang, S.C. Yang, in *Development and Experimental Research of Machining Medium in Barrel Finishing*. Proceedings of the International Conference on Surface Finishing Technology and Surface Engineering, Frontiers of Design and Manufacturing (2006)
57. Y.W. Wang, S.Q. Yang, S.C. Yang et al., Effects of chip performance parameters on the effectiveness in barrel finishing. J. Taiyuan Univ. Technol. **26**(4), 46–52 (1995)
58. G.L. Liu, T. Lv, Mechanism analysis and practice of surface barrel finishing. Mach. Manuf. Eng. **4**, 13–15 (1999)
59. W.H. Li, H.L. Chen, S.Q. Yang, Surface integrity research on barrel finishing of Crankshafts' Part. Key Eng. Mater. **392–394**, 655–660 (2009)
60. GB/T 10095-2008, Cylindrical Gears-System of Accuracy. China Standards Press, Beijing (2008.9.1)
61. M. Massarsky, D.A. Davidson, Turbo-abrasive machining and finishing. Met. Finish. **95**(7), 29 (1997)
62. Z. Yue, H.B. Jiang, J.Q. Yu et al., Experimental research of techniques for few-hour test run in medium-and small-power diesel engine. Intern. Combust. Engines **1**, 25–29 (2002)
63. Z. Yue, H.B. Jiang, L.F. Min, Application finishing technology to improve diesel engine running test specification. Shandong Intern. Combust. Engine **1**, 6–8 (2001)

# Chapter 3
# Magnetic Abrasive Finishing Technology

## 3.1 Introduction

Magnetic abrasive finishing (MAF), also known as magnetic grinding, was first formally presented by the Soviet engineer Kargolow in 1938 years. Since the 1960s, many Soviet scholars have studied and popularized this technology subsequently. They mainly have done a lot of research work for the preparation process of magnetic abrasive particles (MAPs/MAGs), also known as magnetic abrasive, including applying for many invention patents, publishing a large number of papers and several monographs, and they also have hosted many conferences about magnetic abrasive finishing technology. Bulgarian researchers began to study and generalize this technology in the mid-1970s and achieved certain results. Relevant international thematic conferences have been held regularly in Bulgaria. In addition, researchers in Korea, the USA, Germany, China, Taiwan regions, and other countries have also studied the related fields of this technology or complex aspects of other process in-depth [1–4].

Our country began studied the magnetic abrasive finishing technology since the late 1980s, currently the research still in the pilot study stage and rarely used in the practical production [5]. All the time, the study range of this technology has been limited for finishing the internal and external cylindrical surface and flat. In recent years, many research institutes and universities have made progress in the studies of its application in the actual production and have obtained a number of national invention patents. They have also analyzed the preparation of magnetic abrasive finishing, the development of processing apparatus, the processing technology of complex parts, the application of permanent magnet, the composite of magnetic abrasive finishing and other processing methods, and other aspects. Last but not least, they have obtained some valuable data and important conclusions, thus established a good foundation for further studies in the future [6].

 
S. Yang and W. Li, *Surface Finishing Theory and New Technology*,
https://doi.org/10.1007/978-3-662-54133-3_3

The magnetic abrasive finishing technology can finish the cylindrical surfaces, planes, free-form surface, and all kinds of surfaces by different magnetic field generators, relative motion modes, and other methods.

### 3.1.1 Basic Meaning

Magnetic abrasive finishing is a process in which a certain amount of magnetic abrasive particles that contains magnetic properties and grinding ability filled in the reasonable machining gap between the workpiece and the magnetic poles is exerting a certain force via the magnetic field generator (using the electromagnetic or permanent magnetic field to generate a magnetic field that having a certain magnetic field intensity in the machining area), and making the workpiece and magnetic abrasive particles has a complex relative motion through a certain motion mode. Under the action of force and relative motion, the magnetic abrasive particles have the effect of micro-grinding, extrusion, abrasive wear, and electrochemical wear on the workpiece surface [7]. Thus, the geometric characteristics of the workpiece surface are changed (the surface roughness values are decreased, the surface profile support rate is increased, the burrs are removed, and so on), the physical and mechanical properties of surface layer are improved (the deteriorative layer is formed on the surface, and the stress state on the surface is improved), the surface quality of the workpiece is improved, the surface integrity of the workpiece is enhanced, the service performance, the life of workpiece, and products are increased, and the purpose of finishing workpiece surface is achieved [8].

The magnetic abrasive finishing technology can finish the cylindrical surfaces, planes, free-form surface, and all kinds of surfaces by different magnetic field generators, relative motion modes, and other methods. The processing principles are shown in Figs. 3.1, 3.2, 3.3, and 3.4.

The magnetic abrasive finishing technology can be classified by the magnetic field generate mode, the magnetic properties of finished materials, the finishing range, and other factors. The specific classification methods are shown in Fig. 3.5.

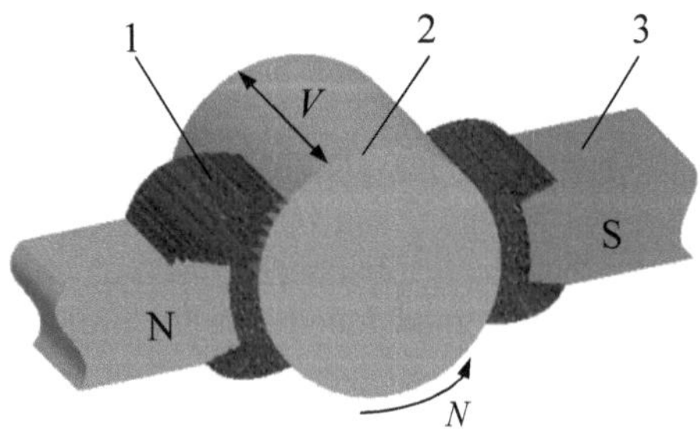


**Fig. 3.1** Schematic diagram of magnetic abrasive particle finishing the outer surface of the rotation. 1—Magnetic abrasive particles; 2—workpiece; 3—magnetic field generator

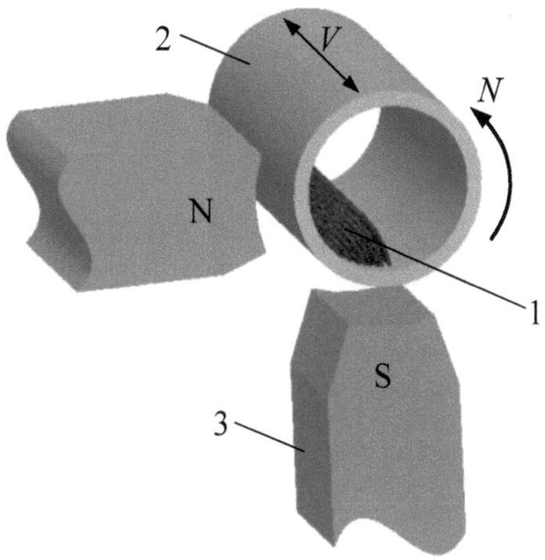


**Fig. 3.2** Schematic diagram of magnetic abrasive particle finishing the inner surface of the rotation. 1—Magnetic abrasive particles; 2—workpiece; 3—magnetic field generator

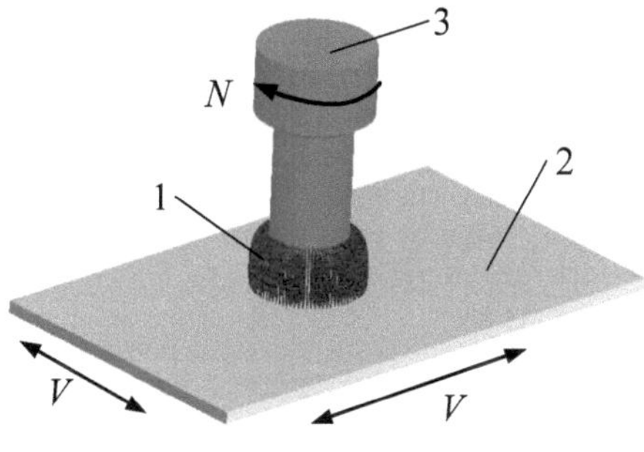


**Fig. 3.3** Schematic diagram of magnetic abrasive particle finishing the plane. 1—Magnetic abrasive particles; 2—workpiece; 3—magnetic field generator

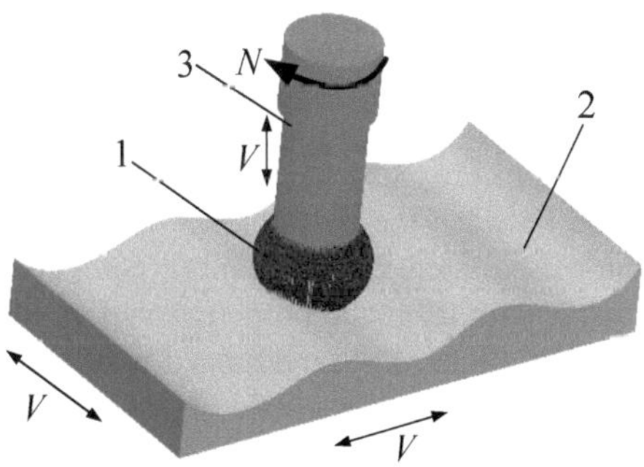


**Fig. 3.4** Schematic diagram of magnetic abrasive particle finishing the free-form surface. 1—Magnetic abrasive particles; 2—workpiece; 3—magnetic field generator

### 3.1.2 Force Analysis

The magnetic abrasive finishing is a new process to achieve workpiece surface finishing under the force generated in the magnetic field from the magnetic abrasive particles, and the relative motion between workpiece and magnetic poles [9]. As

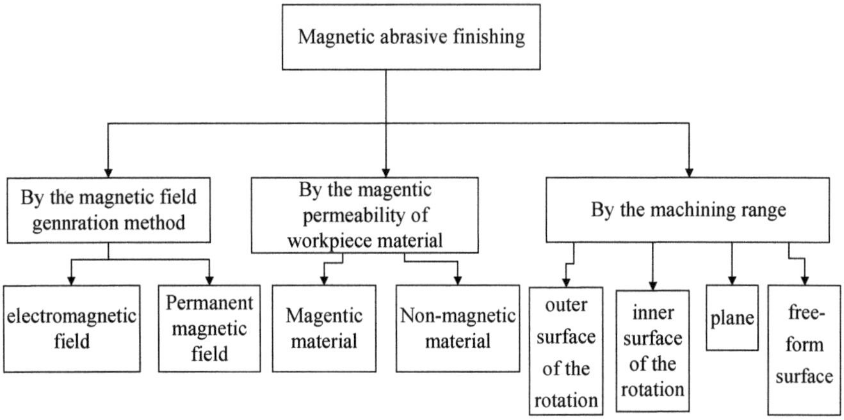


**Fig. 3.5** Classification of magnetic abrasive finishing

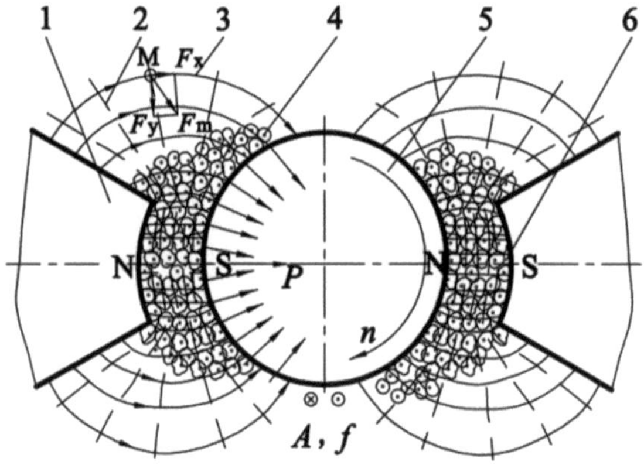


**Fig. 3.6** Sketch of force analysis of magnetic abrasive particles. 1—Magnetic poles; 2—magnetic equipotential lines; 3—magnetic force lines; 4—magnetic abrasive particles; 5—workpiece; 6—machining gap

shown in Fig. 3.6, there is a pair of magnetic poles N and S, forming a magnetic field between the two poles, filling of magnetic energy, and generating magnetic induction intensity. If a kind of complex abrasive, such as magnetic abrasive particles, with magnetic properties and grinding ability, is filled in the magnetic field, these particles will be arranged tightly and regularly under the action of magnetic force, forming a magnetic abrasive brush. Putting the finished workpiece in the magnetic abrasive brush, the force will exert on the workpiece surface in the form of pressure. When finishing the workpiece surface, the workpiece should have relative motion, which is rotational motion and axial vibration. Under the action of relative motion between the workpiece and magnetic abrasive particles, the workpiece surface is grinded and finished. Since the rough surface, burrs, edges, cusps, and other locations of the workpiece have large magnetic induction intensity causing the magnetic field lines are centralized, these parts are finished primarily to achieve the purpose of removing burrs and finishing surface [10].

1. The force analysis of single magnetic abrasive particle

Taking the finished external cylindrical surface of magnetic material workpiece as an example, the force condition of a single abrasive particle within the machining region is shown in Fig. 3.6. After the single magnetic abrasive particle within the machining region is magnetized, it is arranged in the magnetic brush along the magnetic field lines. By putting the finished workpiece in the magnetic abrasive brush, the workpiece and the magnetic abrasive particles in the front end of brush attract each other, under the action of magnetic force the magnetic abrasive particles are pressed on the workpiece surface [11]. Assuming the direction of the line of magnetic force is $X$ direction and the direction of magnetic equipotential lines that are always perpendicular to the magnetic force lines is $Y$ direction, the magnetic forces of magnetic abrasive particles at a point $M$ in the magnetic field are

$$F_X = \left(\pi D^3/6\right)\chi H \frac{\partial H}{\partial x} \tag{3.1}$$

$$F_y = \left(\pi D^3/6\right)\chi H \frac{\partial H}{\partial y} \tag{3.2}$$

The resultant force always points to the workpiece surface, so it is called the magnetic holding force, that is,

$$\vec{F}_{\mathrm{m}} = \vec{F}_X + \vec{F}_Y \tag{3.3}$$

where $D$ is the diameter of magnetic abrasive particles (μm); $\chi$ is the magnetic susceptibility of the magnetic abrasive particles; $H$ is the magnetic field intensity in the machining region (A/m); $\partial H/\partial X$ is the change rate of the magnetic field intensity along the $X$ direction; and $\partial H/\partial Y$ is the change rate of the magnetic field intensity along the $Y$ direction that is also called the gradient of magnetic field intensity.

In the finishing process, under the action of magnetic holding force, each magnetic abrasive particle of the particle group that filled in the machining gap is stably maintained in the gap and acting on the workpiece surface along the fixed direction. At the same time, under the action of magnetic holding force, these particles that are in the external of machining region will come together automatically into the machining region and fill in the gap between the magnetic poles and workpiece to process, ultimately forming a whole and continuous machining process [12].

2. The force that magnetic abrasive particles acting on workpiece surface

In the machining process, due to the action of magnetic holding force $F_{\mathrm{m}}$, each abrasive particle that is filled in the machining gap is adjoined together to constitute the magnetic cluster along the direction of magnetic force lines, thus forming the magnetic abrasive brush. The resultant force $P$ that points to the workpiece surface is generated and it is the force that the magnetic abrasive particles have been acting

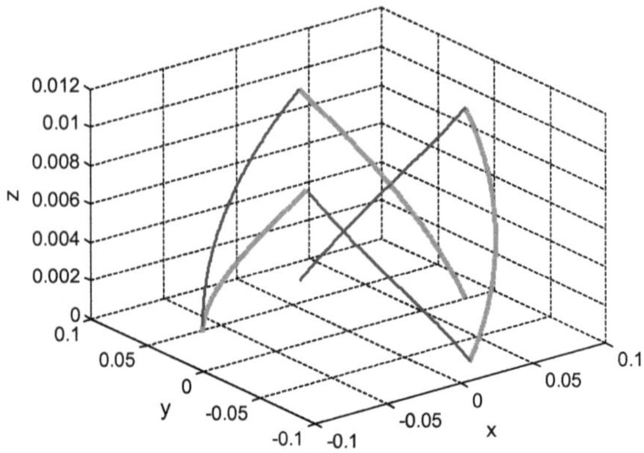


**Fig. 3.7** Simulated results

on the workpiece surface, as shown in Fig. 3.6. Assuming that the shape of single magnetic abrasive particle is spheroid, the force $P$ that magnetic abrasive particles acting on the workpiece surface when finishing the external cylindrical surface is

$$P = \frac{B^2}{4\mu_O} \times \frac{3\pi(\mu_r - 1)\omega}{3(2 + \mu_r) + \pi(\mu_r - 1)\omega} \tag{3.4}$$

where $P$ is the finishing pressure that magnetic abrasive particles acting on the workpiece surface (N/mm$^2$); $B$ is the magnetic induction intensity (T); $\mu_o$ is the magnetic permeability of air (H/m); $\mu_r$ is the relative magnetic permeability of the magnetic abrasive particles; $\omega$ is the volume percentage of the magnetic medium phase in magnetic abrasive particles (%).

Under certain conditions, by using MATLAB software to simulate the relationship among the magnetic force that applied to the magnetic abrasive particles, the magnetic induction intensity $B$, and the volume percentage of the magnetic medium phase in magnetic abrasive particles $\omega$, the results are shown in Fig. 3.7.

From Fig. 3.7, the magnetic force that has been acting on the workpiece surface is mainly related to the magnetic induction intensity $B$, and by changing the magnetic induction intensity, the force will change. Secondly, it is also related to the volume percentage of the magnetic medium phase in magnetic abrasive particles $\omega$, the number of $\omega$ fully embodying the magnetism characteristics of the magnetic abrasive particles and in the preparation of it, the $\omega$ is a major performance indicator [13].

### 3.1.3 *Trajectory of Magnetic Abrasive*

As a kind of free abrasive tool finishing, the force and the complex relative movement between the abrasive tool and the workpiece surface are two necessary conditions for finishing. For the magnetic abrasive finishing, the force generated between the abrasive particles and the workpiece surface is mainly attributed to the magnetic force, and the relative motion between the abrasive particles and the workpiece surface is mainly achieved by certain motion form of the workpiece or magnetic poles.

Taking the finished external cylindrical surface of magnetic material workpiece as an example, in order to make the magnetic abrasive finishing obtain the better finishing effects, during the finishing process, the workpiece making the rotary movement along its own axis of a certain speed *n*, and the axial vibration of a certain amplitude *A*, and frequency *f* helps to realize the complex relative movement between the abrasive particles and the workpiece surface.

The schematic diagram of the trajectory of magnetic abrasive particles is shown in Fig. 3.8. In the finishing process, the cutting trajectories that generated on the workpiece surface by the cutting edge of magnetic abrasive particles in the terminal of brush are

$$\begin{cases} x = r\cos(2\pi nt) \\ y = r\sin(2\pi nt) \\ z = \pm 2Aft \end{cases} \tag{3.5}$$

The cutting trajectories that generated on the workpiece surface and the rotation direction of workpiece constitute an angle called the cross angle; its magnitude is

$$\theta_{\mathrm{m}} = \arctan(60Af/\pi nr) \tag{3.6}$$

where $\theta_{\mathrm{m}}$ is the cross angle (rad); n is the rotation speed of workpiece (r/min); *r* is the radius of workpiece (mm); *f* is the axial vibration frequency of workpiece (Hz); *A* is the axial vibration amplitude of workpiece (mm).

Due to the flexible magnetic abrasive brush is easily deformed after the force is acted on it, the actual trajectories that generated by the cutting edge of magnetic abrasive particles on the workpiece surface are different with theoretical trajectories, and the cross angle is also changed, denoted by $\theta'_{\mathrm{m}}$ and $\theta'_{\mathrm{m}} < \theta_m$.

For the different rotation speed *n*, the axial vibration amplitude *A*, the frequency *f*, and the actual cross angle $\theta'_{\mathrm{m}}$, the trajectories that generated by the magnetic

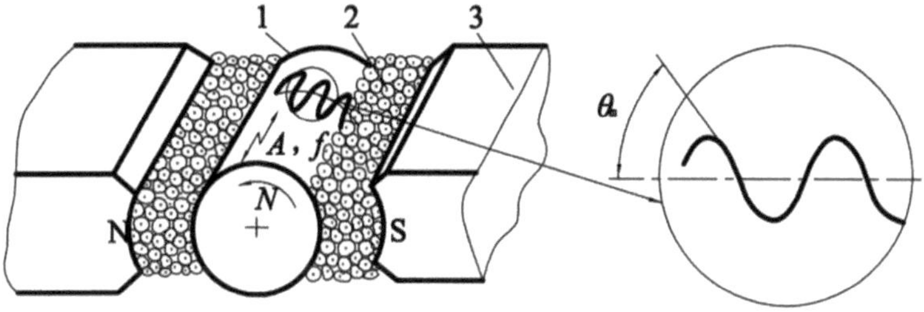


**Fig. 3.8** Motion trajectory of magnetic abrasive particles. 1—Workpiece; 2—magnetic abrasive particles; 3—magnetic pole

**Fig. 3.9** Simulated results

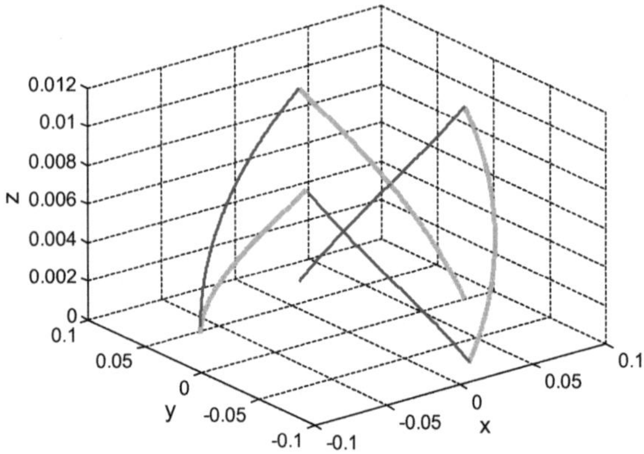


abrasive particles are different and it will affect the finishing effects and efficiency. Due to the vibration of workpiece, the movement distance of the workpiece surface in unit time is increased. At the same time, due to the movement direction of the workpiece surface is changing constantly, the cutting trajectories of workpiece surface generated by the cutting edge of magnetic abrasive particles show the intersecting and non-repetitive reticulate pattern, producing super position effect, and prompting the tumbling and stirring effects in the machining gap of magnetic abrasive particles. The machining locations of magnetic abrasive particles are changed frequently, the self-shaping of magnetic abrasive particles during the finishing process is stronger, and the finishing of the workpiece surface is fully completed.

On the condition of *n* is 600 r/min, *r* is 80 mm, *f* is 25 Hz, *A* is 10 mm, the simulation results of cutting trajectories that generated on the workpiece surface by single magnetic abrasive particle are shown in Fig. 3.9.

From Fig. 3.9, the movement condition of magnetic abrasive particles on the workpiece surface can be known. According to the cross angle formula, the ratio of the circular velocity of the rotary motion and axial vibration velocity should be within a certain range. If the ratio is too large, the texture along the circumferential direction will be more significant, resulting in the isotropic trend to increase on the machined surface and vice versa. In addition, because of the peripheral speed of the rotary motion is not only related to speed, but also to the diameter of the workpiece. Therefore, for the workpieces with different diameters, the choice of motion parameters should be difference.

### 3.1.4 Finishing Mechanism

In the finishing process of magnetic abrasive particles, the magnetic abrasive tool is a particle group that consisted of numerous tiny magnetic abrasive particles. In the magnetic field, under the action of magnetic force, they are pressed against the workpiece surface [14]. The theory of precision cutting and tribology shows that the magnetic abrasive particles have the effects of scratch, extrusion, scoring, and cutting on the workpiece surface in the finishing process to finish the workpiece surface [15–16].

1. The effect of micro-cutting and extrusion

In the finishing process of magnetic abrasive particles, the relative motion (the workpiece or magnetic poles rotation and vibration) between the magnetic abrasive particles and workpiece takes place in a certain way, and the magnetic abrasive particles have the force acting on workpiece surface. Because the hardness of abrasive phase (corundum, silicon carbide, and so on) of the magnetic abrasive particles is higher than the hardness of workpiece, the edge of magnetic abrasive particles will cut the workpiece surface, and the magnetic medium phase of magnetic abrasive particles will have a good extrusion effect on the workpiece surface [17].

The force, magnetic abrasive particles act on the workpiece surface, includes the magnetic holding force and the tangent friction force that generated by the relative motion between the workpiece and the magnetic poles [18]. The magnetic holding force makes the magnetic abrasive particles press against the workpiece surface, and the indentation and the extrusion effect are formed. Thus, the stress state of the workpiece surface is changed. The tangent friction force makes the magnetic abrasive particles move forward, and the abrasive phase likes the cutting edge of the cutter and does the micro-cutting finishing on the workpiece surface. The strength of micro-cutting is related to the shape, size, location, materials, abrasive phase, and working angle of the magnetic abrasive particles and magnetic field characteristics [19].

At the same time, due to the relative motion (rotational movement and axial vibration) between the workpiece and the magnetic poles, and the uneven distribution of the magnetic field, each magnetic abrasive particle of the "flexible magnetic brush" is constantly changing its position to finish, forming a random and discontinuous cutting process. On the other hand, since the irregularity of magnetic abrasive particles, the rake faces of cutting edge mostly have larger negative rake angle, and the cutting depth is very small under the action of the magnetic field force, it generally about 1 μm, or even smaller. Therefore, this process belongs to micro-cutting, the cutting force is small, producing very little cutting heat, the residual stress is not large and is a compressive stress, the elastic deformation region and plastic deformation region on the workpiece surface are very small, and the deteriorative layer is extremely thin. Therefore, the micro-cutting action of magnetic abrasive particles can acquire very good surface quality of the parts, and the surface roughness $R_a$ value of workpiece can reach less than 0.2 μm [20].

2. The effect of multiple plastic deformation wear

The flexible magnetic brush that formed by magnetic abrasive particles group is adsorbed on the workpiece surface by the effect of magnetic field, and the magnetic abrasive particles and workpiece surface are always in the contact state. However, the cutting process is not unique [21]. Under these four situations, first, the shape of magnetic abrasive particles is blunt; second, the arris of abrasive particles rather than the edge facing the movement direction, third, the angle between the edge of magnetic abrasive particles and workpiece is too small; fourth, the material of workpiece surface has very high plasticity, after the magnetic abrasive particles are slipping on the workpiece surface once or more than once, the surface metal will generate plastic deformation or multiple plastic deformation, when the degree of plastic deformation exceeds the plastic deformation limit that allowed by the material. The tiny swarf will fall off from the metal body and become abrasive dust. This is the interaction result of so-called abrasion type plow plane phenomena and "rolling" type rolling scrape phenomena.

Since the magnetic abrasive particles are generally concentrated in the vicinity of the tiny profile peak of surfaces that the magnetic force lines are dense during the finishing process, the asperities of the uneven surface have relatively large plastic deformation wear, so that the roughness declines accelerated. Thus, the effect of multiple plastic deformation wear of magnetic abrasive particles can quickly obtain the smooth workpiece surface.

3. The effects of friction, corrosion, and wear

In addition to the action case above, the effects that magnetic abrasive particles acting on the workpiece surface also include slip and friction phenomena and make the mental surface generate corrosive wear. The corrosive wear is a chemical effect that related to its existent environment. It depends on the environmental conditions and ambient medium at a large extent. The workpiece surface is rubbed by the magnetic abrasive particles, the bare metal surface is corroded by the environment and medium, and then an extremely thin oxide film is formed rapidly. Due to the difference between the expansion coefficients of oxide film and the workpiece material, the temperature changes in the finishing process, and other reasons, the oxide film is falling off in the subsequent slip and friction process. During the continuous finishing process, the metal on the surface layer of workpiece is oxidized and peeled off continuously, so that it improves the finishing effects and enhances the finishing efficiency.

4. The effects of electrochemical wear

Due to the rotation of the workpiece, in the finishing process, the conductive magnetic abrasive particle chain arranges along the magnetic field lines motion and deviates from the magnetic force lines, forming the magnetic field gradient. In addition, both ends of the magnetic abrasive particles chain produce a small electromotive force under the effects of it, so that the workpiece surface produces a small current. The rotating workpiece between the two poles will be got an alternating excitation effect, strengthening the electrochemical process of the metal of workpiece, so that the metal on the surface layer is worn and removed quickly, and further improving the finishing efficiency.

5. The effect of finishing medium

Adding a certain amount of finishing medium (liquid paraffin, stearic acid, silicone oil, etc.) to the magnetic abrasive particles, on the one hand, it can promote the effects of grinding and finishing, and maintaining or improving the surface quality of parts (such as gloss, color, surface roughness), thus reducing the finishing time. On the other hand, the finishing medium can reduce the cutting heat and eddy-current heat during the finishing process, ultimately reducing the impacts on the finishing quality. In addition, the finishing medium has the infiltration effects on the magnetic abrasive particles, thereby reducing the fly off magnetic abrasive particles and improving the utilization rate.

### *3.1.5 Expiration Analysis of Magnetic Abrasive*

From the processing mechanism, the magnetic abrasive finishing enhances the surface quality and improves the surface integrity of the workpiece. This enhancement is caused by micro-cutting and extrusion, friction, and wear, and the electrochemical wear effects on the workpiece surface generated by the magnetic abrasive particles. At the same time, the workpiece also has corresponding effects on the magnetic abrasive particles and makes them wear. However, the serious wear of magnetic abrasive particles can shorten the using time, deteriorate the finishing surface quality, and increase the processing costs. Thus, the wear of magnetic abrasive particles is an important factor that influences the productivity, the processing quality, and the processing costs.

The expiration of magnetic abrasive particles means that due to the wear of these particles leading to the loss of their processing ability, it mainly includes the abrasion wear failure, crushing wear failure, and exfoliation wear failure.

The abrasion wear failure means that under normal circumstances, the cutting edges of abrasive phase of magnetic abrasive particles are dull with the increasing use in time, and the diameter of abrasive phase is greatly reduced due to wear and eventually loses its processing capacity. This failure phenomenon is unavoidable, but there are certain rules to follow. Comparing with the abrasive wheel and other non-free abrasive tool, due to the magnetic abrasive finishing process, existing the non-uniform magnetic field, the vibration phenomenon, and the magnetic abrasive particles are relatively in free states, so that these particles generate the tumbling, transferring, and the changing of position constantly, which reduce the probability of abrasion wear effectively. The disadvantages are those the size of a single magnetic abrasive particle is very small, it cannot be dressed and reused like the abrasive wheel, and it is difficult to ensure that all the magnetic abrasive particles achieve abrasion wear at the same time in once processing. The abrasion wear failure is mainly caused by two factors: the mechanical friction and the thermal effects. By increasing the hardness of the abrasive phase, improving processing technology (such as selecting the suitable processing parameters), and utilizing other methods, it can delay the abrasion wear failure of magnetic abrasive particles and improve the service life.

The crushing wear failure means that due to the improper using of manufacturing processor under the influence of external force, the magnetic abrasive particles can exceed their ultimate strength and fatigue strength and therefore can be smashed. The exfoliation wear failure means that due to the manufacturing process itself or the influence of external force, the magnetic abrasive particles can exceed the bonding strength of adhesives and can result in exfoliation. The crushing wear failure and the exfoliation wear failure are both non-normal failures, which include the smashing of magnetic abrasive particles, the abrasive phase, the separation of abrasive phase, and the magnetic medium phase.

Since the magnetic abrasive particles are the composites that are composed of several different particles, by the influence of the physical compatibility and chemical compatibility between the various components of the composite materials (such as the thermal expansion coefficient, the bonding conditions of the interface, the chemical potential of each element, and the surface energy of multi-component mixture) and other factors, the bonding strength between the magnetic medium phases and among the magnetic medium phase, abrasive phase, and adhesives is relatively weaken (comparing with a single particle). When the bonding strength is reduced to a certain extent, the crush wear becomes serious. Through the manufacturing process (choosing the processing parameters reasonably, improving the crushing process, and so on), enhancing the bonding strength, the toughness of the abrasive phase itself, the service environment, and other measures effectively helps reduce the crushing wear failure probability of magnetic abrasive particles.

The above several abrasion phenomena may occur simultaneously, it is a stochastic process. For the abrasion wear, the wear loss is gradually expanding with the increase in processing time, and it can be divided into the early wear stage, normal wear stage, and rapid wear stage. For the crushing wear and exfoliation wear, they can be analyzed by probability theory.

In addition, the magnetic abrasive particles may also adsorb a certain amount of wear debris, thus affecting the micro-grinding effect that the abrasive phase of magnetic abrasive particles acting on the workpiece, and reducing the processing results. When the effect has a certain influence, the quality of machined surface is obviously deteriorating, and the magnetic abrasive particles are invalid.

Since the magnetic abrasive finishing is a precision surface finishing technology, the failure of the magnetic abrasive particles can be measured like the wear standards of ordinary tool and be determined by the processing time $T$. The finishing time can be used by magnetic abrasive particles under the premise of ensuring the processing quality.

## 3.2 Magnetic Abrasive

Magnetic abrasive particles are the grinding tool for magnetic abrasive finishing [22]. During the finishing process, magnetic abrasive particles are filled in the machining gap between the workpiece and the magnetic poles. Under the action of magnetic force, the particles are arranged in a magnetic brush along the magnetic field lines to achieve the finishing. Therefore, magnetic abrasive particles not only have the grinding ability, but must also have positive magnetic properties and the properties of water resistance, oil resistance, and other as well [23–24].

### *3.2.1 Composition*

Magnetic abrasive particles are iron-based compound abrasives that have magnetization and grinding ability. Their sizes commonly are 60–200 μm, and they are mainly composed of the magnetic medium phase (usually iron matrix), abrasive phase (abrasive), and adhesive. The structure and grain sizes are shown in Fig. 3.10.

1. Magnetic medium phase

Magnetic medium phase is an important part of the magnetic abrasive particles, and it directly affects the acting force that magnetic abrasive particles exert on workpiece during the magnetic abrasive finishing. In order to ensure that the magnetic abrasive particles have higher shear stress, strong controllability, and good stability, the magnetize materials that have good electrical conductivity are usually chosen [25], which are commonly iron, steel, aluminum–nickel alloy, barium ferrite, magnesium–barium alloy, and others.

2. Abrasive phase

Abrasive phase is a kind of high-hardness abrasive which has the action of grinding, lapping, and polishing. Its composition and ratio have directly impacted

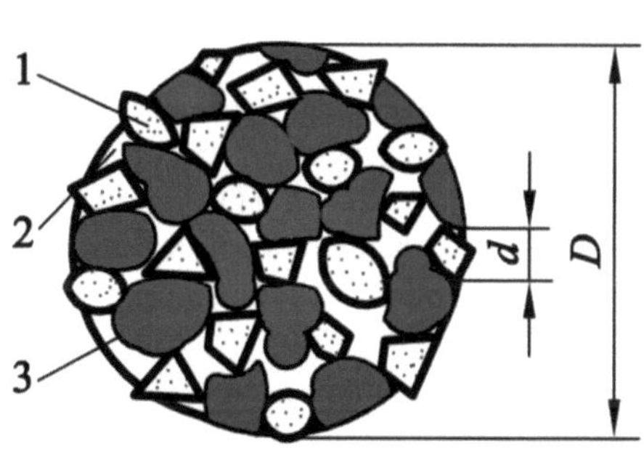


**Fig. 3.10** Structure of magnetic abrasive particles. 1—Magnetic medium phase; 2—adhesive; 3—abrasive phase

the finishing effects and efficiency. It is usually including corundum, silicon carbide, diamond, cubic boron nitride, and others [26].

The basic requirements for abrasive phase are as follow:

(1) It should have higher hardness than the material of workpiece, and its Vickers hardness is usually 1.2–1.5 times of workpiece's.
(2) It should have certain strength and cannot be easily broken, come off, or wear, under the action of external force and in the process of cutting into the workpiece surface [27].
(3) It should have stable physical and chemical properties. Under the action of vortex heat and grinding heat, it is not softened or chemically decomposed and does not react chemically with other materials.

3. Adhesive

The role of adhesive is to bind two components together, making the abrasive phase uniformly distributed in the magnetic medium phase around, and cannot fall off and be separated in the process of smashing and finishing. Thus, the performances of adhesive are the key factors for preparation, and it directly affects the properties of magnetic abrasive particles. The common adhesive includes epoxy resin, polyamide resin, furan resin, metal filling glue, high-temperature-resistant structural adhesive, and others.

The basic requirements for adhesive are as follow:

(1) The adhesive can bind well with the magnetic medium phase and abrasive phase.
(2) The coating of adhesive should be hard and has a certain elasticity. In order to ensure the magnetic abrasive particles have good grinding performance, the adhesive should have certain hardness and strength after solidifying and drying. In order to ensure the magnetic abrasive particles have certain strength, the adhesive must also have certain elasticity.
(3) The adhesives should have good heat resistance. During finishing, magnetic abrasive particles are continuously producing cutting heat, and the vortex heat of the magnetic field generator is also transmitting to the magnetic abrasive particles through magnetic poles, which makes the transient temperature during grinding process become even higher. Using the surface thermocouple to measure, at the range of grinding, the surface temperature of workpiece is usually lower than 120 °C. In order for the adhesive to not become soften or decomposed, the temperature resistance should be higher than 200 °C.
(4) The adhesive should have stable physical and chemical properties. During the finishing process, the magnetic abrasive particles are sometime necessary to add a certain amount of additive, which required the adhesive not to react chemically with the ingredients of additive so as not to affect the bonding strength of the magnetic abrasive particles.

### 3.2.2 Classification and Model

The properties of magnetic abrasive particles are mainly decided by abrasive phase, particle size, ratio, adhesive, and other factors [28]. In order to distinguish, use, and manage the magnetic abrasive particles, it needs to be classified and typed.

1. Abrasive phase

The abrasive phase is a main ingredient of magnetic abrasive particles and also known as abrasives. It plays an important role in grinding process. The code name and application ranges of abrasive that are commonly used are shown in Table 3.1.

2. Particle size

The particle size of magnetic abrasive particles includes two parameters: *D* represents the grain diameter of magnetic abrasive particles and *d* represents the grain diameter of abrasive phase, and they have a great influence on the finishing performance of magnetic abrasive particles. According to the theoretical analysis and a lot of test results, the types and code name of particle size of magnetic abrasive particles that commonly used are shown in Table 3.2.

**Table 3.1** Common types of abrasive

| Category | Name | Code | Remarks |
|---|---|---|---|
| Corundum | Brown corundum | A | Processing carbon steel, alloy steel, malleable cast iron, hard bronze, etc. |
| | White corundum | WA | Processing hardened steel, high-speed steel, high-carbon steel, etc. |
| Silicon carbide | Black silicon carbide | C | Processing cast iron, brass, aluminum, nonmetal, etc. |
| Silicon carbide | Green silicon carbide | GC | Processing cemented carbide, gem, ceramic, glass, etc. |
| High hard classes | Diamond | MBDRVD JR | Processing cemented carbide, gem, ceramic, and other hard and brittle materials |
| | Cubic boron nitride | CBN | Processing stainless steel, high-carbon steel, and other materials that difficult to process |

**Table 3.2** Common types and code name of particle size of magnetic abrasive particles

| | | | | | | | |
|---|---|---|---|---|---|---|---|
| D | Mesh | 40 | 60 | 80 | 100 | 120 | 140 |
| | Code | 1 | 2 | 3 | 4 | 5 | 6 |
| d | Mesh | 325 | 400 | 425 | 500 | 1000 | W5 |
| | Code | 1 | 2 | 3 | 4 | 5 | 6 |

3. Ratio

The ratio of magnetic abrasive particles mainly refers to the mass ratio of the magnetic medium phase and abrasive phase, and different ratio directly affects the magnetic properties and working ability of magnetic abrasive particles. The code name and ratio that commonly used are shown in Table 3.3.

4. Adhesive

The adhesives directly affect the binding strength, finishing effects, service life, finishing cost, and other parameters of the magnetic abrasive particles. The adhesive and its code name that are commonly used are shown in Table 3.4.

In addition, different preparation process of magnetic abrasive particles will also have a certain influence on finishing effects, finishing costs, and so on. The types and code name of preparation process that have wide application and can be used in the production are shown in Table 3.5.

According to the above classification, we can type the magnetic abrasive particles so that users can choose freely, and the specific method for type of establishments is shown in Fig. 3.11.

For example, WA41P4HZ-TG represents the magnetic abrasive particles that are researched and developed by Taiyuan University of Technology, the abrasive phase is white corundum, *D* represents the grain diameter of magnetic abrasive particles is $100^{\#}$, *d* represents the grain diameter of abrasive phase is $325^{\#}$, mass ratio is 4:1, the adhesive is epoxy resin and polyamide resin, and it is prepared by bonding method.

**Table 3.3** Common ratio and code name of magnetic abrasive particles

| Ratio | 2:1 | 3:1 | 4:1 | 5:1 |
|---|---|---|---|---|
| Code | P2 | P3 | P4 | P5 |

**Table 3.4** Common types and code name of adhesive

| Adhesive | Epoxy resin + polyamide resin | Epoxy resin + polyamide resin + furan resin | Metal filling glue | High-temperature-resistant structural adhesive | Liquid paraffin | Zinc stearate |
|---|---|---|---|---|---|---|
| Code | H | HF | JT | GJ | YL | SZ |

**Table 3.5** Common type and code name of preparation process of magnetic abrasive particles

| Preparation process | Sintering method | Vacuum sintering method | Hot-pressing sintering method | Bonding method |
|---|---|---|---|---|
| Code | S | ZS | RS | Z |

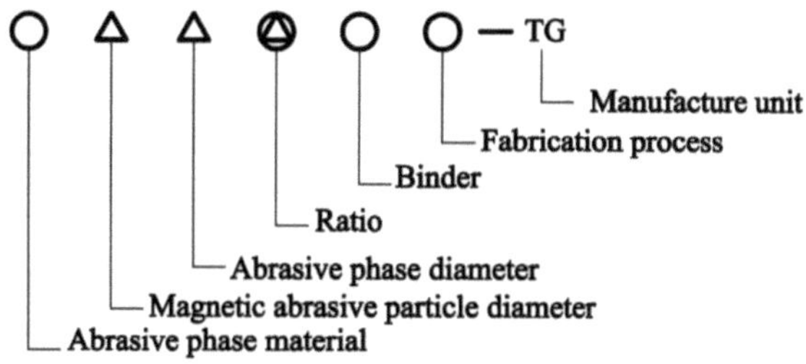


**Fig. 3.11** Type establishment method of magnetic abrasive particles

### 3.2.3 *Preparation*

The magnetic abrasive particles are kinds of enhanced composite materials which are iron-based particles, and their preparation is crucial to magnetic abrasive finishing [29]. With the rapid development of composite materials science, preparation process of magnetic abrasive particles has been developed correspondingly. However, due to the complexity of its preparation process, the complete preparation process is far from being completed, which makes the application of magnetic abrasive finishing process limit to some extent. At present, there are three preparation processes that have many applications, mainly including bonding method, hot-pressing sintering method, and vacuum sintering method, in which the "magnetic grinding tool and its preparation" (ZL00 1 35667.4.2003) that based on bonding process and "hot-pressing sintering preparation for magnetic abrasive particles" (ZL00 1 36573.8.2003) have obtained national invention patent.

1. Vacuum sintering method

The vacuum sintering method preparation for magnetic abrasive particles can avoid the oxidation of pure iron powder and the oxide film that is generated on the surface of the iron particle. Meanwhile, before sintering the oxide film that forms on the surface of pure iron powder particles in each procedure can also be reduced partially or entirely during the sintering process, so that the pure iron powder particles and abrasive particles contact each other closely to achieve complete sintering. This helps to improve the strength and abrasion resistance of the magnetic abrasive particles. The preparation process is shown in Fig. 3.12.

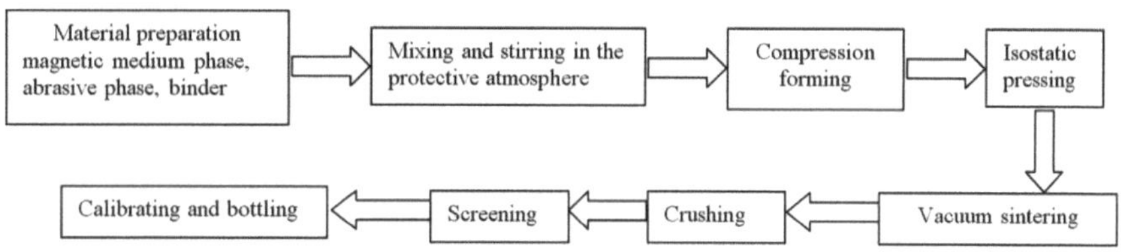


**Fig. 3.12** Vacuum sintering preparation process of magnetic abrasive particles

(1) Material preparation

The ingredients of magnetic abrasive particles mainly include magnetic medium phase (pure iron powder and its size is about 200#), abrasive grains phase (corundum, silicon carbide, its particle size is 325#–400#, 1000#), adhesive (liquid paraffin, zinc stearate).

During the blending process, the ratio and particle size have a great impact on the shaping of magnetic abrasive particles, so the particle size and the ratio of magnetic medium phase and abrasive phase should be chosen reasonably. The larger the particle size, the more uniform the powder is mixed, the more pressing and sintering the performance will be. The more content in the abrasive phase, and the more difficult to press, results more sintering process as it needs higher temperature and longer holding time.

The ingredient ratio of magnetic abrasive particles can be chosen as the mass ratio of magnetic medium phase and abrasive phase is 3:1, 4:1, the adhesive is about 1%.

(2) Mixing and stirring

The mixing and stirring device was shown in Fig. 3.13. Putting the material and adhesive into the V-shaped mixer, and by making it rotate at low speed to achieve the purpose of mixing and stirring. Mixing should be kept uniformly; therefore, not only it is beneficial for pressing and sintering, but also the distribution of abrasive in the sintered magnetic abrasive particles is more uniform [30].

(3) Press forming

After, the adhesive is added into mixed powder, according to the capacity of the molding-die loading a certain quantity of powder, using a press (around 8 MPa) to press it, and then making it vacuums packing.

(4) Isostatic pressing

Using an isostatic press (11–15 MPa) to press the pressed green body, making the particles is further densified.

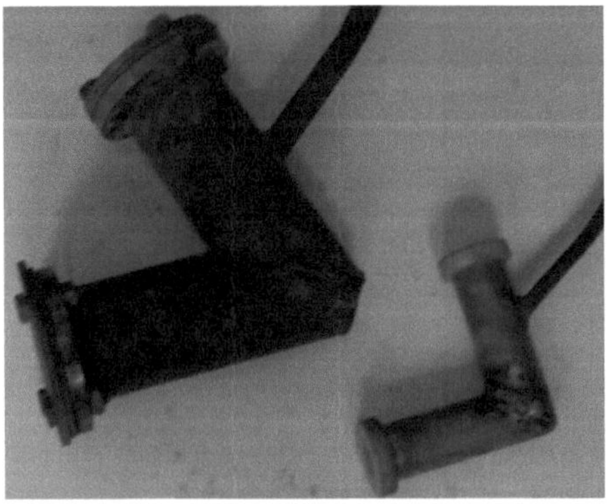

**Fig. 3.13** Photograph of mixing and stirring device

(5) Sintering

Putting the isostatic pressed green body into a groove, according to different test requirements, the green body can be placed in different positions. Due to the sintering temperature, sintering time, holding time, and cooling time, it will have a great impact on the sintering effect, so these parameters should be determined according to the specific test requirements.

(6) Smashing, screening, calibrating, numbering

After, the green body is smashed by the crusher at twice times (Fig. 3.14), using the standard screen (Fig. 3.15) screening and calibrating.

The experiment example of preparation process is as follows: Material are pure iron powder, brown corundum, adhesive A (liquid paraffin) or B (zinc stearate), sintering temperature is 1210 °C, sintering time is 5 h, and holding time is 2 h. The experiment phenomenon: After sintering, the green body is smooth, the end is light

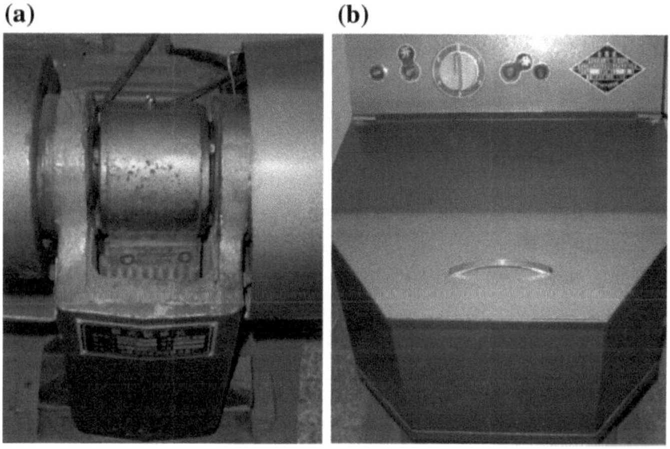

**Fig. 3.14** Photographs of crusher. **a** Rough crusher; **b** fine crusher

**Fig. 3.15** Photograph of standard screen that confirming the particles diameter *D* of magnetic abrasive particles

yellow, and the rest is silver gray; these results in the magnet to having a good absorption effect on the green body. After smashing the section that is dark black, particles distribute uniformly on its surface, and it has metal highlights and molten iron bead particles, which is caused by higher temperature. Because the green body is harder, it is more difficult to crush. Magnet has a good absorption effect on sintered magnetic abrasive particles. After using magnetic abrasive particles to finish the workpiece of the magnetic permeability for nine minutes, the surface roughness $R_a$ value of workpiece which is finished by magnetic abrasive particles has a large-size abrasive phase decreases from 1.95 to 0.25 μm. The surface roughness $R_a$ of a workpiece that finished by magnetic abrasive particle has a small size abrasive phase decreases from 1.95 to 0.18 μm, and the magnetic abrasive particle is prepared by adding liquid paraffin has better finishing effect than adding zinc stearate.

At the conditions of mass ratio of magnetic medium phase and abrasive phase is 4:1, sintering temperature is 1200 °C, sintering time is 5.5 h, holding time is 2 h, and three kinds of magnetic abrasive particles are prepared: The abrasive phase is brown corundum that has grain diameter $D$ of $1000^{\#}$, brown corundum that has $D$ of $325^{\#}$, green silicon carbide that has $D$ of $325^{\#}$, respectively. Figure 3.16 represents SEA photographs of these magnetic abrasive particles at different magnifications.

From Fig. 3.16a, b, when the abrasive phase turns into green silicon carbide, the properties of sintered magnetic abrasive particles are different due to the difference in adhesives, the magnetic abrasive particles that are prepared by adding adhesive B are finer, and have more abrasive particles in the processing and thus the finishing effects are better than the adhesive A. From Fig. 3.16c, d, when the abrasive phase is brown corundum, the magnetic abrasive particles that are prepared by adding adhesive A are finer and the finishing effects are better than the adhesive B. Figure (e) shows that when the particle diameter of abrasive phase is reduced, the formed magnetic abrasive particles are finer and it is beneficial to achieve lower values of surface roughness and improve finishing effects.

(7) Finishing effects

Using the A31P4YLZS-TG magnetic abrasive particles to do magnetic abrasive finishing experiment on the outer circle surface of workpiece (45 steel), the diameter is 30 mm, on conditions that the machining gap is 3 mm, workpiece's rotational speed is 610 r/min, axial vibration amplitude is 15 mm, and frequency is 2.3 Hz. The influence curves of magnetic abrasive particles by vacuum sintering method at different sintering temperatures on the surface roughness are shown in Fig. 3.17.

From Fig. 3.17, the magnetic abrasive particles are prepared when the holding time is the same, and the sintering temperatures 1210, 1200, 1180 °C should have better finishing effects; The magnetic abrasive particles that prepared at 1210 °C have slightly better finishing effects than at the other temperature. It is also discovered in the experiment: In the selected parameter range, the higher the sintering temperature, the longer the holding time; the fewer pores of green body, the better abrasive phase which is wrapped by the magnetic medium phase, the higher bonding strength of both, and the longer service life of magnetic abrasive particles.

According to the test results, in addition to the characteristic parameters of magnetic abrasive particles, sintering temperature, sintering time, and holding time are the key factors to determine sintering effects. In general, these factors should be chosen as sintering temperature 1180–1210 °C, sintering time 5–6 h, holding time 2–3 h. Principles that should be followed are the sintering temperature: being higher, the sintering time increasing, and the holding time should be shorten correspondingly; the sintering temperature is lower, and the sintering time and holding time can be increased correspondingly.

2. Hot-pressing sintering method

Densification is considered to be critical for improving the quality of powder metallurgy materials and products [31]. Although the method that magnetic abrasive particles are sintered after press forming, it can improve the degree of densification. It improves the properties of powder metallurgy material limitedly. Full densification process is an approach to exert pressure and temperature to eliminating pore, namely the hot-pressing sintering process.

Using the hot-pressing sintering process to prepare magnetic abrasive particles is another new approach to apply this process [32]. Since this method can make the iron particles and abrasive particles even finer and closely linked, the strength and hardness of magnetic abrasive particles are high, which can effectively prevent and reduce separation phenomenon of particles and extend the service life of magnetic abrasive particles even longer [33]. Additionally, because the magnetic abrasive particles need to be smashed, the accuracy of mixture amount and stability of die size do not have much impact on the hot-pressing sintering process of magnetic abrasive particles. But the magnetic abrasive particles are not needed to be smashed, and it needs a larger smashing force, so it inevitably leads to some abrasives separate. On the other hand, the hot-pressing sintering needs to die and the decreased size limits the production with many magnetic abrasive particles; it also increases the production cost. The preparation process is shown in Fig. 3.18.

(1) Material preparation

Material: magnetic medium (pure iron powder, and its size is about $200^{\#}$), abrasive phase (corundum, silicon carbide, and its size is $325^{\#}$–$400^{\#}$), Mass ratio of magnetic medium phase and abrasive phase is 2:1, 3:1 or 4:1.

(2) Material weighting

The calculation formula is:

$$G_{\mathrm{F}} = (V_{\mathrm{M}} - G_{\mathrm{A}}/\rho_{\mathrm{A}})\rho_{\mathrm{C}} \tag{3.7}$$

where $G_{\mathrm{F}}$ is mass of magnetic medium phase (g); $V_{\mathrm{M}}$ is volume of mold container ($cm^3$); $G_{\mathrm{A}}$ is mass of abrasive phase (g), for the experimentally aware, the $G_{\mathrm{A}}$ can

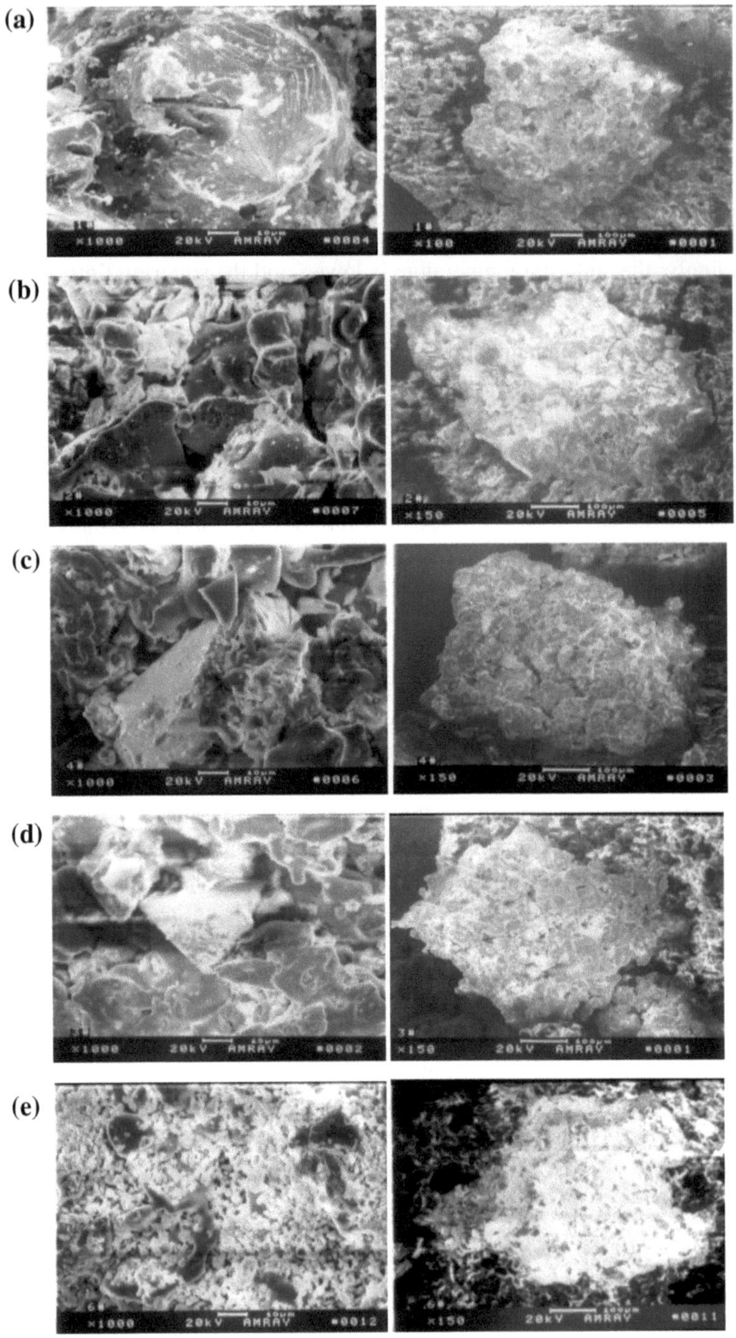
(a)
(b)
(c)
(d)
(e)

◂**Fig. 3.16** SEA photograph of magnetic abrasive particles that prepared by vacuum sintering. **a** GC31P4YLZS-TG; **b** GC31P4SZZS-TG; **c** A31P4YLZS-TG; **d** A31P4SZZS-TG; **e** A35P4YLZS-TG

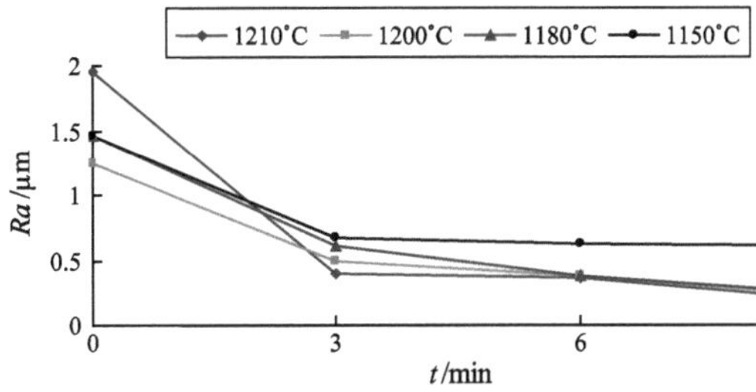


**Fig. 3.17** Influence curves of magnetic abrasive particles by vacuum sintering method at different sintering temperatures on the surface roughness

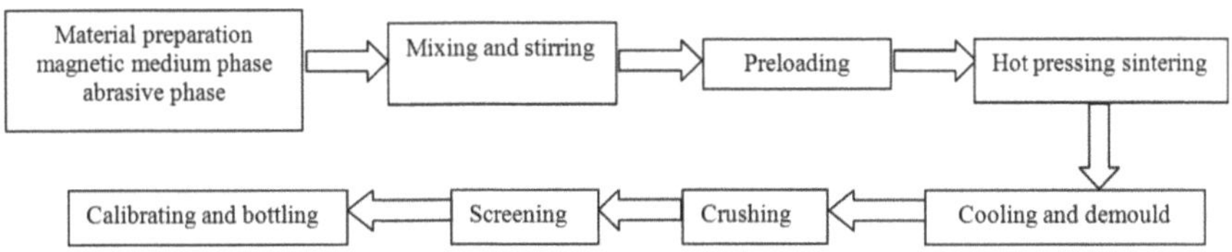


**Fig. 3.18** Hot-pressing sintering preparation process of magnetic abrasive particles

be 1/4, 1/3, or 1/2 of $G_F$; $\rho_A$ is theoretical density of abrasive phase (g/cm$^3$); $\rho_c$ is compact density of mixture material (g/cm$^3$).

(3) Mixing

Using the balance to weight these materials, stirring it manually in these few minutes is off to hell. Loading weighed materials into the V-shaped mixer, mixing it on the lathe with rotational speed of 10–20 r/min in 6–8 h, taking out it and loading in the mold container.

(4) Hot-pressing sintering

The graphite mold frame is located between the upper and lower platens of hot-pressing machine, and atmospheric short sintering is done with selected sintering process parameters: Sintering time is 15–20 min (including heating time is 10–12 min, holding time is 5–8 min), sintering pressure is 5 MPa ± 0.5 MPa, sintering temperature is (1010–1050 °C) ± 10 °C, the curve that hot-pressing sintering temperature changes with time is shown in Fig. 3.19.

(5) Smashing, screening, calibrating, numbering

The green body is smashed twice, and then it is screened and calibrated. Figure 3.20 represents SEA photographs of A31P4RS-TG magnetic abrasive

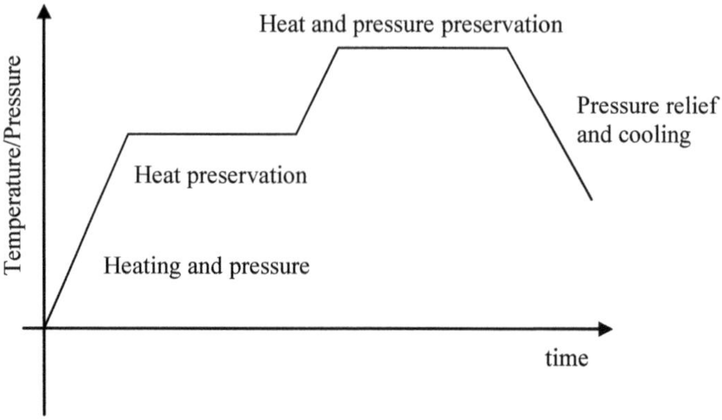


**Fig. 3.19** Schematic diagram of hot-pressing sintering process

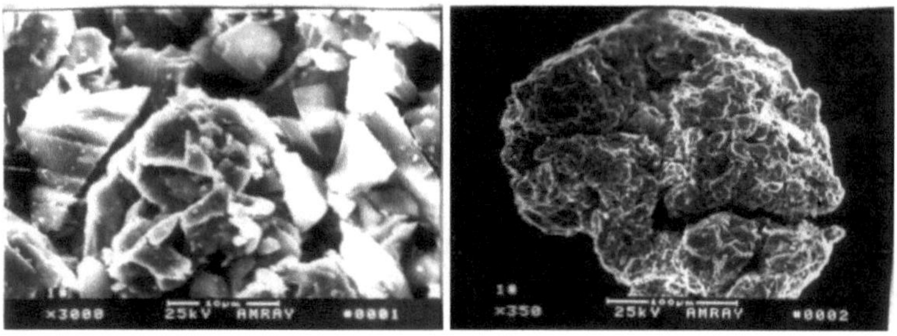

**Fig. 3.20** SEA photograph of magnetic abrasive particles prepared by hot-pressing sintering

**Table 3.6** Finishing effect of magnetic abrasive particles prepared by hot-pressing sintering method

| Workpiece material | Hardened steel | Cast iron | Brass |
|---|---|---|---|
| Workpiece surface roughness $R_a$ value before finishing/μm | 0.82 | 1.7 | 0.83 |
| Magnetic abrasive particles | A31P4RS-TG | WA31P4RS-TG | |
| Magnetic induction intensity *B*/T | 1.15 | 0.97 | 0.42 |
| Workpiece surface roughness $R_a$ value after finishing/μm | 0.20 | 0.58 | 0.48 |

particles that prepared by this process at different magnifications. Table 3.6 represents the finishing effect for different materials of the workpiece.

The difference between the hot-pressing sintering and vacuum sintering is that It does not require cold press molding, but requires exertion of high temperature and high pressure simultaneously to make the molding and sintering at the same time. After, the magnetic abrasive particles prepared by this process have high strength, large density, less pore, and other advantages. Practice shows that using the magnetic abrasive particles prepared by this process to finish, the finishing effect is good but the production cost is also relatively higher.

3. Bonding method

The bonding method is to make a certain percentage of uniformly mixed magnetic medium phase and abrasive phase adhere to the whole by using adhesives, then by pressurized solidification, smashing, screening, and calibrating in order to prepare the magnetic abrasive particles [34]. The preparation process is shown in Fig. 3.21.

(1) Material preparation

Material: magnetic medium phase (pure iron powder, and its size is about 200#), abrasive phase (corundum, silicon carbide, and its size is 325#–400#) and its particle size should be uniform, adhesive (epoxy resin and polyamide resin or epoxy resin, polyamide resin and furan resin or metal filling glue), mass ratio of magnetic medium phase and abrasive phase is 2:1, 3:1, 4:1, 5:1.

(2) Mixing

According to a certain mass ratio regarding weigh materials, stirring it manually helps the loading of the V-shaped mixer, mixing it on the lathe with rotational speed of 10–20 r/min in 6–8 h.

(3) Adding the adhesive

First, according to the mass ratio of adhesive and mixed materials, which is 2:15, it is used to calculate the total mass of the adhesive. Then, according to the mass ratio of 2:1, the weight of the two components of epoxy resin and polyamide resin (or according to the mass ratio) is 2:1:0.5 to weigh the three components of epoxy resin, polyamide resin, and furan resin. Finally, mix it and stir it uniformly.

(4) Uniformly mixing again

Pouring the uniformly mixed adhesive and mixed materials into a large vessel, stir it evenly until the adhesive and mixed materials are completely moist and become the entirety of the object. It helps birth of the lump with a diameter of 30–50 mm.

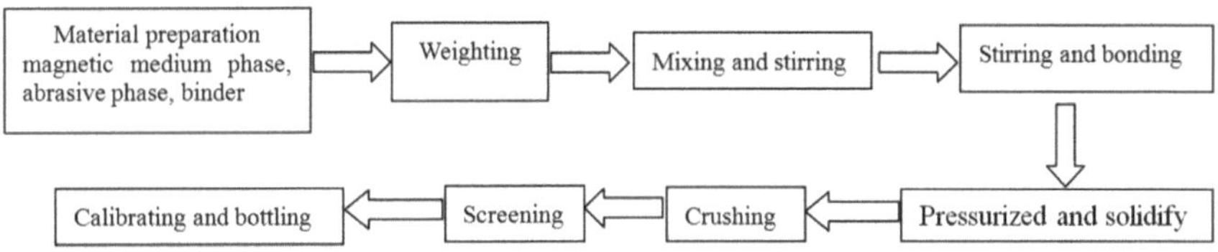


**Fig. 3.21** Bonding preparation process of magnetic abrasive particles

(5) Drying and solidifying

Placing the agglutinate lump composed of adhesive and mixed materials into the dry and dark environment, exerting 0.1–0.3 MPa pressure on it, and solidifying at the room temperature about 24 h.

(6) Smashing, screening, calibrating, numbering

The green body is smashed twice, then using the standard screen according to the requirements, sort it into the size range of $60^{\#}$, $80^{\#}$, $100^{\#}$, $120^{\#}$, $140^{\#}$ for meeting the requirements of rough machining and finish machining.

The experiment example of preparation process is as follows: The material is pure iron powder and brown corundum powder, and the adhesive is metal filling glue. The experiment process: The adhesive has many components; first using the portion of components to mix with materials, after two days of drying and solidifying process, they are still not solidified, and then by adding another component to them and stirring uniformly, it should be able to solidify into a lump that has certain hardness to it after 24 h. The maybe reasons can be analyzed that the ratio of several different components are chosen improperly, and that it does not be pressurized in the preparing process. During the process of adjusting the ratio many times, we increase pressure and get the satisfied results. The experimental results: The bonded body is hard and dense, so it is difficult to be smashed, and the separation phenomenon is rare when smashing. By the magnetic abrasive finishing experiment, we find that the effect is good and it does not have separation phenomenon. After finishing the value of surface roughness, the result is decreased in a well-known manner, and the surface gloss is better than the abrasive particles that were prepared by sintering method and hot-pressing method. The main reasons are the adhesive included resin, a polishing material and a good polishing effect. In order to test its temperature resistance, taking part of the products into the oven, we can use to determine its temperature. After taking them in, take them out in a bit to observe their temperature: 120 °C until their temperature rise every other 30 °C, and after 1.5 h, the temperature rises to 250 °C, preserving heat about 20 min, and remove them to check again. If they are still not softened and deformed, it indicates that the adhesive has good heat resistance. In order to further verify its performance, continuous finishing tests can be done, in which the workpiece has no vibration and feed, the rotating speed is 960 r/min, and finishing time is 8 min. The results show the adhesive still not soften.

Figure 3.22 represents the SEA photographs of the magnetic abrasive particles prepared by bonding method at different magnifications. From Fig. 3.22a, b, when the abrasive phase is brown corundum and silicon carbide, the abrasives phase of magnetic abrasive particles prepared by bonding method is relatively independent and their cutting edge angles are obvious relatively.

Using the magnetic abrasive particles prepared by different adhesives A31P4HZ-TG and A31P4HFZ-TG, the magnetic abrasive finishing experiments on the outer circle surface (45 steel has a 40 mm diameter) are carried out. This is only conditioning that the machine gap is 3 mm, workpiece's rotating speed is 610 r/min, axial vibration amplitude is 15 mm, and frequency is 2.3 Hz, using the

(a)

(b)

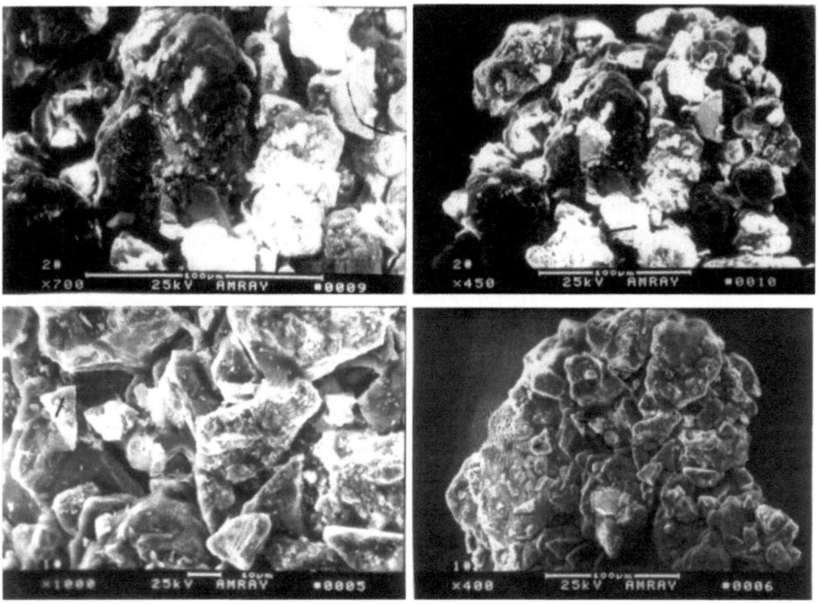

**Fig. 3.22** SEA photograph of magnetic abrasive particles that prepared by bonding method. **a** A31P4HFZ-TG; **b** GC31P4HFZ-TG

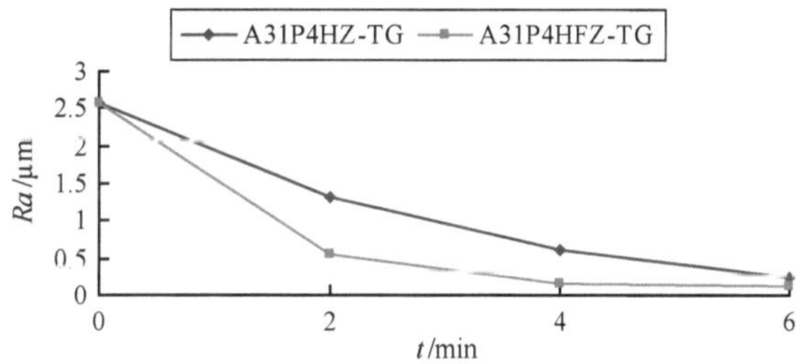


**Fig. 3.23** Influence curves of magnetic abrasive particles with different adhesives on the surface roughness

liquid paraffin as additive. The influence curves of magnetic abrasive particles with different adhesives on the surface roughness are shown in Fig. 3.23.

From Fig. 3.23, when adding a certain amount of adhesives, the workpiece surface roughness $R_a$ decreases exponentially, resulting in an effective improvement in the gloss and brightness of the workpiece surface. Its effect conditions relate to the type and amount of the adhesives. Comparing the adhesive of epoxy resin, polyamide resin, and furan resin with that of epoxy resin and polyamide resin, the former has lower surface roughness $R_a$ value, which indicates a relatively better finishing effect.

Using the magnetic abrasive particles A41P4HFZ-TG, on conditions that the machining gap is 3 mm, finishing time is 3 min, workpiece's rotating speed is 610 r/min, axial vibration amplitude is 15 mm, and frequency is 2.3 Hz, the liquid paraffin is additive, and the magnetic abrasive finishing experiment on the outer circle surface (quenched steel) is 40 mm diameter. Finally, finishing the workpiece surface roughness $R_a$ value decreases from 1.1 to 0.22 μm, and the surface is light and has good physical and mechanical properties.

Using the magnetic abrasive particles GC41P4HFZ-TG, on conditions that the machining gap is 3 mm, finishing time is 3 min, workpiece's rotating speed is 610 r/min, axial vibration amplitude is 15 mm, and frequency is 2.3 Hz, the liquid paraffin as additive as it is, and it deburrs on the edges of those different sizes grooves (total of six places, the groove dimensions: $B \times L = 2 \times 30$ mm, $4 \times 15$ mm, the groove ends have small holes and their diameter is 4 mm) on the aluminum precision outer circle surface (Ø15 × 50 mm). After finishing, the part surface is smooth and light, the burrs of each grooves and holes surrounding are all removed.

The bonding method is an economical and practical preparation method with simple process, but we should pay attention to choose adhesive, especially its temperature resistance should be considered. The magnetic abrasive particles prepared by this process have higher finishing performance, but the binding force between the magnetic medium phase and abrasive phase is weaker than the magnetic abrasive particles prepared by sintering method, which has influence on the service life. But this process has become mature because it has low production cost, easy preparation, and wider application.

4. High-energy ball milling

In this section, the magnetic abrasive is fabricated by using high-energy ball-milling technology. The formation mechanism is listed as follows. During the milling process, the original large-size particles were crushed into smaller ones. Under the condition that high frequency and high stress induced by milling process, the fiercely collision was happened between raw material and grinding balls, and then the larger particles were deformed and broken down. The newly formed metal surfaces can be bound together due to the cold welding effect. Then, the atom diffusion becomes easier between particles. As a result, micro-powder or composite powder can be obtained with this ball-milling technology. The preparation process of magnetic abrasive particles by using ball-milling technology is shown in Fig. 3.24.

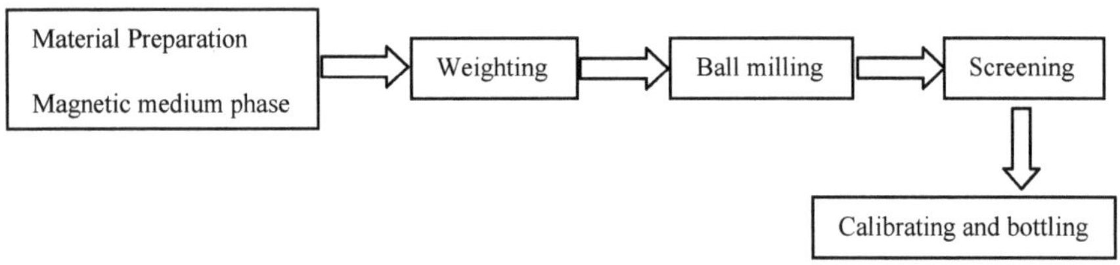


**Fig. 3.24** Preparation process of magnetic abrasive particles by using ball-milling technology

(1) Material preparation

Material: magnetic medium (pure iron powder, 60 meshes), abrasive phase ($Al_2O_3$, 80 meshes), volume ratio of magnetic medium phase and abrasive phase is 4:1, 1:1 or 3:7.

(2) Ball-milling parameters

GCr15 balls with diameter of 3 mm were used for grinding balls. The ball to power weight ratio is 30:1. The loading coefficient is 0.5. The ball-milling time varies in the range of 1–4 h. During the whole milling process, the inside of the tub keeps a vacuum environment.

(3) Screening, calibrating, numbering

The milled powder was screened and calibrated. Figure 3.25 shows the SEM microstructures of magnetic abrasive with different ball-milling times.

The prepared magnetic particles were used to surface polish of 45# steel. Figure 3.26 shows the effect of magnetic particles fabricated by different milling times on the surface roughness of 45# steel after machining. From the figure, we can see that with the increasing of ball-milling time from 1–3 h, the fabricated magnetic abrasive was used to surface finishing of 45# steel, and the surface roughness is decreased obviously. It is indicated that with the increasing of ball-milling time, the magnetic abrasive will have a better machining ability. It is because with the increasing of ball-milling time, the size of the magnetic abrasive is decrease. Then, the number of cutting edges of abrasive phases was increased rapidly. At the same time, the binding energy and cold welding uniformity between two phases, magnetic medium phase and abrasive phase, are improved. Thus, the grinding property of magnetic abrasive was enhanced. As a result, the surface roughness of 45# steel is decreased. When the milling time reaches to 3 h, the surface roughness of workpiece is no longer decreased. If the magnetic abrasive was milled more than 3 h, the surface roughness of the workpiece will increase. The probable cause is the size of the magnetic abrasive is too small to grind the workpiece. It means that the abrasive phase lost the ability to machine resulting in the surface roughness of workpiece increased.

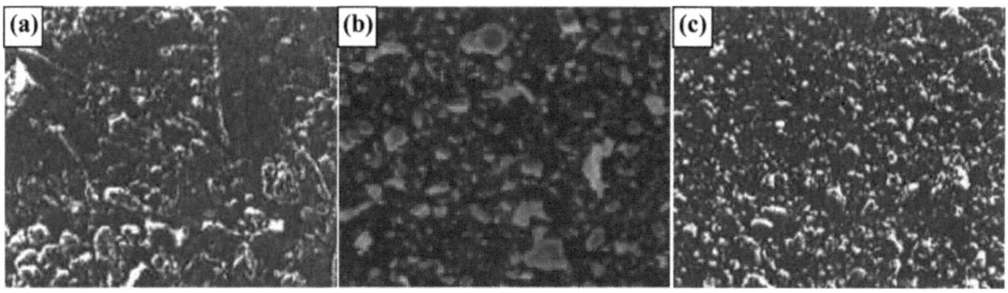


**Fig. 3.25** Dependence of microstructures on the ball-milling time. **a** 2 h; **b** 3 h; **c** 4 h

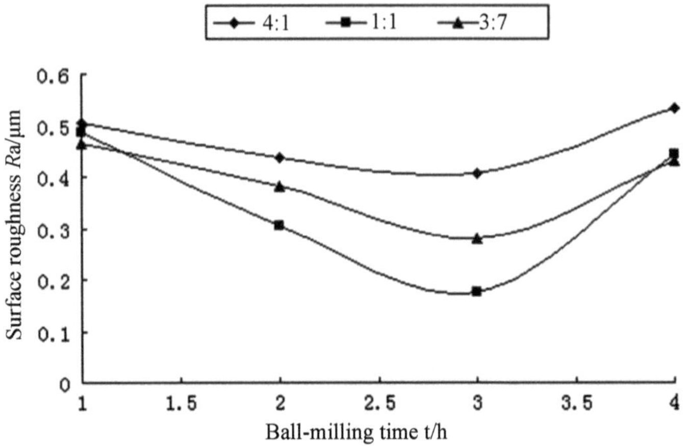


**Fig. 3.26** Influence curves of different ball-milling times with surface roughness

In the experiment, the ratio of iron powder to $Al_2O_3$ powder is 1:1, the inside of the tub keeps a vacuum environment, the ball-milling time is 3 h, and under these conditions the fabricated magnetic abrasive has well surface morphologies, ideal size distribution, and well machine ability. The surface roughness of the workpiece reaches to 0.2 μm by using the above abrasive (Fig. 3.27). At the same time, both the smoothness and glossiness are all improved obviously under the machining condition that the rotation speed of workpiece is 600 r/min, the machining gap is 3 mm, the machining time is 6 min, the magnetic induction strength is 1 *T*, and the lubricant is liquid paraffin wax.

The $Fe/Al_2O_3$ magnetic abrasive has been fabricated successfully by high-energy ball-milling method. This method has the following advantages: lower raw materials cost, simple in preparation, short preparation period, uniformly mixed with magnetic medium phase and abrasive phase and can be produced easily in

**Fig. 3.27** Macro-photograph of workpiece before and after finishing

industrial scale. Therefore, the high-energy ball milling is a magnetic abrasive preparation method with good prospects for practical application.

5. Introduction of other methods

(1) Plasma powder melting method

Plasma powder melting method is mixing the hard abrasive particles with iron powder beforehand, and then ejecting them from the plasma spraying device and melting them together under the high temperature of plasma arc [35]. Figure 3.28 shows the device diagram of plasma powder melting method preparing the magnetic abrasive particles. This method solves the compatible problem of the abrasive and iron matrix at a certain degree, but not suitable for all hard abrasives as $Al_2O_3$, SiC, WC, in which it requires the density distinguish between the abrasive and iron must be very small, otherwise the abrasive will float on the surface of molten state iron matrix rather than dispersing among it uniformly in the spraying process, so this method has some limitations.

(2) Composite coating method

It is called composite coating that uniformly mixes some kind of abrasive with the metal coating to form a special coating by the composite electroplating or composite chemical plating. This method has been successfully applied in many fields of science and process nowadays after 30 years' development. There are still two problems that need to be noticed when applying composite coating process to the preparation of the magnetic abrasive particles: ① The decontamination activation of original pure iron powder particles; ② The abrasive particles are uniformly deposited on the surface of the pure iron powder particles. At present, the second problem can be solved by magnetic stirring, which can easily finish the problem of hardness ratio, but its service life is related to coating thickness. The mass production is difficult to achieve, and there still exists the problem of how to deal with the plating solution and production cost.

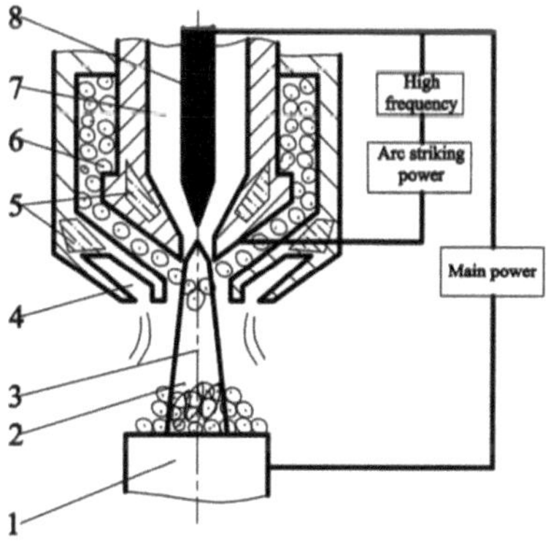


**Fig. 3.28** Device schematic diagram of plasma powder melting method preparing the magnetic abrasive particles. 1—Copper plate; 2—magnetic abrasive particles; 3—plasma arc; 4—protective gas; 5—cooling water; 6—mixture; 7—plasma gas; 8—tungsten electrode

(3) Sand-planting method

When the abrasive melting point is lower than 900 °C, the sintering method cannot be used, but we can plant sand on the iron matrix surface to get the abrasive layer which is the magnetic abrasive particles truly.

(4) Laser Sintering

Magnetic abrasive particle preparation of laser sintering is a new process, which widens the application field of the laser [36]. This process mixes a certain particles size of iron powder with abrasive according to appropriate mass ratio. Laser sintering must be conducted in a reducing atmosphere; therefore, we should try to discharge active gas in the container before the test to prevent reacting between the gas and mixed powder when the laser radiates. During the test, the intake valves and exhaust valves are opened to make the high-pressure nitrogen gas flow within the tube, forming negative pressure in the other branch of Y-shaped tube, pumping out the mixed powder and spraying it into the closed chamber, meeting the laser near the focal spot of the laser beam, and thus occurring sintering phenomenon. Finally, it generates the magnetic abrasive particles. Figure 3.29 shows the device diagram of magnetic abrasive particles preparation of laser sintering. This method is difficult to be applied to the practical production for the expensive cost.

### 3.2.4 Main Performance Parameters

In the process of magnetic abrasive particle finishing, workpiece surface will be finished by a certain interact force and relative motion velocity between magnetic abrasive particles and workpiece surface. This will allow the performance of magnetic abrasive particles to have direct influence on finishing effects, finishing efficiency, processing costs, and other factors. According to the inherent characteristics, work environment, application scope, and others of magnetic abrasive particles, the main performance parameters of magnetic abrasive particles include grain diameter, ingredient and ratio, saturated magnetic induction intensity, density, service life, which are discussed as below.

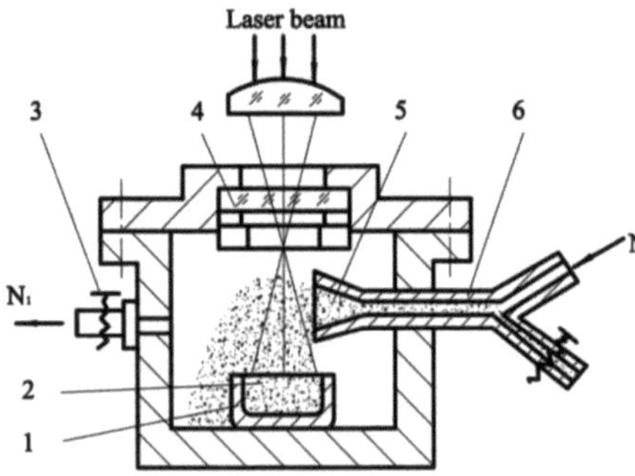


**Fig. 3.29** Device structure schematic diagram of laser sintering method preparing the magnetic abrasive particles. 1—Collector; 2—magnetic abrasive particles; 3—valve; 4—glass sheet; 5—mixture; 6—Y-shape tube

1. Grain diameter

As shown in Fig. 3.10, grain diameters of magnetic abrasive particles include grain diameter of magnetic abrasive particles called $D$ and grain diameter of abrasive called d. From the aforementioned said, the size of magnetic abrasive particles has nothing to do with action force, but the grain diameter of magnetic abrasive particles directly affects the filling effect and agglomeration ability of the magnetic abrasive particles group. Thus, the force (magnetic holding force) of workpiece is influenced by magnetic abrasive particles; therefore, the parameters $D$ and d are closely related to the surface roughness value and material removal rate of workpiece after finishing.

Assuming the magnetic abrasive particles are arranged to the regular square, and the acting force of magnetic abrasive particles on workpiece is set as $P$, the acting force of magnetic abrasive particles of diameter $D$ is

$$f = p\frac{\pi}{4}D^2 \tag{3.8}$$

In the above equation, $f$ is acting force of a single magnetic abrasive particle grain (unit is N); $D$ is grain diameter of the magnetic abrasive particle (unit is mm). For each magnetic abrasive particle grain, the size of its acting force is proportional to square of the grain diameter $D$. Assuming the surface roughness value of workpiece is determined by the cutting depth of each magnetic abrasive particles grain, the larger the grain diameter $D$, and the larger the generated acting force, and so the stronger the effect of friction, extrusion, scoring to the workpiece surface. Under the same acting force $P$, comparing the large grain diameter magnetic abrasive particles with the small grain diameter grains, the former can obtain the larger surface roughness value. If one magnetic abrasive has m abrasive particles, which work at the same time, the acting force of each abrasive particle is

$$\Delta f = \frac{f}{m} = \frac{p\pi D^2}{4m} \tag{3.9}$$

In the above equation, $\Delta f$ is acting force of a single abrasive particle (unit is N); $m$ is the number of abrasive particles.

The smaller the grain diameter, the larger the value of $m$, and it will get smaller with the decrease in $d$; thus, the surface roughness value of workpiece will decline. The influence curves of grain diameter of magnetic abrasive particles on the finishing effect obtained by the test methods are shown in Fig. 3.30. It can be seen when the value of $D$ is the same, the material removal rate increases accordingly but slowly with the increase in $d$; after finishing, the surface roughness value $R_a$ of workpiece is gradually increasing; when the value of $d$ is constant, the material removal rate increases gradually and the surface roughness $R_a$ value also increases due to the increase of $D$. Thus, according to the processing requirements, the grain

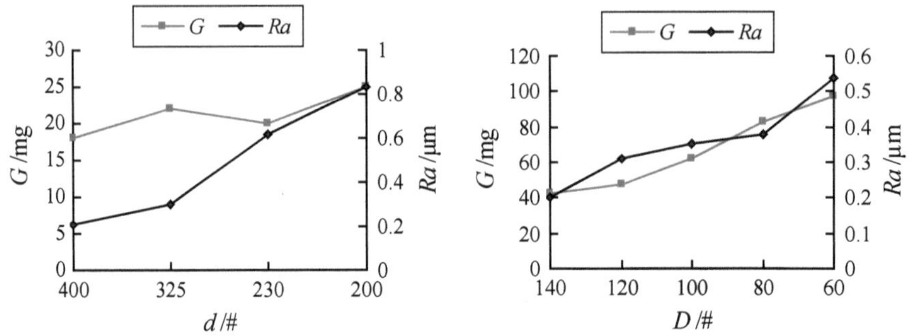


**Fig. 3.30** Influence curves of grain diameter of magnetic abrasive particles on the finishing effects

diameter of magnetic abrasive particles $D$ and the grain diameter of abrasive $d$ should be carefully chosen to obtain optimal finishing effects and efficiency when choosing the type of magnetic abrasive particles.

The influence curves of grain diameter of magnetic abrasive particles when finishing the outer surface of the rotation on the surface roughness $R_a$ value are shown in Fig. 3.31. The influence curve of grain diameter of magnetic abrasive particles when finishing the inner surface of the rotation on the surface roughness $R_a$ value is shown in Fig. 3.32. The influence curve of grain diameter of magnetic abrasive particles when finishing the plane on the surface roughness $R_a$ value is shown in Fig. 3.33.

It is shown in Figs. 3.31, 3.32, and 3.33 that no matter what type of surface, the effect laws of finishing on the grain diameter of magnetic abrasive particles are consistent. Because its impact on the finishing efficiency and finishing quality is not uniform, in actual, the grain diameter $D$ and $d$ have optimal values to measure and select. A large number of experiments have shown that when it is used for rough machining, we can choose larger $D$, $d$ to improve finishing efficiency. The recommended values are as follows: The range of grain diameter of the magnetic abrasive particles is $40^{\#}$–$80^{\#}$, the range of grain diameter of the abrasive is $325^{\#}$–

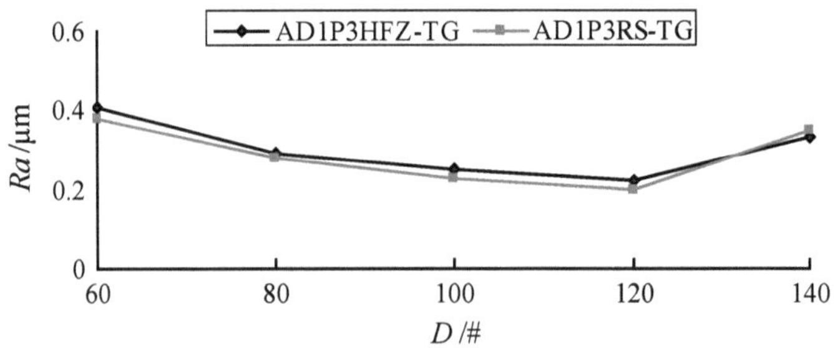


**Fig. 3.31** Influence curves of grain diameter of magnetic abrasive particles when finishing the outer surface of the rotation on the surface roughness $R_a$ value

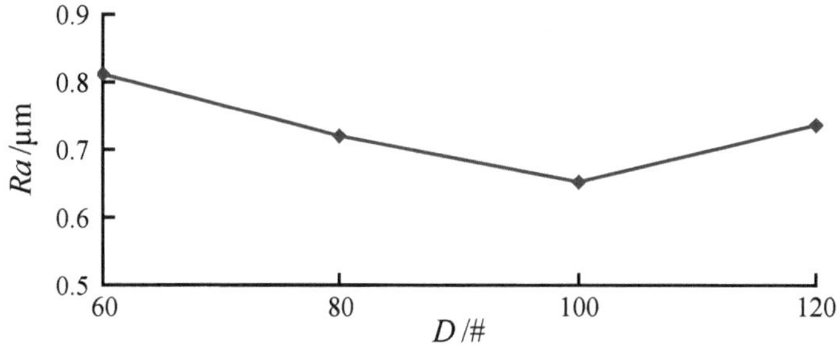


**Fig. 3.32** Influence curve of grain diameter of magnetic abrasive particles when finishing the inner surface of the rotation on the surface roughness $R_a$ value

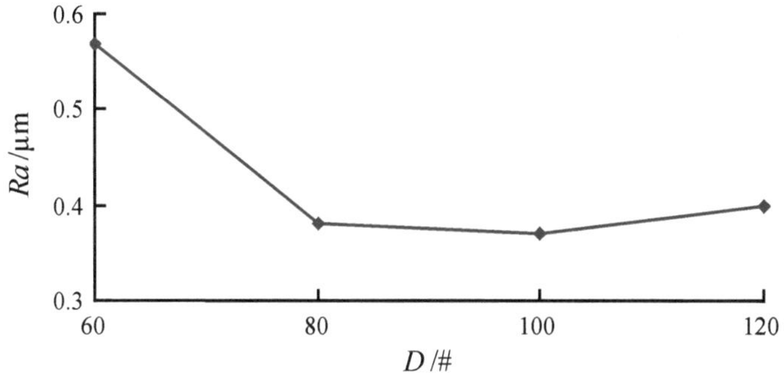


**Fig. 3.33** Influence curve of grain diameter of magnetic abrasive particles when finishing the plane on the surface roughness $R_a$ value

$400^{\#}$; when it is used for finish machining, and we can choose smaller *D*, *d* to obtain the lower surface roughness values. The recommended values are as follows: The optimal range of grain diameter of the magnetic abrasive particles is $80^{\#}$–$140^{\#}$ and the optimal range of grain diameter of the abrasive is $400^{\#}$–W10.

Generally, determination of the size of the grain diameter of abrasive phase (abrasive) *d* can get through the label that has already been given at the time of purchase; for the grain diameter *D*, it can be calibrated by the standard screen (Fig. 3.15) filtering method.

2. Ratio

The ingredients of magnetic abrasive particles include magnetic medium phase, abrasive phase (abrasive), adhesive. The ratio mainly refers to the mass ratio of magnetic medium phase and abrasive phase. Different ingredient and ratio are important parameters to measure the performance of magnetic abrasive particles. And it directly impacts the magnetic properties and processing capability of the magnetic abrasive particles, thereby generating a direct influence on the finishing effects, finishing efficiency, processing costs, and so on [37].

Magnetic medium phase is usually pure iron powder, and its content determines the size of the magnetic holding force. If the ratio is too large, then the action force will increase and the finishing efficiency will improve, but the surface roughness value is larger after finishing and it affects the quality of finishing; If the content reduces, the action force also reduces, the finishing time increases, at this time, the content of abrasive phase increases relatively, and the surface roughness value is lowered. But if the content of pure iron powder is diminishing, and due to the magnetic holding force is decreasing greatly, the magnetic abrasive particles fly away from the machining area therefore losing the machining effects.

The choice of abrasive phase is different due to the difference of materials of the machined workpiece, shown in Table 3.1.

Generally, the ratio (mass ratio) of magnetic medium phase and abrasive phase is usually 5:1, 4:1, 3:1, 2:1. According to the observation of experiment, when the ratio is gradually changed from 5:1 to 2:1, the change of the surface roughness value of workpiece has some difference after finishing, but they can all reduce 1–2 grade. The influence curves of the ingredient ratio of magnetic abrasive particles on the surface roughness value are shown in Fig. 3.34, and the finishing effect comparison photographs of different ratio of the magnetic abrasive particles are shown in Fig. 3.35. During the practical application, choosing the reasonable ingredient and ratio according to the specific requirements, it is usually chosen 3:1 or 4:1.

3. Saturated magnetic induction intensity

Due to the interaction of the molecules of magnetic substance, the molecular current forms many sequential arrays in local area and shows the magnetism, and these small regions are called magnetic domains. When the magnetic substance does not have an external magnetic field, each magnetic domain will arrange in a confusing way, and so the magnetic fields cancel with each other; under the action of external magnetic field, the magnetic domains gradually rotate to the consistent direction of the external magnetic field, it can be said that it is producing a magnetizing field that has consistent direction with the external field, and so the magnetic induction intensity in the magnetic substance increases dramatically that

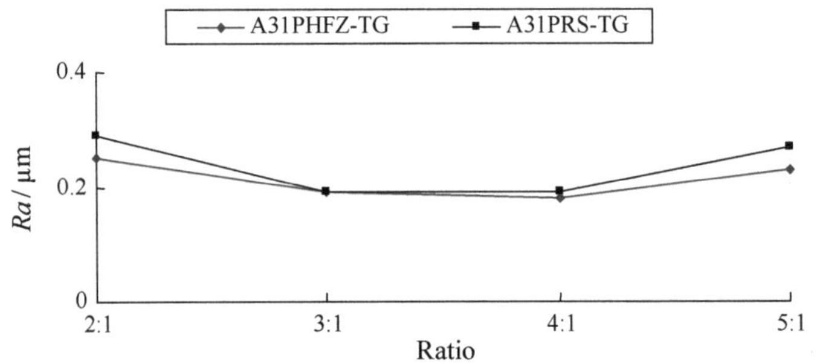


**Fig. 3.34** Influence curves of the ingredient ratio of magnetic abrasive particles on the surface roughness value

**Fig. 3.35** Finishing effect comparison photographs of different ratios of the magnetic abrasive particles. **a** Pre-machining; **b** 2:1; **c** 3:1

namely is magnetized. With the increase of the external magnetic field, the magnetic domains that can provide turning become fewer, the increasing speed of magnetic field of the magnetic substance becomes slower, and finally when the external magnetic field (the current of excitation magnetic field) is increasing to a certain extent, all the magnetic domains turn to the consistent direction with external magnetic field. At this time, the magnetic induction intensity that produced by magnetization of the magnetic substance will not continuously increase with the improvement of the external magnetic field strength that namely the magnetic induction intensity at this time reaches its saturation value which is the saturated magnetic induction intensity of magnetic substance. From Fig. 3.36, when the magnetic field strength $H$ increases to a certain value, the magnetic induction intensity $B$ increases very slowly, and the magnetization is close to saturation, and in Fig. 3.36, $B_{sat}$ represents the saturated magnetic induction intensity, and $H_{sat}$ represents the magnetic field strength that reached at magnetic saturation.

When the entire magnetic domains of magnetic abrasive particles have consistent direction with the external magnetic field, the magnetic induction intensity of the magnetic abrasive particles will reach saturation value. And once it reaches

**Fig. 3.36** *B*–*H* curve of soft iron material

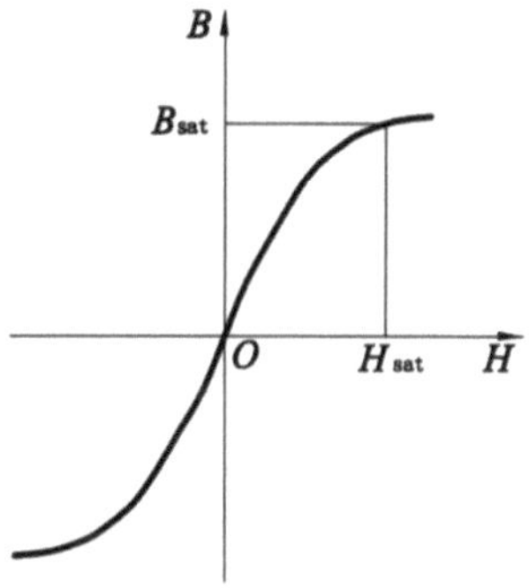

saturation, the external magnetic field strength will increase further (or the current of excitation magnetic field), and the magnetic induction intensity of magnetic abrasive grains will remain the same.

From the formula (3.4), the magnetic induction intensity is the key indicator to determine the size of magnetic properties and action force of magnetic abrasive particles. Under the action of strong external magnetic field, the magnetic abrasive particles along the direction of magnetic field will form a magnetic brush, and when it reaches the saturated induction intensity, the force that magnetic abrasive particles acting on the workpiece surface will achieve the maximum material removal ratio when the finishing is much higher. Therefore, the saturated magnetic induction intensity is not only a very important parameter, but also the basis of designing the magnetic field generation device in the magnetic abrasive particle finishing equipment.

To enhance the magnetic induction intensity of magnetic abrasive particles, there are two main ways: ① trying to select the magnetic medium phase that has higher saturated magnetic induction intensity; ② increasing the proportion of magnetic medium phase in the magnetic abrasive particles. The three magnetization curves named *a*, *b*, *c* that obtained by using the same magnetic medium phase all have different proportions (the proportion is $a > b > c$) are shown in Fig. 3.37. Because the ratio of magnetic abrasive particles to be limited by many factors, choosing the magnetic medium phase with higher saturated magnetic induction intensity has extremely important significance to improve the performance of the magnetic abrasive particles.

Due to the magnetic abrasive particles belonging to a complex with a variety of materials, its saturated magnetic induction intensity is connected with its ingredient, ratio, particle size, and other factors and needs to be obtained by experiment. The model of magnetic abrasive particles used in the experiment is WA21P4HZ-TG.

(1) Measuring the magnetic properties of magnetic abrasive particles with digital storage oscilloscope.

The magnetic abrasive particles that are in the un-magnetized state, which are those under the action of magnetic field magnetizing from weak to strong, we can measure a cluster of hysteresis loop in which its area is from small to large, as shown in Fig. 3.38. Connecting the apex of these hysteresis loops will form magnetization curves of the magnetic abrasive particles.

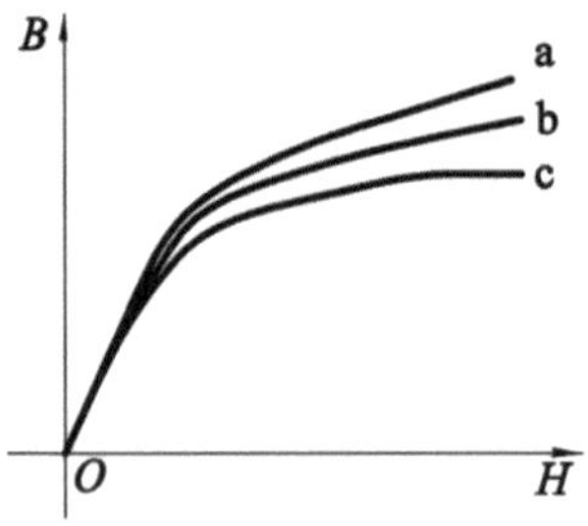


**Fig. 3.37** *B–H* curves of magnetic medium phase having different ratios

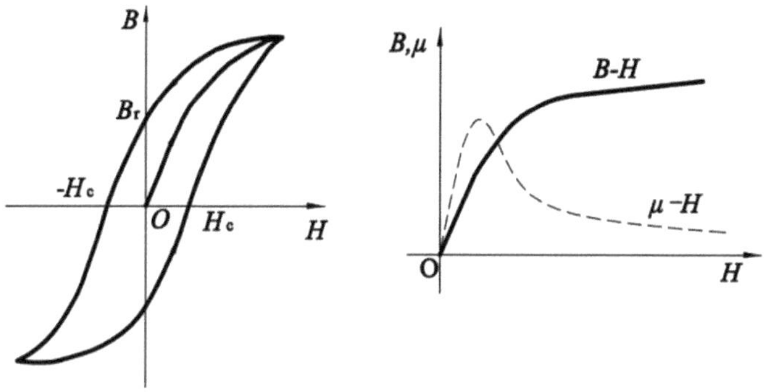


**Fig. 3.38** Magnetization curve schematic diagram of magnetic abrasive particles

From the concept of magnetic permeability, the slope of cables that from one point of the magnetization curves to the origin O is defined as the magnetic permeability at this magnetization state $\mu = B/H$. So the magnetization curves are not linear, and when $H$ starts to increase from O, $\mu$ also increases gradually and reaches at a maximum value; when $H$ increases again, due to the magnetic induction intensity is saturated, $\mu$ starts to decrease sharply. The curves that $\mu$ changes with $H$ are shown in Fig. 3.38.

The testing device is shown in Fig. 3.39; putting specimen boxes that filled with magnetic abrasive particles into the skeleton holes, after switching on the power source, the display screen of oscilloscope will appear hysteresis loop as shown in Fig. 3.40 [38].

By derivation calculations and measurements, the magnetic field strength that generated by the excitation current is

$$H = \frac{N_1 A_x}{L \cdot R_1} X \tag{3.10}$$

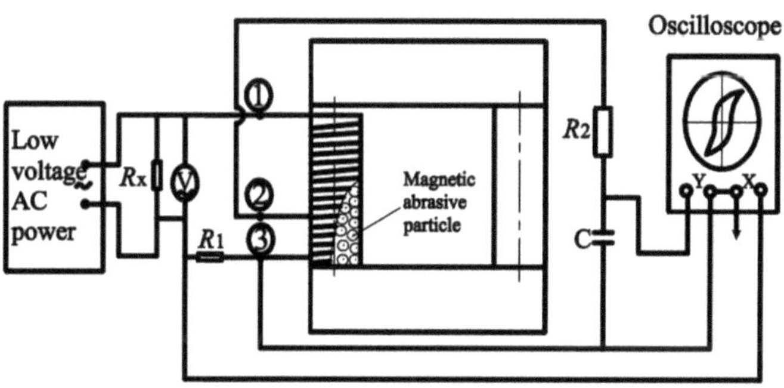


**Fig. 3.39** Schematic diagram of experiment device. ① low-voltage alternating current power; ② magnetic abrasive particles; ③ oscilloscope

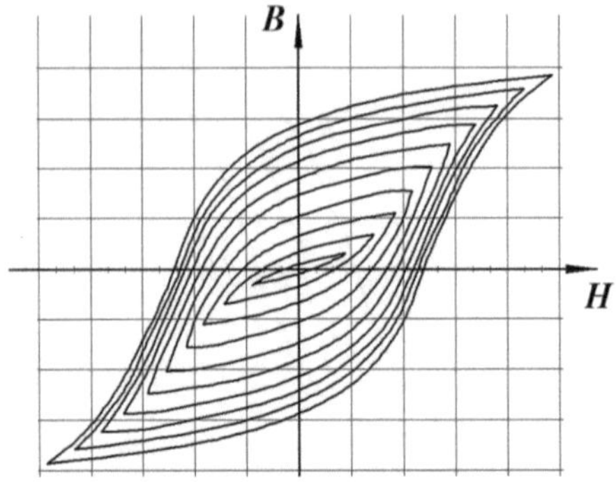


**Fig. 3.40** Hysteresis loop of magnetic abrasive particles

where $H$ is magnetic field strength (A/m); $N_1$ is the number of turns of primary coil (turns); $R_1$ is current sampling resistance (Ω); L is the length of the test box (m); $X$ is x-axis coordinate of the points on the oscilloscope screen; $A_x$ is x-axis deflection sensitivity of oscilloscope (V/lattice).

Accordingly, the magnetic induction intensity that generated by magnetic abrasive particles is

$$B = \frac{CR_2A_y}{N_2S} Y \times 10^3 \tag{3.11}$$

where $B$ is the magnetic induction intensity (T); $C$ is integral capacitor in the circuit (μF); $R_2$ is the integral resistance (Ω); $N_2$ is the number of turns of secondary (turns); $S$ is the sectional area of magnetic circuit of the tested material (mm$^2$); $Y$ is y-axis coordinate of the points on the oscilloscope screen; $A_y$ is y-axis deflection sensitivity of oscilloscope (V/lattice).

According to the calibration proportion of curves that are displayed on the oscilloscope screen and the Eqs. (3.10) and (3.11), the hysteresis loop of magnetic abrasive particles is quantified and *B*–*H* magnetization curve is drawn as shown in Fig. 3.41. From the figure, the magnetic induction intensity of magnetic abrasive particles is gradually increasing with the increase of magnetic field strength. When the magnetic induction intensity increases to a certain level, it will no longer go beyond the limit and thus will reach saturation state.

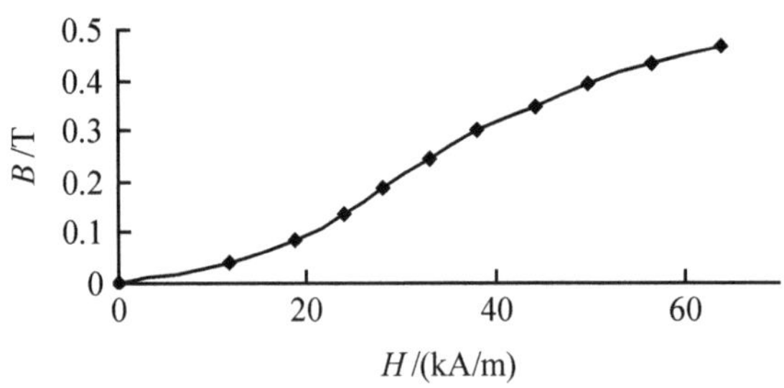


**Fig. 3.41** *B*–*H* magnetization curve of magnetic abrasive particles

When using the alternating current power to test and generate larger eddy-current heat, the magnetic field strength is relatively smaller, so it is generally difficult to achieve saturation state; however, this method can be used to accurately measure its initial magnetization state curve.

(2) Using magnetic circuit to measure the magnetic properties of magnetic abrasive particles

The magnetic circuit schematic of the iron core coil with air gap is shown in Fig. 3.42, and the magnetic potential which generated in the magnetic circuit can be analyzed by magnetic circuit ampere theorem.

According to ampere magnetic circuit theory, assume that

$$\oint H \cdot \mathrm{d}l = N_0 I \tag{3.12}$$

where $H$ is the magnetic field strength (A/m); $l$ is the length of segmented magnetic circuit (m); $N_0$ is the number of turns of coil (turns); $I$ is current that acting on the coil (A). Applying the magnetic circuit theorem to piecewise dispose the magnetic circuit that shown in Fig. 3.38, we can acquire that

$$N_0 I = H_c L_c + H_y L_y + H_p L_p + H_f L_f + H_g L_g \tag{3.13}$$

where $H_c$ is the magnetic field strength in iron core (A/m); $L_c$ is the strength of magnetic core (m); $H_y$ is the magnetic field strength in magnetic yoke (A/m); $L_y$ is the strength of magnetic yoke (m); $H_p$ is the magnetic field strength in magnetic pole (A/m); $L_p$ is the strength of magnetic pole (m); $H_f$ is the magnetic field strength of magnetic abrasive particles (A/m); $L_f$ is the height of magnetic abrasive particles (m); $H_g$ is the magnetic field strength of air gap (A/m); $L_g$ is the height of air gap (m).

Assuming the magnetic circuit does not have magnetic leakage, the relationship between the magnetic induction intensity and magnetic field strength can be expressed as

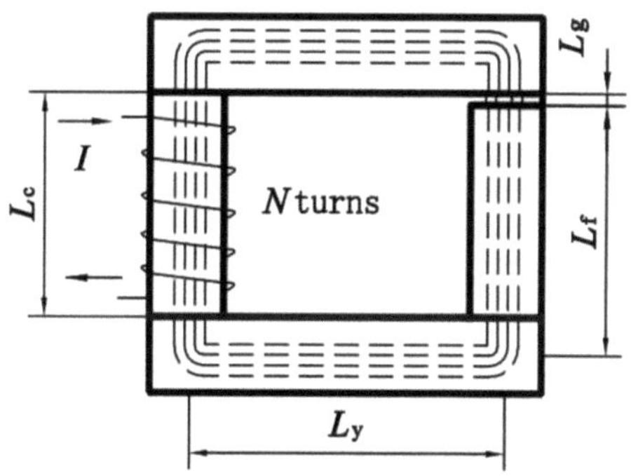


**Fig. 3.42** Magnetic circuit schematic of the iron core coil with air gap

$$B = \mu_o \mu_r H \tag{3.14}$$

where $B$ is magnetic induction intensity (T); $\mu_o$ is vacuum permeability ($\mu_o = 4\pi \times 10^{-7}$ H/m); $\mu_r$ is relative permeability; $H$ is magnetic field strength (A/m). Equation (3.14) is substituted in Eq. (3.15) and obtained relation between the magnetic induction intensity and the current that acting on the coil is

$$N_o I = \frac{B_c L_c}{\mu_o \mu_r} + \frac{B_y L_y}{\mu_o \mu_r} + \frac{B_p L_p}{\mu_o \mu_r} + H_f L_f + \frac{B_g L_g}{\mu_o} \tag{3.15}$$

where $B_c$, $B_y$, $B_p$, $B_g$ are, respectively, the magnetic induction intensity in magnetic core, magnetic yoke, magnetic pole, and air gap.

The photograph of testing device that made according to its principle is shown in Fig. 3.43. Electrifying the adjustable direct current in coils, and then placing the Tesla meter probe into the air gap, we can measure the magnetic induction intensity $B$ of each point in the air gap and further establish the magnetization curves of magnetic abrasive particles.

Since the magnetic abrasive particles that need to be tested are placed in the position of magnetic pole position, Eq. (3.15) can be changed as

$$N_o I = \frac{B_c L_c}{\mu_o \mu_r} + \frac{2 B_y L_y}{\mu_o \mu_r} + H_f L_f + \frac{B_g L_g}{\mu_o} \tag{3.16}$$

By deducing, we can get the magnetic field strength of magnetic abrasive particles which is

$$H_f = \frac{N_o I - \left( \frac{B_c L_c}{\mu_o \mu_r} + \frac{2 B_y L_y}{\mu_o \mu_r} + \frac{B_g L_g}{\mu_o} \right)}{L_f} \tag{3.17}$$

According to the Gauss law of magnetic field, the normal component of magnetic induction intensity continuously runs the boundaries of two medium, that is

$$B_g = B_f = B_c = B_y \tag{3.18}$$

**Fig. 3.43** Photograph of test device

Under normal circumstances, the saturated magnetic induction intensity of magnetic abrasive particles is about 1 T, so the value of $B_g$, $B_f$, $B_c$, $B_y$ also is about 1 T, but the $\mu_r$ is much larger than $\mu_0$, *and* the item $\frac{B_C L_C}{\mu_0 \mu_r}, \frac{2B_y l_y}{\mu_0 \mu_r}$, in Eq. (3.17) can be ignored. After simplifying and compiling, the relation equation between the magnetic field strength of magnetic abrasive particles and magnetic induction intensity is

$$H_f = \frac{N_o I - \frac{B_g L_g}{\mu_o}}{L_f} \tag{3.19}$$

In Eq. (3.19), the number of turns of coils $N_0$, the height of air gap $L_g$, the vacuum permeability $\mu_0$, and the height of test box $L_f$ are known as constant values, and the value of $I$ can be determined by ampere meter, the value of $B_g$ can be read by Tesla meter, so we can calculate the value of $H_f$ correspondingly. In order to establish the relationships between the magnetic field strength of magnetic abrasive particles $H$ and magnetic induction intensity $B$, $B_y$ the actual measurement, we can draw the *B–H* curve of magnetic abrasive particles, as shown in Fig. 3.44.

In the magnetic abrasive particle finishing, in order to facilitate the design of magnetic formation part, we should use the saturated magnetic induction intensity of the abrasive particles to measure the properties of magnetic abrasive particles.

Using the testing device as shown in Fig. 3.43, measuring the magnetic properties of magnetic abrasive particles by the circuit, the changing curves of the magnetic induction intensity of magnetic abrasive particles with different ratio on these fields strength are shown in Fig. 3.45. The influence curves of the magnetic abrasive particles with different magnetic induction intensity on surface roughness $R_a$ valuc of workpiece are shown in Fig. 3.46. The influence curves of the magnetic abrasive particle with different particle diameters on saturated magnetic induction intensity are shown in Fig. 3.47.

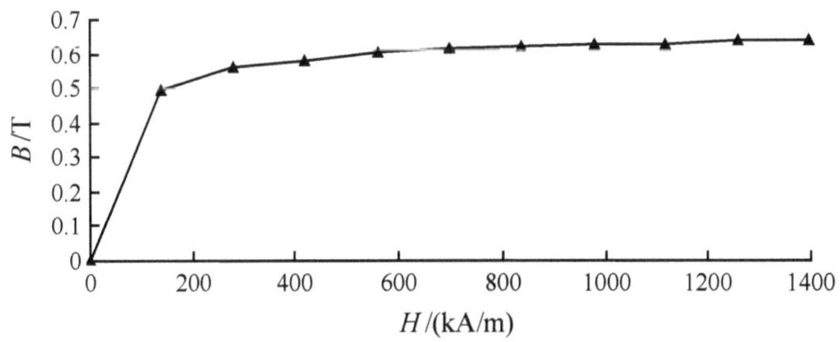


**Fig. 3.44** *B–H* magnetization curve of magnetic abrasive particles

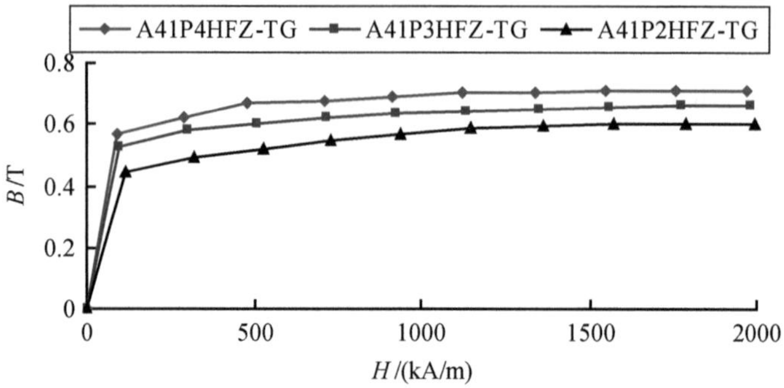


**Fig. 3.45** Changing curves of the magnetic induction intensity of magnetic abrasive particles with different ratios on magnetic field strength

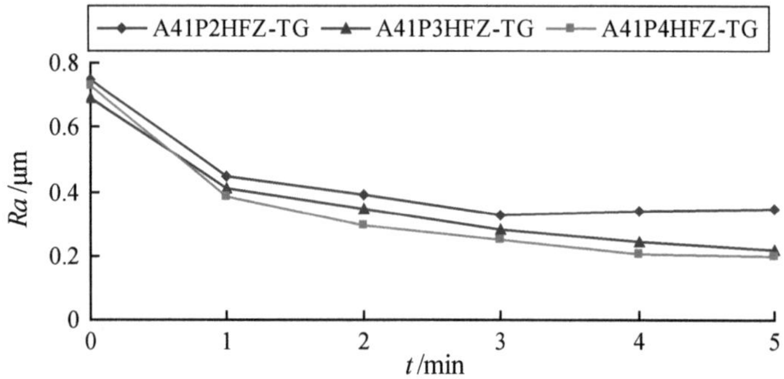


**Fig. 3.46** Influence curves of the magnetic abrasive particles with different magnetic induction intensities on surface roughness $R_a$ value of workpiece

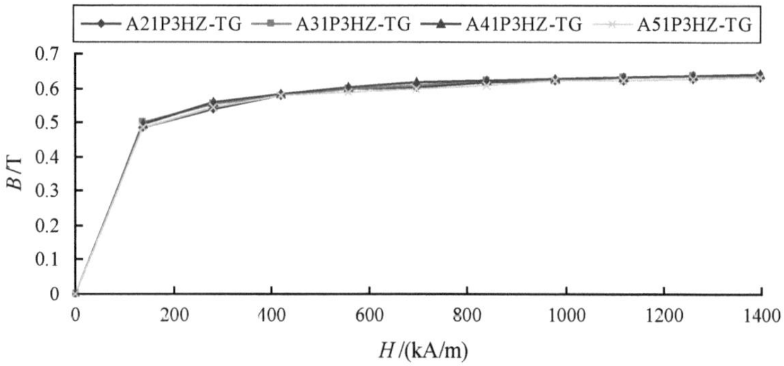


**Fig. 3.47** Influence curves of the magnetic abrasive particle with different particle diameters on saturated magnetic induction intensity

Due to the influence of magnetic resistance and magnetic flux leakage, the magnetic induction intensity when designing the magnetic formation part is increasing, and it can be expressed as

$$B = k_f k_r B_{sat} \tag{3.20}$$

where $B$ is the magnetic induction intensity of designed magnetic formation part (T) and $k_f$ is the magnetic flux leakage coefficient. There is a large difference in the structure of the different magnetic circuits (different size measure, the magnetic potential distribution) and in magnetic circuit of magnetic abrasive particle finishing, and it can be smaller and take 1.2–5; $k_r$ is the magnetic resistance coefficient (magnetic potential loss coefficient, usually is 1.05–1.45); $B_{sat}$ is the saturated magnetic induction intensity of magnetic abrasive particles (T).

In conclusion, according to the finishing workpiece, finishing requirements, and so on, the magnetic abrasive particle finishing process are all different, and the magnetic induction intensity that required by magnetic formation part generally takes about 0.6–1.4 T.

4. Density

The mass per unit volume of substance is called density. The density of magnetic abrasive particles not only is important data in the magnetic abrasive particle finishing, but can also be used to calculate the content of magnetic medium phase and other component particles in the magnetic abrasive particles. The density of magnetic abrasive particles also is related to its magnetic induction intensity, the force that the magnetic abrasive particles has is different than the density exerts on the surface of workpiece, and it can lead to different finishing effects and efficiency. So the density of magnetic abrasive particles is a main performance parameter of magnetic abrasive particles.

The density of magnetic abrasive particles is

$$\rho = \rho_s \varphi_s + \rho_a(\varphi_h - \varphi_s) + \rho_c(1 - \varphi_h) \tag{3.21}$$

In the above equation, $\rho_s$ is the density of magnetic medium phase (g/cm$^3$); $\rho_a$ is the density of bonder (g/cm$^3$); $\rho_c$ is the density of abrasive phase (g/cm$^3$); $\varphi_s$ is the volume percentage of magnetic medium phase (%); $\varphi_h$ is the sum of volume percentage of adhesive and magnetic medium phase (%).

When knowing the volume percentage of various components, we can determine the density of magnetic abrasive particles. The formula (3.21) is used to estimate the density of the magnetic abrasive particles. For example, the loose density of magnetic medium phase (pure iron powder) of magnetic abrasive particles is 2.7 g/cm$^3$; the density (80$^{\#}$) of abrasive phase (brown corundum) is 1.53 g/cm$^3$; the density of adhesive (epoxy resin + polyamide resin) is 1.25 g/cm$^3$. According to the mass ratio of configured magnetic abrasive particles is converted into the

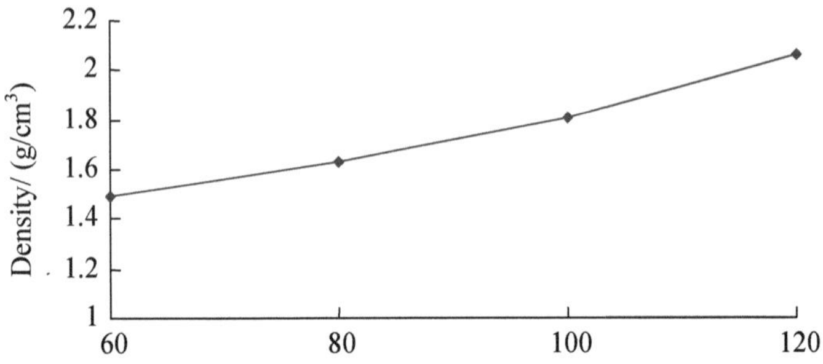


**Fig. 3.48** Influence curve of the magnetic abrasive particles with different particle diameters on density

volume ratio, the volume ratio of magnetic medium phase is 51.8%, the volume ratio of abrasive phase is 47.9%, and the volume ratio of adhesive is 0.3%. The formula (3.21) shows that the density of magnetic abrasive particles is estimated to be 2.13 g/cm$^3$.

The influence curve of the magnetic abrasive particles with different particle diameters on density is shown in Fig. 3.48. From Fig. 3.48, with the particle diameter of magnetic abrasive particles decreasing, the density is showing an increasing trend mainly due to the gap between the magnetic abrasive particles being extremely small and relatively more dense, and that's what causes the density to be increasing.

In addition, according to the theoretical analysis, for a single magnetic abrasive particles grain, different densities have different force on the surface of workpiece, and thus it has an influence on the finishing effects and efficiency of magnetic abrasive particle finishing, but now we cannot measure the density of single magnetic abrasive particle.

5. Service life

In the magnetic abrasive particle finishing, because the requirements of the rigidity and precision of processing equipment are not high, the usage cost of magnetic abrasive particles became a very important index of economy of magnetic abrasive particle finishing, and it is the main problem of the industrialization of this process.

The service life of magnetic abrasive particles refers to the length of time usage of magnetic abrasive particles that under the condition of maintaining its finishing effects, it is closely related to the failure of magnetic abrasive particles.

At present, there is still no report on evaluation of service life of the magnetic abrasive particles, and its evaluation index has no data to reference. In the actual production, the magnetic abrasive particles usually are used only one time, and it is mainly because the abrasive dust that formed during the process of magnetic abrasive particle finishing is also adsorbed in those particles by the magnetic field, forming three-body abrasion and affecting finishing effect. But the service life of magnetic abrasive particles depended on how long on earth must the particles have a quantitative index that can be determined by finishing tests.

The service life of magnetic abrasive particles can be measured by three parameters: the surface roughness $R_a$ value of workpiece after finishing; the material removal rate MRR of processed workpiece; the wear rate of magnetic abrasive particles $\Psi$. The evaluation index can uniformly be assessed by the finishing time of magnetic abrasive particles after these three parameters reach a certain extent.

Because the magnetic abrasive particle finishing itself belongs to the field of precision surface finishing, metal removal thickness is tiny and it is usually under the order of 1 μm, sometime is only 0.2–0.3 μm or smaller. So in the actual production, using the material removal rate MRR as the parameter to measure the service life of magnetic abrasive particles is not convenient to operate, and it is not suitable for the requirements of actual production situation.

The wear rate $\Psi$ of magnetic abrasive particles belongs to the change of microstructure. The abrasive dust that formed during the process of magnetic abrasive particle finishing is also absorbed in magnetic abrasive particles, and it is difficult to distinguish entirely; when detecting especially in the actual using process, it is not easy to be observed, mastered, and evaluated exactly. Therefore, the wear rate $\Psi$ of magnetic abrasive particles is not suitable as the quantification parameter of assessing the service life of magnetic abrasive particles.

The magnetic abrasive particle finishing process is aimed to improve surface quality of workpiece, and surface roughness value is an important parameter for evaluating the surface quality. It greatly affects the performance of the parts, using the surface roughness $R_a$ of the finished workpiece as a parameter to evaluate the service life of magnetic abrasive particles is easy to operate and assess, and it is extremely feasible. As a compound index, $R_a$ somewhat reflects the wear degree of magnetic abrasive particles and material removal rate. Therefore, we can use the surface roughness $R_a$ value of finished workpiece to define the evaluate parameter of service life of magnetic abrasive particles.

After the above theoretical analysis, repeated research, and test, the service life of magnetic abrasive particles is defined as follows: Under the given working conditions, a workpiece is continuously finished by magnetic abrasive particles of certain weight. If surface roughness $R_a$ value of workpiece after abrasive finishing cannot attain the value of original $R_a$ decreased by one level, according to the priority series of R10/3, the magnetic abrasive particles will be thought to lose machinability. Meanwhile, the processing time of magnetic abrasive particles for abrasive finishing is just about its service life $T$ with minute as the unit.

In the process of testing the service life of magnetic abrasive particles, the workpiece is divided into nine sections, the length of each section is 50 mm, each section is finished by a certain amount of magnetic abrasive particles, and the finishing time is 3 min by using the Mahr surface roughness meter to measure each section, conducting relevant analysis and comparison. During the process, the mass

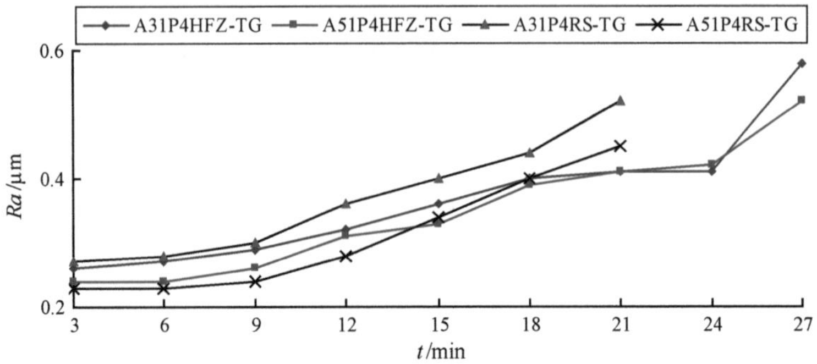


**Fig. 3.49** Service life curves of magnetic abrasive particles

of magnetic abrasive particles that added in the process of testing is also determined by test and quantization, and the service life curves of magnetic abrasive particles are shown in Fig. 3.49.

## 3.3 Magnetic Abrasive Finishing Device

### *3.3.1 Composition of Magnetic Abrasive Finishing Device*

Magnetic abrasive finishing technology can achieve processing of various surfaces. The devices of the magnetic abrasive finishing have a variety of different forms since there are many effects of workpiece's size, shape, material, and so on. The device mainly considers the complex relative motion formed between magnetic grains and the workpiece, and generally it does not need high precision, stiffness, so the device design has its own characteristics. But generally the machining gap and the magnetic field strength are required to be adjustable to accommodate the demand of different workpieces machining under different conditions, respectively. Magnetic abrasive finishing device not only can be converted from the old equipment but also can be produced the special equipment according to the workpiece.

Figure 3.50 represents schematic diagram which represents the magnetic abrasive finishing equipment. Magnetic abrasive finishing equipment utilizes the horizontal lathe to process some rotary surfaces. Rotary motion and axial vibration of the workpiece is achieved through the structure of the machine. The magnetic field generating means will have to be designed according to the size and shape of the workpiece and the processing requirements. Figure 3.51 shows the photograph of processing equipment. In addition, the rotor surface processing can also be implemented through renovation drilling and milling machine. Figure 3.52 represents a photograph of magnetic abrasive finishing equipment which can process the outer surface of the rotary by transformation drilling [39].

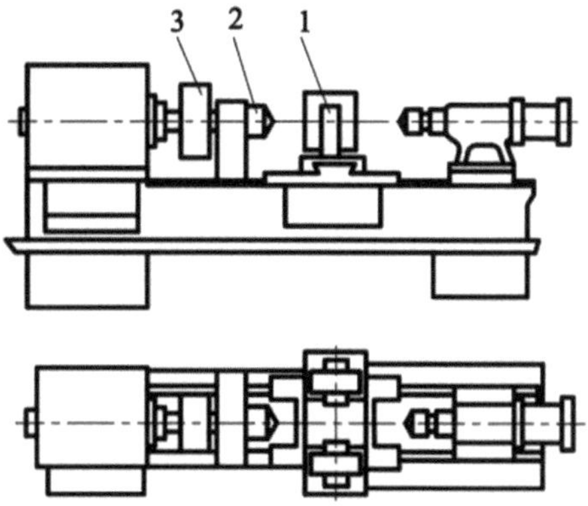


**Fig. 3.50** Surface of rotary magnetic abrasive finishing equipment diagram. 1—Magnetic field source; 2—rotary motion driving mechanism; 3—axial vibration driving mechanism

**Fig. 3.51** Photograph of magnetic abrasive finishing equipment on the surface of rotating body (First)

**Fig. 3.52** Photograph of magnetic abrasive finishing equipment on the surface of rotating body (Second)

Figure 3.53 shows the photograph of magnetic abrasive finishing equipment which processes the inner surface of the rotary using horizontal lathe machining. Figure 3.54 shows the photograph of magnetic abrasive finishing equipment which can process the inner surface of the rotary by transformation drilling. Figure 3.55 shows schematic diagram of magnetic abrasive finishing equipment. Pole head does the rotary motion in a certain speed and axial vibration in the $z$ direction, and the table does the feed movement in the $x$, $y$ direction; under these conditions, the relative movement between the workpiece and the magnetic grains can be achieved.

**Fig. 3.53** Photograph of magnetic abrasive finishing equipment in revolving body surface (First)

**Fig. 3.54** Photograph of magnetic abrasive finishing equipment in revolving body surface (Second)

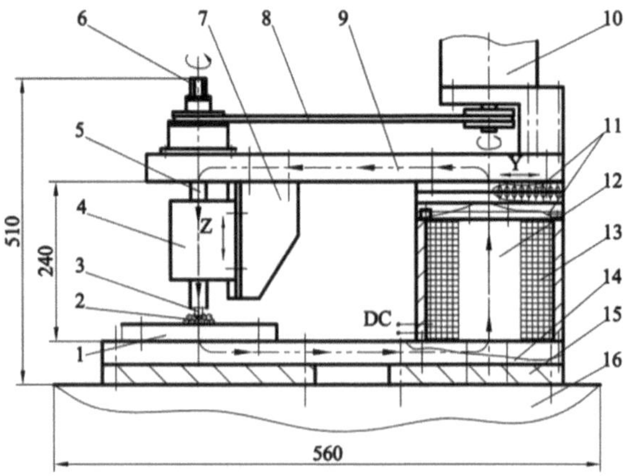


**Fig. 3.55** Schematic diagram of plane magnetic abrasive finishing equipment. 1—Workpiece; 2—magnetic abrasive grain; 3—pole; 4—support sleeve; 5—principal axis; 6—spline shaft; 7—bracket; 8—transmission belt; 9—steel plate; 10—motor; 11—plane bearing; 12—iron core; 13—coil; 14—magnetic base; 15—non-magnetic separator; 16—workbench

Magnetic field generating means can use a permanent magnet source. An electromagnetic source can also be used. In addition, the plane processing can also be achieved through the transformation of drilling machine, milling machine, etc. Figure 3.56 shows the photograph of processing equipment transformed from universal milling machine. The magnetic abrasive finishing of the free-form surface can be realized by adding the magnetic pole head in the motion mode of the CNC drilling and milling machine [40]. Figure 3.57 shows schematic diagram of the magnetic abrasive finishing equipment which can process the free-form surface. Figure 3.58 shows the photograph of finishing equipment which is transformed from CNC drilling and milling machine.

**Fig. 3.56** Photograph of plane magnetic abrasive finishing equipment

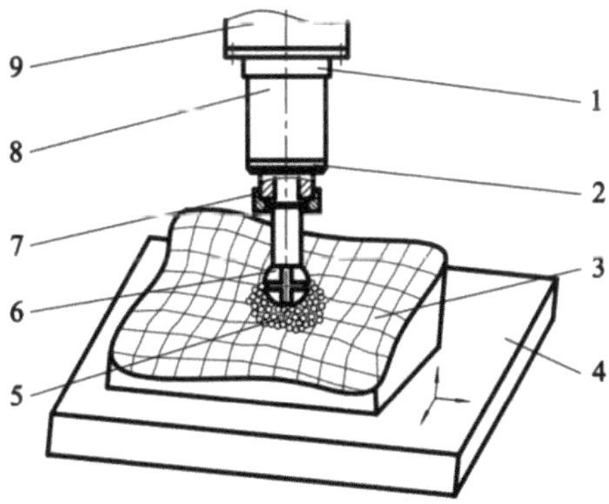


**Fig. 3.57** Sketch of free-form surface magnetic abrasive finishing equipment. 1—Flange plate; 2—seal cover; 3—workpiece; 4—workbench; 5—magnetic abrasive particles; 6—pole; 7—lock nut; 8—winding; 9—NC spindle unit

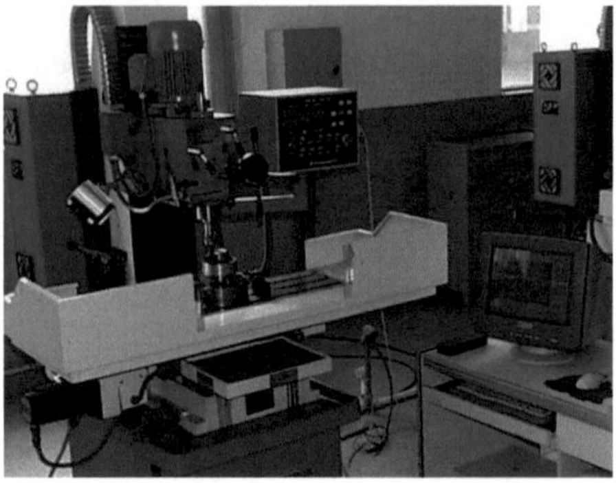

**Fig. 3.58** Photograph of the free-form surface magnetic abrasive finishing equipment

### *3.3.2 Design of the Magnetic Field Generator*

The interaction and the relative velocity between the magnetic grains and the workpiece are two necessary conditions for the realization of magnetic abrasive finishing. The interaction force is generated by magnetic force between the magnetic grains in the formation of N–S poles. Therefore, a suitable magnetic field generating equipment must be designed first, and it is also known as the source of magnetic field (The designs of magnetic field source include circuit design and pole design). Source of magnetic field can be divided into electromagnetic source and permanent source [41].

Electromagnetic source can produce large magnetic flux density, and the size of magnetic flux density can be adjusted by adjusting the output of the current. When machining large workpieces, rotation is needed for the magnetic field source. Rotation can fix the magnetic field source and may achieve by a pulse current [42]. But the pulse current needs windings. So not only it is complex and bulky, but also heat sensitive. The heat is produced by an eddy current which causes the rise of workpiece temperature, affecting the processing results, while it is not suitable for connection and precision equipment. The structure of permanent magnet sources is simple, and it is portable equipment. Compared with the electromagnetic source, the permanent magnet sources have the smaller the magnetic flux density and less convenient adjustment of magnetic flux density (Although it can be achieved by the design of magnetic circuit, but it not as good as the electromagnetic source convenience.).

1. The basic theory of magnetic circuit design

In motors, transformers, and various ferromagnetic elements, the common magnetic material is made out of a certain shaped core. Permeability of core is higher than the surrounding air or other matter a lot, so a closed loop can be formed with the majority of the magnetic flux through the core. This man-made magnetic flux loop is called the magnetic circuit.

Circuit design can be divided into electromagnetic circuit design and permanent magnet magnetic circuit design. The key of two circuit designs is different, but their methods are similar. First, they all need to brief the magnetic circuit, forming an equivalent magnetic circuit, and then calculate magnetic field intensity. Combing with the electromagnetism theory, an example of electromagnetic magnetic circuit can be seen as follows.

Magnetization curve of ferromagnetic material (Fig. 3.59) shows that with the increase of magnetic field intensity *H*, the magnetic induction *B* increases but permeability $\mu$ decreases. When the magnetic field intensity *H* is increased to a certain extent, magnetic induction *B* tends to be saturated. The purpose of the magnetic circuit design is to make the magnetic induction *B* of the ferromagnetic material in a reasonable range. If *B* is too small, it will result in waste material.

If *B* is too large, then the oversize magnetic field intensity *H* is needed because it needs to increase the weight of the winding and consume more power and also excessive MMF to generate incentives.

Figure 3.60 shows schematic diagram of electromagnetic source. Figure 3.61 shows the equivalent magnetic circuit. Design of the electromagnetic excitation coil is the key of the circuit design, and this means that under the condition of the certain maximum magnetic flux density, ampere-turns of the coil is calculated [43].

In order to meet the magnetic flux density required in machining gap, section of the magnetic conductor must avoid magnetic saturation, while the flux should flow

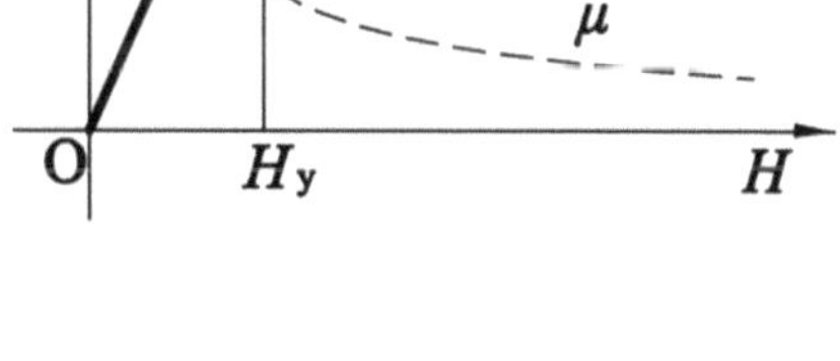


**Fig. 3.59** Magnetization curve of ferromagnetic material

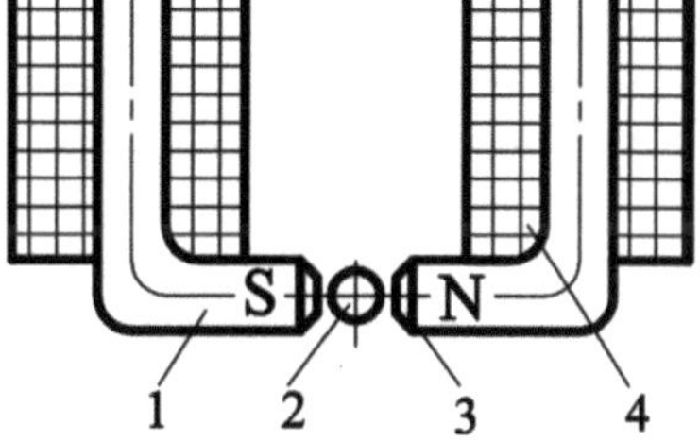


**Fig. 3.60** Schematic diagram of electromagnetic source

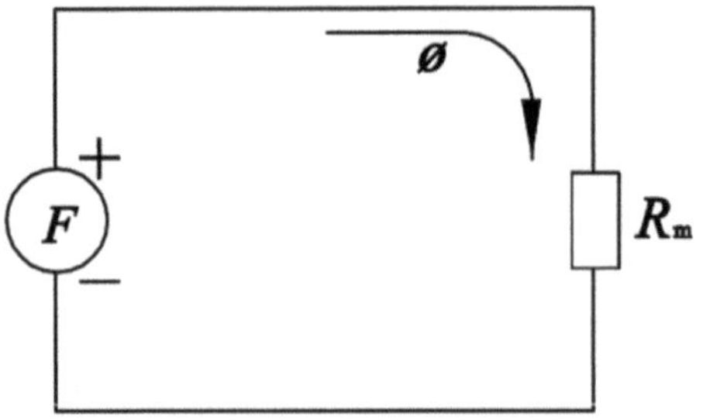


**Fig. 3.61** Electromagnetic source equivalent magnetic circuit

through all of unsaturated magnetic conductor. Each part of the magnetic circuit will weaken the flow of the magnetic flux, which is called magnetoresistance. Magnetic permeability of the material $\mu$, length l, and cross-sectional area s will affect the size of magnetoresistance [44], which can be represented by the following formula:

$$R_{\mathrm{m}} = \frac{L}{\mu s} \tag{3.22}$$

where $R_{\mathrm{m}}$ represents the reluctance ($\mathrm{H}^{-1}$); $\mu$ represents the permeability (H/m); $L$ represents the length (mm); $S$ represents the cross-sectional area ($\mathrm{mm}^2$).

Magnetic properties of the magnetic material are nonlinear. Different magnetic materials have different magnetic permeability. Therefore, the size of the magnetoresistance will change as the change of magnetic permeability. In order to facilitate the calculation, the reciprocal value of the magnetoresistance is used. It means that permeability can be expressed as the following formula:

$$\Lambda_{\mathrm{m}} = \frac{1}{R_{\mathrm{m}}} = \frac{\mu s}{L} \tag{3.23}$$

According to magnetic circuit Ohm's law, the magnetic flux passing the core and yoke can be expressed as the following formula:

$$\Phi = \frac{F}{R_{\mathrm{m}}} = \frac{IN_0}{R_{\mathrm{m}}} \tag{3.24}$$

where $\Phi$ represents magnetic flux (Wb) passing the core and the yoke; $R_{\mathrm{m}}$ represents magnetoresistance ($\mathrm{H}^{-1}$); $I$ represents the current (A); $N_0$ represents the number of turns (turns); $F$ represents total MMF which is required to overcome the magnetoresistance of all magnetic circuit.

Ignoring the magnetoresistance of other parts, the MMF of the magnetic circuit is determined by the magnetoresistance. If the permeability is obtained, the magnetoresistance can be calculated, and then MMF will be solved. In addition, when the current in the magnetic circuit is constant, the number of turns of the coil can be determined.

2. Engineering Design of magnetic circuit

(1) The calculation of Permeability

In the magnetic circuit, the total magnetoresistance of each series-wound segment can be computed by following formulas:

$$R_{\mathrm{m}} = R_{\mathrm{m1}} + R_{\mathrm{m2}} + \ldots + R_{\mathrm{m}i} \tag{3.25}$$

or

$$\frac{1}{\Lambda_{\mathrm{m}}} = \frac{1}{\Lambda_{\mathrm{m1}}} + \frac{1}{\Lambda_{\mathrm{m2}}} + \ldots + \frac{1}{\Lambda_{\mathrm{m}i}} \tag{3.26}$$

The total magnetoresistance of each shunt-wound segment can be computed by following formulas:

$$\frac{1}{R_{\mathrm{m}}} = \frac{1}{R_{\mathrm{m1}}} + \frac{1}{R_{\mathrm{m2}}} + \ldots + \frac{1}{R_{\mathrm{m}i}} \tag{3.27}$$

or

$$\Lambda_{\mathrm{m}} = \Lambda_{\mathrm{m1}} + \Lambda_{\mathrm{m2}} + \ldots + \Lambda_{\mathrm{m}i} \tag{3.28}$$

Figure 3.62 shows machining gap which is the shortest distance between the pole head and workpiece surface. Machining gap is very common in the process of magnetic abrasive finishing.

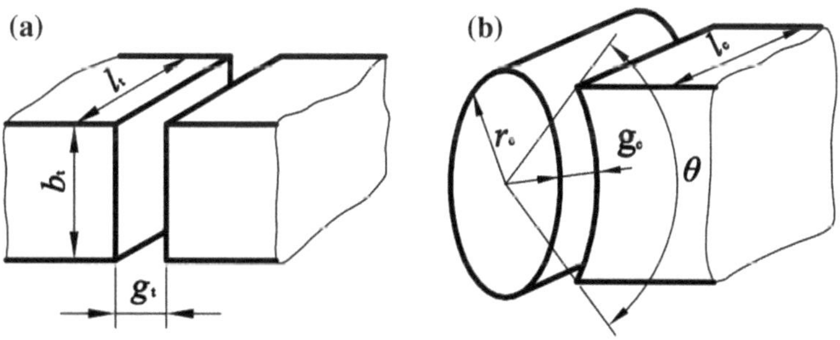


**Fig. 3.62** Shape and parameter of machining gap. **a** The rectangular surface machining gap; **b** The annular surface machining gap

(1) Permeability in the machining gap between rectangular pole surface. As shown in Fig. 3.58, assuming that the flux in the machining gap is uniform, the following formula can be obtained.

$$\Phi_t = B_t b_t l_t \tag{3.29}$$

$$F_t = H_t g_t \tag{3.30}$$

Thus, the permeability can be obtained by the following formula:

$$\Lambda_t = \frac{\Phi_t}{F_t} = \mu_0 \frac{b_t l_t}{g_t} \tag{3.31}$$

where $\mu_0$ represents the magnetic permeability; $l_t$ represents the width of the magnetic poles; $b_t$ represents the height of the magnetic poles; $g_t$ represents the machining gap.

(2) Permeability in the machining gap between ring-type pole surfaces.

Inside the cylinder, the magnetic flux density in the radial direction is not uniform, but it is uniform in the axial direction. It is shown in Fig. 3.58 that in ignoring the diffusion of the lines of magnetic force at the edge of the pole, the total magnetic flux passing either radius of the cylinder surface is equal. Therefore, the following formula can be obtained.

$$\Phi_c = r_c \theta l_c B(r_c) = r_c \theta l_c \mu_0 H(r_c) \tag{3.32}$$

In addition, the following formula can be obtained from the definition of MMF.

$$F_c = \int_{r_c}^{r_c + g_c} H(r)\mathrm{d}r = \frac{\Phi_c}{\mu_0 \theta l_c} \ln \frac{r_c + g_c}{r_c} \tag{3.33}$$

Thus, the permeability can be obtained by the following formula:

$$\Lambda_c = \frac{\Phi_c}{F_c} = \frac{\mu_0 \theta l_c}{\ln(1 + g_c / r_c)} \tag{3.34}$$

When $g_c \le r_c$, the formula 3.34 can be simplified as the following formula:

$$\Lambda_c = \mu_0 \frac{l_c \theta}{g_c} \tag{3.35}$$

In considering the diffusion of the lines of magnetic force at the edge of the pole, the permeability can be obtained by the following formula:

$$\Lambda_c = \frac{\mu_0}{g_c}\left[\theta\frac{(2r_c + g_c)}{2} + 2g_c\right](l_c + 2g_c) \tag{3.36}$$

where $r_c$ represents the radius of the workpiece; $\theta$ represents the angle between the magnetic poles; $l_c$ represents the width of the magnetic poles; $g_c$ represents the machining gap.

In the case of the same shape and size, the machining gap has great influence on the machining of magnetic materials and non-magnetic materials. Therefore, it should be discussed separately.

① Workpiece material is a magnetic material

When machining a cylindrical surface, the design drawing of the magnetic circuit is shown in Fig. 3.63.

The magnetic circuit includes two magnetoresistances. They form respectively the unilateral gap between magnetic pole and workpiece. In this condition, the magnetoresistance of the magnetic circuit can be calculated by the following formula:

$$R_{cm} = R_{cm1} + R_{cm2} = 2R_{cm1} = 2R_{cm2} \tag{3.37}$$

Then, the permeability of the magnetic circuit can be calculated by the following formula:

$$\Lambda_{cm} = \frac{\Lambda_{cm1}}{2} = \frac{\Lambda_{cm2}}{2} \tag{3.38}$$

According to magnetic circuit Ohm's law, the magnetic flux can be expressed as the following formula:

$$\Phi_{cm} = \frac{F}{R_m} = \frac{IN_0}{R_m} = \Lambda_{cm} IN_0 \tag{3.39}$$

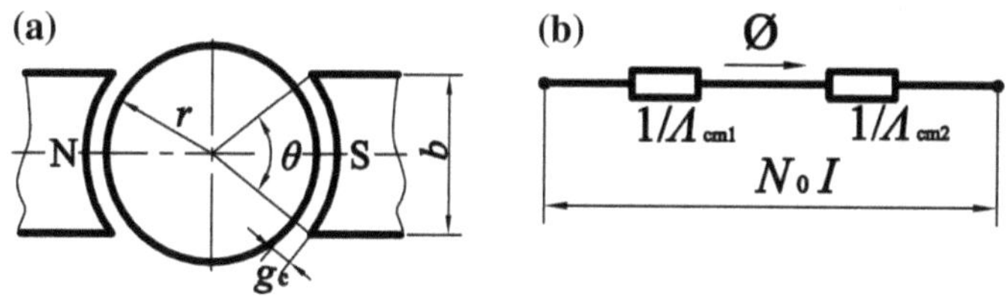


**Fig. 3.63** Magnetic circuit design for the processing of magnetic conduction workpieces. **a** Pole shape; **b** equivalent magnetic circuit

Therefore, the magnetic flux density within the processing region can be calculated by the following formula:

$$B_{\mathrm{cm}} = \frac{\Phi_{\mathrm{cm}}}{S} = \frac{\Phi_{\mathrm{cm}}}{\theta l_{\mathrm{cm}}(r_{\mathrm{cm}} + g_{\mathrm{cm}})} = \frac{\Lambda_{\mathrm{cm}} I N_0}{\theta l_{\mathrm{cm}}(r_{\mathrm{cm}} + g_{\mathrm{cm}})} \tag{3.40}$$

In the formula, permeability can be calculated by formula 3.36.

Engineering calculation examples: These are to be informed that the diameter of the workpiece is 30 mm, length is 45 mm, the width of the magnetic poles is 40 mm, and height is the two-thirds of the diameter that is 20 mm. The scope of the machining gap is 1–3 mm. Permeability can be calculated from above. When the magnetic flux density reaches to 1.4 T in the processing region, total MMF is required to overcome the magnetoresistance of all magnetic circuit. This can be calculated with the formula 3.40. The results are shown in Table 3.7. Wires have different diameters and maybe used in the solenoid coil. When the diameter is 1.5 mm, the excitation current can be 1.5 A. In this condition, turns in different machining gap are obtained out in the magnetic circuit. The results are shown in Table 3.7. Taking into account the leakage magnetic phenomena in magnetic circuit, the actual application should appropriately increase the number of turns. For example, when the machining gap is 3 mm, 3000 turns should be selected.

② Workpiece material is a non-magnetic material

When machining the cylindrical surface, the design drawing of the magnetic circuit is similar as Fig. 3.59.

Because the workpiece is a non-magnetic material, the magnetoresistance in the magnetic circuit is composed of the diameter of the workpiece and the machining gap. Obviously, the magnetoresistance of the non-magnetic materials is much larger than magnetic materials. Therefore, the magnetic flux density will be reduced a lot in the processing region. In order to facilitate the calculation, surface of the pole is approximately regarded as a rectangle. The following formula can be obtained.

$$B_{\mathrm{cf}} = \frac{\Phi_{\mathrm{cf}}}{S} = \frac{\Phi_{\mathrm{cf}}}{l_{\mathrm{cf}} b_{\mathrm{cf}}} = \frac{\Lambda I N_0}{l_{\mathrm{cf}} b_{\mathrm{cf}}} = \frac{I N_0}{g_{\mathrm{cf}}} \tag{3.41}$$

**Table 3.7** Calculation results

| Gap/mm | Magnetic permeance/H | Magnetomotive force/A | Turn ratio/Turn |
|---|---|---|---|
| 1 | $6.04 \times 10^{-7}$ | $2.01 \times 10^{-7}$ | 1400 |
| 2 | $3.34 \times 10^{-7}$ | $3.59 \times 10^{-7}$ | 2400 |
| 3 | $2.45 \times 10^{-7}$ | $4.85 \times 10^{-7}$ | 3300 |

**Table 3.8** Calculation results

| Gap/mm | Turn ratio/Turn |
|---|---|
| 1 | 9000 |
| 2 | 10,100 |
| 3 | 11,120 |

In the formula

$$\Lambda_{cf} = \frac{l_{cf} b_{cf}}{g_{cf}} \mu_0 \tag{3.42}$$

Engineering calculation examples: These are to be informed: The diameter of the workpiece is 30 mm, length is 45 mm, the width of the magnetic poles is 40 mm, and the height is the two-thirds of the diameter that is 20 mm. The magnetic flux density in the processing region is 1.4 T, current is 1.5 A, and the scope of the machining gap is 1–3 mm. Therefore, the number of turns is calculated. The results are shown in Table 3.8.

From the above calculation, it can be drawn that if the magnetic properties of the workpiece are different, the same source of magnetic field cannot be used because otherwise the desired effect cannot be achieved. When the machining gap of the non-magnetic material is 1 mm, permeability in this gap can be calculated by the following formula

$$\Lambda_{cf} = \left[\frac{l_{cf} b_{cf}}{g_{cf}} + \left(0.528 + \frac{1.28}{g_{cf}/x + 1}\right)(l_{cf} + b_{cf} + 0.308 + x)\right] \mu_0 \tag{3.43}$$

where $x$ means the edge length of the pole.

The answer can be calculated by the formula 3.43. In the same condition, permeability of the magnetic material is $6.04 \times 10^{-7}$ H. This shows that permeability of non-magnetic material of the workpiece is only a quarter of the permeability of magnetic materials.

When machining a non-magnetic workpiece on the source of magnetic designed for processing of magnetic materials, it is necessary to increase the current of about four times the original to achieve the same magnetic flux density. Conversely, when machining a magnetic workpiece by using the source of the magnetic design for processing of non-magnetic materials, magnetic saturation will occur.

For the permanent magnet, the critical of the design in magnetic circuit is the selection of the structure shape and size of the permanent magnet. The rest is similar to the electromagnetic source [45].

(2) Design of core and yoke

When the workpiece is a magnetic material and the machining gap is only 1 mm in the magnetic circuit, which was shown in Fig. 3.59, permeability can be calculated by using formula 3.36. The result is $6.04 \times 10^{-7}$ H. At this time, the magnetic flux through the core and the yoke of the frame is $2.08 \times 10^{-3}$ Wb.

If the cross section of the iron core and the yoke are square and when the saturation magnetic flux density is 1.6 T, it can be calculated that its unilateral length must be more than 36 mm according to the formula $B = \Phi/S$. As a result, the length of iron core and yoke is selected as 40 mm.

3. Finite element analysis of magnetic field

Take the electromagnetic field generated by DC as examples, which is used to machine the outer surface of the rotary in magnetic abrasive surface finishing. ANSYS software is used for the finite element analysis. After the adoption of software analysis, the distribution and movement of magnetic flux in a machining area can be determined. This provides the theoretical support for the analysis of processing mechanism, which provides a theoretical basis for the experimental study.

(1) Different workpiece materials

For non-slotted pole, the distribution of the magnetic lines of non-magnetic and magnetic in the machining region is shown in Fig. 3.60. It is shown in Fig. 3.64a that the lines of magnetic force are almost all going through the magnetic workpiece. If abrasive grains are placed into the gap between the head of the magnetic poles and the workpiece, the abrasive grains will be aligned along the lines of magnetic force, forming the magnetic brush in the gap. It is shown in Fig. 3.64b that permeability of non-magnetic materials is very small. Therefore, there is no line of magnetic force going through the workpiece, and it is distributed along the upper and lower sides of the workpiece. Through the contrast of the distribution of the magnetic force line and the formation of the magnetic brush, it can be concluded that the processing effect of the permeable material is better than that of the non-permeable material [46].

(2) Changes in the structure of the magnetic pole head

When machining a permeable workpiece with a different magnetic pole head structure, the contour map of the magnetic flux density in the machining area is shown in Fig. 3.65. The figure shows that the "magnetic flux" is formed at the edges of the slot portions. Here, the distribution of the lines of magnetic force is relatively dense, and the density of magnetic flux is larger. Due to the magnetic brush being dense, the stiffness is stronger. In addition, after the pole head is slotted, in the machining area the uniform magnetic field becomes non-uniform magnetic field, and the magnetic flux density is increased correspondingly. This situation is more conducive to improving the processing effect [47, 48]. Therefore, in order to improve the processing effect and improve the processing efficiency, the structure of the pole head with a discontinuous surface should be used. Figure 3.66 shows the photograph of the magnetic brush which formed under the slotted pole head.

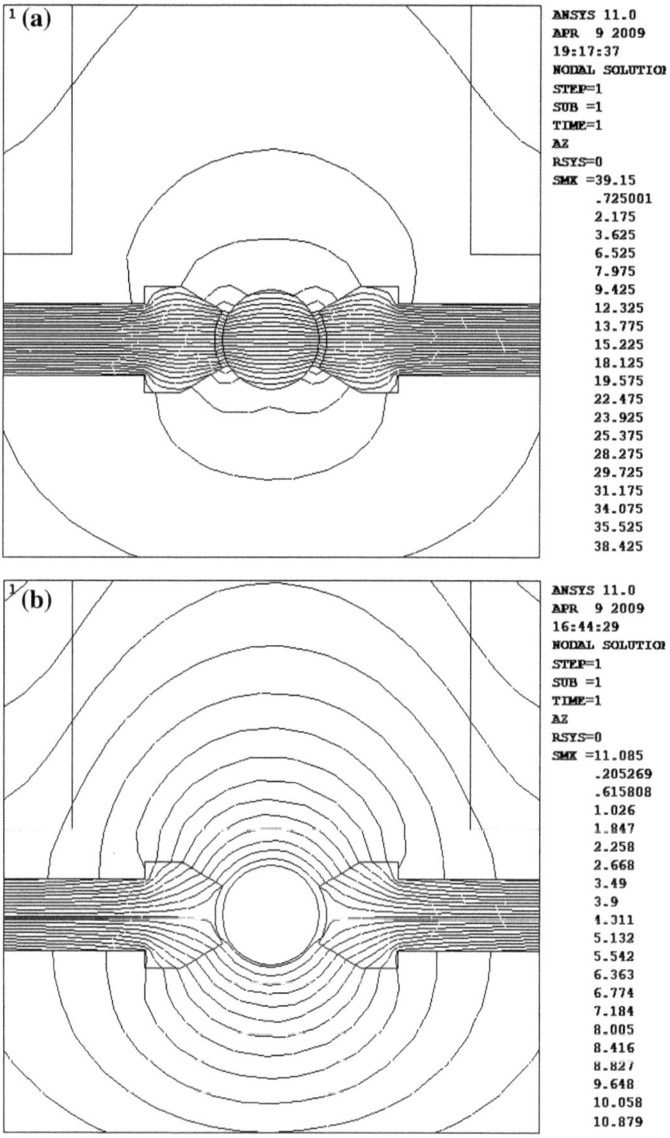


**Fig. 3.64** Distribution map of regional processing lines. **a** Magnetic conducting workpiece; **b** non-magnetic conducting workpiece

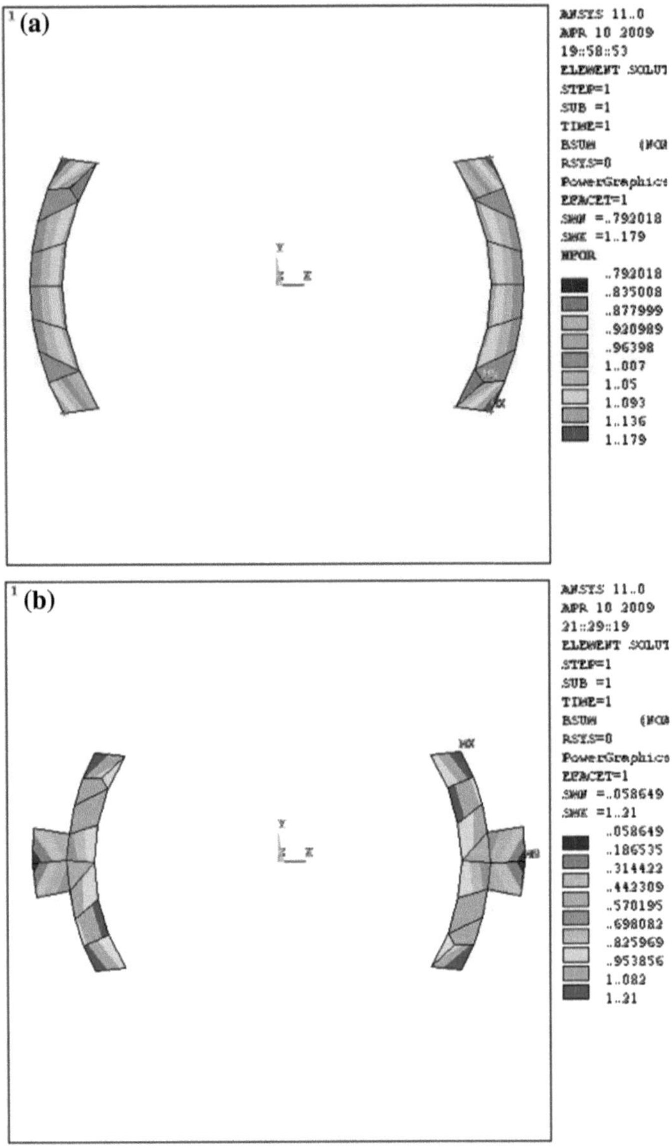


**Fig. 3.65** Magnetic flux density contour map of processing area in different magnetic pole head structures. **a** Non-slotted; **b** slotted

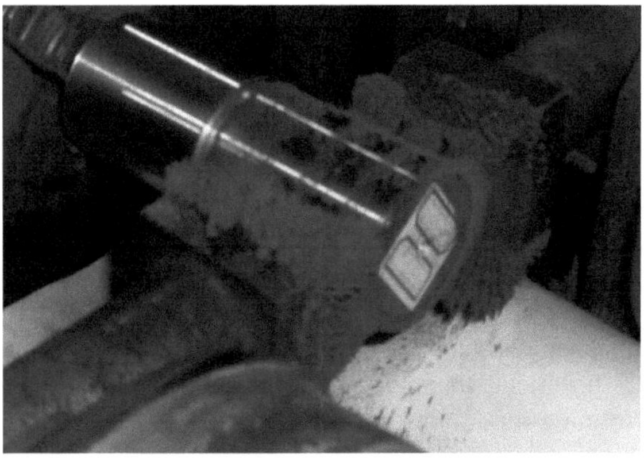

**Fig. 3.66** Magnetic brush photographs

### 3.3.3 *Design of Magnetic Pole Head*

In the magnetic abrasive finishing, the magnetic abrasive force on the workpiece is generated by the magnetic field, and processing methods are irrelevant. Magnitude of the force is affected by the changing magnetic flux density, machining gap, and the arrangement of the magnetic field generator.

In the case of the magnetic flux density is a certain value, the force is mainly determined by the shape of the magnetic pole and the space between the two poles. Thus, the structural design of the pole head is very important [49].

1. The surface of the magnetic pole head is a revolving body.

In general, the structural dimensions and shape of the magnetic pole head are related to the shape and size of the engineering parts. In many cases, the magnetic pole head is the matching parts of the machined part. Figure 3.67 shows a magnetic pole head for machining the outer surface of the workpiece. The shape of the magnetic pole head is the circular arc inner surface [50].

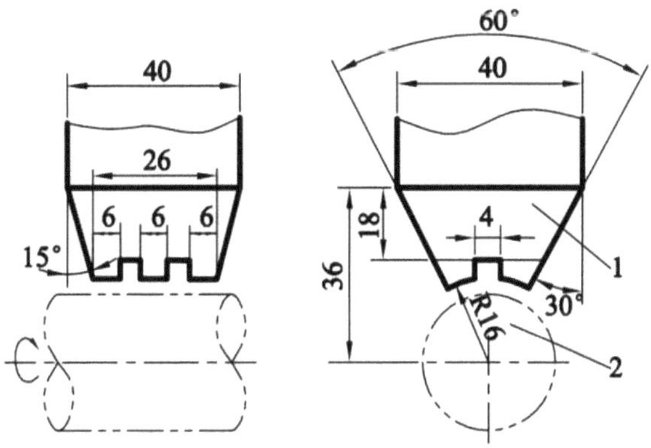


**Fig. 3.67** Shape of magnetic pole head. 1—Pole; 2—workpiece

Figure 3.67 shows that the surface of the magnetic pole is neither flat nor smooth. There are a lot of crisscross grooves. This can be explained by the formulas 3.3 and 3.4. The size of magnetic abrasive in magnetic field is not only related to the magnetic susceptibility and the shape of magnetic abrasive particles, but also related to the rate of change of the magnetic field intensity. The rate of change of the magnetic field intensity is also called the magnetic field gradient.

If the surface of the magnetic pole is provided with a groove, the gap between the pole and the workpiece will be changed. In large gap region, the magnetoresistance is larger, so magnetic field intensity is reduced. In the small one, the magnetoresistance is smaller; therefore, magnetic field intensity is strengthened. It is beneficial to improve the processing effect of magnetic abrasive finishing by forming a non-uniform magnetic field [51].

In the magnetic abrasive finishing process, when the magnetic pole surface is slotted, the change of magnetic field lines and magnetic flux intensity can be simulated by Maxwell 2D. Here, the simulation of magnetization process for the permanent magnet along its axial direction is an example. Figure 3.68 shows a map of the distribution of magnetic field lines when the non-magnetic material is processed. Figure 3.69 shows a map of the distribution of magnetic field lines when the magnetic material is processed.

As shown in Figs. 3.68 and 3.69, it can be learned that whether it is processing of magnetic materials or processing of non-magnetic materials, the lines of magnetic force and magnetic flux density will be uneven when the surface of the magnetic pole is provided with a groove. In the edge region of the groove, lines of magnetic force are most concentrated, the magnetic flux density is the biggest, and the processing capacity of the magnetic brush formed here increased significantly. The above are beneficial to improve the processing results. In addition, in the processing of magnetic materials, lines of magnetic force on the surface of the workpiece are relatively dense, therefore, the magnetic flux density is stronger. Results have shown that when the force is large, the surface quality after processing is good. The experimental results show that when the magnetic flux density on the smooth surface of the magnetic pole is 1.1 T, the phenomenon of focus, magnetic intensity, and gradient will increase as it will appear on the edge and corner of the groove. This is due to the drastic change of the flux. Thus, the magnetic field intensity and the magnetic field intensity gradient of these positions are greatly enhanced. The magnetic force will also increase greatly. The magnetic flux density at the edge portion can be obtained of 1.5 T.

In the process of machining the surface of a rotating body with magnetic abrasive finishing technology, various factors should be considered when designing the specific structure of the magnetic pole head, including the shape, size, and magnetic conductivity of the workpiece [52, 53]. Figure 3.66 shows a schematic diagram, which shows several different structures of the magnetic pole head.

Figure 3.70a–d shows that the basic shape of the poles is the same. Since the widths of grooves are different, the effective processing area of magnetic pole is

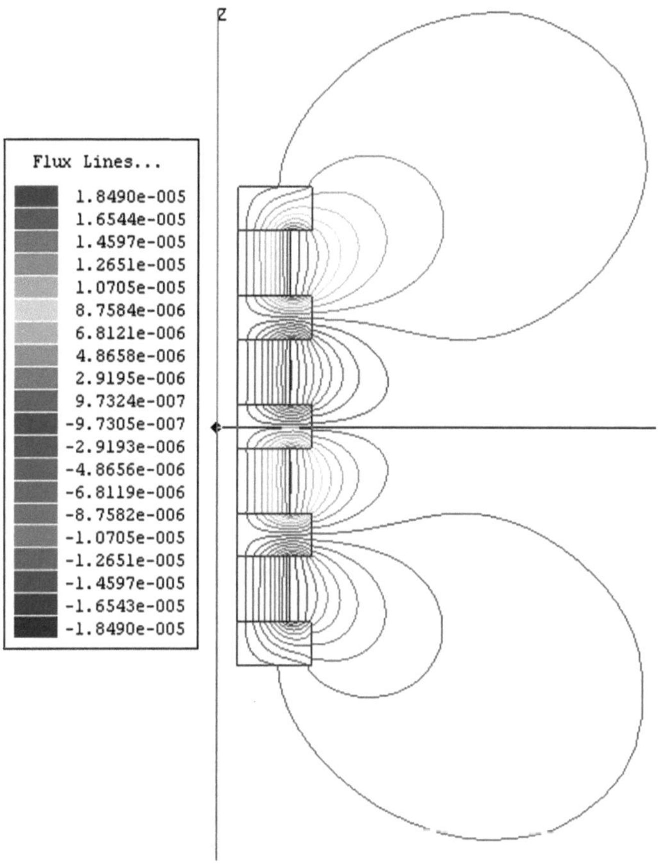


**Fig. 3.68** Processing of non-magnetic flux distribution map when the axial magnetization

different. As shown in Fig. 3.70e, b, the effective processing area of magnetic pole is the same, and the difference is the b provided with a groove but e with no groove; As shown in Fig. 3.70f, g, the shape and the effective area of the magnetic poles are equal, but working position is different.

When the workpiece is a magnetic material, the gap in the magnetic circuit is roughly equal to the machining gap. Most of the magnetic flux is emitted from a magnetic pole, flows into the workpiece through the machining gap, and finally flows into the other pole [54]. In order to minimize the magnetic flux leakage outside the machining area, the design of the magnetic poles can refer to some of the shapes shown in Fig. 3.70.

Through a large number of experimental studies, the influence of the geometry size of the magnetic pole head on the processing effect is summarized, and the data are listed in Table 3.9.

**Fig. 3.69** Processing of magnetic flux distribution map when the axial magnetization

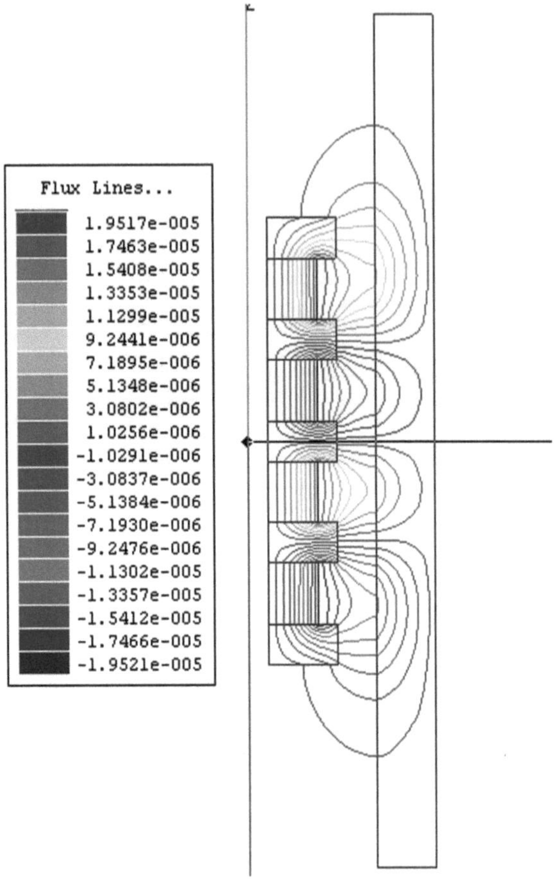


The following conclusion is supported by experimental results obtained in the table:

(1) The geometries of the magnetic poles are the same, and effective working area is shown in Fig. 3.70a–d. The metal removal is decreased with the decrease in working area.
(2) The effectiveness of the working area for the magnetic poles b and e is equal, and the geometries for both of them are different. In this case, a strong metal removal is obtained on the surface of the magnetic pole with a groove. Grooved pole has the uneven surface, so the machining gap can be changed; thus, the non-uniform magnetic field in the processing area will be formed. The magnetic field lines at the edge of the groove are dense and the magnetic flux density is large, under the action of strong magnetic force. The magnetic abrasive grains can be concentrated in the small area of the machining gap, but they are relatively less in the grooves are. Thus, a plurality of different sizes of the magnetic brush is formed, and the processing ability is enhanced. Therefore, in

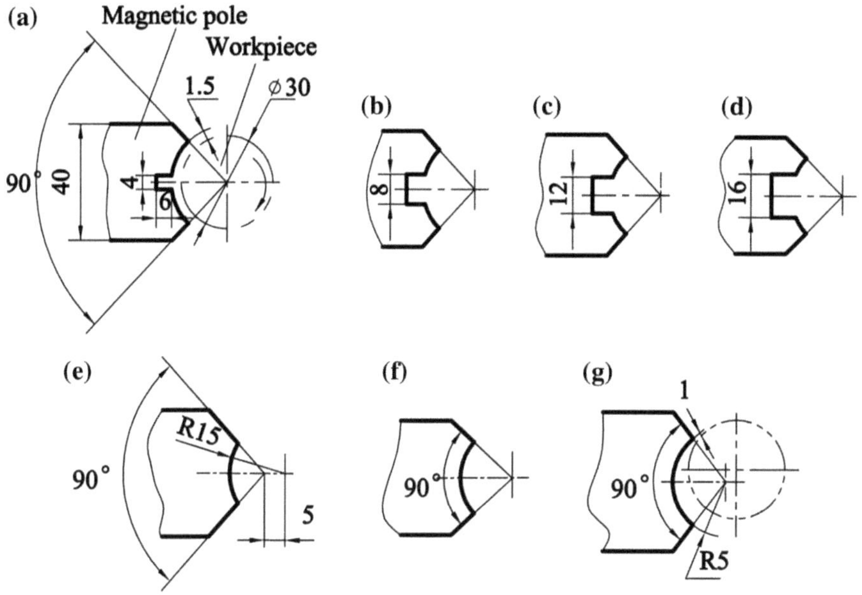


**Fig. 3.70** Magnetic pole head with different structures

**Table 3.9** Effect of geometry size of magnetic pole head on machining effect

| Pole head shape (Fig. 3.66) | a | b | c | d | e | f | g |
|---|---|---|---|---|---|---|---|
| Surface area $S$/cm$^2$ | 17.5 | 16 | 11.6 | 8 | 16 | 19 | 19 |
| Metal removal $G$/mg | 15 | 15.3 | 11.6 | 8.4 | 11 | 18 | 25 |
| Surface roughness $R_a$/μm | 0.45 | 0.48 | 0.65 | 0.75 | 0.58 | 0.41 | 0.58 |

the slotted poles, the surface area in the processing is reduced, but the amount of metal removal has become larger. Meanwhile, the different shape, size, and density of the groove will have a greater impact on machining quality and efficiency. It needs to be taken into consideration during the design process.

(3) The effective area and the geometry of the magnetic poles f and g are approximately equal, but the centers of those two are different from the center of the workpiece, and metal removal rate is also different. Due to the center of the pole g is different from the center of the workpiece when the pole is installed, the metal removal of the magnetic pole g is larger, and the extrusion effect on the surface of the workpiece is generated during the process. With the gap between the workpiece and the magnetic pole being narrowed, the magnetic flux density is increased, lines of magnetic force become dense, and the magnetic abrasive will flow along dense lines of magnetic force. Due to the rotary motion of the workpiece, in the smaller gap region, the arrangement of the magnetic abrasive grains is dense and compressed, so that the surface of the

workpiece can be processed. But too strong extrusion effect will scratch the surface and increases the surface roughness, so the shape of the magnetic pole is commonly used in rough machining [55].

In this section, the pole head with the shape as shown in Fig. 3.70 is used for processing the non-magnetic material.

The following results can be obtained by comparing with the processing effects between non-magnetic materials and magnetic materials.

(1) The effect from the increase of the working area of the magnetic poles on the metal removal is not obvious. The reason is that the workpiece is non-magnetic material; it has no magnetic domain structure and does not have magnetic properties. In the magnetic circuit, the phenomena of serious magnetic flux leakage of the non-magnetic material make it difficult to obtain more than 1 T of the magnetic flux intensity in machining area. And actually, it can at most obtain only 0.5 T, thereby reducing the metal removal ability.
(2) In the magnetic pole shape g, the center of the pole and the center of the workpiece are biased which allow an un-uniform gap to be formed. At the smaller gap, the magnetoresistance is small and the magnetic flux intensity is larger, so that more magnetic abrasive grains are concentrated here, resulting in the increasing of the metal removal capacity and the enhancement of the processing ability.
(3) Compared with the magnetic pole b and e, the metal removal of b is greater. This once again shows that regardless of whether the workpiece is a magnetic material or not, the processing capacity of the pole shape with uneven and discontinuities surface is better than continuous and smooth surface.

2. Magnetic pole head of plane and free surface

When smooth pole head processes the plane and curved surface, the following phenomena are discovered. It is found that the delamination and disconnection phenomenon appear between the surface of the magnetic pole and the magnetic abrasive near the workpiece surface during the process of plane machining. This causes the rapid wear of the magnetic pole surface and the reduction of the service life of the magnetic abrasive particles; when the normal direction of the contact point of the machining surface is coincident with the rotation axis in the process of machining surface, the situation is similar to that of the plane processing. However, when a certain angle is formed between them (Fig. 3.71), the phenomenon of the magnetic abrasive grains remaining on the surface of the workpiece will reappeared; at this time, the object to be processed is likely to be reversed, so that the surface of the workpiece becomes the surface of the pole, and the surfaces of the magnetic pole will become the surface of the workpiece. Figure 3.72 shows a comparison of the removal of metal from the surface of the smooth ball magnetic pole with the grooved ball one in the process of curved surfaces machining.

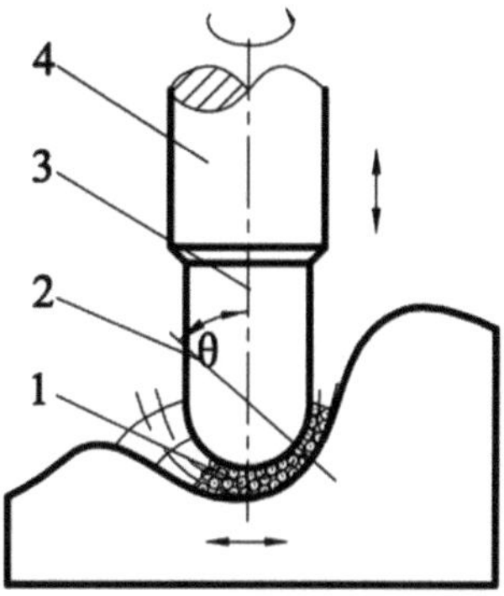


**Fig. 3.71** Schematic diagram of machining free surface of smooth spherical pole head. 1—Magnetic abrasive particles; 2—normal contact point; 3—axis of rotation; 4—pole

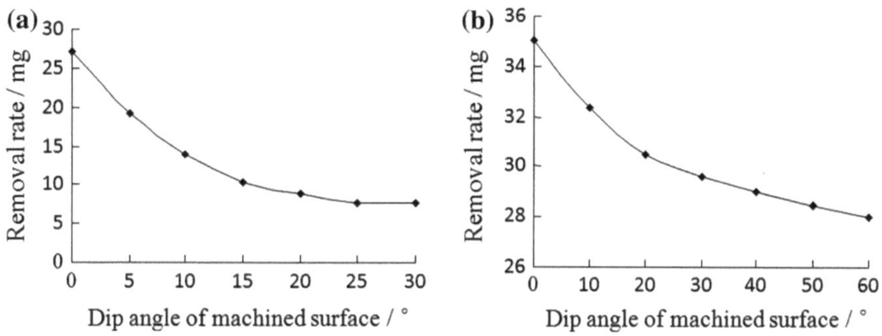


**Fig. 3.72** Relationship between the tilt angle of the machined surface and the metal removal. **a** Smooth spherical magnetic pole head; **b** slotted spherical magnetic pole head

From Fig. 3.72, conclusion can be drawn that with the inclination angle of the machined surface changes from 0° to 30°, the metal removal rate of the ball head pole from the smooth surface is decreased by 64%. For the ball head pole of the grooved surface, the metal removal rate of the workpiece surface only decreased by 17% when the inclination angle of the machined surface changed from 0° to 60°. Thus, at the edge of the groove on the surface of the magnetic pole, the magnetic field gradient will turn from a uniform field to a non-uniform field. This can effectively improve the working conditions of the magnetic abrasive particles in the magnetic field, increase the magnetic force of the abrasive on the workpiece, and improve the processing effect.

The complex free surface (such as the mold cavity surface) is always composed of horizontal, vertical, inclined plane, inner and outer arc surface, and so on. Corresponding to these basic elements, the geometry of the magnetic pole head can be designed into some typical shapes.

According to the above discussion, it is found that the working surface of the magnetic pole should not have smooth continuous transition. So the magnetic head working surface shall be a discontinuous grooved surface, to form a non-uniform magnetic field, aggregate the magnetic lines of force, improve the magnetic flux intensity, and further obtain a stronger the cutting ability and a higher machining efficiency.

Figure 3.73 shows some common geometric shape of the pole head.

Figure 3.74 shows the combined type planar permanent magnetic pole head, and Fig. 3.75 shows the combined type curved surface permanent magnetic pole head. The magnetic field source and the magnetic pole can be made into one combined pole, which is more simple and applicable, and has the following advantages:

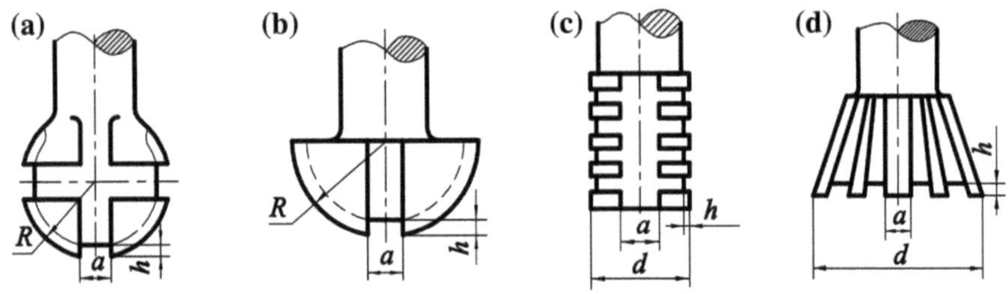


**Fig. 3.73** Magnetic pole head of common geometry shape. **a** Sphere; **b** hemisphere; **c** cylinder; **d** discoid

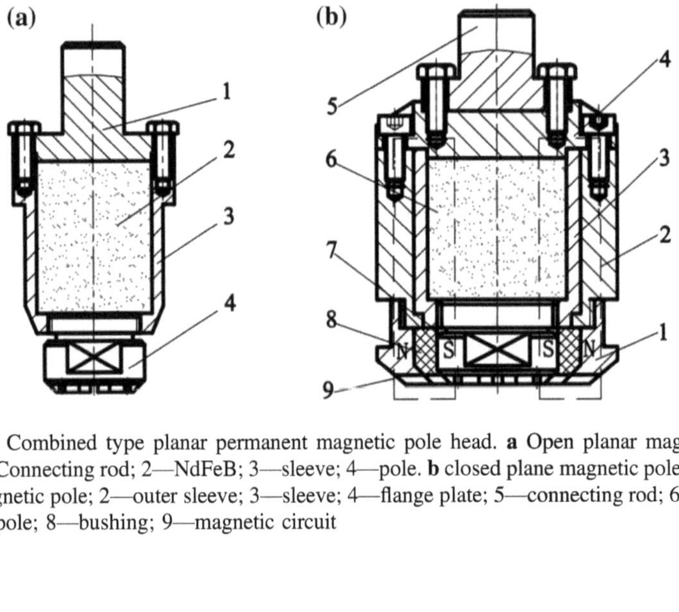


**Fig. 3.74** Combined type planar permanent magnetic pole head. **a** Open planar magnetic pole head. 1—Connecting rod; 2—NdFeB; 3—sleeve; 4—pole. **b** closed plane magnetic pole head. 1—Outer magnetic pole; 2—outer sleeve; 3—sleeve; 4—flange plate; 5—connecting rod; 6—NdFeB; 7—inner pole; 8—bushing; 9—magnetic circuit

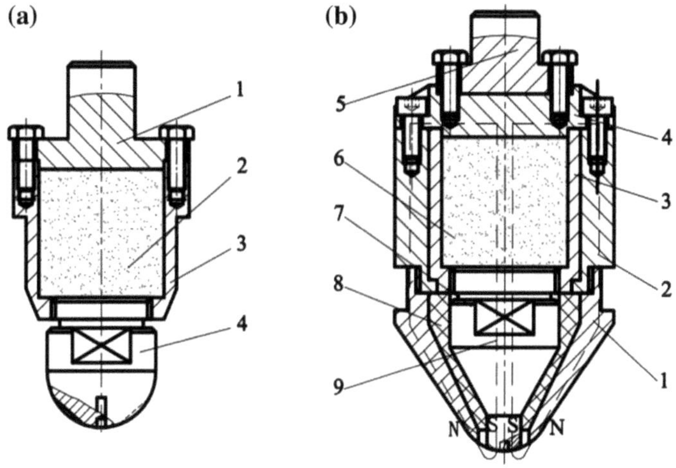


**Fig. 3.75** Combined surface permanent magnetic pole head. **a** Open circuit surface magnetic pole head. 1—Connecting rod; 2—NdFeB; 3—sleeve; 4—pole. **b** Closed circuit surface magnetic pole head. 1—Outer magnetic pole; 2—outer sleeve; 3—sleeve; 4—flange plate; 5—connecting rod; 6—NdFeB; 7—inner pole; 8—bushing; 9—magnetic circuit

(1) It not only avoids the influence of the eddy-current heat on the machining quality, but also integrates the magnetic field source and the magnetic pole head into a whole, so that the device has the advantages of small size, light weight, and flexibility. The connection with various devices is easier and more convenient, and reliable for future operations.
(2) The structure of the permanent magnetic pole head with convenient manufacture and high processing efficiency is provided.
(3) It has changed the original structure of the magnetic pole; it also made the deburring and the finishing of some parts became realized. These parts include various molding surfaces, special-shaped curved surfaces, internal and external circular surfaces with different sizes of tiny grooves and cross-holes, and the surface cannot be finished precisely using traditional methods. It solves the technical problems of practical production.
(4) Since a plurality of permanent magnets, the processing efficiency is significantly improved [56]. The surface roughness of those parts are effectively reduced, edge burrs are removed, the distribution of the surface stress of the parts is improved, the surface hardness is improved, and these will lead to improvements in coordination, conductivity, sealing, and service life of the parts.

The magnetic pole head mentioned above realizes the processing of some surface by means of various processing equipments such as lathes, vertical milling machine, CNC machine tools, and NC machining centers, which proven their usefulness to be positive. These surfaces from the parts include the following: the inner and outer circular surfaces of the part, forming surface, the profile surface, the surface of small parts with grooves in different sizes and shapes, edges, and all kinds of root of the fillet.

The part's surface roughness after processing can drop a lot based on the original, and $R_a$ value can reach 0.2 μm. The burr removal effect for the edge of the groove, pitted surface, and the cross-holes is also very good, and it has the advantages of simple structure and convenient manufacture.

Figure 3.76 shows photographs of the several developed pole heads in different states.

The following is the finishing process and the effect of the two kinds of workpiece:

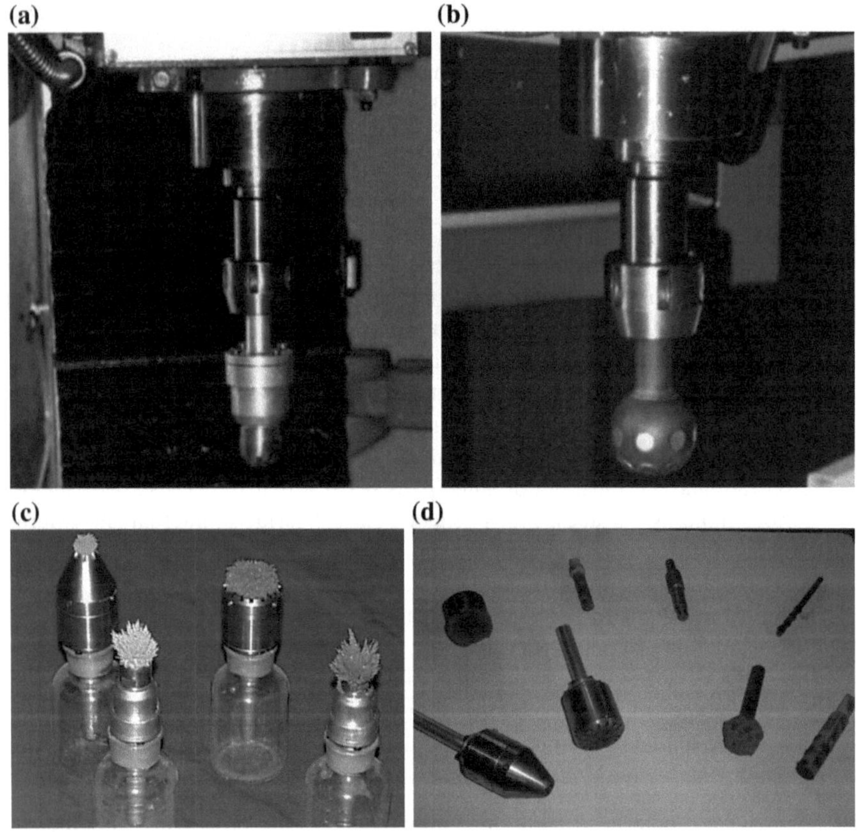


**Fig. 3.76** Photographs of the magnetic pole head. **a** Closed circuit surface magnetic pole head; **b** mosaic-type magnetic pole head; **c** the state of the magnetic pole of the abrasive particles; **d** some kinds of magnetic pole head

(1) The 3D surface is processed by mosaic type of permanent magnet pole head. After machining, the surface roughness $R_a$ value decreased to 0.21 μm from 0.85 μm. The forming process of permanent magnet pole head is as follows. First, a diameter of 40 mm Bakelite ball is processed, and then 17 cylindrical holes with 8 mm in diameter and 8 mm in depth are processed, $\theta = 45°$. Then, 17 cylindrical permanent magnets (diameter: 8 mm, height: 8 mm) were put into the cylindrical holes, respectively.

(2) The embedded permanent magnet pole head is used for finishing and removing the surface of the inner hole with cross-hole. The $R_a$ value of the surface roughness after machining is decreased from the original 0.64 μm down to 0.19 μm, and the entire burr is removed. The forming process of permanent magnet pole head is as follows. First, 8 mm in diameter, length of 90 mm brass cylinder is processed, 8 mm of the center distance, 60° in the interlaced axial direction, and 5 mm in diameter eight cylindrical holes are processed, and then 5 mm in diameter and height of 8 mm cylindrical permanent magnets are load to processed holes. Finally, the round hole and the cylinder are bonded together by using the epoxy.

## 3.4 Factors Affecting Finishing Effects

There are three main evaluation indexes of magnetic abrasive finishing: material removal rate; surface roughness value of the processed workpiece, and physical and mechanical properties of the surface layer of the workpiece [57, 58].

The main factors that affect the processing effect and efficiency of the magnetic abrasive finishing machining are the following: motion parameters, process parameters, and equipment parameters. When the surface of the revolving body is processed, the motion parameters include rotational speed of the workpiece, frequency of the axial vibration and amplitude, feeding speed. When the plane is processed, the motion parameters include the rotational speed of the pole head, the velocity of the longitudinal, and transverse of the workpiece. When the free-form surface is processed, the motion parameters include rotation speed of the magnetic pole head and the speed of the composite forming of the workpiece. Process parameters include the magnetic flux intensity, processing time, the type of magnetic abrasive particles, and the processing medium. Equipment parameters include machining gap, structure shape of the magnetic pole head, etc. Other parameters include the material of the workpiece to be machined, the initial surface roughness value, etc. This section mainly studies the effects of various parameters on the processing effect and processing efficiency and further determines the effect of magnetic abrasive finishing machining on the basis of the optimal matching. The process of the study and method is shown in Fig. 3.77 [59, 60].

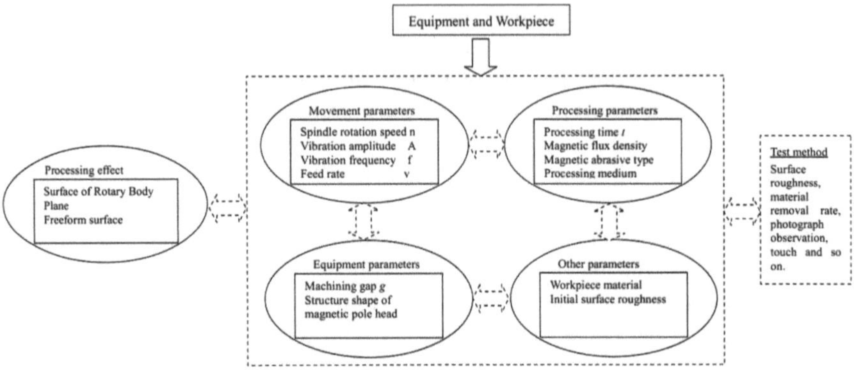


**Fig. 3.77** Research process and methods

## *3.4.1 Motion Parameters*

1. Relative rotation speed

For the surface of the rotating body parts, the relative speed of movement of the workpiece is usually determined by relative linear speed of rotation due to the different diameters of the workpiece. The choice of the parameter depends on the magnetic flux intensity, the shape of the magnetic pole, the gap, the processing time, and so on [61].

The outer surface of the exterior circle (workpiece material is 45 steel), has a diameter of 30 mm, is processed by the magnetic abrasive finishing technology. The conditions are as follows: Machining gap is 3 mm, axial vibration amplitude is 15 mm, the frequency is 2.3 Hz, lubricant is liquid paraffin, processing equipment is shown in Fig. 3.51, and the grade of the magnetic grinding grains is A31P4HFZ TG. The influence curve of different rotational speeds on the surface roughness of the workpiece is shown in Fig. 3.78.

Figure 3.78 shows that the higher the relative speed between magnetic abrasive and the workpiece, the greater the relative displacement per unit of time, and the

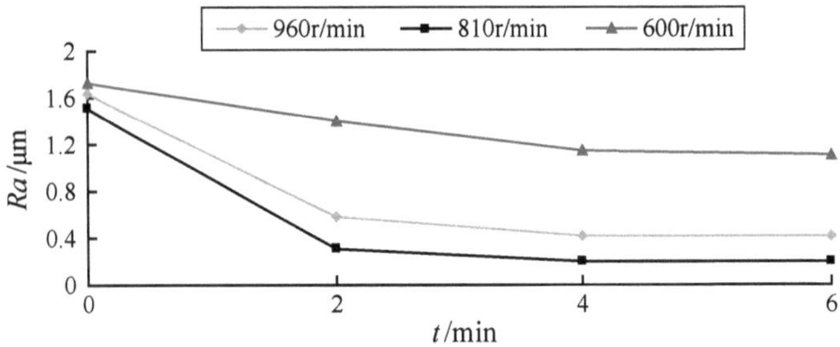


**Fig. 3.78** Influence of different rotating speeds on the surface roughness value

higher the efficiency. At the same time, in a certain range of relative motion speed, the value of the surface roughness of the workpiece will decrease with the increase in the relative motion speed. But when the relative motion speed exceeds a certain range, the magnetic abrasive particle will get rid of the bounds of the magnetic field and ejects due to the increases in the centrifugal force which loses the processing capacity; on the other hand, for the pole of the electromagnetic source, it is necessary to consider the influence of the eddy-current heat and the cutting heat on the surface quality of the workpiece.

The magnetic abrasive finishing technology is used to processes the 45 plates of steel. The conditions are as follows: Machining gap is 2 mm, lubricant is liquid paraffin, processing equipment is shown in Fig. 3.56, and the grade of the magnetic grinding grains is A31P4HFZ TG. The influence curve of different rotational speeds of the surface roughness of workpiece is shown in Fig. 3.79.

Figure 3.79 shows that when the rotation speed of the pole head is 649 r/min, the drop of the surface roughness of the workpiece is the fastest, and the processing efficiency is the highest.

However, when the rotation speed reached 1028 r/min and machining time is 3 min, the rise temperature of the workpiece is apparent with the smell of the melting binder. And the surface has a very fine powder. After removal of the powder, it has been found that the workpiece surface has been blackened by oxidation, while the magnetic grains appear agglomeration.

The relative rotational speed between the magnetic abrasive and the workpiece is also related to the saturation of the magnetic flux density, the increase in density of the magnetic abrasive, and the increase in magnetic properties of the processed workpiece. When other processing conditions are the same, the saturation magnetic flux density is higher and the density of magnetic grains is smaller, and this allowed the relative rotational speed to be higher.

A large number of experiments have shown that when the relative speed of movement of the workpiece reaches 150 m/min with the increase of processing time, the surface temperature will rise due to the increase of the cutting heat and

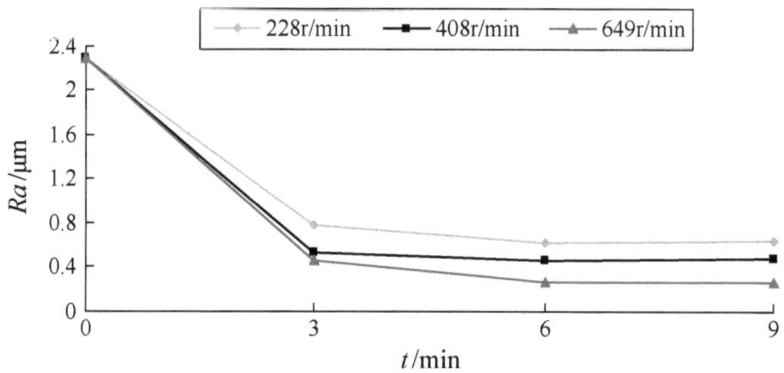


**Fig. 3.79** Influence of different rotating speed on the surface roughness value

eddy heat, and it will affect the quality of the surface. So the physical and mechanical properties of the surface are damaged. For example, the surface of the quenched steel will be annealed, and the surface hardness will decrease.

In short, the choice of the relative movement speed of the workpiece is very important, and it need to be determined according to the specific circumstances. Usually, the relative rotation speed is 20–100 m/min.

2. Relative axial vibration

In order to improve the processing effect and the machining efficiency, an axial vibration should be attached to the circular motion. It will not only improve the "self-sharpening" of magnetic abrasive grains, but also can effectively prevent the generation of the phenomenon of "the hole plugging of abrasive tools [62]."

The 30 mm surface of the exterior circle (workpiece material has 45 plates of steel) is processed by magnetic abrasive finishing technology. The conditions are as follows: Machining gap is 3 mm, the rotating speed of the workpiece is 610 r/min, lubricant is liquid paraffin, processing equipment is shown in Fig. 3.46, and the grade of the magnetic grinding grains is A31P4HFZ TG. The influence curve of axial vibration on machining effect is shown in Fig. 3.80 [63].

As shown in Fig. 3.80, the metal removal of the workpiece with axial vibration is much larger than the workpiece without axial vibration. This will shorten the processing time and improve the processing efficiency. At the same time, having the axial vibration than without the vibration is more conductive to the decrease in the surface roughness value. The value of the surface roughness after superimposed vibration is reduced from 1.6 to 0.2 μm.

The vibration of the workpiece makes the machining distance of the surface increase with each unit of time. The cutting path showed in workpiece surface is reticulated, and resulting in a superposition effect. It is also very beneficial to reduce the surface roughness value.

In addition, the synergistic effect will lead to stirring effect of the magnetic abrasive grains in magnetic abrasive machining gap to increase the strength and make frequent changes in the processing position. The effect is conducive to the

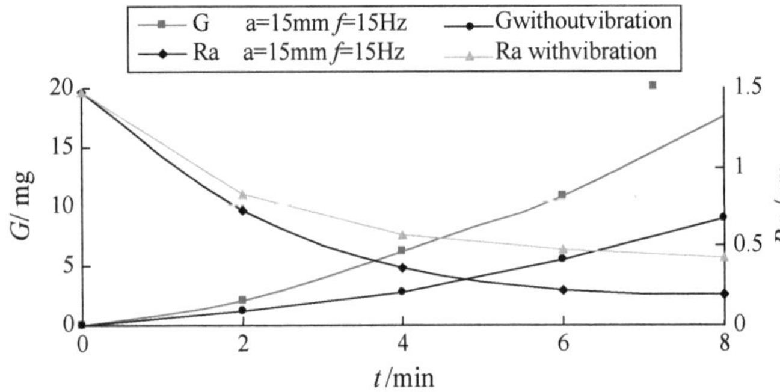


**Fig. 3.80** Curve of the effect of different axial vibration on the processing

emergence of new cutting edge persistence. It can be better on the surface processing to obtain a good processing effect.

The amplitude and frequency of the axial vibration should be determined by the experimental study.

When the amplitude is smaller and frequency is larger, the contact between the magnetic abrasive grain and workpiece surface is not sufficient. Therefore, the cross- and non-repeated reticulate trajectory is difficult to be produced in the tip of "magnetic brush" and the workpiece surface. The magnetic abrasive grain cutting edge in the workpiece surface cannot show the mixing function of the multi-direction cutting effect, thus affecting the processing effect.

In order to obtain the satisfactory processing effect, the amplitude and frequency of a certain size are required. The desired size of amplitude is 10–15 mm, and 2–5 Hz for frequency.

3. Feed movement speed

As the size of the magnetic pole head has a certain range, the feeding speed of the pole or the workpiece should be considered when machining the workpiece. For the determination of the feed motion speed, the maximum feed rate should be less than the amplitude of the vibration motion in order to ensure the machining effect.

## 3.4.2 Process Parameters

1. The magnetic flux density

The magnetic flux intensity is one of the main factors that affect the processing effect. With the change of the magnetic flux intensity, the processing efficiency and quality of the magnetic abrasive finishing machining can be changed, so that the different processing results can be obtained [64].

In determining the size of the magnetic flux density, we must pay attention to "magnetic saturation" phenomenon. The magnetic flux density in the core and magnetic circuit should be in no saturated state. Figure 3.81 shows the relation curve of the magnetic flux intensity and the current, which is obtained from the experimental study. The magnetic field generating device in the processing equipment is shown in Fig. 3.51.

Figure 3.81 shows that when the current is in the range of 0.2–1.2 A, the magnetic flux density will increase with the incensement of the current. When the current exceeds 1.4 A, magnetic flux density would not increase, even a modest reduction. This is called a magnetic saturation phenomenon.

The surface of the outer circle (workpiece material is 45 steel and the surface roughness of the unfinished workpiece is 1 μm), has a diameter of 30 mm, is processed by the magnetic abrasive finishing technology. The conditions are as follows: Machining gap is 3 mm, processing time is 3 min, the rotation speed of the workpiece is 960 r/min, axial vibration amplitude is 15 mm, the frequency is

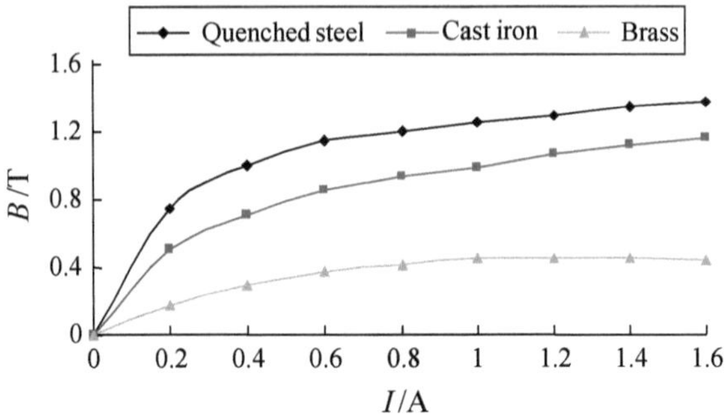


**Fig. 3.81** Relationship between the magnetic induction intensity and the current size

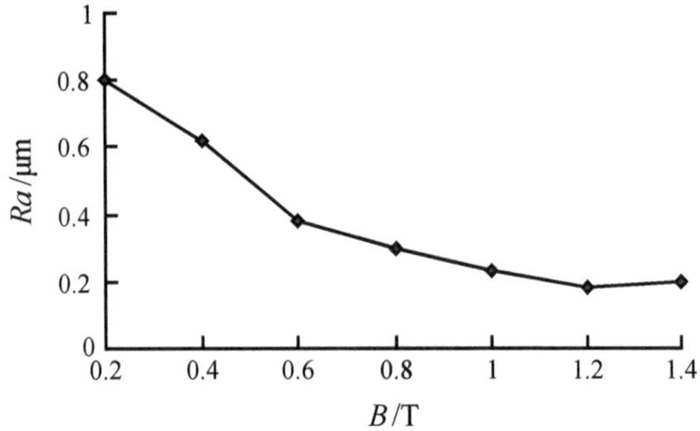


**Fig. 3.82** Influence of different magnetic induction intensity on the surface roughness value

2.3 Hz, lubricant is liquid paraffin, processing equipment is shown in Fig. 3.46, and the grade of the magnetic grinding grains is A31P4HFZ TG. The influence curve of different magnetic flux density on the surface roughness is shown in Fig. 3.82.

Figure 3.82 shows that within the same processing time, the drop of the workpiece surface roughness will increase as the magnetic flux density increases, and so the processing efficiency will also increases; When the magnetic flux density is greater than 0.6 T, processing 3 min can make the surface roughness value decreased to level 1–2, but if the magnetic flux density exceeds 1.4 T, the surface roughness value will rise. From the formula (4.3), it is known that the force on the surface of the workpiece by magnetic abrasive is proportional to the quadratic magnetic flux density. When the magnetic flux density is weak, the force becomes small, and the holding force between the “magnetic strings” formed by magnetic grains connected along magnetic field lines is also weak. When the workpiece and the magnetic abrasive grains at the end of a magnetic string have a mutual movement, the magnetic abrasive grain is easy to fall off from the “magnetic string,” so the surface roughness value of the workpiece is decreased slowly. With the increase in the magnetic flux density, magnetic holding force, and the holding force of the “magnetic strings,” processing capacity has been strengthened, and the surface roughness value is decreased more quickly.

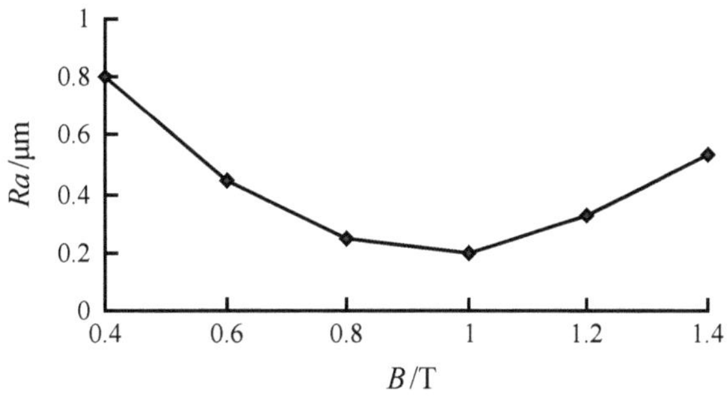


**Fig. 3.83** Influence of different magnetic induction intensity on the surface roughness value

When the magnetic flux density continues to increase, it will allow the holding force of the "magnetic string" to further increase. Also with the rigidity of the "magnetic brush" significantly enhanced, the force is further increased. Therefore, the magnetic abrasive particles near the surface are attached to the surface, and the "magnetic brush" does not generate a tumbling phenomenon, reducing the processing capacity. It is possible to make the magnetic grains with rotary motion of the workpiece lose cutting ability by flying away from processing areas; meanwhile, with the increase of the rigidity of the magnetic brush, the scratch on the surface of the magnetic abrasive grain is increasingly serious, and the value of the surface roughness also increases.

The 45 plates of steel (the surface roughness of the unfinished workpiece is 0.95 μm) are processed by magnetic abrasive finishing technology. The conditions are as follows: Machining gap is 2.5 mm, processing time is 3 min, the rotation speed of the workpiece is 649 r/min, lubricant is liquid paraffin, processing equipment is shown in Fig. 3.56, and the grade of the magnetic grinding grains is A31P4HFZ TG. The influence curve of different magnetic flux density on the surface roughness is shown in Fig. 3.83.

Figures 3.81, 3.82, and 3.83 show that there is an optimal range of the magnetic flux density. When machining magnetic materials, the magnetic flux density should be selected in the 0.8–1.2 T, usually not more than 1.4 T.

2. Processing Time

In order to accurately determine the optimal processing time of magnetic abrasive, a new method of replacing the same quality is carried out in each time period, taking into account the effect of the loss and wear debris of the magnetic abrasive grains. Figure 3.84 shows the change law of the value of surface roughness of the workpiece with time. Figure 3.84 shows that the surface roughness value declined sharply in the early stages of processing, but 4 min later, it is small change and finally basically stable at a certain value. The value is about 0.48 μm. It can be drown from the above that there is a limit value of the surface roughness. At this time, the processing time is called the optimum processing time.

Figure 3.85 shows the influence curve of all the different sizes of the magnetic abrasive grains on the surface roughness of the workpiece under different

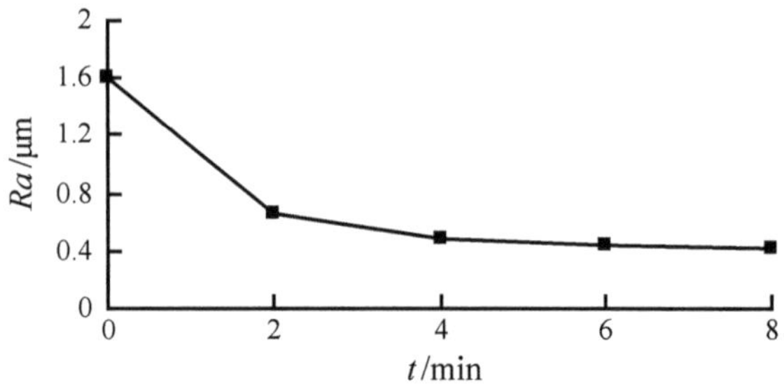


**Fig. 3.84** Influence of finishing time on the surface roughness

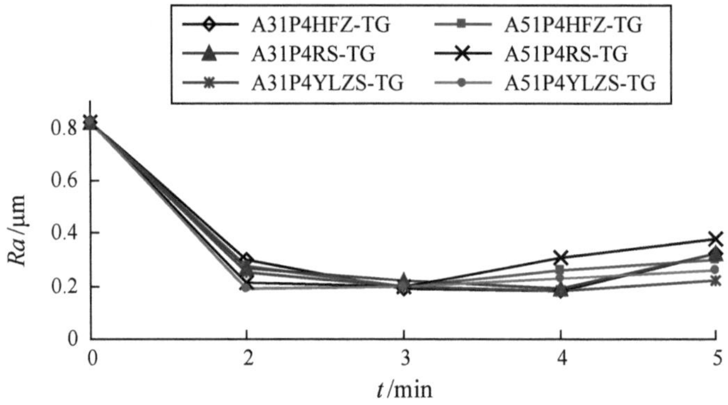


**Fig. 3.85** Influence of different magnetic abrasive particles on the surface roughness value

processing time. The magnetic abrasive grains are prepared by vacuum sintering method, hot-pressing sintering method, and bonding method. Figure 3.85 shows that the surface roughness of the workpiece is decreased with the increase in the processing time, whether or not the magnetic abrasive grains are prepared by the vacuum sintering method, hot-pressing sintering method, or the bonding method. In a period of time at the beginning of the process, the surface roughness (shown in the figure is 2 min) decreased rapidly, and then it becomes gradually smooth. However, when the processing time is over a certain range, the surface roughness value of the workpiece will not decrease, and it is likely to increase. It also shows that there is an optimum processing time in magnetic abrasive finishing.

According to the above test results, the best processing time of the magnetic abrasive finishing is generally 3–4 min.

2. Magnetic abrasive grain

The choice of magnetic abrasive grains mainly considers the size of the magnetic abrasive particles, the composition of the magnetic abrasive particles, and the ratio of the magnetic abrasive particles. The law of influence can be found in the relevant content of the second section.

4. Processing medium

Theoretical analysis and extensive testing showed that adding a certain amount of process media in the magnetic abrasive, on the one hand, it may play the role of infiltration in the magnetic particles, reducing the scattering of the magnetic abrasive grains; on the other hand, it can reduce the impact of cutting heat and eddy-current heat on the processing quality (such as the emergence metamorphic layer, crack, residual stress, etc.), and also it can effectively improve the gloss and brightness of the workpiece surface. In the processing of magnetic abrasive finishing, the common processing medium is liquid paraffin, stearic acid, silicon oil, and so on.

The surface of the outer circle (workpiece material are hardened steel), has a diameter of 40 mm, is processed by the magnetic abrasive finishing technology. The conditions are as follows: Machining gap is 3 mm, the rotation speed of the workpiece is 610 r/min, axial vibration amplitude is 15 mm, the frequency is 2.3 Hz, lubricant is liquid paraffin, processing equipment is shown in Fig. 3.51, and the grade of the magnetic grinding grains is A31P4HFZ TG. The influence curve of different processing medium on the surface roughness is shown in Fig. 3.86.

As shown in Fig. 3.86, if magnetic abrasive grain does not add any processing medium, the surface roughness values will gradually decline as time passes b. But the processing effect is poor, and the gloss and brightness of the workpiece surface is not ideal. After adding a certain amount of processing medium, the grinding effect of the magnetic abrasive action on the workpiece becomes strong, and it will leave more machining marks on the machined surface. The value of the surface roughness will decrease more significantly, and it can effectively improve gloss and brightness of the surface. Its influence law is related to the type and quantity of the processing medium. The test results show that the effect of processing of liquid paraffin is better than that of stearic acid. The main reason is that the temperature resistance of stearic acid in the process is not balanced, so it does affect the processing.

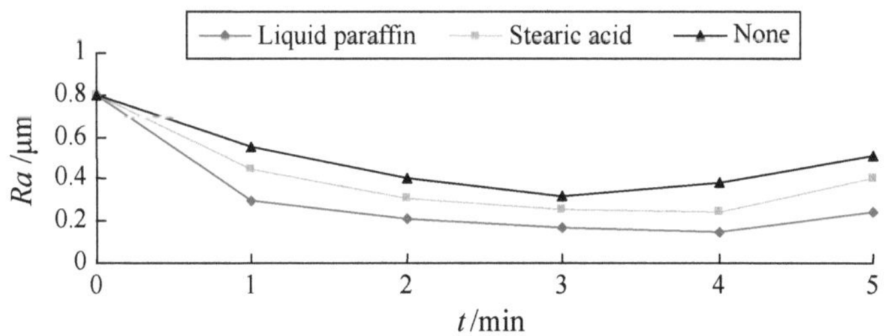


**Fig. 3.86** Influence of processing medium on the surface roughness value

## *3.4.3 Equipment Parameters*

1. Machining gap

For the processing of magnetic materials, the machining gap is the distance between the magnetic pole and the workpiece (shown in Fig. 3.6).

On the one hand, it is closely related to the size of the magnetic field in the magnetic circuit. With the increase in the machining gap, the magnetic reluctance of the magnetic circuit increases, the magnetic flux leakage increases, and the magnetic induction intensity decreases; this led to the decrease in the processing ability of the magnetic abrasive (affecting the processing results). On the other hand, it is closely related with the filling amount of magnetic abrasive grains. It can be known that the choice of machining gap is very important, and it has a direct effect on the efficiency and quality in the magnetic abrasive finishing. The size of the machining gap can be determined by experiment.

The surface of the outer circle (workpiece material is 45 steel), has a diameter of 30 mm, is processed by the magnetic abrasive finishing technology. The conditions are as follows: The rotation speed of the workpiece is 960 r/min, axial vibration amplitude is 15 mm, the frequency is 2.3 Hz, lubricant is liquid paraffin, processing equipment is shown in Fig. 3.51, and the grade of the magnetic grinding grains is A31P4HFZ-TG. The curve of machining gap and the surface roughness value of the workpiece obtained when the magnetic abrasive grains are filled with enough quality is shown in Fig. 3.87.

Figure 3.87 shows that when the machining gap is 3 mm, the surface roughness value decreases from 1.61 to 0.67 μm; while the gap is 2.25 mm or 4 mm, the surface roughness value will have only a little decrease. That is because there is a problem about the size of the rigidity of "magnetic brush" which formed in the machining gap. Under certain processing conditions, when the machining gap is relatively small, the "magnetic string" formed in the machining gap is shortened, magnetic induction intensity becomes stronger, the flexibility of the magnetic brush decreases near to nothing; therefore, the action of the tumbling and stirring of the magnetic grains cannot be sufficiently played, and the scratches will form easily on the work surface that results in the roughness on the workpiece which does not improve significantly. At the same time, due to the decrease in the machining gap, it

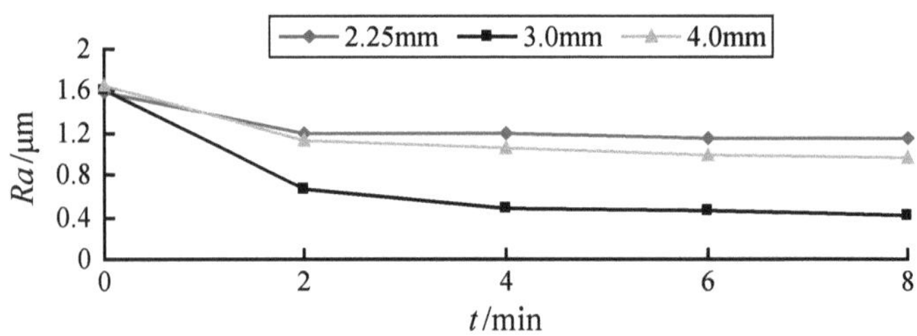


**Fig. 3.87** Curves of surface roughness value with different processing times

is not easy to fill enough magnetic abrasive grains, which will also affect the processing results. And when the machining gap is relatively large, the following conditions will appear: Magnetic flux leakage will increase, the magnetic induction intensity will be weakened, the force of the magnetic abrasive action on the surface of the workpiece will be reduced, and the flexibility of the "magnetic brush" is relatively large. These conditions make the magnetic abrasive grains cannot be well maintained in the machining gap, resulting in decrease in the processing capacity, and the roughness of the workpiece surface cannot be significantly improved.

The magnetic abrasive finishing technology processes the plate of the 45 steel. The conditions are as follows: The rotation speed of the workpiece is 649 r/min, lubricant is liquid paraffin, processing equipment is shown in Fig. 3.56, and the grade of the magnetic grinding grains is A31P4HFZ-TG. The influence curve of different machining gaps on the surface roughness is shown in Fig. 3.88.

Through the experimental study, when the permeable material or non-permeable material is processed, in order to obtain a better processing effect, the generating device of the magnetic field may take on different arrangements, and the machining gap can generally be taken to about 3 mm.

2. Filling amount

After the machining gap is determined, the filling amount of magnetic abrasive grain will also affect the processing effect. Filling amount is the amount of magnetic abrasive grains in the process of processing. The selection of the filling amount of magnetic abrasive grains will directly affect the surface quality and the cost of processing, because too much amount will cause the dispersion of magnetic abrasive grains to the area outside the processing. It is only when the filling quantity is appropriate, the magnetic abrasive particles in the magnetic field can form the ideal magnetic brush, the rolling movement phenomenon is generated, and the optimal processing results can be obtained.

The surface of the outer circle (workpiece material is 45 steel), has a diameter of 30 mm, is processed by the magnetic abrasive finishing technology. The conditions are as follows: The rotation speed of the workpiece is 960 r/min, machining gap is 3 mm, processing time is 3 min, axial vibration amplitude is 15 mm, the frequency is 2.3 Hz, lubricant is liquid paraffin, processing equipment is shown in Fig. 3.51,

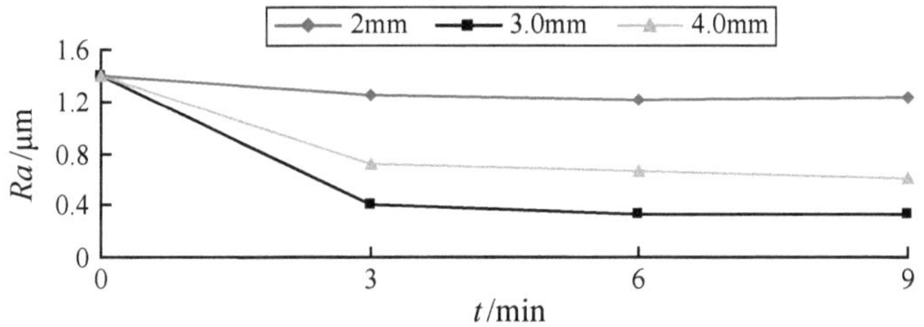


**Fig. 3.88** Curves of surface roughness value with different processing times

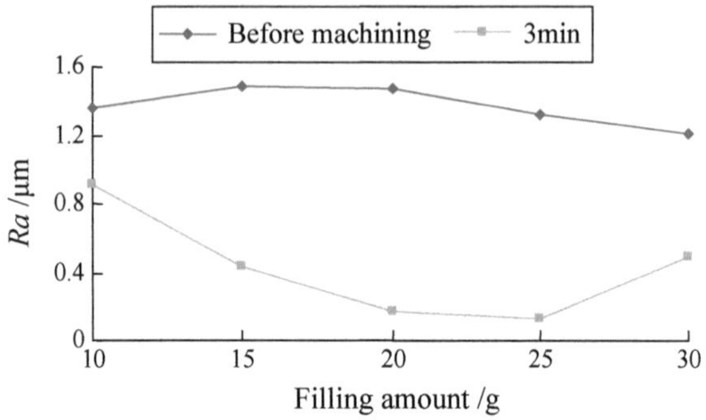


**Fig. 3.89** Influence of surface roughness on the surface roughness of machined surface

and the grade of the magnetic grinding grains is A31P4HFZ-TG. Influence curve of different filling amount of magnetic abrasive grain on the surface roughness is shown in Fig. 3.89.

Figure 3.89 shows that in certain conditions, with the increase of filling amount, the number of grinding abrasive grains in each unit area is increased, the surface roughness value is decreased more obviously, the processing effect is better, and the processing efficiency is higher. When the filler amount is 20 g, the value of the surface roughness decreases from 147.7 to 0.166 μm; when the filler amount is 25 g, the value of the surface roughness decreases from 1.317 to 0.127 μm; when the filling amount increases to 30 g, the decrease in the value of the surface roughness is obviously slowed down, and the value is reduced from 1.21 to 0.488 μm. Excessive amounts can cause scattering of magnetic abrasive to the area of the machining outside, while the magnetic grains will accumulate in a local. Therefore, it causes processing unevenness and affects the processing results.

The amount of filling is related to the size of the workpiece, the machining gap, and the structure of the magnetic pole head. In order to better measure or choose the filling amount, only from the point of view of considering the quality is not comprehensive, measurement should be based on the volume between the workpiece and the magnetic pole. Through theoretical analysis and a large number of experimental observations, it can be learned that it is suitable to fill the gap between the workpiece and the magnetic pole with magnetic abrasive grains. Due to the magnetic field, the magnetic pole will absorb a certain amount of magnetic abrasive grains, and the actually used quality of the magnetic abrasive grain should be slightly increased. That can be expressed by the following formula:

$$m = k\rho V_{\mathrm{g}} \tag{3.44}$$

where $m$ represents the filling amount of magnetic abrasive grains (g); $k$ represents the correction factor; $\rho$ represents the magnetic abrasive density (g/cm$^3$); $V_{\mathrm{g}}$ represents the volume between the workpiece and the magnetic pole (cm$^3$).

According to the results, when the filling amount is determined, the correction factor can be taken as 1.2–1.4. In order to verify the validity of the method, a corresponding magnetic abrasive finishing experiment for plane was carried out.

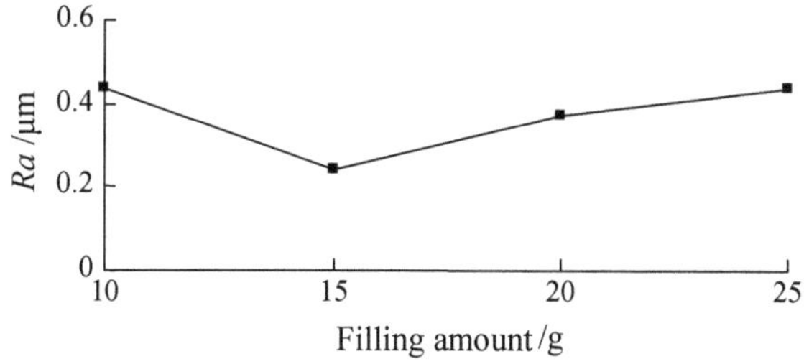


**Fig. 3.90** Influence of filling amount on surface roughness of machined surface

The plate of the 45 steel (the surface roughness of the unfinished workpiece is 1.2 μm) is processed by the magnetic abrasive finishing technology. The conditions are as follows: Machining gap is 3 mm, processing time is 3 min, the rotation speed of the workpiece is 649 r/min, lubricant is liquid paraffin, processing equipment is shown in Fig. 3.56, and the grade of the magnetic grinding grains is A31P4HFZ-TG. Influence curve of different filling amounts on the surface roughness is shown in Fig. 3.90.

Figure 3.90 shows that the influence of the filling amount on effect of the plane and the surface of the outer circumferential is the same. When the filling amount is less than or equal to 15 g, with the increase of the filling amount, the number of grinding particles is increased in per unit area, the value of the surface roughness is significantly reduced, and the minimum value can reach 0.24 μm. When the filling amount is larger than 20 g, the local accumulation phenomenon will occur, the grinding is not uniform, and the surface roughness value is increased.

3. Pole head

Discussion of the relevant content in the third section shows that in addition to the design of magnetic source, the properties of the magnetic grain and machining gap, and other factors, the shape and size of the magnetic head is the most important factor because it can change the magnitude and distribution of the magnetic field, and it directly affects the processing efficiency and the processing quality of the workpiece.

In general, for the surface of the rotating body, plane, free surface, and other types of surfaces, the design or choice of the structure of the magnetic pole head can refer to the relevant content of the third section.

Machining parts with special structures, it is necessary to design a special magnetic pole to get better a processing result. For example, when processing the large non-magnetic sleeve parts and the change in diameter, in order to reduce the machining gap and increase the magnetic flux intensity of the processing area, a magnetic head can be arranged inside the workpiece (it is also called as slotted magnetic spindle).

The long diameter non-magnetic sleeve-type part (The material is aluminum alloy, the diameter of the workpiece is 105–90 mm, the wall thickness is 2 mm, and the length is 78 mm.) is processed by magnetic abrasive finishing technology. The conditions are as follows: The range of machining gap is 2–12 mm, the rotation speed of the workpiece is 500 r/min, lubricant is liquid paraffin, processing equipment is shown in Fig. 3.58, and the grade of the magnetic grinding grains is A41P4HFZ-TG, and separately using the structure of magnetic pole head with no spindle installed, a smooth magnetic spindle installed, and slotted magnetic spindle to processing workpiece (These three poles is shown in Fig. 3.91). The influence curve of different structure of magnetic pole head on the surface roughness is shown in Fig. 3.92.

Figure 3.92 shows that when the spindle is not loaded, because of the larger diameter of the workpiece (which leads to large machining gap), the magnetic flux density in machining area is only 0.16 T. In the process, most of the magnetic abrasive grains are adhered to the inner wall of the workpiece, and there is no relative motion between the magnetic abrasive grains and the workpiece, so the surface roughness value is almost no change.

**Fig. 3.91** Magnetic pole structure photographs. **a** Outer magnetic pole head; **b** inner magnetic pole head (Slotted magnetic spindle); **c** processing state

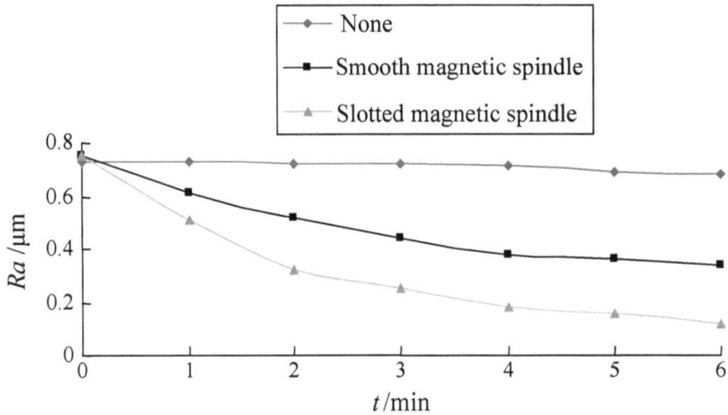


**Fig. 3.92** Influence of the pole structure on the surface roughness of the surface of the inner hole

When the smooth magnetic spindle is loaded, due to the existence of the spindle, it changes the machining gap and the distribution of the magnetic field, thus increasing the magnetic flux density in the processing region. In the machining process, the magnetic abrasive is no longer adhered to the inner wall of the workpiece. At this point, the spindle is equivalent to a magnetic pole. Because there is no slot, non-uniform magnetic field is formed, so the value of the surface roughness is only decreased by one level after processing.

When slotted magnetic spindle is loaded, the value of the surface roughness of the workpiece decreased from 0.75 to 0.12 μm. It is a good processing result. This is mainly because when the magnetic spindle is provided with a groove, the magnetic field gradient changes from the uniform field to non-uniform field. It can effectively improve the force of magnetic abrasive particles in the magnetic field. Small magnetic brushes can be formed at each edge of grooves; therefore, the force of magnetic abrasive particles on the surface of the workpiece is greatly increased, and processing results can be significantly improved.

### 3.4.4 Other Parameters

1. Material of the workpiece

The magnetic properties of the workpiece are very important properties to the effect of the processing and the processing efficiency. If the magnetic material workpiece is processed, the size of the magnetic material will not have a greater impact on the processing results. The magnetic flux density in the machining area is about 1 T in general when there is no-load, so it can obtain a satisfactory processing effect in a relative short amount of time. However for the non-magnetic materials, the size of the workpiece will directly affect the strength of the magnetic flux density in the machining area. For example, a non-magnetic material with a diameter of 30 mm, the magnetic flux density in the machining area can only reach about 0.5 T, so it is difficult to achieve a good processing effect. Through the test, under the same condition, the magnetic flux density of the magnetic material workpiece is three times to four times higher than that of the non-magnetic material workpiece.

(1) Magnetic material workpiece

Figure 3.93 shows the distribution of the force on the surface of the workpiece by magnetic abrasive in the machining. Lines of magnetic force generated by the magnetic pole N flow to the surface, through the workpiece, and then flow out from the surface, reach to the pole S. The magnetic brush composed of magnetic abrasive grains will be formed along the lines of magnetic force and acts on the surface of the workpiece in a certain size. At the same time, the magnetic flux density in the machining area is much higher than that in the machining area, forming a complete inhomogeneous magnetic field. This makes the magnetic abrasive grains located

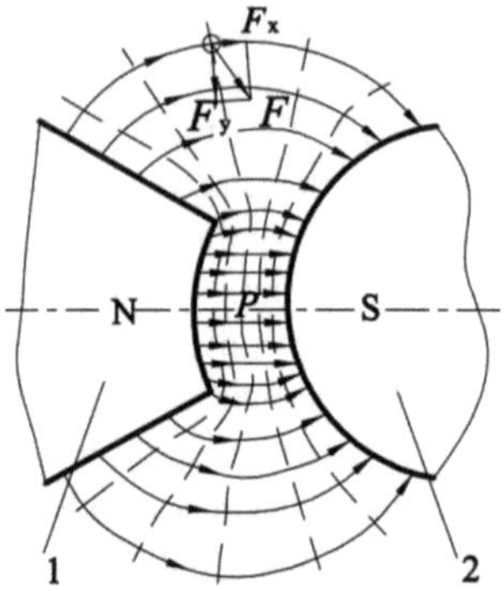


Fig. 3.93 Schematic diagram of the force distribution of the magnetic conducting workpiece. 1—Pole; 2—workpiece

outside the machining area to be concentrated and filled in the machining area, forming a strong magnetic effect. This also causes the magnetic abrasive folded, which effectively avoids the scattering of the magnetic abrasive grains caused by the rotary workpiece. "Magnetic brush" in the process of processing constantly focuses on and folded under strong magnetic force and great force, so the processing efficiency is high. For magnetic materials, as long as the machining gap remains unchanged, they will still be the same kind of processing effect even if the workpiece diameter increases.

(2) Non-magnetic material workpiece

If the workpiece material is a non-magnetic material, the situation is very different. As shown in Fig. 3.94 At this time, the magnetic flux density of the magnetic field is only related to the two magnetic poles of S–N, the formation of "magnetic brush" can also be determined only by these two poles of S–N that makes the reluctance of the magnetic circuit increases rapidly, and lines of the magnetic force will disperse to result in the magnetic flux density decreased significantly. And with the increase in the diameter of the workpiece, the magnetic flux density gradually decreases. It is different from the magnetic material parts that there are two kinds of pressure in the process. One kind of force is the positive

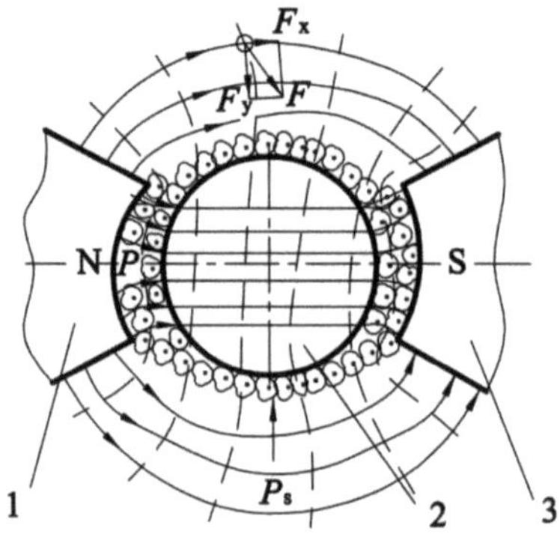


Fig. 3.94 Schematic diagram of the force distribution of non-magnetic conducting workpiece. 1—N-pole; 2—workpiece; 3—S-pole

pressure acted on the surface of the workpiece, which is produced by "magnetic brush" neatly arranged along lines of magnetic force between two poles. The other is the side pressure generated by the magnetic brush on the side surface of the workpiece. These two kinds of pressure act on the workpiece surface together, both participating in the processing. But the side pressure is much smaller than the positive pressure, and the pressure is not possible to be produced in the machining of the magnetic material. It is a special kind in machining process of the non-magnetic material. Besides the magnetic field intensity and the gradient of it, the factors affecting the positive pressure and the side pressure are also related to the shape and size of the magnetic pole head.

The plates (the materials of the plate are, respectively, 45 steel and aluminum alloy) are processed by magnetic abrasive finishing technology. The conditions are as follows: The machining gap is 3 mm, the rotation speed of the workpiece is 649 r/min, lubricant is liquid paraffin, processing equipment is shown in Fig. 3.56, and the grade of the magnetic grinding grains is A31P4HFZ-TG. The curve of different materials on the surface roughness is shown in Fig. 3.95.

Figure 3.95 shows that when processing a non-magnetic conducting material (likes aluminum alloy), the value of the surface roughness decreases slowly with the increase of processing time due to the decrease of the magnetic flux density. When the processing time is 2 min, the surface roughness value of the workpiece decreases from 2.0 to 0.9 μm. However for the magnetic material (likes 45 steel), it dropped to 0.48 μm by the 2.0 μm.

In summary, under the same processing conditions, the effects of the magnetic material and the non-magnetic conducting material are different. In order to achieve the ideal processing effect, it can be realized by selecting different magnetic field generators and the magnetic poles with specific structure.

2. The initial state

The surface of the outer circle (workpiece material is 45 steel) with the diameter of 40 mm is processed by magnetic abrasive finishing technology. The conditions are as follows: Machining gap is 3 mm, the rotation speed of the workpiece is

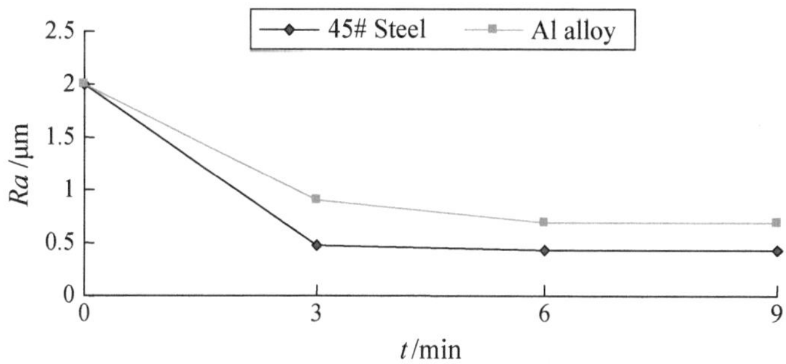


**Fig. 3.95** Influence of magnetic materials and non-magnetic materials on the surface roughness value

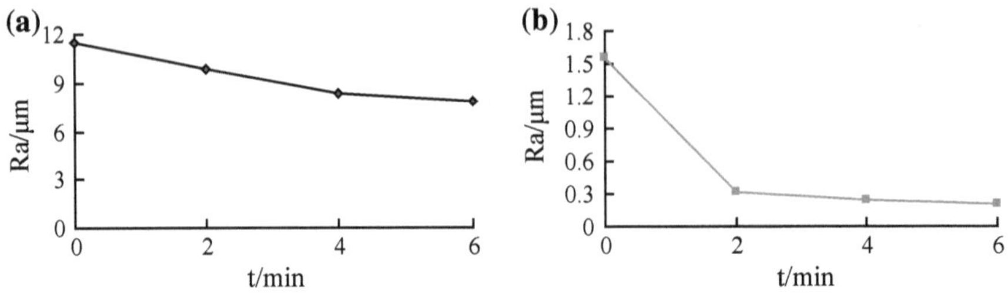


**Fig. 3.96** Effect of surface roughness on the processing results. **a** 1#workpiece; **b** 2#workpiece

610 r/min, axial vibration amplitude is 15 mm, the frequency is 2.3 Hz, lubricant is liquid paraffin, processing equipment is shown in Fig. 3.46, and the grade of the magnetic grinding grains is A31P4HFZ-TG. Influence curve of different surface roughness values of the unprocessed workpiece on the machining effect is shown in Fig. 3.96. Figure 3.96 shows that for the workpiece with different surface roughness values, these values will decrease by the same law with the increase in processing time. However after machining, the difference of value of the surface roughness is obvious. After processing 2 min, the surface roughness value of the workpiece is decreased from 11.5, 1.56 to 9.8, 0.31 μm, respectively. This also shows that the magnetic abrasive finishing is micro-machining, it has certain requirements for the original state of the workpiece, and it is a kind of surface finishing technology of precision parts.

## 3.5 Application Example of Magnetic Abrasive Finishing

### *3.5.1 Introduction*

When the magnetic abrasive finishing technology is applied, the prior-processing of the parts, the finishing of the parts, and the post-processing of the parts are mainly included.

The purpose of the prior-processing of the parts is to remove the oil slops, which was remained on the surface of the parts. The oil slops will enable the parts surface to be gloomy and also weaken the processing capacity of the magnetic abrasive particles. The effect of the magnetic abrasive finishing is reduced, and the method to deal with the oil slops is mentioned in the second chapter.

According to the structure, shape, size, the processing requirements, the processing ways, equipment type, and processing parameters (motion parameters, process parameters, equipment parameters, etc.) are selected to determine the result in the magnetic abrasive finishing of the parts.

The following process is to remove the rust and magnetization. Found through a large number of experiments, the surface becomes smooth and bright after the magnetic abrasive finishing, the active metal molecule exposed in the air soon will be oxidized to make the surface black, the rusts will occur at the end, and the derusting must be implemented; the workpiece with the special requirements needs to remove them magnetization.

### *3.5.2 Finishing Effect*

1. The surface geometry feature

In order to further study the surface geometry feature after the magnetic abrasive finishing, the surface characteristic parameter of the rough is tested by the surface roughness measuring instrument. Figure 3.97 shows the surface roughness value of the test curve under different processing time, and the main characteristic parameter variation curve of surface roughness with processing time is shown in Fig. 3.98.

From Figs. 3.97 and 3.98, the surface roughness of the parts was reduced from 0.6 to 0.19 μm; the surface profile support rate is increased significantly.

2. Formation of the surface compression stress

During the finishing process, due to certain pressure affected by the magnetic abrasive particles on the working surface, the surface plastic deformation makes the surface residual compression stress appear, and the deteriorative layer with a few microns in the depth is formed. The residual compression stress is related to the parameters such as the rotational speed, vibration frequency, amplitude, magnetic induction, magnetic abrasive particles types, and the filling amount.

The depth of the surface deteriorative layer after the magnetic abrasive finishing becomes different due to the different materials. For the magnetic material, the depth is about 2.7 μm. For the non-magnetic material, the depth is about 2 μm. The surface stress testing is exerted after the magnetic abrasive finishing. For the shafts for 45 steel, when the magnetic induction intensity is 1.2 T and the machining gap is 3 mm, the axial surface stress is −198 Mpa, and the radial surface stress is −230 MPa; for the inner surface of these parts which the material is made with brass, when the magnetic induction intensity is 0.9 T, the machining gap is 3 mm, the axial surface stress is 160 MPa, and the radial surface stress is −154 MPa. It can be seen that the magnetic abrasive finishing can greatly improve the stress state of the surface and improve the service life of the parts.

3. The deburring

The magnetic abrasive finishing can get rid of the burr produced through the cutting processing on the workpiece surface in all directions. Because the magnetic lines distribution in the magnetic field at the workpiece edge is not uniform. The place where the magnetic lines are dense will have the higher magnetic induction, so the magnetic force is larger. Whether the workpiece or the pole device has an axial vibration, the vibration direction is roughly perpendicular to the magnetic

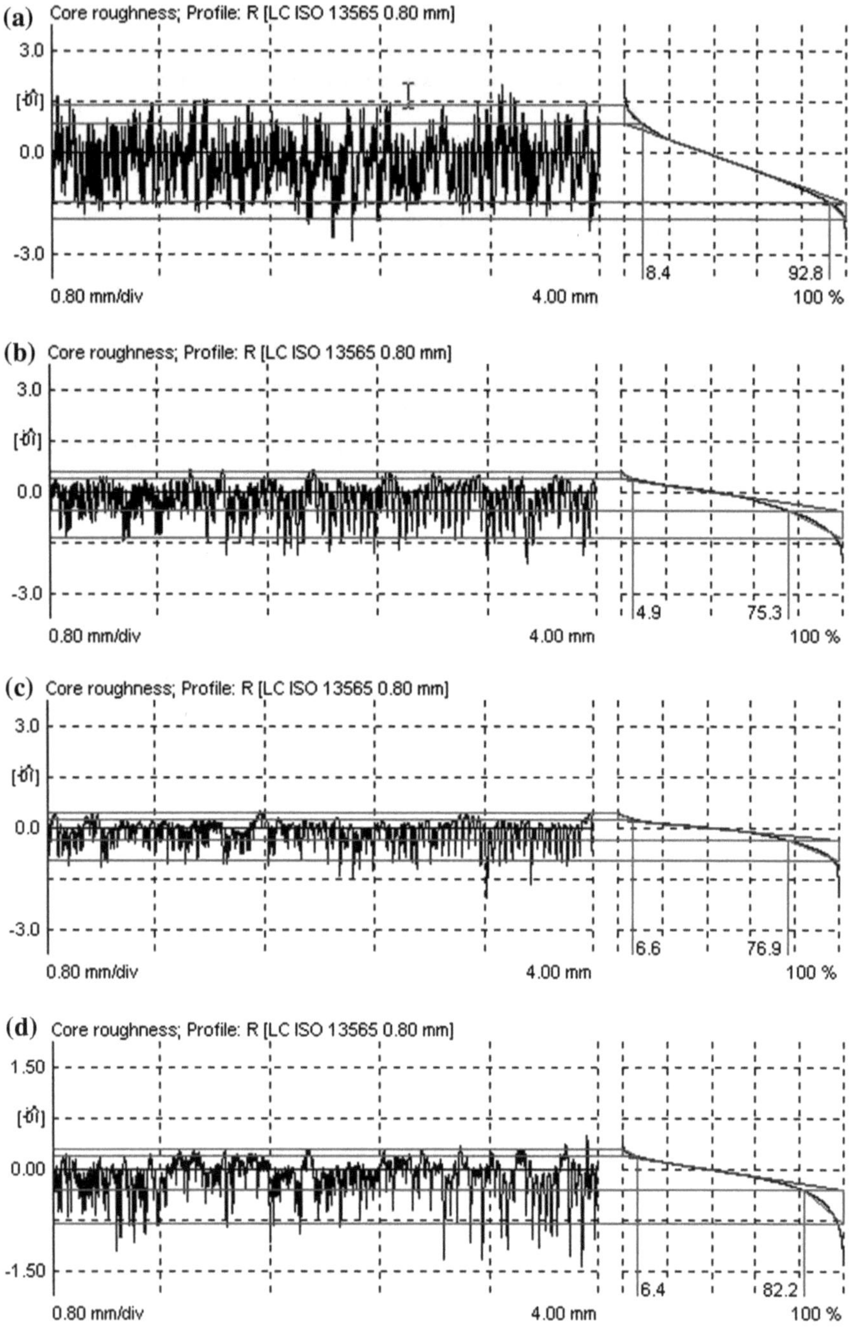


**Fig. 3.97** Test curve of surface roughness value with processing time. **a** Before processing; **b** 1 min; **c** 2 min; **d** 3 min

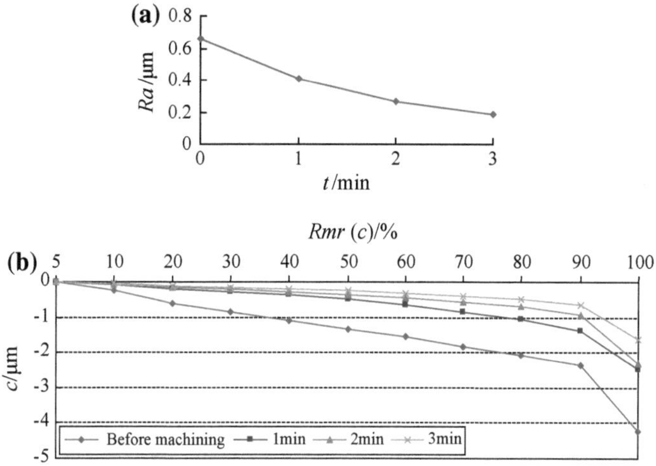


**Fig. 3.98** Main characteristic parameter variation curve of surface roughness with processing time. **a** *Ra*; **b** *Rmr* (*c*)

brush, and the end of the magnetic brush has stronger processing opportunities. Therefore, the burr generated along the radial direction that is vertical to the processing direction can be completely removed after the 1 min; for the burr generated along the axial direction, which is parallel to the machining direction, the time to removal the burr is relatively longer.

### 3.5.3 Application Examples

1. The outer surface

As mentioned before, known from the principle of the magnetic abrasive finishing, in addition to the external force, there is relative motion between the workpiece and magnetic pole along the circumferential and axial direction [65], which is determined by the diameter and length of the workpiece. In Fig. 3.99a, the working length of the magnetic pole is equal to the length of the workpiece, so all of the surface can be processed at the same time. The magnetic outer surface of which the diameter is 15–150 mm and the length is equal to or less than 200 mm magnetic can be processed in this way.

This method is also suitable for the processing of the non-magnetic outer surface of the sleeve having the outer diameter of 15–150 mm. The magnetic central spindle should be installed into the sleeve, so a closed circular magnetic circuit can be formed between the non-magnetic sleeve and the magnetic central spindle. But the total reluctance is obviously the bigger than that of the solid magnetic workpiece. According to $B = \Lambda NI/S$, the magnetic motive force and the coil number can be calculated.

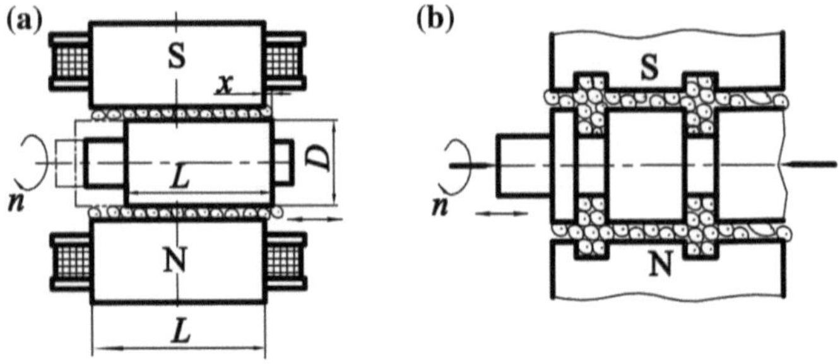


**Fig. 3.99** Diagram of continuous and discontinuous surfaces of the outer surface. **a** The continuous surface; **b** the discontinuous surface

Figure 3.99b shows the processing of the discontinuous surface. The shapes of magnetic pole and the workpiece are made of discontinuous surface. In the process of finishing, the workpiece is still required to be rotating and axial vibrating simultaneously. When the length of the workpiece is longer than 250 mm, there will be a series of difficulties in the processing as follows: ① The longer workpiece will lead to an increasing working surface of the magnetic pole and the filling amount of magnetic abrasive particles; ② with the increasing of the working area of the magnetic pole that the coil number must be increased to ensure the magnetic induction intensity enough should make the machine bigger; ③ increasing the magnetic motive force can make the magnetic force stronger for the rigid requirements of the workpiece; ④ with the increase in working length of the pole, the geometric deviation of the magnetic pole (roundness, cylinder, etc.) is reflected in the surface of the workpiece which will lead to the geometric shape deviation of the workpiece; ⑤ it is also difficult to produce axial vibration for the longer workpiece which will cause vibration of the magnetic pole and reduce the productivity. In view of the above situation, the structure shown in Fig. 3.99 is changed to the structure shown in Fig. 3.100. The length of the magnetic pole is still determined by the length of the workpiece which is less than or equal to 200–250 mm, and it is only to increase the reciprocating motion along the axial direction and can realize the total processing of the workpiece. But the new question occurs. When the workpiece rotates, the friction between the abrasive particle and the workpiece surface makes the magnetic abrasive particle move in the opposite

**Fig. 3.100** Diagram of the processing of the outer surface with large diameter

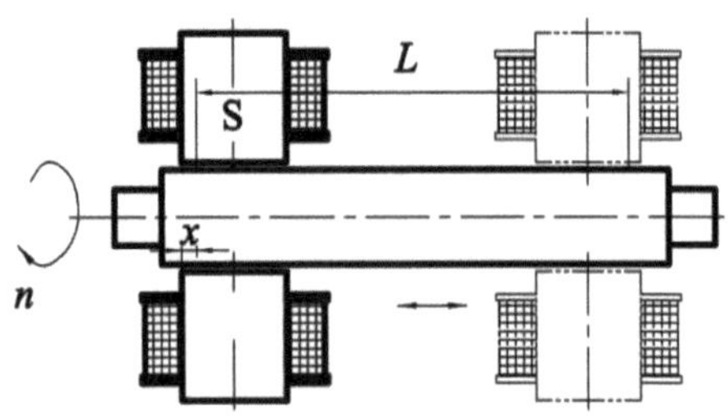

direction of the workpiece lead to the falling of abrasive particle behind the pole, so to increase the run-length of the magnetic pole to ensure the end of the workpiece fully processed, the moving length is shown by the double dotted line in Fig. 3.101. Without the magnetic spindle between two poles, the increasing gap will make magnetic induction intensity decrease a lot, which leads some of the magnetic abrasive particles to fly out of the gap, so the processing capacity is poor for the poor efficiency and quality of the processing. In order to obtain satisfactory processing effect, the inhomogeneous magnetic field lines can be artificially created. The structure shown in Fig. 3.102a is for the continuous smooth surface; the structure shown in Fig. 3.98b is for the discontinuity surface.

When the diameter of the workpiece is longer than 200 mm, because it is increasing, the increased processing area will weaken the magnetic induction intensity. In order to get the equivalent amount of the magnetic induction intensity, the number of coils must be increased and the size of the winding coil contour must also be increased at the same time, but the magnetic circuit becomes correspondingly large and bulky.

The device shown in Fig. 3.103 is used to solve these problems. When the workpiece rotates, the magnetic pole moves to-and-forward along the axis of the workpiece and the relative axial vibration between the workpiece and the magnetic pole can be generated, so the processing effect can be further improved.

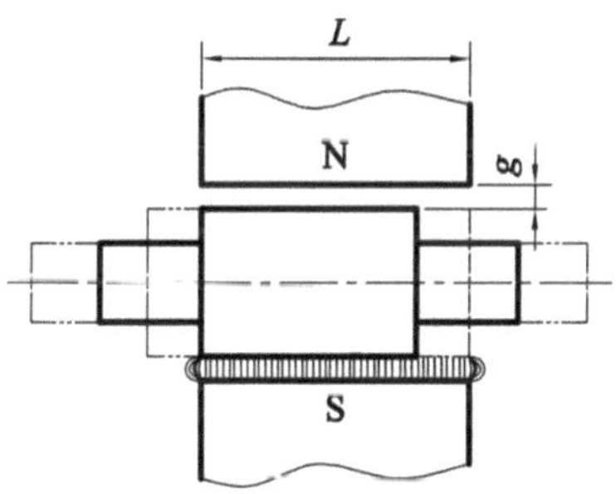


**Fig. 3.101** Processing diagram of increasing the magnetic pole moving length

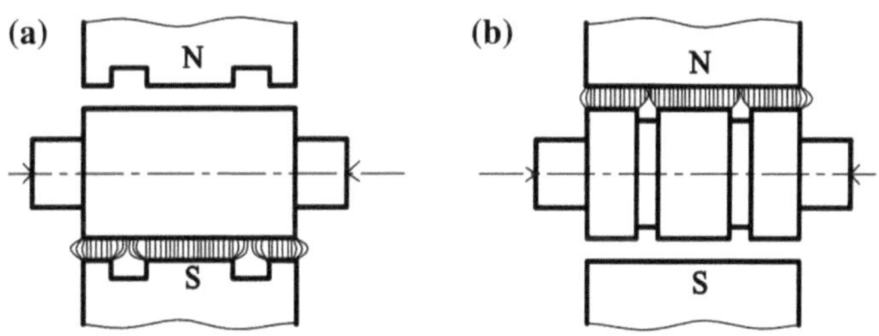


**Fig. 3.102** Processing diagram of different magnetic poles

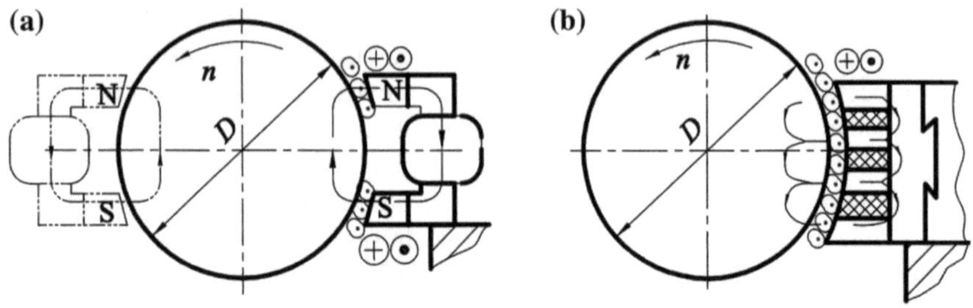


**Fig. 3.103** Diagram of the shape of the pole

For the device in Fig. 3.99, the reduction of the area of the pole decreases the number of the coil and the whole magnetic circuit has characteristics of small size, compact structure, and small energy consumption. Considering the single attraction between the workpiece and magnetic pole, two pairs of poles in place of a pair are designed in the symmetrical form, as Fig. 3.103b shows the processing effect is further improved.

When the diameter of the workpiece is less than 15 mm, the method shown in Fig. 3.104 can get a good processing effect. If the shape of the pole is in the form of an arc with unchanged width of the pole, the large surface area of the pole should form a magnetic short circuit in the tip of the pole where the magnetic abrasive particles gathered, instead of in the area of processing. The width of the pole is too small to hold a sufficient number of magnetic abrasive particles, and it will also significantly reduce the processing effect. In Fig 3.104, the surface of the two magnetic poles, with the distance of $D$ + 2–4 mm, is a plane which is full of the magnetic abrasive particles. The workpiece with the circular motion and axial vibration at the same time also has a reciprocating linear motion which will further accelerate the processing of the surface of the workpiece.

In this way, the force is related to the magnetic abrasive particles, which hinder the linear motion of workpiece workpiece. When the magnetic abrasive particles go through the minimum distance between the poles and the surface, the surface of the workpiece should be squeezed for better processing, because the workpiece is completely surrounded among the magnetic abrasive.

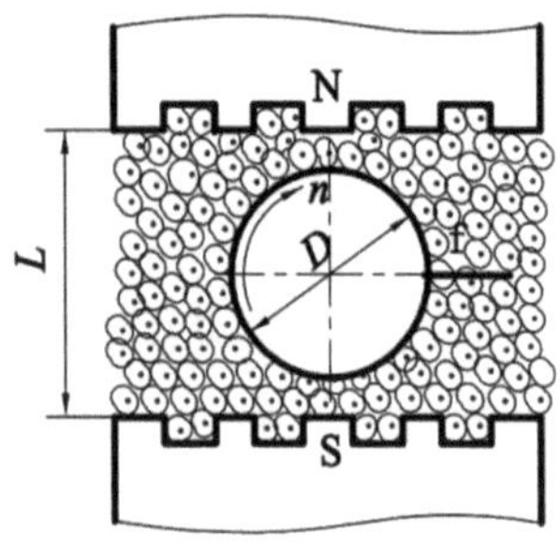


**Fig. 3.104** Diagram of processing for the workpiece that measures less than 15 mm diameter

**Fig. 3.105** Effect photograph of the magnetic abrasive finishing for the outer surface

**Fig. 3.106** Contrast photographs for a thin sleeve before and after the magnetic abrasive finishing

The photograph of processing effect of the magnetic abrasive finishing for the outer surface is shown in Fig. 3.105. Figure 3.106 shows the contrast photographs for a thin sleeve before and after the processing. The finishing effect photograph for the piston by using magnetic abrasive is shown in Fig. 3.107.

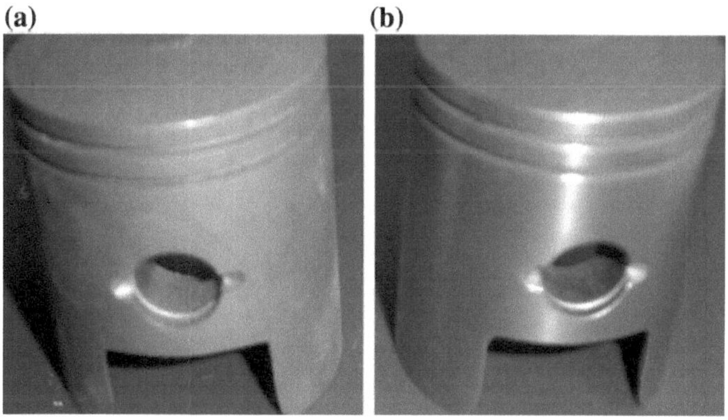


**Fig. 3.107** Effect photograph of the magnetic abrasive finishing for the piston

2. The inner surface

Compared to outer surface, the inner surface will be more difficult to process, especially for some vessels and holes with the small caliber. The flexible magnetic abrasive brush is constituted by numerous magnetic abrasive particles that can vary with the shape of the workpiece. The inner surface of the workpiece can be put into the smaller caliber and hole, so the magnetic abrasive finishing is provided for the finishing of the inner surface of the special precise workpiece such as the thin-walled sleeve, cylinder, hydraulic valve, deep hole, larger pores, and so on. The chemical container with small inlet, big cavity, inner surface of the medical stainless steel utensils, and all other kinds of mold and three-dimensional surface can also be processed in this way, and other special-shaped parts such as the piston, camshaft, and blade can also be polished by this way [66–69].

In general, the equipment for the outer surface is fit for the inner surface finishing [70], but because the abrasive particles are loaded into the internal of the workpiece, the movement of the abrasive particles is influenced not only by the magnetic induction intensity, but also by the diameter of the inner hole. At the same time, the filling amount is closely related to the diameter of the inner hole. Compared with the magnetic abrasive finishing for the outer surface, the main parameters affecting the machining results are the same as those for the inner surface, but the changes and the choices of these parameters are quite different. The processing for the magnetic and non-magnetic materials is also different [71], and the most common non-magnetic material is the thin-walled sleeve. The diagram of the magnetic abrasive finishing for the thin-wall stainless steel sleeve is shown in Fig. 3.108.

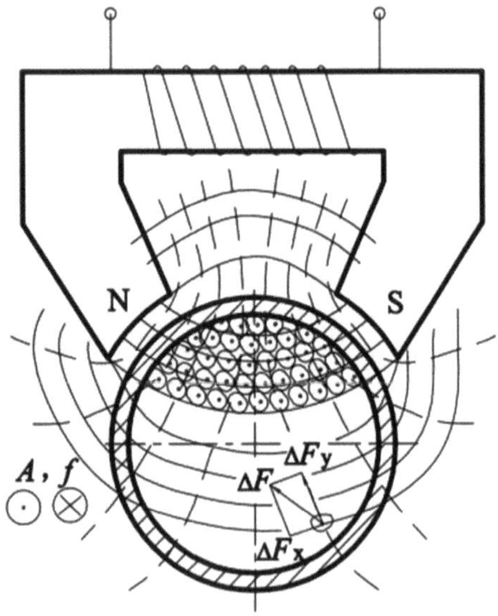


**Fig. 3.108** Diagram of the magnetic abrasive finishing for the thin-wall stainless steel sleeve

For the different materials and shapes, the inner surface can be further improved by the following ways:

(1) With the increasing of the diameter of the workpiece and the machining gap for the non-magnetic sleeve, the magnetic induction intensity is drastically reduced to greatly affect the processing quality [72]. In order to reduce the influence of the size of the workpiece for the magnetic induction intensity, a permanent magnet is put into the sleeve serves to reduce the machining gap, and to change the distribution of the magnetic field; the magnetic induction intensity is also increased to improve the processing capacity. According to configuration of magnetic pole, the shape of the permanent magnet needs to change. The common shape is shown in Fig. 3.109.
(2) Instead of the purely magnetic abrasive particles, the mixture of these particles and the electrolytic iron powder are used for the magnetic abrasive finishing in order to increase the grinding pressure in order to improve the processing quality. The quality percentage of the magnetic abrasive particles and the electrolytic iron powder is needed to determine the result. According to the experimental determination, when the diameter of the magnetic abrasive particle is 80 μm and the diameter of the pure iron powder is about 300 μm which percentage is 70%, the mixture is used to process stainless steel thin-walled sleeve, and the effect is the best so that the surface roughness value is decreased from 0.8 μm to about 0.2 μm.
(3) For some small sleeve parts, the permanent magnetic pole used for the magnetic abrasive finishing make the magnetic source and the magnetic pole together as one in order to get a simple structure, and a convenient processing. For example, the magnetic pole shown in Fig. 3.110a is used for the processing of the aluminum alloy sleeve in which the size is 35 × 5 × 130, and not only the inner hole surface roughness value is less than 0.2 μm, but also the burrs located at the entrance of the hole is removed.
(4) The rotating magnetic field is fit for the inner surface of the sleeve comprising an elbow pipe with the longer thin-walled that can make the inner surface of the

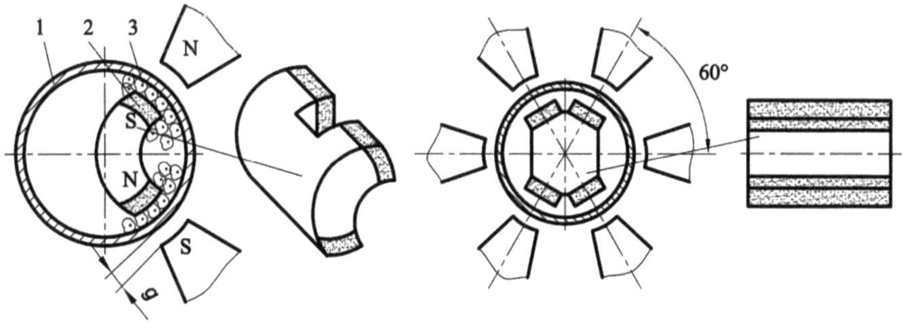


**Fig. 3.109** Array of the permanent magnet for the inner surface of the non-magnetic thin sleeve. 1—The workpiece; 2—the permanent magnet; 3—the magnetic abrasive particles

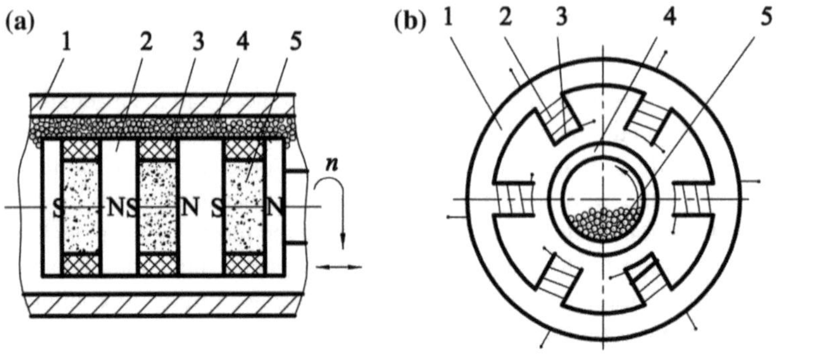


**Fig. 3.110** Diagram of the special processing method for the inner surface. **a** The combination of the multiple permanent magnet source: 1—the workpiece; 2—the guide bar; 3—the non-magnetic filler; 4—the magnetic abrasive particles; 5—permanent magnetic source; **b** the rotating electromagnetic source: 1—the magnetic yoke; 2—the magnetic coil; 3—the magnetic pole; 4—the workpiece; 5—the magnetic abrasive particles

tube like a vacuum tube. The medical stainless steel pipe is polished easily, as shown in Fig. 3.110b. Without the moving parts, the reliable operation and the simple device can be obtained, but it is only for the revolving parts [73, 74].

(5) For the magnetic workpieces, the magnetic field lines are more concentrated near the thick wall; the internal magnetic induction intensity is too small to the poor effect. In order to pass through the thick wall into the interior, the permanent magnet is put into the sleeve so as to change the distribution of the magnetic lines to improve the processing effect. The diagram for special processing method of the magnetic cylindrical inner surface is shown in Fig. 3.111.

3. The flat

The difference of the magnetic abrasive finishing between the flat and the outer surface lies in: ① the shape of the magnetic pole, which is flat instead of the curved surface; ② the workpiece or the magnetic base under the workpiece is taken as the

**Fig. 3.111** Diagram of special processing method for the magnetic cylindrical inner surface

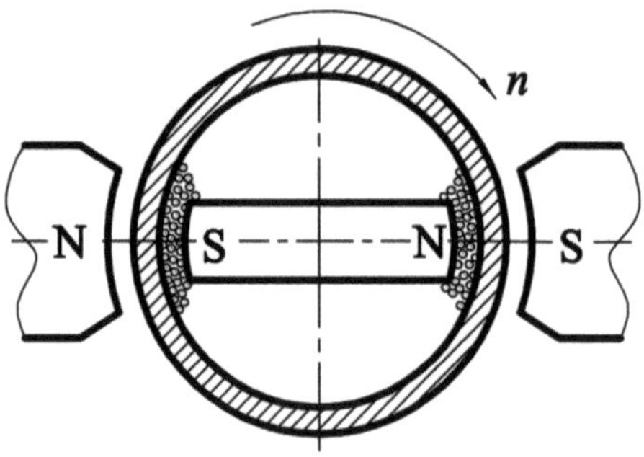

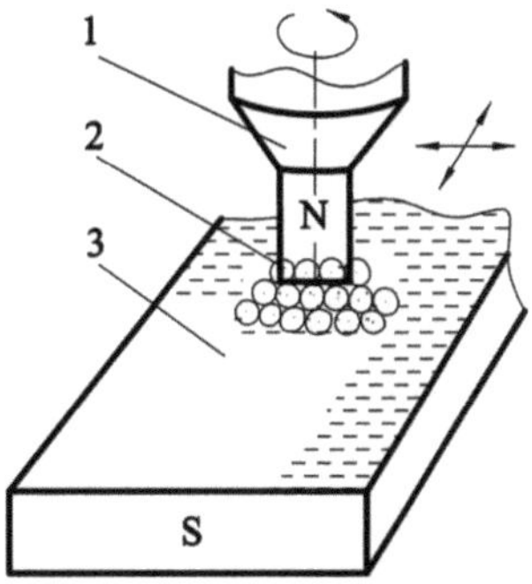


**Fig. 3.112** Magnetic abrasive finishing for the flat. 1—The rotating magnetic pole; 2—the magnetic abrasive particles; 3—the workpiece

**Fig. 3.113** Processing state of the steel sheet

end of the magnetic pole and the rotating tool is taken as the other end of the magnetic pole. The magnetic abrasive finishing for the flat surface is shown in Fig. 3.112 [75].

The processing state of the steel sheet is shown in Fig. 3.113. The effect photograph of the steel and the copper sheet after the magnetic abrasive finishing is shown in Fig. 3.114. Table 3.10 shows the contrasts of the roughness for several different magnetic workpiece.

4. The complicate surface

Known from the principle of the magnetic abrasive finishing, under the action of magnetic force, the magnetic abrasive particles that are filling in the machining gap will form a flexible magnetic brush which is fully consistent with the shape of the workpiece bringing the convenience for the processing of the complicate surface [76]. For example, the tableware like the spoon and fork, decoration like earrings and necklace, medical instruments like pliers and tweezers, industrial products like the inside and outside raceways of the bearing, the turbine blades, the mold, all are fit for the magnetic abrasive finishing in which the molding tool and the special processing equipment as well as the profiling device need not to manufacture to reduce the production cost and improve production efficiency [77].

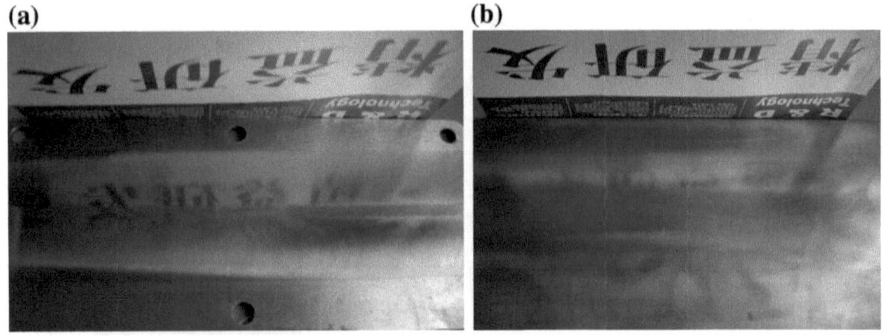

Fig. 3.114 Effect of the magnetic abrasive finishing for the flat: **a** the steel sheet, **b** the copper sheet

**Table 3.10** Contrast of the roughness for several different magnetic workpiece

| Material | $R_a$/μm | |
|---|---|---|
| | Before finishing | After finishing |
| Steel sheet | 2.0 | 0.2 |
| High-carbon steel sheet | 2.0 | 1.0 |
| Quenching steel sheet | 1.0 | 0.3 |
| Ceramic sheet | 2.0 | 0.3 |

(1) The finishing of the complex cylindrical outer surface between two magnetic poles

The processing of the complex rotating cylindrical outer surface is shown in Fig. 3.115. The shape of the magnetic pole is the simplest in which the machining gap reversely changes with the diameter of the workpiece. Mentioned earlier, the gap has a great influence on the processing quality. This simplest shape is only suitable for the complex cylindrical surface in which the difference between the

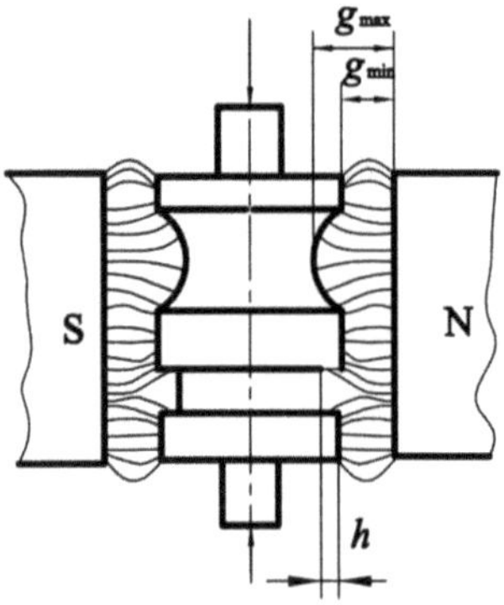


**Fig. 3.115** Diagram of the processing of the common magnetic pole for the complex magnetic cylindrical outer surface

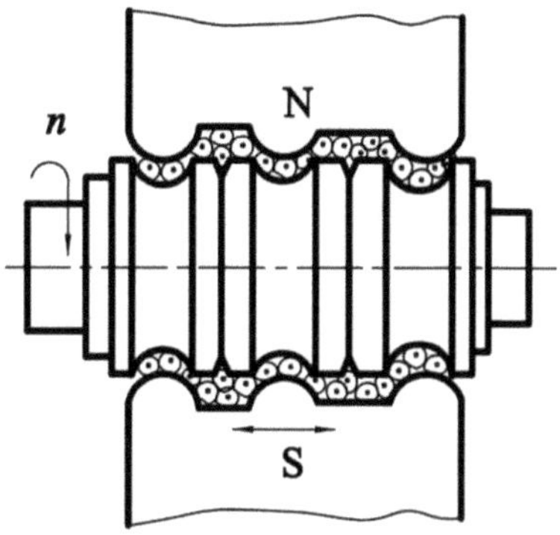


**Fig. 3.116** Diagram of the processing of the special magnetic pole for the complex cylindrical outer surface

maximum gap and the minimum gap is less than or equal to 2 mm. When the difference is more than 2 mm, the shape of the magnetic pole needs to be redesigned according to the shape of the workpiece, as shown in Fig. 3.116. But sometimes, it will affect the axial vibration.

(2) The finishing of the complex cylindrical outer surface in the cylindrical groove

For the non-magnetic materials, the complex surface and the larger size workpiece are especially difficult to process. The larger gap weakens the magnetic induction intensity and the magnetic field force, so the magnetic abrasive particles are not well maintained in the gap. The way shown in Fig. 3.117 can change the situation to process this kind of workpiece. The workpiece is fixed on the shaft of the non-magnetic cylindrical sleeve with a pair of magnetic poles placed outer, and the magnetic abrasive particles are placed at the bottom of sleeve and can be promoted to the gap between the poles and fully surround around the surface of the workpiece. If the workpiece can rotate and axial vibrate, the workpiece can be completely processed.

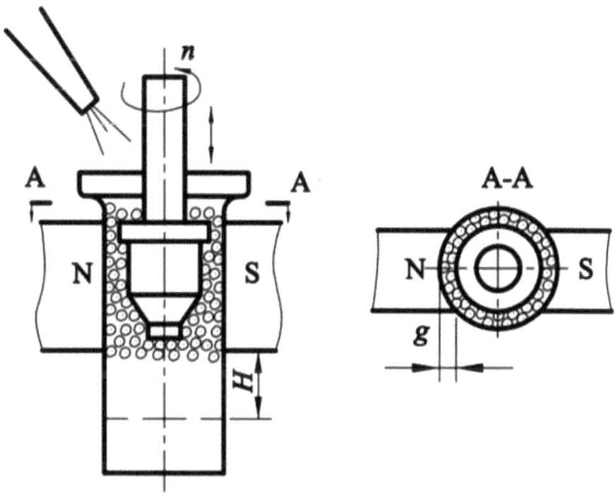


**Fig. 3.117** Diagram of processing for the non-magnetic cylindrical outer surface

In order to reduce the influence of magnetic induction intensity in the processing region exerted by the non-magnetic thin-walled sleeve, the following measures can be adopted; the first thing needed to be done is to try to choose a sleeve with the thinner wall; the second is to dig the grooves in the wall of the sleeve where the magnetic pole situated and flushed with the inner wall of the sleeve.

This processing way will prevent the magnetic abrasive particles flying. Before processing, the magnetic field must be connected to make the magnetic abrasive particles fly from the bottom to the gap between the poles. But the height is limited because of certain magnetic induction intensity. The experimental results show that when the diameter of the sleeve is 20 mm and the magnetic induction intensity is 1 T, the magnetic abrasive particles can only rise to $H = 13$ mm; when $H$ is 22 mm, only 25% of the magnetic abrasive particles required can rise; when $H$ is 30 mm, only fifty percent of the magnetic abrasive particles required can rise. Therefore, in the actual processing, the magnetic field is usually connected and the sleeve moves along the axial direction at the same time to ensure that there is enough magnetic abrasive particles arise from the bottom to the processing gap.

If the diameter of the workpiece is larger, the magnetic pole shown in Fig. 3.118 can be used effectively to prevent the gap from becoming larger and prevent the magnetic induction intensity from being weakened; by doing this, we can achieve a satisfactory processing effect. The special magnetic pole is shown in Fig. 3.118a, b shows the number of magnetic poles increased.

(3) The surface of shaft and disk with the medium or small size

The way shown in Fig. 3.119 can be used to process the surface of the shaft and the disk with medium- or small-size units. The workpiece clamped by the vertical fixture and controlled by the planetary mechanism either can be rotated around itself or rotated around the center axis. The workpiece is placed on a non-magnetic annular groove that is filled with numbers of magnetic abrasive particles, and six pairs of electromagnetic source are arrayed along it so as to make the magnetic lines pass through the radial direction of the annular groove. There is no essential difference from the outer surface, but because many workpieces can be processed together, the production efficiency is greatly improved.

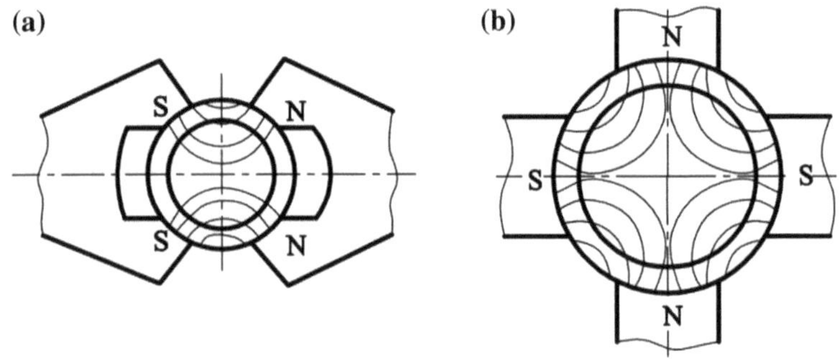


**Fig. 3.118** Special-shaped magnetic pole and distribution diagram of the magnetic line

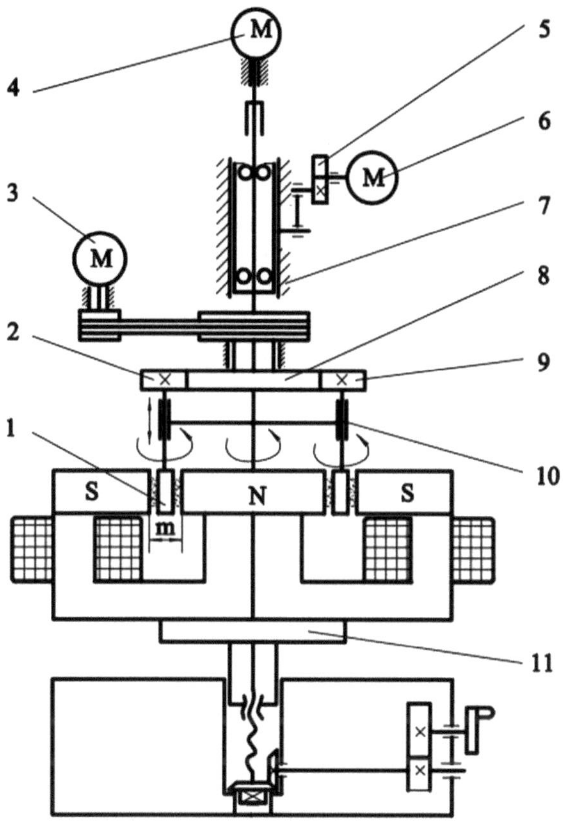


**Fig. 3.119** Diagram of the processing for the surface of shaft and disk with the medium or small size. 1—The workpiece; 2, 5, 8, 9—the gears; 3, 4, 6—the motors; 7—the rolling bearing; 10—the shaft; 11—the support disk

(4) The surface of the micro-workpiece with the complex shape

The way shown in Fig. 3.120 can be used for the finishing of the surface of the complex-shaped micro-workpiece. The workpiece randomly placed in the annular groove cannot only rotate but also vibrate along the axial direction, so a complex trajectory is formed to ensure that the surface of the workpiece completely processed. The material may be magnetic or non-magnetic.

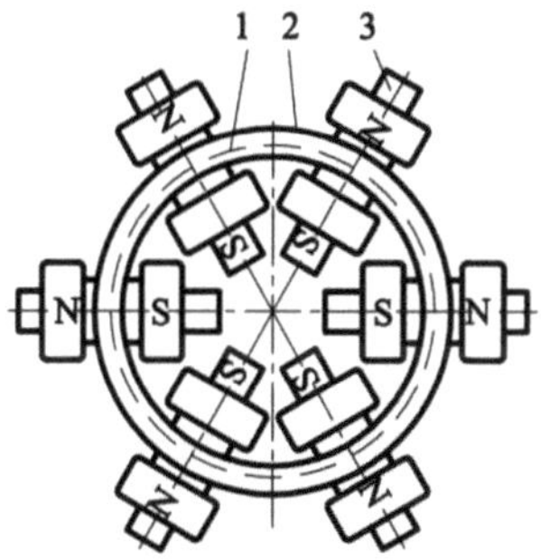


**Fig. 3.120** Diagram of the processing for the surface of the micro-workpiece with the complex shape. 1—The workpiece; 2—the annular groove; 3—the magnetic pole

(5) The complex surface

The magnetic brush is composed of magnetic string, which contains numerous magnetic abrasive particles that can make use of magnetic energy and change its shape according to the shape of the workpiece. The plasticity of the magnetic abrasive particles has a special characteristic that can realize the finishing of the uneven surface or free surface. As long as the parameters (such as the rotary speed of the magnetic pole, the feed rate of the magnetic pole, the gap between the poles are certain, or the magnetic brush) can match with the three-dimensional surface to achieve similar profiling, it can ensure that the processing of all the parts of the workpiece and automation will achieve a great machining process [78]. At the same time, the small metal removal will not damage the dimensional accuracy nor the form's accuracy.

① The device is shown in Fig. 3.57. The central part of the magnetic field source mandrel is the same as that of the NC machining central spindle. The magnetic pole fixed on the spindle of the magnetic field source is convenient enough to connect and disassemble to achieve the finishing for a variety of complex 3D surfaces or the surface of the mold cavity.

② The formation of the force for the magnetic abrasive finishing of the complex surface is shown in Fig. 3.121. The magnetic brush will change with the surface of the workpiece which is two-dimensional or three-dimensional as the changing of the curvature.
Whether the horizontal or the vertical surfaces, or the convex and the concave, the size of the force will change slightly, but the direction of the force is always pointed to the surface. The flexibility and plasticity of the magnetic brush can realize the magnetic abrasive finishing of the mold cavity surface successfully.

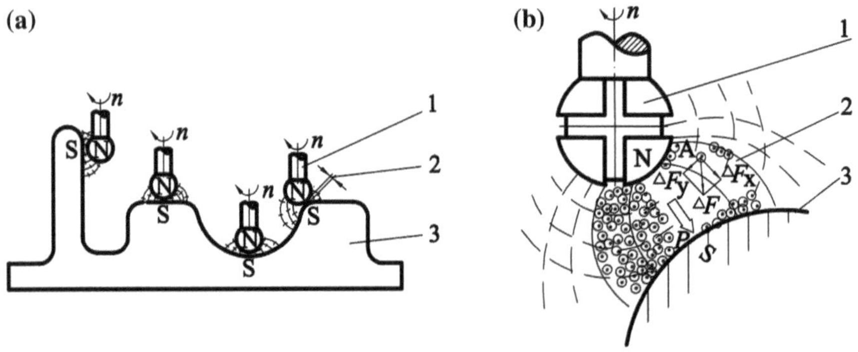


**Fig. 3.121** Formation of the force of the magnetic abrasive finishing for the curved surface. **a** Two-dimensional surface: 1—the magnetic pole; 2—the gap; 3—the workpieces; **b** three-dimensional surface: 1—the magnetic pole; 2—the magnetic abrasive particles; 3—the surface of the workpieces

③ The geometry shape of the magnetic pole should be designed on the basis of the surface that is maybe horizontal, vertical, inclined, arc, or freedom. The basic elements of the surface are complex and variable, and thus the shape of the magnetic pole is different, which is shown in Figs. 3.73, 3.74 and 3.75 [79].

The machining state diagram for the free surface is shown in Fig. 3.122; the processing effect photograph from different angle is shown in Fig. 3.123.

Known from Fig. 3.123, the magnetic abrasive particles sucked around the magnetic pole make the contact area between the magnetic pole and the workpiece larger for the concave so as to get the better effect; but for the convex surface, that same magnetic abrasive particles sucked at the bottom of the magnetic pole are fewer, which can lead to the poor processing effect.

In a word, the magnetic abrasive finishing is similar to a practical and precision technology that has many advantages such as wider processing range, better processing quality, the lower processing cost, the higher processing efficiency, the

**Fig. 3.122** Machining state diagram for the free surface

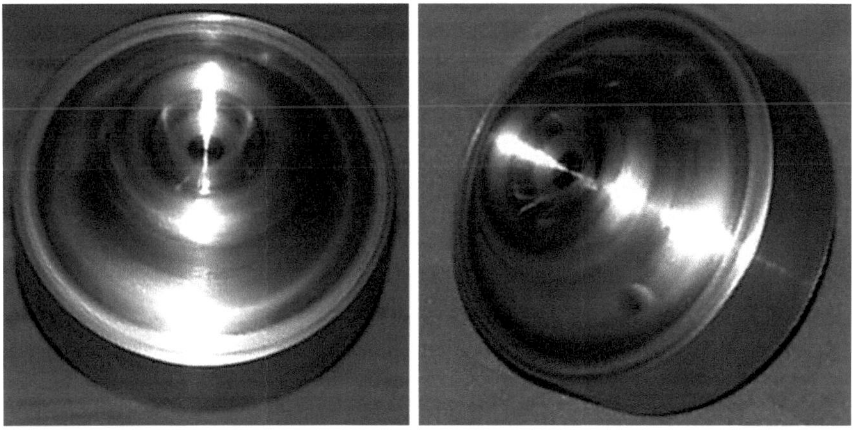

**Fig. 3.123** Effect diagram of the magnetic abrasive finishing for the free surface

simpler processing equipment, and so on. After using the magnetic abrasive finishing for various surfaces, the burr, the rust corrosion, and the oxide skin are completely removed, the sharp corners and edge become well polished, and the overall surface is smooth and bright without deformation.

## References


1. S. Jha, V.K. Jain, Design and development of the magnetorheological abrasive flow finishing (MRAFF) process. Int. J. Mach. Tools Manuf. **44**, 1019–1029 (2004)
2. N. Umehara, T. Kobayashi, T. Kato, Internal polishing of tube with magnetic fluid grinding. J. Magn. Magn. Mater. **149**, 188–191 (1995)
3. W.I. Kordonski, D. Golini, Fundamentals of magnetorheological fluid utilization in high precision finishing. J. Intell. Mater. Syst. Struct. **10**, 683–689 (2001)
4. J.D. Kim, Y.H. Kang, Y.H. Bae et al., Development of a magnetic abrasive jet machining system for precision internal polishing of circular tubes. J. Mater. Process. Technol. **71**(3), 384–393 (1997)
5. S.C. Yang, H.F. Li, Application and development of surface finishing and deburring technology in China. New Technol. New Process **3**, 35–36 (1995)
6. Z.Y. Xiao, Experimental research on the magnetic abrasive finishing technology. Die Mould Ind. **12**, 47–50 (2004)
7. B.H. Yan, G.W. Chang, R.T. Hsu, Electrolytic magnetic abrasive finishing. Int. J. Mach. Tools Manuf. **43**, 1355–1366 (2003)
8. S.C. Yang, Z.M. Wang, Y.X. Zhang, *Surface Quality and Finishing* (China Machine Press, Beijing, 2000)
9. Y.H. Zhou, X.J. Zheng, A general expression of magnetic force for soft ferromagnetic plates in complex magnetic fields. Int. J. Eng. Sci. **35**(15), 1405–1417 (1997)
10. H.L. Chen, *Practical Technology Investigation of Magnetic Abrasive and Magnetic Abrasive Finishing* (Taiyuan University of Technology, Taiyuan, 2009)
11. N.I. Wakayama, Utilization of magnetic force in space experiments. Adv. Space Res. **24**(10), 1337–1340 (1999)
12. Y. Wang, D.J. Hu, Grinding mechanism and preparation technology of magnetic abrasive grains. Tool Eng. **6**, 32–34 (2003)
13. J.D. Kim, M.S. Choi, Simulation of the prediction of surface-accuracy in magnetic abrasive machining. J. Mater. Process. Technol. **53**, 630–642 (1995)
14. T. Shinmura, H. Yamaguchi, Study on a new internal finishing process by the application of magnetic abrasive machining. Int. J. JSME Ser C **38**(4), 798–804 (1995)
15. T. Mori, K. Hirota, Y. Kawashima, Clarification of magnetic abrasive finishing mechanism. J. Mater. Process. Technol. **143–144**, 682–686 (2003)
16. H. Yamaguchi, T. Shinmura, Study of an internal magnetic abrasive finishing using a pole relation system discussion of the characteristic abrasive behavior. Precision Eng. **24**, 237–244 (2000)
17. H.L. Chen, Y.X. Zhang, Grinding mechanism study on magnetic abrasive grains. J. Taiyuan Univ. Technol. **31**(5), 562–565 (2000)
18. B.M. Li, B. Zhao, *Modern Grinding Technology* (China Machine Press, Beijing, 2003)
19. Y.Y. Guo, Y.X. Zhang, Research on magnetic abrasive finishing for cylindrical surface. China Mech. Eng. **8**(3), 16–78 (1997)
20. Y.Y. Guo, Y.X. Zhang, S.C. Yang, Research on magnetic abrasive finishing for inner surface. China Mech. Eng. **8**(6), 23–25 (1997)

21. K. Suzuki, T. Makizaki, Finishing of Small Edges Magnetic Abrasive Finishing Method Utilizing Centrifugal Force, in *Proceedings of ICPE'96 and 6th SJSUT* (1996), pp. 1111–1114
22. G.W. Chang, B.H. Yan, R.T. Hsu, Study on cylindrical magnetic abrasive finishing using unbonded magnetic abrasives. Int. J. Mach. Tools Manuf. **42**, 575–583 (2002)
23. H.L. Chen, Y.Y. Guo, Preparation and Performance of Magnetic Abrasive Particles, in *Proceedings of CMES'2001* (2001) pp. 467–470
24. S.C. Yang, S.Q. Yang, The development direction of free abrasive tool finishing technology. Machinist Metal Cutting **1**, 11–12 (2002)
25. V.E. Oliker, A.F. Zhornyak, V.Y. Shlyuko, Structure formation in magneto-abrasive materials from iron-carbon melts alloyed with carbide-forming elements. Poroshkovaya Mettelurgiya **22**(2), 76–84 (1983)
26. H. Yamaguchi, T. Shinmura, Internal finishing process for alumina ceramic components by a magnetic field assisted finishing process. Precision Eng. **28**, 135–142 (2004)
27. M.D. Krymsky, Magneto-abrasive materials produced by means of inner nitrating. Poroshkovaya Mettelurgiya **22**(9), 77–82 (1983)
28. Y.X. Zhang, H.L. Chen, Y.H. Ding, *Magnetic Grinding Tool and Its Preparation*. China, ZL00 1 35667.4.2003 (2003)
29. H.L. Chen, *The development and experimental research on finishing characteristic of magnetic abrasive grains* (Taiyuan University Technology, Taiyuan, 2000)
30. Y.P. Sun, *Study on the Free Form Surface Magnetic Abrasive Finishing Based on 5-DOF Parallel Machine Tool* (Northeastern University, Shenyang, 2004)
31. L.J. Xu, W. Wang, C. Yang, Review of the magnetic abrasive finishing. Modular Mach. Tool Autom. Manuf. Tech. **1**, 41–43 (2003)
32. Y.X. Zhang, Y.H. Ding, H.L. Chen, *Hot Pressing Sintering Process of Preparing Magnetic Abrasive*. China, ZL00 1 36573.8.2003 (2003)
33. H.L. Chen, Y.X. Zhang, Study of technical parameters of magnetic abrasive finishing. J. Taiyuan Univ. Technol. **33**(1), 16–18 (2002)
34. H.L. Chen, Y.X. Zhang, Y.Y. Guo, Research on preparation of magnetic abrasive grains with binding method. J. Taiyuan Univ. Technol. **32**(5), 533–535 (2001)
35. K. Masaki, T. Nakagawa, Fabrication of magnetic abrasive using plasma powder melting method. Powder Metall. **38**(4), 472–476 (1991)
36. M.Y. Liao, J.J. Zhou, Laser sintered magnetic abrasive particles and magnetic abrasive finishing. Manuf. Technol. Mach. Tool **5**, 36–38 (1999)
37. V.E. Oliker, A.F. Zhornyak, T.Y. Gridasove, Functional role of structural components for magnetic-abrasive powder particles. Poroshkovaya Mettelurgiya **24**(9), 71–75 (1985)
38. Y. Zhang, S.L. Lu, Using digital milliteslameter measuring ferromagnetic hysteresis loop and curve of magnetization. Res. Explor. Lab. **20**(5), 48–51 (2001)
39. X.Y. Jia, M. Chen, C. Lu, Refit of magnetic grinding device on vertical milling machine. Mach. Tool Hydraul. **4**, 316–317 (2003)
40. Y.P. Lu, J. Ma, J.W. Cai et al., Study of intelligent control system in magnetic abrasive finishing on molds free surface process. Trans. Chin. Soc. Agric. Mach. **37**(7), 163–166 (2006)
41. 進村武男. 磁気研磨法加工機構の解明と装置の開発応用に関する研究[M]. 日本精密工学出版社 (1995)
42. J.C. Fang, Z.J. Jin, W.J. Xu et al., Research on magnetic abrasive machining by applying a rotating magnetic field. China Mech. Eng., **12**(11),1304–1306, 1326
43. M.R. Cao, Y.X. Zhang, Magnetism and circuit design for magnetic polishing set. J. Taiyuan Univ. Technol. **33**(3), 257–259 (2002)
44. K.D. Shen, G.Z. Zhang, M. Chen et al., The electromagnetic inductor and coils calculating on the magneto-abrasive machining. Coal Mine Mach. **1**, 25–27 (2004)
45. S.J. Cao, S.L. Fu, Q. Zhang, Research on app1ication of permanent magnet loop in magnetic force grinding. J. Shenyang Univ. **15**(4), 62–64 (2003)

46. Z.J. Jin, C.R. Wang, The application of finite element method in magnetic abrasive finishing using rotating magnetic field. Electro Mach. Mould **1**, 46–49 (2002)
47. J.D. Kim, M.S. Choi, Development and finite element analysis of the finishing system using rotating magnetic field. Int. J. Mach. Tools Manuf. **36**(2), 245–253 (1996)
48. Z.Y. Xiao, Y.G. Zhao, W.Q. Wu, The numerical analyzing for the effects of magnetic pole shape to the process of magnetic abrasive finishing. Mod. Manuf. Eng. **10**, 10–12 (2003)
49. B.F. Feng, Q.F. Gai, W.S. Zhao et al., Impact of magnetic abrasive tool's shapes on abrasive effects. Chin. J. Constr. Mach. **6**(1), 101–104 (2008)
50. X.B. Cheng, *The Exploitation of Strong Permanent Magnetic Poles on Plane Surface and the Experiment Research on its Working Character* (Taiyuan University of Technology, Taiyuan, 2002)
51. X.X. Wang, *Development of Permanent Magnetic Poles Based on Freeform Surface Finishing and Experimental Studies on Their Working Performance* (Taiyuan University of Technology, Taiyuan, 2003)
52. Z.Y. Zhou, X.M. Ye, M. Chen, Two kinds of electromagnetic vibrating magnetic pole-head for magnetic skiving. Precise Manuf. Autom. **2**, 43–45 (2006)
53. 木下直治, 高哲孝哉. 精度設計と部品仕上システム技術[M]. 日經技術圖書株式會社 (1990)
54. 田中義信, 津和秀夫, 井川直哉. 精密工作法[M]. 共立出版 (1995)
55. 研磨布紙加工技術研究会. 新しい研磨技術[M]. ォーム社, (1992)
56. H.L. Chen, S.C. Yang, J.M. Wang, Internal magnetic abrasive particles surface finishing based on permanent magnetic field. Adv. Mater. Res. **53–54**, 65–68 (2008)
57. H. Yamaguchi, T. Shinmura, Study of the surface modification resulting from an internal magnetic abrasive finishing process. Wear **225–229**, 246–255 (1999)
58. L.Y. Zhang, *Modern Processing Technology* (China Machine Press, Beijing, 2002)
59. V.K. Jain, P. Kumar, P.K. Behera et al., Effect of working gap and circumferential speed on the performance of magnetic abrasive finishing process. Wear **250**, 384–390 (2001)
60. Z.J. Yuan, X.K. Wang, *Precision and Ultra-Precision Manufacturing Technology* (China Machine Press, Beijing, 2005)
61. Z.Y. Xiao, The analyses on the technology parameter of magnetic abrasive finishing. Mod. Manuf. Eng. **12**, 61–63 (2004)
62. S.H. Yin, T. Shinmura, Vertical vibration-assisted magnetic abrasive finishing and deburring for magnesium alloy. Int. J. Mach. Tools Manuf. **44**, 1297–1303 (2004)
63. S.H. Yin, T. Shinmura, A comparative study: polishing characteristics and its mechanisms of three vibration modes in vibration-assisted magnetic abrasive polishing. Int. J. Mach. Tools Manuf. **44**, 383–390 (2004)
64. K.S. Dhirendra, V.K. Jain, V. Raghuram, Parametric study of magnetic abrasive finishing process. J. Mater. Process. Technol. **149**, 22–29 (2004)
65. P.X. Yao, H.L. Chen, J.M. Wang et al., The principle of magnetic abrasive finishing under rotating electromagnetic field and its application in surface finishing of heavy crankshaft. Key Eng. Mater. **304–305**, 379–383 (2005)
66. J.D. Kim, Polishing of ultra-clean inner surfaces using magnetic force. Int. J. Adv. Manuf. Technol. **21**, 91–97 (2003)
67. B.P. Shang, S.J. Jiang, Experimental study on magnetic abrasive machining of inner surface of small diameter hole. Diamond Abras. Eng. **1**, 57–59 (2006)
68. H.L. Chen, W.H. Li, S.C. Yang, Study of magnetic abrasive finishing to thin sleeve of Al alloy. Key Eng. Mater. **407–408**, 565–568 (2009)
69. J.C. Fang, H.O. Zhang, Z.J. Jin, Research on magnetic-abrasive finishing machining for inner surface of stainless steel tube by rotating magnetic field. Mech. Sci. Technol. Aerosp. Eng. **89** (1), 92–93 (2001)
70. L. Wu, J.M. Luo, Study on magnetic abrasive machining for the smaller hole. Mach. Tool Hydraul. **2**, 97–98 (2006)
71. Y. Wang, D.J. Hu, Study on the inner surface finishing of tubing by magnetic abrasive finishing. Int. J. Mach. Tools Manuf. **45**, 43–49 (2005)

72. H.L. Chen, Y.X. Zhang, W.L. Yan, The technology of finishing process of die space based upon magnetic abrasive finishing. Key Eng. Mater. **259–260**, 657–661 (2004)
73. H. Yamaguchi, Development of a new precision internal machining process using an alternating magnetic field. Precision Eng. (27), 51–38 (2003)
74. J.D. Kim, M.S. Choi, Simulation of an internal finishing processing of rectangular tube using a magnetic field. Wear **184**, 67–71 (1995)
75. X.F. Wang, *Study on Ball Milling Preparation Technology and Finishing Experiment of Fe/* $Al_2O_3$*Magnetic Abrasive* (Taiyuan University of Technology, Taiyuan, 2013)
76. Y.G. Zhao, S.C. Jiang, J.J. Zhou, New kind of machine tool for magnetic abrasive finishing complex surface. J. Mech. Eng. **36**(3), 100–102 (2000)
77. S.Q. Yang, W.H. Li, S.C. Yang, Experimental Research on Improving Performance of the engine by surface finishing technology. Trans. Chin. Soc. Agric. Mach. **37**(2), 17–20 (2006)
78. Y.P. Sun, S.X. Yuan, B.F. Feng et al., Research on magnetic abrasive finishing for the free form surface of mould by 5-DOF parallel virtual axis machine tool. Key Eng. Metering **259–260**, 631–635 (2004)
79. J.D. Kim, M.S. Choi, Study on magnetic polishing of free-form surfaces. Int. J. Mach. Tools Manuf. **37**(8), 1179–1181 (1997)

# Chapter 4
# Fluid Magnetic Abrasive Finishing Technology

## 4.1 Introduction

The fluid magnetic abrasive (FMA) is a new type of precision surface finishing abrasive. It is a viscous suspension which is prepared by dispersing the magnetic particles, surfactants, non-magnetic abrasive particles, and anti-rust agent in a carrier liquid. In the absence of magnetic field, the FMA is a stable suspension with low processing capability. When an external magnetic field is applied, these rheological properties of FMA show the rapid changes, and viscosity and shear yield stress increase apparently. These magnetic particles in the FMA aligned in the external magnetic field direction form a stable chain-like structure, and the non-magnetic abrasive particles are embedded in these chain-like structures. Then, the FMA changed into something like half-solid abrasive with the processing capability. A flexible grinding layer is formed at the contact surface of the workpiece, and the precise surface finishing on the workpiece with intricate or complex shape will be realized as soon as the relative movement occurs between the workpiece and the grinding layer [1–3].

### *4.1.1 Presentation of the Fluid Magnetic Abrasive Surface Finishing Technology*

FMA is developed based on the rheological property of the magnetorheological fluid and the magnetic fluid in the magnetic field. The magnetorheological (MR) fluid is a new smart material, which is developed by an American Jacob Rabi now in 1949. It is a suspension of soft ferromagnetic particles, surfactants, and carrier liquid. Normally, in the absence of magnetic field, MR fluid is free-flowing liquid with all properties of the Newtonian fluid. However, when an external magnetic field is applied, the rheological properties of the MR fluid will change



S. Yang and W. Li, *Surface Finishing Theory and New Technology*,
https://doi.org/10.1007/978-3-662-54133-3_4

dramatically, and the viscosity and the shear yield stress will increase apparently over to 100 kPa. At this time, the MR fluid is looked like changed from liquid into solid. When the external magnetic field is removed, the MR fluid returns to its original state. The MR fluid was extensively concerned because of its smart properties. With the development of the stability, especially since the 1990s, it was widely used in mechanical engineering, such as the damping element (automobile shock absorber, building shock absorption), the agility control element (clutch, brake rigging device), and the controllable valve. The ferrofluid and the ER fluid have the similar rheological properties [4–6]. The development of the ER fluid is far earlier than the MR fluids. It is prepared by dispersing the electric polarized particles in an insulating carrier liquid. When an external electric field is applied, the viscosity and the yield stress will increase rapidly as well. However, many shortages limit the usage of the ER fluid in engineering, such as the obtaining of the lower yield stress (about 3 kPa), the demand of the higher voltage (over 4 kV), and the sensitivity to the temperature will change. The ferrofluid was developed in the mid 1960s. The compositions and properties are similar with ER fluids. The ferrofluid is also prepared of soft ferromagnetic particles (nano-sized), surfactants, and carrier liquid. Because the nano-sized soft ferromagnetic particles can be suspended in carrier liquid steadily, the ferrofluid is a colloid, and the sedimentation stability of the ferrofluid is better than the MR fluids. As the nano-sized soft ferromagnetic particle is a single magnetic domain structure particle, its coercivity is very high. So the nano-sized soft ferromagnetic particles are commonly used as magnetic recording material, not as the MR fluid particles. When the radius of the ferromagnetic particle is smaller than the critical radius, the ferromagnetic particle shows the paramagnetic property, which we called superparamagnetism property. When an external magnetic field is applied, because of the weak interaction force of the ferromagnetic particle, although the viscosity of the ferrofluid increases rapidly, the shear stress increases rarely (nearly 0, $\mu_y = 0$). At present, the ferrofluid is used in the sealing element, the lubrication system, the damping element of loudspeaker coil, and the radiator element [7]. The main properties of the MR fluid, the ferrofluid and the ER fluid are compared in Table 4.1.

Based on the rheological property of the MR fluid and the magnetic fluid in the magnetic field, the Japanese Y. Tain and K. Kawata developed a magnetic-field-assisted polypropylene lenses polishing technology in 1984. Umehara and Kato developed the magnetic float polishing (MFP) technology realized by the rheological property, and then they polished the ceramic balls by using MFP technology. In the early 1990s, [8, 9] W.LKordonski cooperated with Jacobs of the Center for Optics Manufacturing (COM) of the University of Rochester to put forward and verify the deterministic magnetorheological finishing (MRF) technology to process the aspheric optical lens. Based on the MRF technology, the commercialized MRF polishing equipment can be obtained by the manufacture of the QED Technologies Inc. of USA in 1998 (Fig. 4.1) [10]. In fact, both in the MRF and in the MFP, the abrasive particles are dispersed in the magnetorheological (MR) fluid or the ferrofluid. Then, the finishing process will be realized when the abrasive particles rise to the top layer of the fluid in the role of the

**Table 4.1** Comparisons of the MR fluid, the ferrofluid, and the ER fluid

| Property | MR fluid | Ferrofluid | ER fluid |
|---|---|---|---|
| Type of magnetic particle | Iron, ferrite, etc. | Iron, ferrite, and magnetic ceramic, etc. | Silicate, polymer, $SiO_2$ and $BaTiO_3$ |
| Size of magnetic particle | 0.1–10 μm | 2–10 nm | 0.1–10 μm |
| Carrier liquid | Non-polar oil, polar liquid, and water, etc. | Polar and non-polar oil, and water, etc. | Insulating oil |
| Density (g/ml) | 3–5 | 1–2 | 1–2 |
| Initial viscosity (MPa•s) | 100–1000 | 2–500 | 50–1000 |
| Excitation field | Magnetic field | Magnetic field | Electric field |
| Shear yield stress | 100 kPa | $\Delta\eta(B)/\eta(0) \approx 2$ | 10 kPa |

**Fig. 4.1** Magnetorheological finishing machine for optical glasses

magnetic levitation force. The details of the MFP technology will be seen in the next section.

Based on the rheological property of the MR fluid in magnetic field, the fluid magnetic abrasive (FMA) is developed. By using the fluid magnetic abrasive, the finishing process on the workpieces made of variety of materials with intricate or complex shapes is realized, and the machining results are satisfied [11, 12].

### *4.1.2 Characteristics of the Fluid Magnetic Abrasive Surface Finishing Technology*

As a new type of precise surface finishing technology, the FMA technology has advantages in precisely machining the workpieces with intricate or complex shapes. The characteristics of it were shown below.

- Strong adaptability to the shape.
  The finishing process on the both outer surfaces and the inner surfaces of the workpieces with the irregular shapes can be achieved by the dynamic phase transition. During the finishing process, the FMA will automatically adapt to the surface of the workpiece, and then a flexible grinding layer is formed at the contact surface. If the chain-like structures of magnetic particles are destroyed by the movement of workpiece during the finishing process, it may be reconstructed immediately by the external magnetic field. In the other word, the FMA can dynamically adapt the surfaces of the workpieces. Therefore, a continuous finishing process on both the outer surfaces and inner surfaces of the workpieces with the irregular shapes will be achieved. Especially, excellent results may be obtained in the rotary surface finishing and the small-sized hole surface finishing.
- Strong adaptability to the materials.
  Both the metallic materials and the non-metallic materials (include some fragile materials) can be finished by the FMA.
- Better controllability.
  By adjusting the ingredient of the FMA, the material removal rate and the final surface roughness can be controlled. During the finishing process, by adjusting the magnetic field intensity and the magnetic field direction, the process can be controlled. In the initial phase, a high-intensity magnetic field is applied to achieve a high material removal rate; in the later phases, a low-intensity magnetic field is applied to achieve a low surface roughness.
- Better processing quality.
  The surface roughness can be reduced 10–1000 times after the finishing process. However, the machining precision is the same as before.
- Simple processing.
  Firstly, only general equipment is needed to prepare the FMA; secondly, the preparation processing is relatively simple; lastly, during the finishing process, the necessary machining equipment includes the magnetic field generator, the drilling machine, or the milling.
- Innate lubricating function, cooling function, self-sharpening function, and abrasive dust clean function.
  Firstly, the water, as the carrier liquid in the FMA, will lubricate and cool the workpiece during the finishing process; secondly, the magnetic particle's chain-like structures are continuously destroyed by the workpiece movement, and they may be persistently reconstructed. The obtuse particles and the abrasive dusts will be taken away at the same time.
  In conclusion, the FMA is one excellent precision finishing abrasive (both the plastic material and the fragile material can be finished by the FMA), and it can be widely used in manufacturing. Especially, it has advantages in precisely finishing the workpieces with intricate or complex shapes.

# 4.2 Fluid Magnetic Abrasive Finishing Mechanism

## *4.2.1 Fluid Magnetic Abrasive Rheological Properties Mechanism*

1. FMA composition and microstructure

FMA is mainly composed of ferromagnetic particles, non-magnetic abrasive particles, surfactants, rust inhibitor and base fluid, and others. The ferromagnetic particles mainly achieved the function of interaction with the magnetic field, which is a function particle of creeping effect in FMA. It can control the viscosity and shear yield intensity of the abrasives by changing the applied magnetic intensity and direction, which to a certain extent dynamically control the material removal rate of fluid magnetic abrasives finishing. Ferromagnetic particles can choose one kind or make multi-kinds of different types and sizes into one-kind soft ferromagnetic particle, which accounted for 10–60% of the weight of FMA. The non-magnetic abrasive particle is a machining function particle, which mainly completes the workpiece surface finishing processing, and it can be made by one-kind or multi-kind particles, and its mass percent is 30–60%. Surfactants mainly improve the dispersing power in the base fluid and slow down their settlement and cohesion. By the adsorption, reaction, and coating on the surface of the magnetic and non-magnetic abrasive particles, it can choose one kind based on the situation of surface active agent, and it can also be combination of both, and even a variety of surfactants to use, in fluid magnetic abrasives, the dosage of surfactants is determined according to the proportion of magnetic particles and abrasive particles, which accounted for 0.5–6% of the weight. The main function of the base fluid is to dispersing medium. Theoretically, base fluid can choose any fluid with chemical stability, but based on the characteristic about finish machining, it generally is water and kerosene, which its percentage is 20–60%.

On microcosmic view, when shown without an external magnetic field, the magnetic particles and abrasive grains are random distribution in base fluid, which are uniform distribution, and stand independent of each other. The state to be presented is free-pouring suspending liquid in a macro scale (Fig. 4.2a) [13]; when going with an external magnetic field, the magnetic particles form chains that align themselves with the field lines where convoys of non-magnetic particles in magnetic chain (Fig. 4.2b), thus it forms the abrasive tool, magnetic particles would move to the direction of the high magnetic field intensity and push the non-magnetic particles to the direction of the less magnetic field intensity, resulting the abrasive percentage in the lower magnetic field strength to be higher.

2. Rheological properties mechanism

The rheological properties of FMA refer to the apparent change of its viscosity and shear yield with the applied magnetic field. When the magnetic field strength is above the critical value, FMA will lose fluidity and be solidified in order to keep its

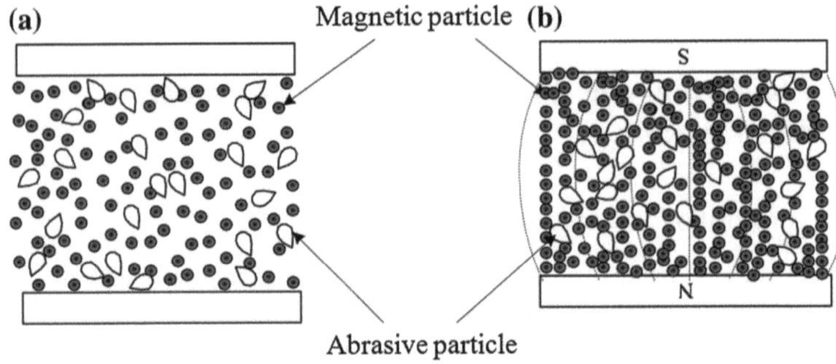


**Fig. 4.2** Schematic diagram of microscopic structure of fluid magnetic abrasives. **a** Without applied external magnetic field; **b** After applying an external magnetic field

shape or have a certain ability of skid resistant, and it will show the yield phenomenon, peculiar to the solid [14].

The rheological properties mechanism of MR fluids, magnetic liquid, and magnetic abrasives are similar. But because the occurrence of MR effect is a complicated magnetism, thermodynamics and dynamic process, although after a long-term research, which was also discussed by scholars both at home and abroad, there are no conclusion and theories for the model [15]. The rheological properties mechanism and theoretical model are built on the base of the simplification of the actual situation; nowadays, there exists two typical theories: the theory of field-induced dipole moment and phase transition nucleation theory.

The theory of field-induced dipole moment: The magnetic particles dispersed in base fluid freedom without an applied magnetic field. The MR fluid, magnetic liquid, and magnetic abrasives presents as suspension or colloid state, respectively. When there is an external magnetic field, these magnetic abrasives are polarized, with each particle is polarized into magnetic dipole. Then, the magnetic dipole will attract to each other's form in a chain formation along the direction of the magnetic field (Fig. 4.2). Owing to the chain structure, the viscosity and shear yield stress are improved. The intensity of MR effect depends on the size of the force between the magnetic dipole. With the improvement of magnetic field intensity, the polarized degree of magnetic particles increases, and the force between the magnetic dipole also increases. Therefore, the viscosity and shear yield strength increase. After the magnetic particles are driven into saturation, the viscosity and shear yield strength will not change obviously with the increase of the magnetic field strength. Magnetic particles will quickly return to its original state of freedom without the magnetic field, so MR fluid, magnetic liquid, and magnetic abrasives will return to the standard of suspension liquid or gel. Meanwhile, this theory believes that there are three critical fields, $H_{c1}, H_{c2}, H_{c3}$, in the process of transition from liquid to solid. When $H < H_{c1}$, the magnetic particles are in random state of suspension, when $H_{c1} < H < H_{c2}$, there is a situation of short chain and particles together, and when $H_{c2} < H < H_{c3}$, it is in a nematic crystal flow state; when $H_{c3} < H$, it is in a steady state of cylindrical solid, namely magnetic saturation state.

Corresponding to the theory of field-induced dipole moment, there is another theory: the phase transition nucleation theory. This theory believes that the magnetic particles dispersed in the base are in a random state, without a magnetic field, which follows Brownian motion law. When applied an external magnetic field, the magnetic particles began to be magnetized, with the magnetic field intensity increased to a certain extent, some particles are beginning to converge into an orderly arrangement. Under the effect of thermal fluctuation and magnetic field, it is known as the ordered phase (or nucleation). With the further increase of magnetic field intensity, the ordered phase is connected to long chain, which then absorbs short chains, so the chain further become larger in diameter, eventually it constitutes solid phase. Phase transition point can explain some phenomena of the MR fluid, but it has not been accepted by most people, so it also needs to be validated.

After applying a magnetic field, the rheological properties of MR fluid, magnetic liquid, and fluid magnetic abrasives change rapidly, which will no longer follow the Newtonian fluid law. When the additional shear force is less than the shear yield strength, sticky stiff MRF is equal to the flexible solid. When the external force exceeded the shear yield strength, the magnetic chain structure is destroyed, and it will start to behave in a liquid form. After removing to the shear force, the damaged magnetic chain structure will rapidly regenerate [16]. By many theories and experiments, people found that the shear stress of MR fluids is in the line with the characteristic of Bingham fluid, and its expression is

$$\tau = \tau_{y,d}(B) + \eta_0 \dot{\gamma} \tag{4.1}$$

where $\tau_{y,d}(B)$ is the shear yield stress, which generated by externally the magnetic flux density $B$, and is proportional to the strength of applied magnetic field; $\eta_0$ is the viscidity of thc liquid, without the applied external magnetic field; $\gamma$ is the shear strain, namely d$v$/d$y$ [17].

Before the magnetic particles in the liquid reach saturation, by the applied external magnetic field, the shear yield stress $\tau$ of FMA can be expressed by Eq. (4.1), the expression of $\tau_{y,d}(B)$ is:

$$\tau_{y,d}(B) = \alpha B^n \tag{4.2}$$

where $\alpha$ is a constant; $n$ is the parameters, which is determined by the liquid property, and the value is 1–2.

Considering the above type, we can see that not only the actual shear stress of MFA is close to its property, but also related to the external magnetic field and shear strain and other factors.

### 4.2.2 Fluid Magnetic Abrasives Finishing Mechanism

1. Microscopic mechanism

FMA is a viscous suspension, which is composed by adding an appropriate proportion of magnetic particles, surfactants, non-magnetic abrasives, and anti-rust agent. These materials are then dispersed in the base fluid, and in large part, have similar properties with MR. In the absence of the applied magnetic field, the magnetic particles in the abrasive and non-magnetic are in a state of freedom (Fig. 4.3a); the entire abrasive appears as a kind of Newtonian flow; but under the effect of magnetic field, the abrasive in the magnetic particles will be quickly polarized and form a chain structure along the magnetic field direction. This leads to the rapid changes in the rheological properties of the liquid. The shear yield stress and viscosity values will also be improved and form into a solid-like Bingham substance. The chain structure of magnetic particles, just as the bond of plain wheel, clamps the non-magnetic abrasives between the magnetic chains (Fig. 4.3b), and if the chain structure in the manufacturing process is destroyed, then the new chain structure will be regenerated rapidly under the action of a magnetic field, so as to form a continuous finishing "soft wheel" [18].

FMA finishing is essentially a complicated physicochemical process; in this process, the friction, swipe, plowing, micro-machining, and corrosion by the abrasive particles may produce finishing effects to the workpiece surface. Among these effects, the main reason for finishing is the mechanical action, especially the micro-machining and swipe of the convex portions, which are the main reason to achieve the finishing. Oxidation and chemical corrosion also contributed to the finishing process.

From the microscopic point of view, after applying an external magnetic field, the magnetic particles form a chain structure; the chain structure of magnetic particles, just as the bond of plain wheel, clamps the non-magnetic abrasives between the magnetic chains, and the abrasive particles can be pushed to the surface by the magnetic levitation force (Fig. 4.3b). The "soft wheel" will automatically form the grinding contact surface according to the shape of the surface of the workpiece to be machined, thus ensuring a sufficient of abrasive particles near the machined surface.

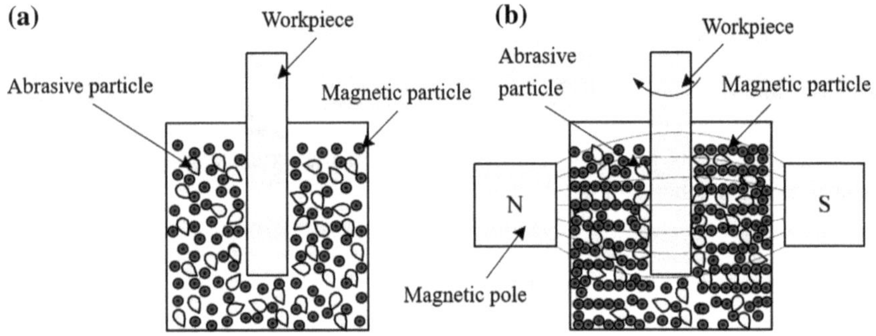


**Fig. 4.3** Fluid magnetic abrasive machining principle. **a** Without magnetic field of fluid magnetic abrasive; **b** Applying an external magnetic field magnetic abrasive fluids

When the workpiece is moving (generally includes rotation and the compound movement in the mobile of the abrasive), the abrasive particles and magnetic particles contact with the workpiece surface and thus are influenced by frictional force. If the frictional force is greater than the clamping force of its magnetic chain, the abrasive particles will damage the magnetic chain and do a relative movement by the force (Fig. 4.4a), that is to say non-magnetic particles must make MFA yield [19]. Assumes that the movement speed of workpiece surface and the direction of the magnetic field are vertical, the forces acting on the non-magnetic grits should be equal to the shear stress of magnetic abrasive. If the tangential component of the forces acting on the abrasive particle is greater than the friction between the abrasive particles and the workpiece surface, the relative sliding will be generated between the abrasive particles and the workpiece surface, then the slip brush, roll, and micro-machining will be obtained (Fig. 4.4b). For smaller convex, abrasive particles can cut it by micro-machining (Fig. 4.4c); if the tangential force is not big enough to cut off the bump, then the particle will slide along the surface. When the abrasive particles will once or several times contact with the workpiece surface (collision, friction, sliding, etc.), one or more tiny plastic deformation and superposition, even crack initiation, will be obtained. When the deformation degree is higher than the plastic deformation limit, the material allows, or after the crack extension, tiny chip can fall off from the surface (Fig. 4.5), so as to realize surface finishing.

2. Material removal model

In addition to the mechanical action of the non-magnetic abrasive particles, in the FMA finishing, the magnetic particles, the abrasive debris mixed in the magnetic particles, surface oxidation, and chemical corrosion may influence the finishing process. To better illustrate the mechanism of FMA finishing, it still assume that the finishing will mainly go through the micro-machining of non-magnetic particles, then the fluid magnetic abrasives finishing model can be deduced from the microscopic view through the analysis to a single abrasive particle [20–22].

Arbitrarily selecting an abrasive particle contacting with the workpiece surface, for the convenience of calculation, the grinding grain can be simplified as cones

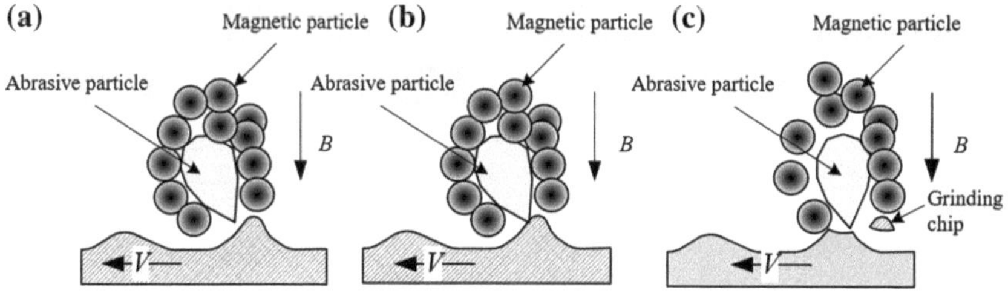


**Fig. 4.4** Micro-machining of fluid magnetic abrasives on the workpiece surface. **a** Abrasive particles close to the surface bump; **b** Micro-machining of abrasive particles on the surface convex; **c** Chip off

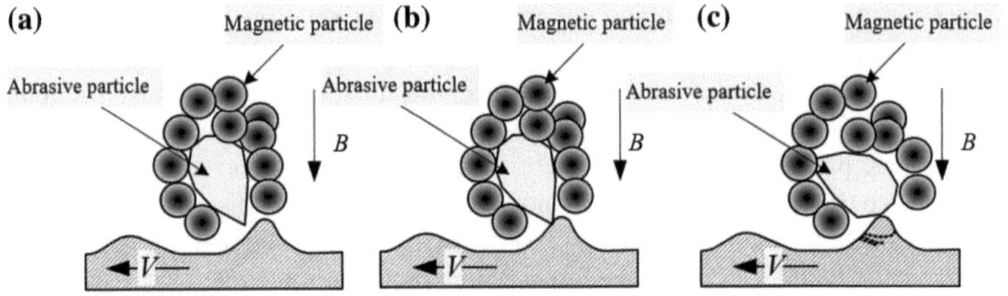


**Fig. 4.5** Collision, friction, and sliding of fluid magnetic abrasives on the workpiece surface. **a** Abrasive particles close to the surface bump; **b** The collision, friction, sliding of abrasive particles to surface convex; **c** On the surface of the metal occurring one or more small plastic deformation and plastic deformation superposition

(Fig. 4.6). When subject to the effect of the movement workpiece, abrasive particles will generate a trend in the same direction of the speed of the workpiece, and then the magnetic particle chain will prevent the movement of the particles. If the particles had a movement, the maximum stress acting on the particles is the shear yield stress when the magnetic particle chain is damaged, so that the maximum possible force of a single particle is

$$F = \frac{\tau h^2 tg\alpha}{2} \tag{4.3}$$

The cutting depth of abrasive on the convex portion is

$$a = \sqrt{\frac{F}{\sigma_s tg\alpha}} = \sqrt{\frac{\tau h^2}{\sigma_s}} = h\sqrt{\frac{\tau}{\sigma_s}} \tag{4.4}$$

Where $F$ is the force to particles subjected by the magnetic particle chains; $\tau$ is the shear stress of fluid magnetic abrasives; $h$ is the height of abrasive particles; $\alpha$ is

**Fig. 4.6** Machining schematic of single particle on the workpiece surface

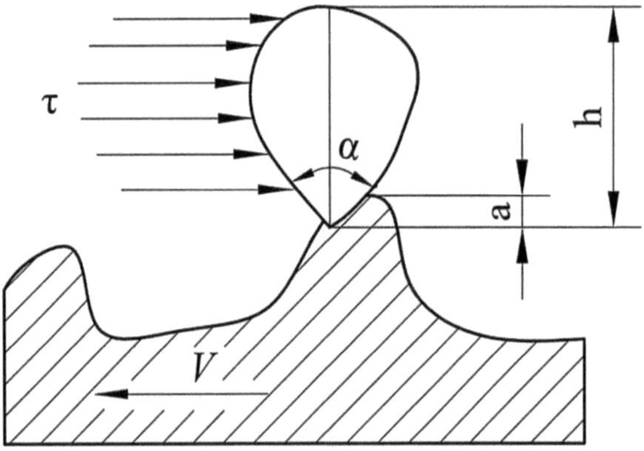

the semi-vertex angle of abrasive particles; $\delta_s$ is the yield stress of the workpiece surface; $a$ is the possible cutting depth on the convex portion by abrasive particles.

The above analysis is based on the micro-machining of the bumps on the surface by abrasive particles, since the trough of the workpiece surface and the bottom of the concave depression will be subjected to the action of abrasive particles, but due to the force and the number of raised surface by these particles are larger than the recessed portion, the material removal rate of the raised surface is higher than the concave part. But when the roughness of the surface decreases to a certain value, the difference of heights between the convex and the concave is not large, and the material removal rate of the convex and the concave part is relatively close. In this case, if the processing time is continuously increased, then there will not be any significant effects to reduce the surface roughness.

The efficiency and the surface roughness value greatly depend on the depth of micro-machining on the workpiece surface. A single particle that has greater cutting depth will result the FMA to have a higher finishing efficiency. Then, the surface roughness will also be greater. From the Eq. (4.4), it shows that the finishing efficiency is proportional to the size of the particles and the 1/2 power of shear stress, and it is inversely proportional to the yield stress of the workpiece surface. The smaller the abrasive particle's diameter and the FMA's shear yield strength, the greater the yield stresses of the workpiece surface, and the smaller the final surface roughness. Whereas if the abrasive particle's diameter and the shear yield strength of FMA are greater, the yield stress of the workpiece surface will be smaller, and the final surface roughness will be greater. The result is commendably validated by the test in Chapter V. Thus, by adjusting the diameter of the particles and the yield shear strength of the FMA, the finishing efficiency and the final surface roughness can be controlled, especially by controlling the shear yield strength of FMA through the change of the magnetic field strength. Also, the balance between the processing efficiency and processing quality can be well made.

From a macro-perspective, the material removal rate of the finishing can be described by Preston equation (2.3). Because in the fluid magnetic abrasives finishing, the material removal is mainly completed by the shear stress [23, 24], so it can be expressed as:

$$R = \mu \Lambda P V \tag{4.5}$$

where MMR is the material removal rate; $K$ is the coefficient, which is due to the combination of factors, including the type of abrasive, particle size, percentage, the types of magnetic particles, the diameter of the magnetic particle, the percentage of magnetic particles. Also, many processing parameters have affect the coefficient $K$, which always gained from experiment results. $A$ represents the finishing area; $\tau$ is the shear stress; $f$ is the friction coefficient; $v$ is the relative velocity between the abrasive and the workpiece surface.

As can be seen from the equation, the relative velocity between the abrasive and the workpiece surface is an important parameter, which has important influence to processing efficiency and the surface quality.

## 4.3 Fluid Magnetic Abrasive

### *4.3.1 Composition*

1. Magnetic particles

The magnetic particles play the most important role in FMA. Generating of magnetorheological (MR) effect is relay on the interaction between magnetic particles. The composition, morphologies, size, and other factors of the magnetic particles can affect the shear stress, controllability, and stability of FMA. A suitable candidate should be a soft magnetic material with many properties as follows: high saturation magnetization values, magnetoconductivity, low residual magnetism, and coercivity. The optimum diameter of magnetic particles is in the range of 1–10 μm and distributed uniformly. A kind of spherical particle with satisfactory ability of dispersion is preferred [25, 26].

(1) The type choice of magnetic particles

The MR effect of FMA is increased with increasing saturation magnetization, and the FMA would have a higher maximum shear yield stress, which is beneficial to improve the material remove rate. The saturation magnetization of FMA is related to magnetic particles and its proportion. As a result, electing the magnetic particle with high magnetic flux intensity is very significant to improve the properties of FMA.

Magnetic permeability is used to evaluate the ability of a material to be magnetized. A material with high magnetic permeability would have higher magnetic dipole moment under a low magnetic field, which also ensure that higher interaction force exists between magnetic particles at the same time. High interaction force makes the FMA has well shear yield strength and it is easier to response to the exterior magnetic field. All of this makes the machining process more controllable.

Residual magnetism is defined as the magnetic flux intensity of a magnetic material when the external magnetic field was removed after the material was magnetized to saturation. A material with high residual magnetism is difficult to be demagnetized after magnetization. It is needed to add a reverse magnetic field for a material with residual magnetism to be demagnetized. The coercivity is used to evaluate the demagnetized resistance of a magnetic material. The magnetic particles with low residual magnetism and coercivity can ensure that the shear yield stress of FMA could be well responded to the change of the magnetic field strength. The magnetic particles will not hold any residual magnetism under the condition that there is no external magnetic field. Then, the aggregation would not occur between magnetic particles, which will help keep its original state. Furthermore, magnetic particles with narrow magnetic hysteresis loop and low cohesive force will induce a decrease in demand of energy consumption for FMA. All of these metrics are beneficial to the application of the FMA.

(2) The diameter of particles

Besides the matrix material, the diameter and morphology are also important for determining the processing characteristics and stability of FMA. The particles size has an important influence on the shear yield strength and stability of the FMA. According to the magnetization law, the magnetic particle with multi-domain structure has a well-magnetized ability. If the magnetic particle is too small to only single domain structure, its magnetism will decrease obviously. Thus, choosing a larger magnetic particle in diameter is advantageous to the shear yield strength of FMA. Many researchers found that when the original particle size is small, and when small particles account for the same proportions in different formulation, the mechanical properties of MR fluid will increase with the increasing in the particle's diameter. When the magnetic flux density is 300 mT and the particle size increased from 0.1 to 1.0 μm, the shear stress will increase from 2.3 to 8.6 kPa over 3 times. However, when the particle size increased to a certain value, the shear stress of FMA will increase slowly with the increasing of particle size. For example, when the particle size increased from 5 to 30 μm, the shear stress of MR fluid would not improve obviously.

The phenomenon is induced by two reasons as follows: (1) The magnetic particles with large diameter have multi-domain structures. Under external magnetic field, these multi-domains will rotate in the same direction of the field. That means a new magnetic field, produced by particles, has the same magnetic field direction with that of the external field. As a result, the magnetic induction inner particles will greatly increase. In other words, the dipolar interaction between magnetic particles is strengthened. By contrast, the smaller magnetic particles have a weaker dipolar interaction, and especially, the nano-scale particles in single domain state have very low magnetized ability. (2) Smaller magnetic particles in diameter have larger specific surface area and lower bulk density. If the particles account for the same volume proportion, magnetic particles will account for a smaller proportion in weight. Then, the shear yield strength of FMA will be affected.

From the point of view of magnetism, it is better to choose the magnetic particles with large diameter. However, too large magnetic particles in diameter cannot ensure its settlement stability in the magnetic fluid. If surfactant is absence in MR fluid and the diameter of magnetic particle larger than 1 μm, these particles are difficult to suspend stably in the aqueous medium and will easily settle. Magnetic particles larger than 20 μm in diameter are difficult to suspend in the aqueous-based media even if the surfactant is added. Thus, selecting the particles with diameters in the range of 1–20 μm is the best choice for the fabrication of the FMA.

(3) The morphology of particles

Particle morphology is an image composed of every point located in the particle's boundary and surface. Particles in nature or used in industry have various morphologies that are defined by the manufacturing process. The morphology has significant influence on the specific surface area, magnetism, flow ability, dispersibility, and isotropy of the particles. As a result, not only the particle diameter is critically selected, but morphology is also selected. Because the sphere-like particles have well flow ability, stability, dispersibility, and very low anisotropy, sphere-like or near-spherical magnetic particle is the optimum choice for FMA.

(4) Introduction for magnetic particles

There are many magnetic particles, such as $\gamma$-$Fe_2O_3$, $Fe_3O_4$, Ni, Co, Fe, FeCo, NiFe, and they are suitable for the fabrication of FMA. The saturation flux density of the material listed as above is increased in sequence; however, their oxidation resistance in air is reduced in turn. $\gamma$-$Fe_2O_3$ is much more stable than the others but its magnetization is low. The situation is reversed in $Fe_3O_4$, as its color is changed quickly from black to brown in medium and its magnetization is also decreased in the oxidation process. The essential characters of $Fe_3O_4$ would not be changed even if its oxidation process can be retarded by the surfactant to some extent. FeCo and NiFe can also be oxidized easily. In these magnetic materials, pure Fe particle has the highest saturation magnetization and magnetoconductivity and is a kind of well magnetic material. But pure Fe is easily oxidized and cannot be used for FMA directly.

At present, ferrite and carbonyl iron powder are always used for the fabrication of FMA. Because the existence of the oxide in ferrite, the shear stress of FMA will decrease to some extent. The decreased value is related to the content of the oxide. The oxide can induce solid particles that account for much more volume than that without of the oxide, when they have the same weight. Inevitably, the FMA fabricated has higher initial viscosity in a zero magnetic field. Iron oxides' saturation flux density is much less than that of the pure iron powder. Thus, the existence of oxide in FMA will result in a decrease in magnetic flux density and shear yield strength. However, when the particle size is a certain value, oxides account for much proportion in magnetic particles, which will lead to an increase in subsidence stability of FMA.

Carbon monoxide (CO) can react with iron to form liquid $Fe(CO)_5$. Carbonyl iron powder will be obtained from liquid $Fe(CO)_5$ through evaporating method. The Fe element content of carbonyl iron powder is higher than 98%. Carbonyl iron powder has a high magnetoconductivity and saturation magnetization as iron, moreover, its lower residual magnetism and coercivity (Hc = 2.76 Oe, Mr = 0.126 emu/g). All of these metrics make the carbonyl iron powder an excellent soft ferromagnetic material. Carbonyl iron powder is in a suitable diameter range of 3–5 μm, and it has a multi-domain structure and a satisfactory dispersibility. It can be used directly without screening. Carbonyl iron powder has a spherical or onion-like morphology and smooth surface, which can ensure the FMA to have a decent flow ability and magnetic isotropy. Furthermore, carbonyl iron powder has well oxidation resistance, even if it is in a higher temperature and humidity environment. At present, carbonyl iron powder is the ideal choice for the fabrication of FMA [27].

Certainly, there are some inadequacies for choosing the carbonyl iron powder. The carbonyl iron powder can easily settle in the carrier liquid, especially for the water carrier liquid, and it is difficult to be dispersed again after settling. For this reason, surfactant is added to prevent the settlement and improve the dispersibility of carbonyl iron powder.

2. Carrier liquid

Carrier liquid is a carrier for magnetic particles, abrasive particles, and surfactant of FMA. The type and property of carrier liquid have significant influence on the fabrication, properties, and cost of the FMA. Generally, FMA should have these characteristics as follows:

(1) The carrier liquid should have well physical and chemical stability and will not react with the other composition of FMA. Moreover, it cannot be corrosive to the workpiece. The physical and chemical properties will not change, such as the decomposition and deterioration under normal service condition.
(2) It should have a higher boiling point, lower freezing point, and lower saturated vapor pressure under the service condition. All of these features are to ensure that the FMA has a wider service of temperature range.
(3) It should have a suitable viscosity. A low viscosity will induce bad settlement stability. Considering from the settlement stability, the carrier liquid should have a high viscosity. But a high viscosity of carrier liquid can restrict the improvement of magnetic particles and abrasive particles in proportion, which will affect the saturate shear yield stress and material remove rate of the FMA.
(4) It should have a low magnetic energy density and magnetic loss, in which will prevent the generation of excess heat.
(5) After being machined and being took out from carrier liquid, the workpiece can be cleaned easily. Cheapness is another advantage for widespread application. According to the requirements listed as above, water, mineral oil, and synthetic oil are always used for carrier liquid. Among them, the most used liquid are water and kerosene. Compared with other types of carrier liquid, water has many merits as follows: (1) low cost, pollution-free, and it makes the abrasive particles dispersed easily. (2) It has a well cooling, washing, and lubricating effect. (3) The machined workpiece can be easily cleaned by using the water carrier liquid. (4) Low viscosity of water leads to more sold particle proportions, which can be added to FMA that will have higher saturation shear yield stress and material remove rate. Thus, water is a better choice to be used for carrier liquid of FMA.

3. Surfactant

If magnetic particles and abrasive particles can be dispersed homogeneously in carrier liquid, the choice of surfactant is one of the deciding factors. The main purpose of adding surfactant is to improve the wetting properties and the prevention of aggregation of solid particles, i.e., the aim of adding surfactant is to improve the stability of dispersed solid particles and prevent subsidence and aggregation. When subsidence occurred, the adding of surfactant aimed to improve the re-dispersibility of solid particles [28].

The carrier liquid of FMA always has a low density. For example, the value of water is 1 g/cm$^3$, and by contrast that of magnetic particle is 7–8 g/cm$^3$. The density of abrasive particle is larger than that of the carrier liquid. The suspended particles are easy to settle in the carrier liquid due to solid particle having a higher density. The magnetic particle has a diameter of 5 μm and a large specific surface area. Particle aggregation occurs easily. Single particle can grow into a larger secondary particle, and then it will settle in the form of aggregation. At present, an effective method for solving this problem is to add several types of surfactants to the carrier liquid.

Generally, small amount of surfactant is needed for FMA. It can decrease the interfacial tension between solid–liquid two phases, make particles to be wetted effectively, and change the surface properties of the suspended particles. It is always used to decrease the interfacial energy between the two incompatible phases. There are many surfactants which can be classified by various methods. From the functionality of application, surfactant can be classified into emulsifying agent, detergent, foaming agent, wetting agent, dispersant, penetrating agent, and so on. According to their electrical properties, surfactant can be classified into two groups: non-ionic type and ionic type. The ionic surfactant contains three types as follows: anionic, cationic, and amphoteric types. Surfactant is composed of two groups, one is called the hydrophobic group and the other is hydrophilic group, from aspect of the molecular structure. Thus, surfactant is an amphipathic molecule. For example, $C_{12}H_{25}SO_4Na$ is an anionic surfactant, its alkyl ($-C_{12}H_{25}$) is hydrophobic group, and its sulfate group ($-SO_4Na$) belongs to the hydrophilic group. After adding a surfactant into the carrier liquid, the hydrophobic group will absorb the solid particles with the positive charge through the physical and chemical adsorption effect. The hydrophilic group will extend to the aqueous phase, which leads to the powder surface wetted quickly. The same situation occurs in magnetic particles and abrasive particles. On one hand, this effect will increase the volume of particles. On the other hand, the interaction between carrier liquid and surfactant belongs to a mixed-effect that will make the free energy of the system increased. Then, the particle aggregation will be avoided effectively.

It must be noted that the proportion of surfactant is not the more the better. If the surfactant on the particle surface reaches saturation, the dispersion effect will not increase with adding of more surfactant. Moreover, more surfactant will lead to the stability of particles to decrease quickly. Thus, the optimal content of surfactant should be gained from many tests, through changing the proportion of carrier liquid, magnetic particle, abrasive particle, and their diameter, respectively. The best dosage of surfactant is in the range of 0.5–5% by weight.

The choice of surfactant is tightly related to the types of carrier liquid, magnetic particle, and abrasive particle. At present, sodium dodecyl sulfate, sodium dodecyl sulfonate, ester, and polyether-based compound are always used as surfactant for water-based FMA.

4. Abrasive particles

Abrasive is a kind of granular material with functions of grinding and polishing. It is the main component of FMA, and it is also the main contributor for cutting and finishing processes. Abrasive type, size, and its percentage account for not only great influence on the removal rate, power consumption, durability, and the quality of the workpiece surface, but also the sedimentation stability of FMA [29].

There are many kinds of abrasives which are used in various areas. Generally, the selection of abrasive must be taken into considerations many of these factors as shown: hardness, strength, toughness, chemical stability, thermal stability, and particle size. The type, particle size, and its percentage of abrasive have important influence on the performance of the FMA. In type selection, the first thing to consider is the properties of the workpiece. For the hardness of abrasive, the principle is that the abrasive has the self-sharpening ability during the machining process and avoid its excessive wear. Workpiece that has a high hardness will result the abrasive to have a low value and vice versa. At the same time, the abrasive should go well with chemical stability to ensure that it will not react with another component of FMA.

(1) The choice of abrasive type
Generally, the abrasive type is mainly determined by the workpiece that is going to be machined. In the field of surface finishing, the boron carbide, silicon carbide, aluminum oxide, and diamond powder are always used as abrasive. In them, silicon carbide and alumina are suitable for processing steel or aluminum parts. In the experiment, two kinds of abrasive materials, including green silicon carbide (SiC) and brown fused alumina (mainly $Al_2O_3$), were chosen. For processing of steel and alloy steel, silicon carbide is not recommended; because it can easily be decomposed and react with iron at high temperature. The temperature of the FMA has almost no change during the usage. Then, the FMA containing silicon carbide have achieved a good processing effect in the processing of steel, alloy steel and hard aluminum [30].
(2) The choice of particle size
How to choose the particle size of the abrasive? The most important factor to be considered is that the roughness of the un-machined surface. First, the particle size has an important influence on the process. From the Eq. (4.4), we can see that the smaller mesh number (the larger the particle size of abrasive), the higher the material removal rate. The surface finishing efficiency increases with the increasing of particle size in a certain range. But when the particle size becomes larger, the final surface roughness obtained is greater. The larger the particle is, the more difficult to suspend in liquid, resulting a decline in the stability of FMA. According to large amounts of experiments, the optimal mesh number is in the range of $46^{\#}$–$180^{\#}$. If the grain size is too large, special surface treatment needs to be carried out in order to achieve a good suspension. When the mesh number is greater than $180^{\#}$, the processing efficiency will become too low. Both machining efficiency and surface roughness requirements should be considered for the selection of particle size. At the same time, the original surface roughness and the size of abrasive should adapt to the machining requirements. If the original surface roughness value is big, but that is small after processing, it will be difficult to meet the requirements with a fine abrasive

through a single process. Then, the processing can be divided into two or multiple steps. The choice of abrasive is according to the original and finial surface roughness of workpiece. In principle, one-pass processing should make the roughness decreased 1–2 grade and reach the balance between the machining efficiency and surface quality requirements.

(3) Percentage of abrasive account for FMA

The content of abrasive particles also has a very important influence on the performance of the FMA. In general, the larger the volume percentage of abrasive particles, the higher the machining efficiency is. Too high percentage of abrasive particles will result in a reduction of shear yield strength of FMA. This condition also has a negative effect on the stability of FMA. According to the experiment results, when the mass ratio of the abrasive particles is in the range of 30–40%, the FMA will have good comprehensive performance.

## 4.3.2 Preparation Technology

1. Fabrication of FMA

(1) Tools for the fabrication of FMA

The preparation of FMA requires many measuring instruments such as weighing tools (such as balance, cylinder), stirring tools (such as electronic stirrer and colloid mill), and heating tools (Fig. 4.7). The stirring tool is mainly used to make each component of FMA uniformly dispersed in the carrier liquid; heating tools can be used in low temperature environment for speeding up the process of dispersion [31, 32].

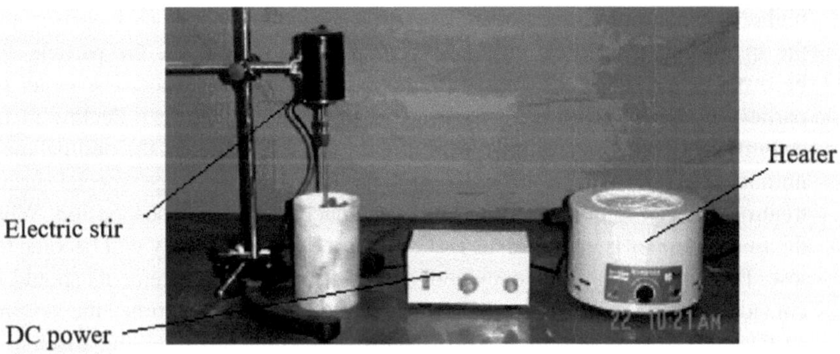


**Fig. 4.7** Measurements used for the fabrication of FMA

(2) Formulation of FMA
Three kinds of formulations are given as follows:

(a) 5.9 g of sodium dodecyl sulfonate was added in 195 ml deionized water, and the solution was heated until it is completely dissolved. Then, add 197 g of carbonyl iron powder and 200 g of green silicon carbide ($80^{\#}$) to the solution. Finally, the solution was mixed 6 h, under the condition that the rotational speed is 1000 r/min. Requirements: The sedimentation rate is less than 5% after standing for 24 h.
(b) 5.9 g of sodium dodecyl sulfonate was added in 195 ml deionized water and the solution was heated until it is completely dissolved. Then, add 270 g of carbonyl iron powder and 300 g of alumina ($80^{\#}$) to the solution. Finally, the solution was mixed for 6 h, under the condition that the rotational speed is 1000 r/min. Requirements: The sedimentation rate is less than 5% after standing for 24 h.
(c) 5.9 g of sodium dodecyl sulfonate was added in 200 ml deionized water, and the solution was heated until it is completely dissolved. Then, add 270 g of carbonyl iron powder, 300 g of green silicon carbide ($80^{\#}$) to the solution, and 2 g of ethanol. Finally, the solution was mixed for 6 h, under the condition that the rotational speed is 1000 r/min. Requirements: There is no precipitation after standing for 48 h.

Figure 4.8 is the digital photograph of the FMA. Figure 4.8a shows the situation of FMA without of the external magnetic field's influence. Figure 4.8b shows the situation of FMA in the external magnetic field [33].

2. Disperse process

Dispersion is one of the most important procedures, which aims to make each group stably dispersed in the carrier liquid during the preparation of the FMA. Generally, dispersion process can be divided into three steps, namely wetting of the solid particles, through the external force enables larger aggregates dispersed to smaller particles, stabilizing the dispersed particles, and preventing them to become aggregate. There are two kinds of dispersion methods: chemical dispersion method and physical dispersion method.

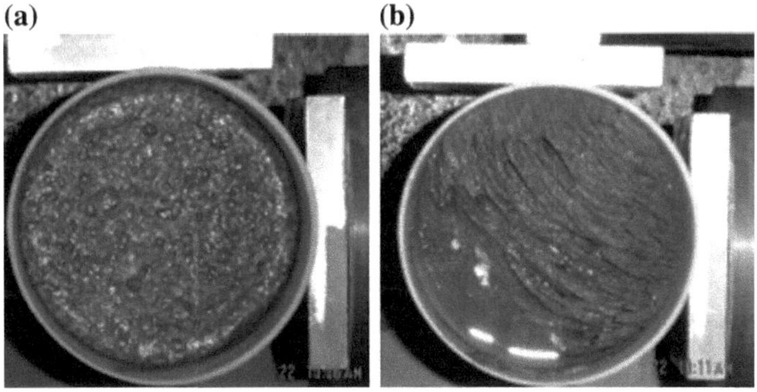

**Fig. 4.8** Digital photograph of the FMA under different magnetic field environment. **a** Without of the external magnetic field; **b** With external magnetic field

Chemical dispersion method is mainly used through the addition of surfactants, by changing the interaction between solid phase particles and liquid phase medium to achieve the purpose of dispersion. In the preparation of FMA, the adding of dodecyl sulfonate sodium belongs to the chemical dispersion. But it is still not enough to only use the chemical dispersion technology to prepare the FMA. It is necessary to use the physical dispersion method to achieve the purpose of uniform dispersion [34].

The physical dispersion includes two kinds of methods; they are mechanical dispersion and ultrasonic dispersion. Mechanical dispersion is a strong turbulent motion of liquid through a strong mechanical agitation, which makes the solid particles dispersed into liquid phase. In fact, the dispersion of mechanical dispersion is a complex physical and chemical process. In these processes, the whole dispersion system will appear in a series of physical and chemical changes. Through dispersion, the aggregated particles in the FMA can be broken into small ones under the effect of impact and shear effect. These particles will be driven by strong turbulence movement and fully contact with the surfactant. Then, the surfactant is attached to the particle surface, through strong turbulence movement, making the particle surface to be sufficiently wet so as to achieve the purpose of dispersion. It is very necessary to increase the dispersion time appropriately in order to make the particles absorb the surfactant and be wetted sufficiently. According to experience, in general, the dispersion time should not be less than 4 h.

There are many kinds of mechanical dispersion equipment. Electronic stirrer is relatively a simple one, and it is suitable for the preparation of less quantity of FMA. If fabricated amount is large, it is best to choose a professional high-speed mechanical mixing machine. But the ball mill and colloid mill with crushing effect are not suitable for an eventually dispersed device for FMA. In addition, an appropriate increase in the temperature of the liquid also helps to accelerate the dispersion process, but if the temperature is too high, it will cause aggregation and then affect the effect of the dispersion.

In order to improve the dispersion effect, it is necessary to support the mechanical dispersion by ultrasonic dispersion technology. Ultrasonic is an effective method for dispersing; just need to put the FMA into the ultrasonic field, and it will appropriately control the ultrasonic frequency and duration. By using the local high temperature, high pressure, strong shock wave, and micro-jet, the generated ultrasonic cavitating effect can achieve the auxiliary dispersion effect. The use of ultrasonic dispersion cannot be too long, otherwise it will lead to overheating, resulting a decline in the stability of FMA.

### *4.3.3 Performance Parameters*

There are many indicators for evaluating the performance of FMA. These indicators include viscosity, density, stability, rheology, and saturation magnetic induction intensity. In addition, particle size and volume proportion of abrasive also have important influence on performance of FMA.

1. Initial viscosity of FMA

(1) Initial viscosity

The initial viscosity, also known as zero field viscosity, is the viscosity of the FMA when the magnetic field is not added. A relative motion exists between the adjacent fluid layers when the fluid is flowing. The frictional resistance between the two fluid layers is called viscous force. Viscosity is used to measure the viscous force; high viscosity means high internal friction. In the fluid flow, due to the adhesion between fluid and solid surface, intermolecular interaction and momentum exchanges between the fluid particles, thus the shear deformation will occur in these particles. The fluid viscosity has deformation resistance ability to shear deformation. Without the magnetic field, the FMA presents the properties of Newton fluid, in accordance with the Newton's law. According to the Newton's inner friction law, the internal friction (shear stress) is proportional to viscosity and is linear with velocity gradient (shear strain rate). The relationship between shear stress and shear strain rate can be described as:

$$\tau = \eta_0 \dot{\gamma} = \eta_0 \frac{\partial u}{\partial y} \tag{4.6}$$

where $\tau$ is the shear stress; $\eta_0$ is the zero field viscosity of the FMA. For Newtonian fluid, $\eta_0$ is a constant; $\dot{\gamma}$ is the shear strain rate and is equal to the rate change of flow velocity along the thickness direction of the fluid.

For FMA, the physical properties, shape, the percentage concentration of magnetic medium particles, non-magnetic abrasive particle, and the carrier liquid and surfactant will directly affect its initial viscosity $\eta_0$. The relationship between liquid viscosity and the temperature is satisfactory, but the relationship between the liquid viscosity and the pressure is small. When the pressure $P$ is lower than 5 MPa, the change in viscosity can be ignored.

The initial viscosity of FMA determines its flow ability under the condition of zero magnetic fields. When the initial viscosity is low, the FMA has good flow ability. Because the FMA is always used for finishing, then a low initial viscosity is required. Thus, magnetic particles and abrasive particles can flow freely in FMA. When the magnetic field is applied, parameter's control range is relatively large. The surface finishing effect for workpiece will be better.

Initial viscosity of FMA is related to the physical properties, shape, volume, concentration ratio of magnetic particle, non-magnetic abrasive particle, the carrier liquid, and surfactant. The concentration of solid phase plays a decisive role in viscosity of FMA. The relationship between the viscosity, volume fraction of solid phase, and the carrier liquid viscosity of extremely dilute suspension was founded by Einstein in 1906 and can be described as:

$$\eta_0 = \eta_c(1 + 2.5\phi_v) \tag{4.7}$$

where $\eta_0$ is the viscosity of carrier liquid and $\phi_c$ is the volume fraction of solid phase.

It can be seen from Eq. (4.7) that the viscosity of the FMA increases as the volume fraction of the magnetic particles and the abrasive particles increases. Einstein's equation can qualitatively describe the relationship between viscosity and volume fraction of solid phase. But there are some errors compared with the calculated results and actual measured value. This is because Eq. (4.7) is only applicable in the condition that the solid volume fraction of single solid phase suspension is less than 0.02. Afterward, Chong improves the equation as follows:

$$\eta_0 = \eta_c\left(1 + 0.75\frac{\phi_v/\phi_{vc}}{1 - \phi_v/\phi_{vc}}\right) \tag{4.8}$$

where $\phi_{\text{vc}}$ is the maximum solid volume fraction, which is a function of particle shape. It is approximately 0.74 in the water carrier. It will increase when the particle size is not uniform due to the small particles will enter the space formed by the large particles. When there is an infinite number of particle sizes distributing randomly, then $\phi_{\text{vc}} < 1$.

(2) Viscosity measurement

Generally, there are three methods to measure the viscosity of liquid: the falling-ball method, the rotating-cylinder method, and the capillary method. The falling-ball method and the rotating-cylinder method are briefly introduced in this chapter.

(1) Falling-ball method

When a small ball with a radius of $r$ and smooth surface moves in an infinite liquid at a low velocity of $v$ with no swirl occurs; then the Stokes' law pointed out that the resistance of ball movement can be described as: $F = 6\pi\eta vr$ ($r$ is the radius of the ball and $\eta$ is the viscous force of liquid).

When the ball drops in the liquid, three kinds of forces will be acting on the ball, namely gravity (mg), buoyancy (ρvg), and viscous resistance ($6\pi\eta vr$). When the ball is just begun to fall, the speed of $V$ is very small; the ball is accelerated with little resistance. As the speed increases, the resistance increases. When the sum of the resistance and buoyancy is equal to the gravity, the ball will began to drop at a constant speed, that is, mg = ρvg + $6\pi\eta vr$. At this time, the ball speed is called final speed. Then,

$$\eta = (m - \rho v)g/6\pi vr \tag{4.9}$$

The falling-ball method is usually used to measure the viscosity of transparent or semitransparent liquid.

(2) Rotating-Cylinder Method
Figure 4.9 is the schematic diagram of the rotating-cylinder method. In Fig. 4.9, A is a rotor that spins slowly, driven by the electromotor spindle through hairspring. B is a beaker.

Liquid that is being measured is placed in the beaker (the liquid level is flushed with the neck of the rotor). When the spindle of the motor rotates at a certain speed of $\omega$, the rotor (A) will rotate through the driving of gossamer.

When the moments acting on rotor A is equal to that on the gossamer, the rotor will rotate at the speed of $\omega$. At this time, the rotor A rotates $\theta$ angle compared with the motor spindle (gossamer tighten $\theta$ angle). The relationship between viscosity and $\theta$ angle for gossamer is:

$$\eta = k\theta \tag{4.10}$$

Where $k$ is the instrument coefficient determined by the torsion coefficient of gossamer, rotor structure, and rotation speed.

The NDJ-1 rotary viscometer was used in the experiment, which work as the principle that is shown in Fig. 4.10. The tested FMA has a good finishing effect on the experiments. It is composed of 5.9 g of sodium dodecyl sulfonate, 200 ml deionized water, 2 g of ethanol, 150 g of RXe carbonyl iron powder, 300 g of 80$^{\#}$ green silicon carbide. Four groups of test were carried out. The experimental results are 630cP, 640cP, 590cP, 590cP; the mean initial viscosity value of the FMA is 612.5cP.

2. Density of FMA

The density of the FMA not only is one of the most important parameters during finishing, but it is also used to calculate the content of magnetic particles and other particles. The density of the FMA has a certain relationship with saturation magnetization. If the volume of the FMA is considered to be the sum of the volume of each component, the density of FMA can be written as:

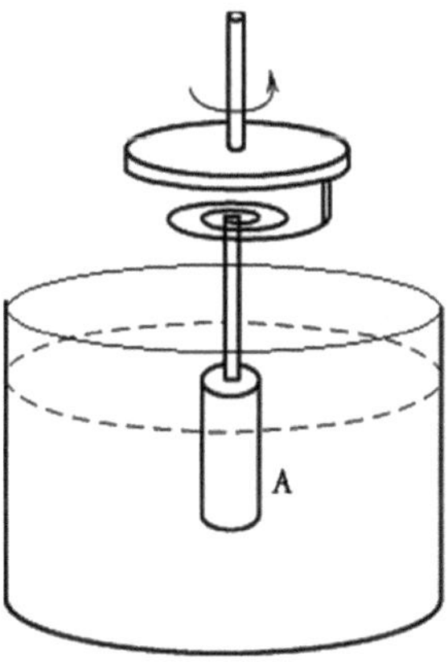


**Fig. 4.9** Principle of rotary viscometer

**Fig. 4.10** NDJ-1-type rotary viscometer

$$\rho = \rho_s\phi_s + \rho_a(\phi_h - \phi_s) + \rho_c(1 - \phi_h) \quad (4.11)$$

Where $\rho_s$ is the density of the solid phase, solid phase including abrasive phase and magnetic particle phase; $\rho_a$ is the density of surfactant; $\rho_c$ is the density of carrier liquid; $\phi_s$ is the volume fraction of solid phase, $\phi_h$ is the sum of volume fraction of surfactant and solid phase.

In the FMA, the loose-packed density of carbonyl iron is 3.32 g/cm$^3$; that of the green silicon carbide loose-loading density is 1.58 g/cm$^{3,}$ and that of surfactant sodium dodecyl sulfonate is 0.578 g/cm$^3$. If computed by volume ratio, carbonyl accounts for 16.9%; green silicon carbide accounts for 39.4%; sodium dodecyl sulfonate accounts for 2.1%; water carrier fluid accounts for 41.5%; ethanol accounts for 0.1%. According to Eq. (4.13), the density of the FMA is about 1.55 g/cm$^3$.

3. Stability of FMA

The stability of the FMA is that it can keep its initial state for a long time, and all the particles are in a state of uniform suspension without settlement. It consists of two aspects: the settlement stability and aggregation stability. Settlement stability, also known as dynamic stability, is used to describe whether the solid particles are easy to settle or not. Many factors can affect the settlement stability, which includes size, shape, density, the deformation ability of particle, fluid density, viscosity, temperature, pressure, and particle concentration. But the determining factor comes to be gravity and the settlement resistance force. When gravity and the resistance force are equal, the particle will settle uniformly. According to Stokes' law, if the particle shape is spherical, the settling velocity is:

$$v = \frac{2r^2(\rho - \rho_0)g}{9\eta_c} \quad (4.12)$$

where $v$ is settling velocity; $r$ is the radius of the spherical particles; $\rho$ is the density of particles; $\rho_0$ is the density of dispersion medium; $\eta_c$ is the viscosity of the dispersion medium; $g$ is the gravitational acceleration.

The settling velocity of particles are proportional to their square of the particle radius and to the density difference between particles and medium. However, the settling velocity of particles are inversely proportional to the viscosity of medium. The particle size has the greatest impact on the settlement velocity. The settling velocity of suspension particle relies on the radius of the particles and the viscosity of the suspension. Therefore, in the preparation of FMA, we should choose particles with smaller size and lower density. The carrier liquid with high viscosity can greatly improve the settlement stability of the FMA.

Aggregation stability refers to whether the solid particles in FMA are easy to aggregate automatically. The single magnetic particle or abrasive particle in the FMA can grow up to become a bigger secondary particle due to the interaction force between particles in the zero magnetic fields.

The aggregation may induce by the interaction between electrostatic, steric stabilization of adsorption layer, lyophobic, and van der Waals forces. In order to prevent the aggregation, surfactants are usually added to improve the settlement stability of solid particles.

Although the requirements on stability of FMA are not as high as the magnetic liquid and the MR fluid, it is necessary to have a certain stability and re-dispersibility. A good FMA requires its magnetic particles and abrasive particles to have a low settling velocity and can be dispersed evenly in the carrier liquid. Moreover, the settlement particle does not aggregate and has a good re-dispersibility.

4. Saturation magnetization of FMA

(1) Saturation magnetization

According to the Sect. 3.4, Chap. 3 which shows that the magnetic flux density of magnetic materials induced by magnetization increase, the magnetic flux density increases to a certain value and will not continue to increase as the external magnetic field increases, i.e., when the magnetic flux density reaches a saturation value and that is the saturation magnetization of a magnetic material.

Because the rheological effect of FMA is generated by the suspended magnetic particles, the directions of all domains of FMA are identical with the external magnetic field; thus the magnetic flux density will reach the saturation value. Once the saturation magnetization has been reached, the increase of external magnetic field strength will not cause the increase in magnetic flux density of FMA.

When the external magnetic field is applied, magnetic particles will form a chain structure along the direction of the field. Abrasive particles are held between the chains. When the FMA reaches the saturation magnetic intensity, abrasive particles are hijacked by magnetic particle chains force, which are also reached to the maximum. At this time, the shear yield stress of FMA will reach to its maximum value. For FMA, the cutting force is mainly contributed by shear yield stress. Therefore, saturation magnetization not only is one of the most important parameters for FMA, it is also the design basis of magnetic field generator of finishing equipment. For FMA, a high saturation magnetization is needed in order to lead to more obvious rheological effect; a greater maximum shear yield stress can be gained, the material removal rate will be higher, the control ability will be stronger when the external magnetic field is applied.

There are two main methods for improving the saturation magnetization of FMA. First, the choice of magnetic particles should have a saturation magnetization as high as possible; second, improving the percentage of magnetic particles account can also improve the saturation magnetization of FMA. Three magnetization curves were obtained by using the same magnetic particles with different content and are shown in Fig. 4.11. The mass percentage of magnetic particles is in accordance with the trends of a > b > c. Due to the increase of the percentage of magnetic particle being restricted by many factors, it is very important to choose the magnetic particles with high saturation magnetization to improve the performance of the FMA.

The saturation magnetization of FMA can be tested by means of a special instrument, and it can be estimated by equations. In general, the saturation magnetization of FMA is equal to the product of the magnetic saturation value of ferromagnetic material used for making magnetic particles and the volume ratio of the magnetic particles. For example, the volume ratio of magnetic particles is 0.2. Assuming that the magnetic particles were obtained through the ground iron, the magnetic saturation value of iron is about 2.1 T (oxidation state is about 0.7), and then the saturation magnetization of this FMA is about $0.2 \times 2.1 = 0.42$ T.

(2) The measurement of saturation magnetization

When all domains' directions in the FMA are the same with the external magnetic field, the magnetic flux density will reach the saturation value, which is the saturation magnetization of FMA. When external magnetic field is applied, the magnetic particles will form columnar chains. Abrasive particles are held in these chains. The hold force will reach the maximum value when the magnetic flux density of FMA reaches saturation. At this time, the FMA has the largest shear stress and the highest machining efficiency. The saturation magnetization is one of the most important parameters for FMA and has significant influence on its machining ability, machining efficiency, controllability, ultimate surface roughness

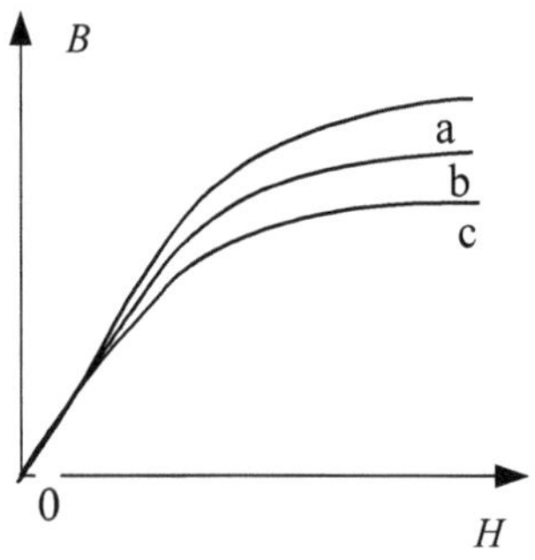


**Fig. 4.11** Response characteristics of magnetic flux density of FMA with different volume ratios vary with the magnetic field intensity

value, and so on. Therefore, it is necessary to obtain the saturation magnetization of FMA for the design of magnetic field generator and energy saving.

Figure 4.12 is a schematic diagram of experiment device for testing the saturation magnetization of FMA. The closed magnetic circuit forms by coil, iron core, upper and lower magnetic yoke, pole, FMA, and air in the FMA. A DC power is used to supply to coil, which is used to generate a magnetic field. Iron core, upper and down magnetic yoke, two poles, FMA, and air in the FMA help form a closed magnetic circuit. According to the magnetic circuit theory:

$$F = NI = \sum H_j l_j \tag{4.13}$$

$$F = R\phi \tag{4.14}$$

Then,

$$NI = R\phi \tag{4.15}$$

$$R = l/\mu S \tag{4.16}$$

$$\phi_j = B_j S_j \tag{4.17}$$

where $F$ is the magnetic potential; $N$ is the number of turns; $I$ is the current flowing in the coil; $R$ is reluctance; $l$ is the length of medium in the magnetic circuit; $B$ is the magnetic flux density; $\phi$ is flux, $S$ is the area of the medium in magnetic circuit; $\mu$ is permeability; $H_j$ is the magnetic field intensity of the $j$ section of magnetic circuit; $l_j$ is the length of the $j$ section of magnetic circuit; $\phi_j$ is the magnetic flux of $j$ segment; $B_j$ is the magnetic flux density of the $j$ section of magnetic circuit.

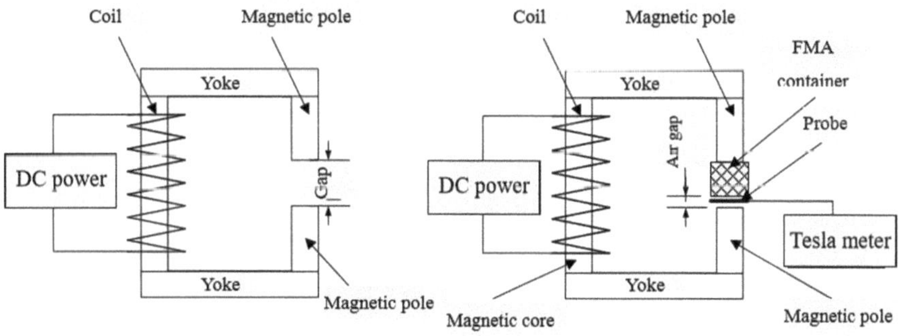


**Fig. 4.12** Schematic diagram of experiment device for testing the saturation magnetization of FMA. **a** state without adding specimen; **b** state during the test

Because $B = \mu H$, according to the Eq. (4.13), then:

$$NI = \sum \frac{B_j l_j}{\mu_j} \tag{4.18}$$

According to Eq. (4.13) and (4.18):

$$H_f = \left[ NI - \frac{B_g l_g}{\mu_0} - \frac{B_x l_x}{\mu_y} - \frac{B_y l_y}{\mu_y} \right] \Big/ l_f \tag{4.19}$$

where $H_f$ is the magnetic field intensity of the FMA; $B_g$ is the magnetic flux intensity of the air; $l_g$ is the length of the magnetic circuit in the air; $\mu_0$ is the air permeability; $l_x$ is the sum length of the iron core and pole; $B_x$ is the magnetic flux intensity of core (pole); $\mu_y$ is the permeability of pure iron, $l_y$ is the total length of upper and down magnetic yoke; $B_x$ is the magnetic flux intensity of magnetic yoke; $l_f$ is the length of FMA in magnetism loop.

The number of turns inside the coil can be learned, the current in the coil provided by the DC power will be shown in ammeter, the Tesla meter is used to measure the magnetic flux density, and the length of the medium can also be measured. Then, the permeability can be computed out. The magnetic flux intensity and the permeability are known, and the magnetic flux is the same during the whole magnetic circuit. According to the Eq. (4.19), the magnetic field intensity $H_f$ of FMA can be calculated.

According to the knowledge of electromagnetism, the interface of FMA and air gap is as followed:

$$B_g = B_f \tag{4.20}$$

When the magnetic flux intensity of air gap was 1 T, $B_x$ and $B_y$ are less than 1 T. At this time, iron core and magnetic yoke were unsaturated (if magnetic circuit formed by pure iron, the saturation magnetization of iron $B_{\text{sat}} = 2.1$ T), so the permeability of iron permeability is far larger than that of air:

$$\mu_y \gg \mu_0 \tag{4.21}$$

Then, Eq. (4.19) can be simplified as follows:

$$H_f = \left( NI - \frac{B_g l_g}{\mu_0} \right) \Big/ l_f \tag{4.22}$$

According to the test method described above, the parameter design of measuring equipment and calculation process are introduced in Table 4.2. According to the experiment results, the saturation magnetization of FMA was estimated and was less than 1 T. The magnetic flux intensity of designed device was also less than 1 T. Initially set: pole radius $R = 0.03$ m; column space $\delta = 0.03$ m.

**Table 4.2** Parameters design and calculation of the measuring device

| Computing project | Equation | Remarks |
|---|---|---|
| Permeability of vacuum | $\mu_0 = 4\pi \times 10^{-7}$ | |
| Magnetic flux intensity of air gap | $B_g = 1$ T | |
| Air gap height | $h_1 = 0.01$ m | |
| Permeability of magnetic liquid | $\mu = 3\mu_0$ | |
| Magnetic intensities of magnetic liquid | $H_1 = B_g/\mu = 2.65 \times 10^5$ | |
| Magnetic intensities of air gap | $H_g = B_g/\mu_0 = 7.96 \times 10^5$ | |
| Magnetic potential | $F = H_1(\delta - h_1) + H_g h_1 = 1.326 \times 10^4$ | |
| Magnetic pressure drop of iron core | $F_{\max} = 1.2F = 1.5912 \times 10^4$ | |
| Current in coil | $I = 2$ A | |
| Coil turns | $N = \frac{F_{\max}}{I} = 7.956 \times 10^3$ | $N = 7.960 \times 10^3$ |
| Current in wire | $J = 4 \times 10^6$ | |
| Wire diameter | $d = \sqrt{\frac{4.2}{\pi J}} = 7.979 \times 10^{-4}$ | Q 0.90-type enameled wire |
| Maximum outer diameter of first-level enameled wire | $d_{\max} = 0.960$ | |
| Duty factor of coil | $k = 0.7$ | |
| Window area of coil | $A_w = \frac{Nd_{\max}^2}{k} = 1.278 \times 10^{-2}$ | |
| Skeleton dimension of coil | Assume that: $H = 140$ mm; wall thickness $\Delta = 2$mm | |
| Inner diameter of coil skeleton | $\phi_1 = 2R + 0.004 = 0.064$ m | |
| Outer diameter of coil skeleton | $\phi_2 = \phi_1 + 2\Delta + \frac{A_w}{H - 2\Delta} = 0.162$ m | |
| Total length of wire | $L_d = \pi\left(\frac{\phi_2 + \phi_1}{2}\right)N = 2825.8$ m | |
| Weight per kilometer of wire | 5.83 kg | |
| Total weight of wire | $W = 5.83$; $L_d = 16.33$ kg | |
| Resistance of wire | $R = 0.0181 \times 10^{-6} \times L_d/(\frac{\pi}{4}d^2) = 80.4$ | |
| DC voltage | $U = 1.4$; RI = 225.12 V | |
| Iron yoke width | $b = 2$; $R = 0.06$ | |
| Iron yoke height | $h_2 = \pi R^2/b = 0.047$ | |
| Iron yoke length | $L_y = 4R + 0.04 + \phi_2/2 = 0.241$ | Set as 240 mm |
| Pole piece dimension | $h_3 = (H + 0.004 - \delta)/2 = 0.057$ | Set as 60 mm |

The rationality of the design should be verified. The calculation results are shown in Table 4.3. Assuming that the cross-sectional area of iron core is $S_y$ ($m^2$), the average length of the iron core is $L_y$ (m), the cross-sectional area of air gap is $S_g$ ($m^2$), and the length of the air gap is $L_g$ (m).

**Table 4.3** Calculation check for the measurement device

| Computing project | Equation | Remarks |
|---|---|---|
| Air gap reluctance | $R_{mg} = L_g/\mu_0 S_g = 2.553 \times 10^6$ | |
| Air gap flux | If $B_g = 1\,T$, $S_g = \pi r^2 = \pi \times (0.063 \div 2)^2 = 3.117 \times 10^{-3}$ then $\phi_g = B_g S_g = 3.117 \times 10^{-3}$ | |
| Flux of iron yoke | $\phi_y = \phi_g$ | |
| Magnetic flux intensity of iron yoke | $B_y = \phi_y/S_y = 1.1$ | |
| Pure iron | when $B_y = 1$ T, $H_y = 120$, then $\mu_y = B_y/H_y = 0.0083$ | |
| Reluctance of iron yoke | $R_{my} = L_y/\mu_y S_y = 2.05 \times 10^4$ | |
| Reluctance of iron core | $R_{mx} = L_x/\mu_y S_{yx} = 1.04 \times 10^4$ | |
| Reluctance of FMA | $R_{mm} = L_m/\mu_m S_{yx} = 1.703 \times 10^6$ | |
| Total reluctance | $R_m = R_{mg} + R_{my} + R_{mx} + R_{mm} = 4.29 \times 10^6$ | |
| Magnetic potential | $F = R_m \cdot \phi_g = 1.337 \times 10^4$ | |

The calculation check results are consistent with the design results; this shows that the designed device was able to meet the saturation magnetization test requirement for FMA.

The test was carried out by using the experimental device above. When the circuit of electromagnet was 0.5, 1.0, and 2.0 A, respectively, the magnetic flux intensity in magnetic circuit is 0.24, 0.31, and 0.355 T, respectively. The magnetic field intensities of FMA were calculated under three kinds of conditions shown in Table 4.4 by using Eq. (4.22).

According to the above calculation results, the *B–H* curves of FMA are shown in Fig. 4.13.

5. Rheological properties of FMA

(1) Influence factor

For FMA, the rheological properties are shown from the aspects of shear yield stress and viscosity.

(1) Effect of shear yield stress on rheological properties

Without adding the external magnetic field, the FMA acts as a standard suspension liquid. It has a property in accordance with that of the Newtonian fluid (Eq. 4.6). Once the external magnetic field is applied, its rheological properties will

**Table 4.4** Magnetic field intensity of FMA

| Current | Magnetic field intensity |
|---|---|
| 0.5 A | $H_f = (NI - B_g l_g/\mu_0)/l_f = 107667$ A/m |
| 1 A | $H_f = (NI - B_g l_g/\mu_0)/l_f = 273323$ A/m |
| 2 A | $H_f = (NI - B_g l_g/\mu_0)/l_f = 637000$ A/m |

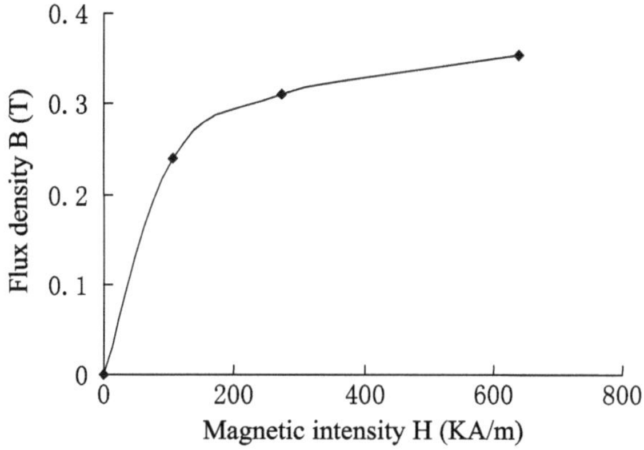


**Fig. 4.13** *B–H* curves of FMA

change rapidly. The FMA's viscosity is increased, and it showed a character similar to the Bingham fluid. If the external force acting on FMA is not enough to overcome the shear yield stress, the abrasive will depict a state like that of a solid, and it will not flow. In case the external force is larger than the shear yield stress, the abrasive will begin to flow. Before the magnetic flux density reaches saturation, the shear yield stress and viscosity will increase with the enhancement of the magnetic field intensity (Fig. 4.14) and vice versa. The relationship between the shear yield stress and the magnetic field should be described as Eq. (4.2). When reaches saturation, the shear yield stress will not change continually due to the increasing of magnetic field intensity. Generally, when the magnetic flux density reaches saturation, the shear yield stress of FMA will reach the maximum value.

The relationship between the shear yield stress of FMA and the saturation magnetization of magnetic particles is described as:

$$\tau_y \propto B_{\mathrm{sat}}^2 \tag{4.23}$$

where $\tau_y$ is the shear yield stress; $B_{\mathrm{sat}}$ is the saturation magnetization of magnetic particles.

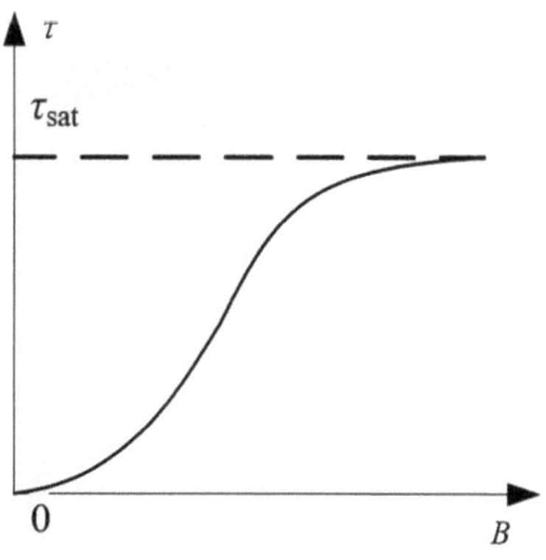


**Fig. 4.14** Response characteristics of magnetic flux density vary with the shear yield stress

For the FMA to contain high concentration of magnetic particles, and on the basis of the analysis on the magnetic particle chain structure, the saturation yield stress $\tau_{sat}$ of FMA can be calculated by following equation:

$$\tau_{sat} = \frac{4}{5^{5/2}} \xi \phi \mu_0 B_{sat}^2 \tag{4.24}$$

where $\xi$ is a constant in range of 1.2–1.5, $\phi$ is the volume ratio of magnetic particles, and $B_{sat}$ is the saturation magnetization of magnetic particles.

From Eq. (4.24), it can be seen that the selection of a magnetic particle with high saturation magnetization and the improvement of the volume ratio of magnetic particles in FMA will lead to an increase in saturation shear yield stress.

$$\begin{cases} \tau = \text{sgn}(\dot{\gamma})\tau_0 + \eta\dot{\gamma}; & |\tau| > \tau_0 \\ \dot{\gamma} = 0; & |\tau| < \tau_0 \end{cases} \tag{4.25}$$

From Eq. (4.25), it can be seen that the shear force of FMA during the finishing process is related to the shear yield stress and shear strain rate. With the increasing of shear strain rate, the shear force will be increased. (Fig. 4.15 shows the shear force vary with different magnetizations.)

(2) Effect of viscosity on rheological properties

Not only the shear yield stress is related to the magnetic field intensity, but also it is related to the viscosity of FMA. Under the effect of magnetic field, the FMA loses its flow ability gradually when changing from Newtonian fluid to Bingham elastic–plastic body. When it has lost all of its ability to flow, its viscosity tends to lean toward infinity. When the external shear stress exceeds its yield stress, the FMA will have the ability to flow. With the increasing of the applied magnetic field intensity, both the shear yield stress and the viscosity will increase. But the relationship between the viscosity and shear strain rate is in opposition; that is to say, the viscosity of FMA has a tendency to become smaller with the increasing of shear strain rate, which is called shear thinning. The response characteristic of viscosity with the change of magnetic field intensity is an important index to reflect its adjustability and controllable properties. Due to the change in the magnetic field

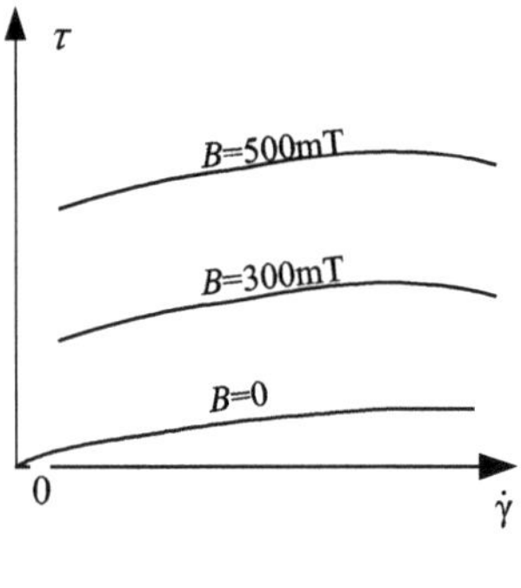


**Fig. 4.15** Dependence of shear force on shear stain rate

intensity, the greater the change will be in viscosity, and it shows a larger adjustable range in its performance [35].

In a word, the rheological properties of FMA can be effectively controlled by the external magnetic intensity. As previously mentioned, the main processing capacity of FMA is contributed by its shear stress. So the shear stress is an important parameter for assessing the processing capacity of FMA under the different magnetic field intensities and shear strain rates conditions. When the shear stress is high, the finishing efficiency of FMA will be high, and the controllable range will become wider. When the composition of FMA is fixed, the magnetic field intensity determines the value of the shear stress in a certain range. The purpose of studying the shear stress is to achieve the maximum shear stress with the most reasonable magnetic field intensity.

(2) The test of shear yield stress

There are no matured commercialization systems for shear stress testing, and there is also no uniform set of standards, although the test principle and test equipment can borrow that of the MRF. But some test equipment is needed to be self-made in the production practice.

Figure 4.16 shows the device used by Felt (1996). Brookfeild LVDV III rotational viscometer was placed in the electromagnetic coil for testing. The viscometer with conical plate structure can produce the uniform flow field. Because the coil produces the magnetic field of viscometer directly, the magnetic flux intensity is very low, as it only reaches 800 mT. However, due to the conical plate structure, the magnetic field distribution in the MR fluid is not uniform, which greatly affects the test results.

Figure 4.17 shows the static measuring instrument for MR fluid developed by Tang and Gorral (1996). S and N are the two magnetic poles of the electromagnet, which can generate an adjustable magnetic field. The magnetic flux intensity $B$ can be measured by the Gauss instrument that connected to the Holzer sensor. When the supported aluminum plate moves at a very small constant velocity of $V$, the MRF is bearing the simple shear. The supported force $F$ for aluminum plate is then

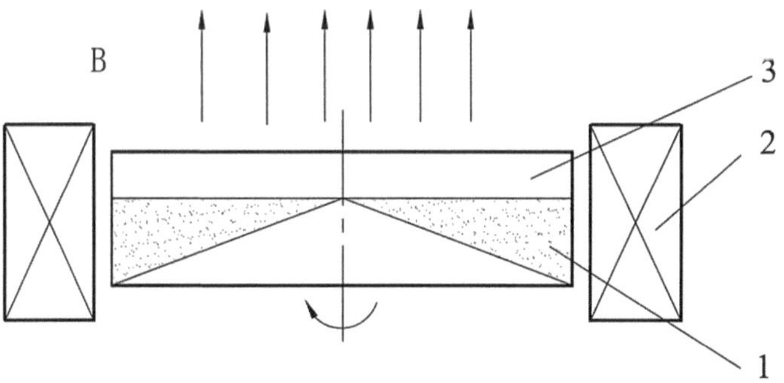


**Fig. 4.16** Schematic diagram of No. 1 test equipment: 1—MRF, 2—Coil, 3—Cone viscometer

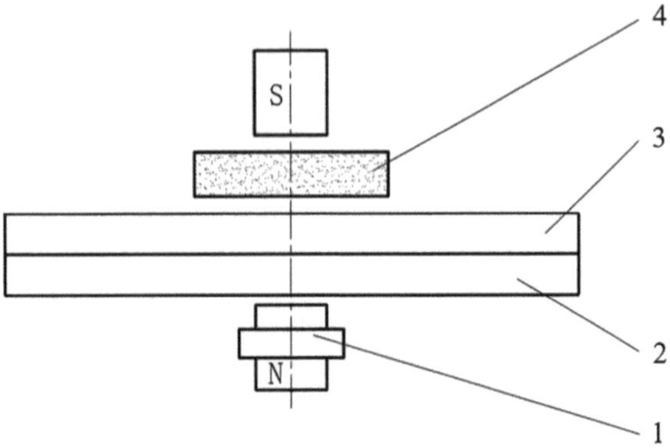


**Fig. 4.17** Schematic diagram of No. 2 test equipment. 1—Holzer sensor; 2—Aluminum plate; 3—Iron plate; 4—MRF

measured. Due to $V$ being very small, it can be considered as a static measurement. The disadvantage of the device is that the force $F$ is not easily and accurately grasped, and the magnetic liquid can easily overflow. The device is also complex.

Figure 4.18 shows the magnetorheological properties test system, developed by the National University of Defense Technology, Dai Yifan, in 2005. The driving speed and rotating speed of the system are controlled through the numerical control machine. The torque is measured by the static torque sensor. A common Tesla meter measures the magnetic field, and a column-shaped electromagnet provides it. In order to not affect the magnetic field, tank and shear disk (ring) are made out of aluminum. When the shear plate and tank are rotated with a relative angular velocity, the shear rate is the linear function of polar radius $r$. Then, the torque $M_N$ required for keeping the MRF tank does not move. It can be described as:

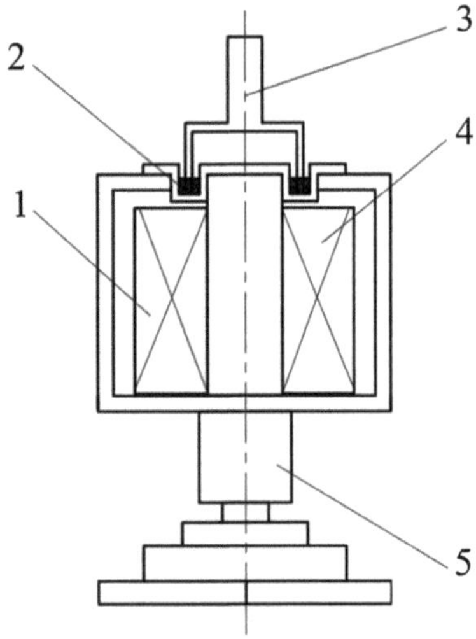


**Fig. 4.18** Schematic diagram of No. 3 test equipment

$$M_N = 2\pi \int_{R_1}^{R_2} r^2 \tau \mathrm{d}r \tag{4.26}$$

where $R_1$ is the diameter of inner loop; $R_2$ is the diameter of external loop. According to Eq. (4.26), then:

$$\tau = \frac{3M_N}{2\pi(R_2^3 - R_1^3)} \tag{4.27}$$

The shear stress can be obtained by the Eq. (4.27) through measuring the torque and rotational speed. This device can accomplish the shear stress measurement of MRF, especially in the measurement of dynamic shear stress. But its testing precision largely depends on the precision of the torque sensor and the operational stability of the whole device. The weight of the electromagnet is large, thus ensuring the stability of the device is very important.

There are some problems from the above devices due to the fact that retrofitting the existing viscometer. They are used to measure the shear stress of FMA and now needed to be further improved. In order to facilitate the measurement of shear stress, a simple instrument was self-designed. Two points are mainly considered as follows: first, simple manufacture, and second, low cost. The device can accurately test the shear stress of workpiece suffered from different magnetic intensities.

This system is mainly composed of bearing, rotating shaft, specimen, support rod, bearing seat, pulley bracket assembly, lifting rope, and pallet as shown in Fig. 4.19.

When the test is in progress, the magnetic field should be turned off. The workpiece is exposed to air. Sand was added to the pallet gradually. When the total weight of sand and pallet reach the ultimate static torque of the shaft, the shaft will

**Fig. 4.19** Shear stress testing instrument of FMA. **a** Workpiece immersed in FMA; **b** Workpiece exposed in air

rotate slightly. At this point, the mass of the sand and the pallet multiplied by the rotating shaft radius is the ultimate static torque of the shaft. The 2/3 of the test workpiece is immersed in the FMA, and then the magnetic field intensity will increased gradually. Under different magnetic field intensities, the sand is added to the pallet until the shaft is slightly rotated. The total mass of the sand and the pallet is weighted and measured. Then, the total mass multiplies by the radius of the shaft will help obtain the different torque values. The ultimate static torque can be obtained under different magnetic field conditions. These ultimate torque values subtract the torque values tested in the air is the static torque values of FMA under different magnetic field conditions.

According to the knowledge of the materials mechanics, the relationship between shear stress and torque at the edge of the circular section is known as:

$$\tau = \frac{T}{W_t} \tag{4.28}$$

where $\tau$ is shear stress; $T$ is the torque of round shaft subjected to; $W_t$ is torsional section modulus.

$$W_t = \frac{I_p}{R} \tag{4.29}$$

where $I_p$ is the polar moment of inertia of round shaft; $R$ is the radius of round shaft; $d$ is the diameter of round shaft.

Then, the polar moment of inertia of round shaft is:

$$I_p = \frac{\pi d^4}{32} \tag{4.30}$$

Combined Eqs. (4.29) and (4.30), then:

$$W_t = \frac{I_P}{\frac{d}{2}} = \frac{\pi d^3}{16} \tag{4.31}$$

According to the Eqs. (4.31) and (4.28), we can get the shear stress of FMA in different magnetic field intensities.

The fabricated FMA was tested by using this device, and Table 4.5 shows the maximum torque force of rotation shaft when it was subjected to different conditions.

The average values of the maximum torque force under six conditions are (the diameter of the workpiece is 0.02 m) 0.1804, 0.229, 0.7378, 0.923, 1.2044, and 1.2898 N, respectively. The shaft diameter is 40 mm, then the torque will be 0.03608, 0.0458, 0.14756, 0.1846, 0.24088, 0.257965 N·m. According to the Eq. (4.28), the calculated results are shown in Table 4.6.

Table 4.5 Maximum torque force of rotation shaft subjected to different conditions (N)

| Test number | Test condition | | | | | |
|---|---|---|---|---|---|---|
| | Air | Zero magnetic field | Current 1 A | Current 1.5 A | Current 2 A | Current 2.5 A |
| 1 | 0.174 | 0.228 | 0.735 | 0.915 | 1.148 | 1.274 |
| 2 | 0.186 | 0.234 | 0.728 | 0.91 | 1.24 | 1.32 |
| 3 | 0.18 | 0.229 | 0.745 | 0.95 | 1.176 | 1.258 |
| 4 | 0.175 | 0.224 | 0.747 | 0.904 | 1.233 | 1.30 |
| 5 | 0.187 | 0.23 | 0.734 | 0.936 | 1.225 | 1.297 |

Table 4.6 Computation results of shear stress (Pa)

| Test condition | Calculation result |
|---|---|
| In air | 2230 |
| Zero magnetic field | 2920 |
| Current 1 A | 9390 |
| Current 1.5 A | 11,800 |
| Current 2 A | 15,300 |
| Current 2.5 A | 16,400 |

The analysis shows that the shear stress of workpiece tested in air is defined as the system damping. The real shear stresses are the ones that the workpiece had suffered from different conditions subtracted from the system damping. When the current reaches 2.5 A (the maximum magnetic flux intensity is about 0.6 T), the shear stress of FMA is 25.4 times as much as that in the zero magnetic fields.

When adding the sand, it is done to help prevent the shaking of pallet, so as not to affect the accuracy of the measurement. This device is simple and accurate for measuring the shear stress of FMA in the static condition.

## 4.4 Fluid Magnetic Abrasive Finishing Device

### *4.4.1 Basic Requirements of Finishing Device*

Fluid magnetic finishing device mainly consists of the magnetic field generator, control device, workpiece motion-driving device, and vessel of FMA. Generally, fluid magnetic finishing device is determined by the machining environment and the workpiece shape. Usually, motion-driving device can use milling machine, drilling machine, numerical control drilling, milling machine, and other equipment. The designed principle of the vessel is to reduce the impacts on the magnetic field on the premise of workpiece movement as much as possible. Its shape is determined by workpiece shape and processing conditions, and it should be manufactured from non-magnetic materials. Magnetic field generator and control device is one of the key equipment of fluid magnetic finishing device. In general, it needs to be designed and manufactured according to production requirements. Therefore, the design is an important part of the fluid magnetic finishing device.

There is a schematic diagram of simple fluid magnetic finishing device that drives the device into numerical control drilling and a milling machine, as shown in Fig. 4.20. The device mainly consists of magnetic field generator, vessel of FMA, numerical control drilling, and milling machine, etc. The workpiece is fixed on the main shaft of numerical control drilling, and the milling machine is fixed close to the scroll chuck. The workpiece has a rotary motion and a reciprocating motion by the main shaft. The magnetic field generator creates this magnetic field after electrifying it. The phase of FMA is transformed to realize surface finishing. The device is suitable for processing the outer surface of rotational parts.

## 4.4.2 Magnetic Field Design and Finite Element Analysis

The magnetic field generator is an important part of fluid magnetic finishing device. Generally, in order to get a good machining effect, the magnetic field intensity of the generator needs to achieve saturation magnetic induction of the fluid magnetic finishing device [31]. Meanwhile, the generator should have the advantages of wide controllable range, stable performance, low heat, and low energy consumption. Magnetic fields are produced by coils, electromagnet, permanent magnet, etc. The coil is difficult to generate high magnetic field intensity due to the limitation of current density in the winding. Coils can produce higher magnetic field strength when they are forced to be cooled. At present, superconducting coils can produce the strongest magnetic field source. However, its price is very expensive. Permanent magnet has the characteristics of small volume, stable performance, and zero power. But the magnetic field intensity is difficult to be adjusted continuously. Electromagnet is an economical magnetic source with high magnetic field intensity

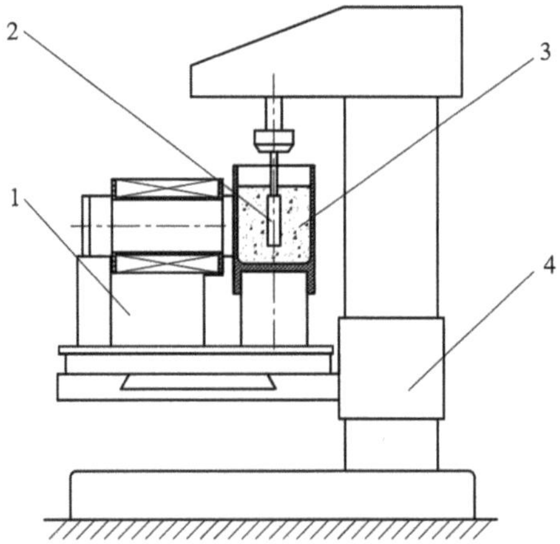


**Fig. 4.20** Overall structure of test device. 1—Magnetic field generator; 2—Workpiece; 3—Fluid magnetic abrasive; 4—Numerical control drilling and milling machine

at room temperature, which has the advantages of being continuously adjustable, well distributed, and other positive movements. Generally, electromagnet is selected to be external magnetic field of fluid magnetic finishing device.

1. An example of engineering design of magnetic circuit

*Example 1* As shown in Fig. 4.21, let cross-sectional area of iron core be $S_y$ ($m^2$), the average value of iron core total length is $L_y$ (m), cross-sectional area of air gap is $S_g$ ($m^2$), length of air gap is $L_g$ (m). If the value of magnetic field intensity of air gap is 1 T, then the required excitation ampere-turns of coil are calculated as below.

*Solution* Method 1, the method of magnetic circuit, is shown in Table 4.7. Method 2 is shown in Table 4.8.

*Example 2* If $F$ (ampere-turns) is known, the design method of coil is shown in Table 4.9.

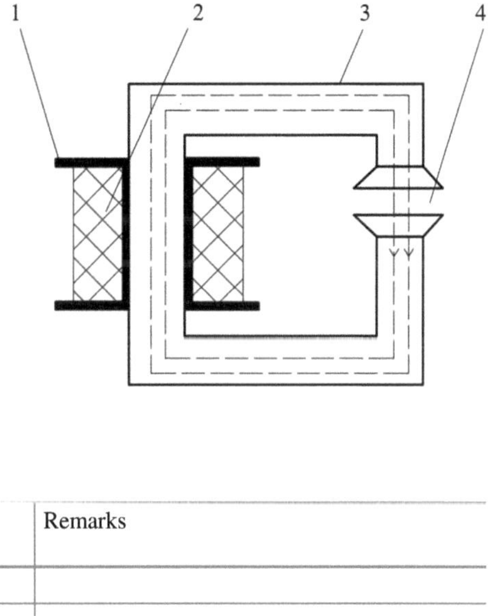


**Fig. 4.21** Electromagnet. 1—Frame; 2—Coil; 3—Iron yoke; 4—Air gap

**Table 4.7** Method 1

| Computing project | Calculation equation | Remarks |
|---|---|---|
| Air gap reluctance | $R_{mg} = L_g/\mu_0 S_g$ | |
| Air gap flux | $\phi_g = B_g S_g$ | |
| Iron yoke flux | $\phi_y = \phi_g$ | |
| Iron yoke magnetic induction | $B_y = \phi_y/S_y$ | If the iron core magnetic induction was not suitable, $S_y$ need to adjust |
| Iron yoke permeability | $\mu_y = B_y/H_y$ | By *B–H*, the corresponding $H_y$ is found out according to $B_y$ |
| Iron yoke reluctance | $R_{my} = L_y/(\mu S_y)$ | |
| Total reluctance of the magnetic circuit | $R_m = R_{my} + R_{mg}$ | |
| Required magnetic motive force | $F = R_m \phi_g$ | The required ampere-turns can be obtained according to $F = NI$ |

**Table 4.8** Method 2

| Computing project | Calculation equation | Remarks |
|---|---|---|
| Air gap magnetic field intensity | $H_g = B_g/\mu_0$ | |
| Corresponding magnetic pressure drop | $U_{mg} = H_g L_g$ | |
| Iron yoke magnetic field intensity | Process is same as Method 1 | |
| Iron yoke magnetic pressure drop | $U_{my} = H_y L_y$ | |
| Required magnetic motive force | $F = U_{my} + U_{mg}$ | |

*Note* As long as the magnetic induction of iron yoke is not in a supersaturated state, the magnetic pressure drop of the iron core is far less than the magnetic pressure drop of the air gap, and the equation of required magnetic motive force is $F \approx U_{mg}$. The "Method 2" only needs the first and second steps, which is more convenient to quickly evaluate

**Table 4.9** Design method of coil

| Computing project | Calculation equation | Remarks |
|---|---|---|
| Selecting appropriate current density | General current density J selects 3–5 A/mm$^2$ | When J is large, coil has the problem of being seriously heated. When J is small, then the copper wires are wasted. Normally, long-term system magnetic circuit selects 2–4 A/mm$^2$, repeated short-term system magnetic circuit selects 5–12 A/mm$^2$, and short-term system magnetic circuit selects 13–30 A/mm$^2$ |
| Required window area for coil frame | $S_w = F/(Jk)$ | $k$ is duty factor, which selects 0.5–0.7 usually |
| Size of frame window | $S_w = H_w b_w$ | $H_w$ is height, $b_w$ is width. Generally, $H_w$ should be greater than $b_w$ |
| Turns N of primary selection coil | | |
| Coil current | $I = F/N$ | |
| Cross-sectional area and diameter of wire | $S_d = I/J$<br>$d = \sqrt{4S_d/\pi}$ | The diameter of enameled wire is discontinuous actually. The model of calculated value $d$ is selected in product manual |
| Coil resistance | $R = NL_{av}/(\gamma S_d)$ | Where $\gamma$ is the conductivity of wire, $L_{av}$ becomes the average length of each turn of the coil. The conductivity of wire is related to the material and temperature. When the coil temperature rises, selection of thermal electrical conductivity should be considered when calculating the coil resistance |
| Required DC supply voltage | $U = RI$ | (1) If the supply voltage value $U_{dc}$ is fixed and non-adjustable, such as $U_{dc} = 150\,V$. If $U > U_{dc}$, $N$ need be reduced. If $U < U_{dc}$, $N$ need be increased, and the above steps are repeated until $U = U_{dc}$.<br>(2) If supply voltage is adjustable, the calculated value of $U$ can be adjusted in the range of $U_{dc}$, otherwise readjusting and calculating the number of turns |

2. Finite element analysis

In order to determine the performance of the magnetic field from the fluid magnetic finishing device, the two magnetic field devices (Figs. 4.22 and 4.23) were analyzed individually based on the electromagnetic module in the ANSYS. Electromagnetic module can be used to analyze two-dimensional static magnetic field: harmonic magnetic field and transient magnetic field. The module can analyze the three-dimensional static magnetic field; the harmonic magnetic field and the transient magnetic field based on the boundary, and help analyze three-dimensional magnetic field (static, harmonic, and transient) and the electric field based on node. In addition, the electromagnetic module can be used to calculate electromagnetic field or high-frequency electromagnetic field of wave radiation. In the experiment, the magnetic field was generated by electromagnet, which is a two-dimensional static magnetic field. The magnetic field of the whole finishing device was analyzed by using the magnetic nodal module, and the results are shown in Figs. 4.24 and 4.25.

As shown in Fig. 4.24, the distribution of magnetic line of the magnetic field generator is consistent with previous theoretical analysis. Therefore, the device can generate an ordered magnetic field in FMA areas in the process. The distribution map of the magnetic line is right-angled along the FMA areas. The intersection of two poles at the center formed a magnetic field that diffuses outward at a quarter of the circle. The distribution of the magnetic line gradually became sparse from the

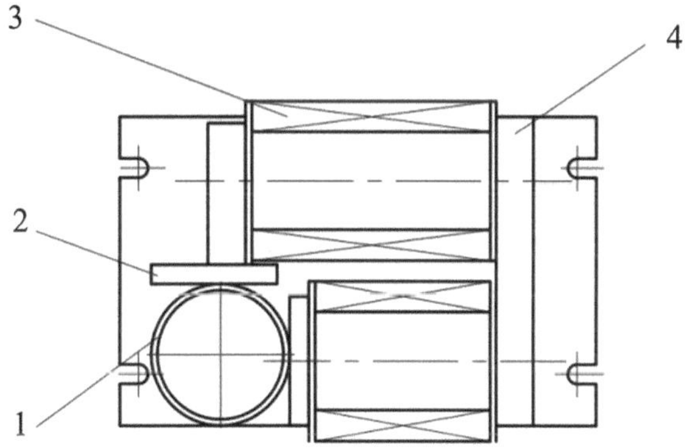


**Fig. 4.22** Electromagnetic field generator 1. 1—Vessel; 2—Poles; 3—Coil; 4—Iron yoke

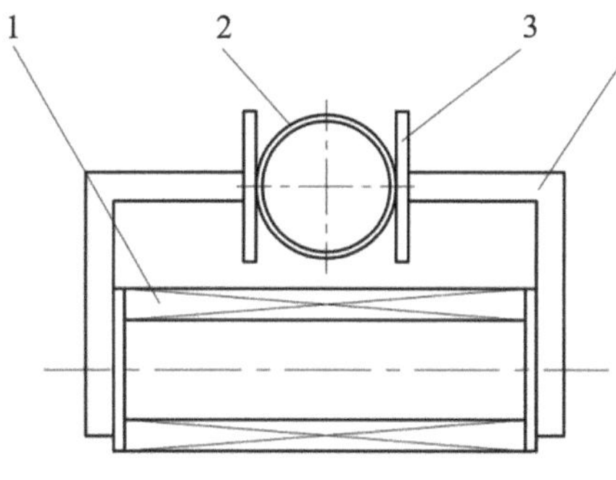


**Fig. 4.23** Electromagnetic field generator 2. 1—Vessel; 2—Poles; 3—Coil; 4—Iron yoke

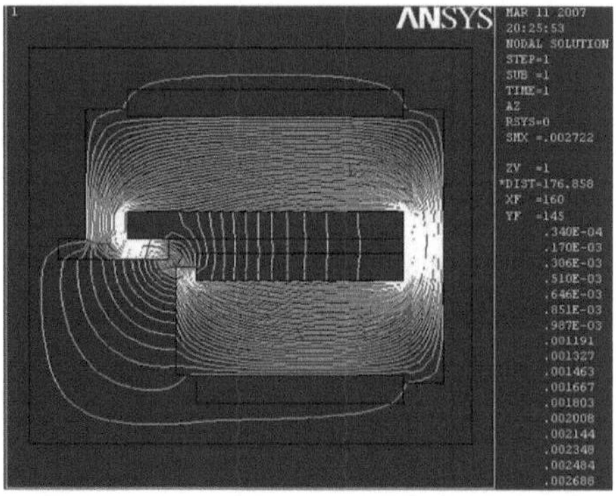


**Fig. 4.24** Distribution of magnetic line of electromagnetic field generator 1

inside out. Therefore, the device is particularly suitable to the process of the rotational parts. Figure 4.25 shows the distribution map of magnetic line of the magnetic field generator 2. This magnetic line of the magnetic field is a symmetrical distribution along the symmetry line of those poles, and the distribution of the magnetic line gradually becomes sparse from the inside out (the densest in the center of the distribution). The maximum magnetic field intensity is definitely lower than the magnetic field generator 1.

When FMA was placed in a round vessel between the poles, the distribution of the magnetic line is formed, as shown in Figs. 4.26 and 4.27. From the figure, soft magnetic material with high saturation magnetic induction is selected to be the magnetic abrasive particles in FMA (magnetic particle usually is produced by iron and ferrous alloy). In response, when the magnetic field is added to the FMA, the FMA closes the magnetic lines. When the magnetic field is formed, it pushes the Newtonian fluid into the Bingham fluid inside the FMA. In this process, the distributions of the magnetic abrasive particles are changed. In the dense areas of the magnetic line, the density of the magnetic abrasive particles is high, and the particles will have high shear force during the process. Therefore, it is reasonable that the motion of the workpiece is concentrated in the dense areas of the magnetic line,

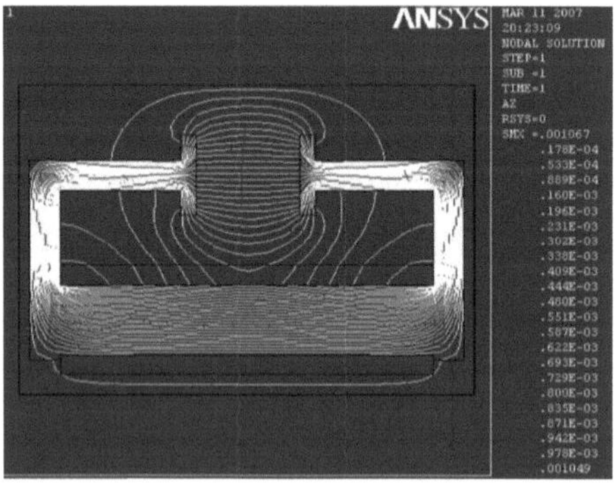


**Fig. 4.25** Distribution of magnetic line of electromagnetic field generator 2

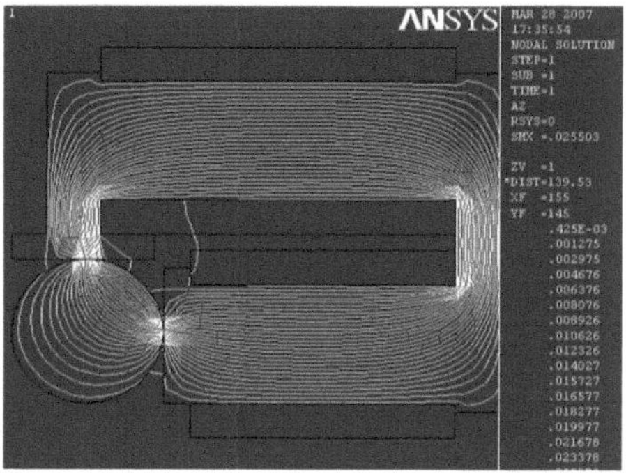


**Fig. 4.26** Distribution of magnetic line of electromagnetic field generator 1 with fluid magnetic abrasive

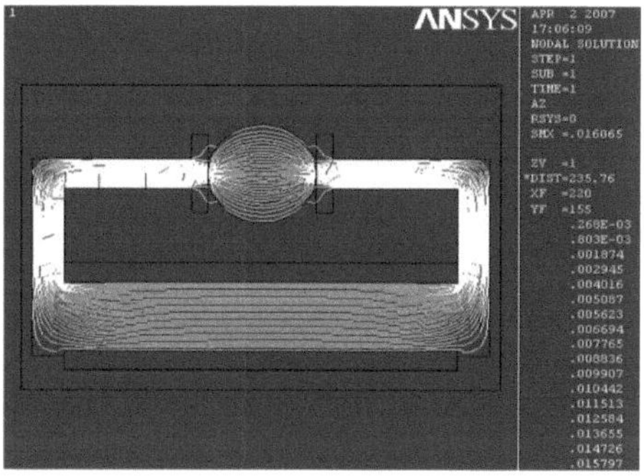


**Fig. 4.27** Distribution of magnetic line of electromagnetic field generator 2 with fluid magnetic abrasive

namely the areas of high magnetic field intensity. It not only makes full use of the efficacy of the magnetic field generator, but it also improves the processing efficiency.

The distributions of the magnetic induction of two kinds of magnetic field generators are, respectively, shown in Figs. 4.28–4.31. Figures 4.28 and 4.29 are the distributions of magnetic induction without FMA. Figures 4.30 and 4.31 are the distributions of magnetic induction with FMA. As shown in the Figure, the FMA is

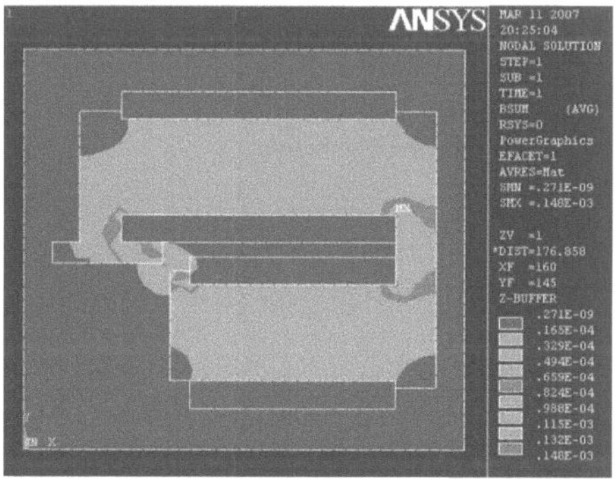


**Fig. 4.28** Distribution of magnetic induction of electromagnetic field device 1

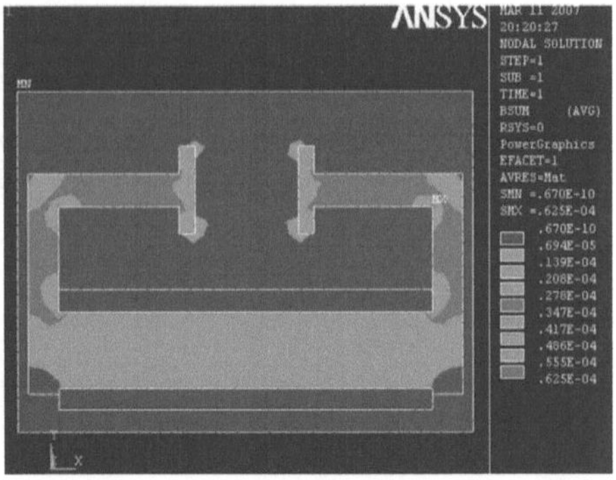


**Fig. 4.29** Distribution of magnetic induction of electromagnetic field device 2

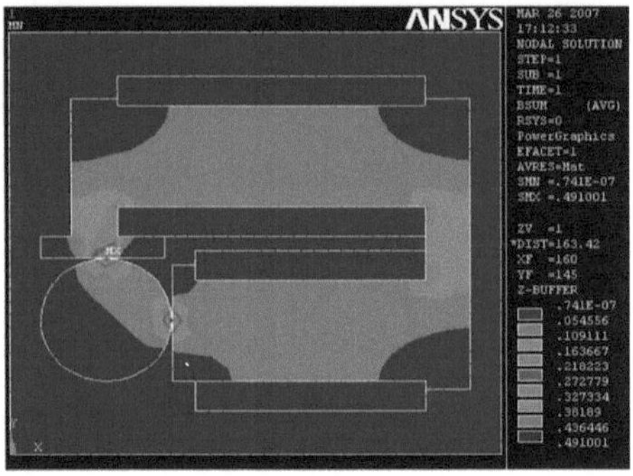


**Fig. 4.30** Distribution of magnetic induction of electromagnetic field device 1 with fluid magnetic abrasive

placed between magnetic field generator 1 and generator 2, and the distribution of the magnetic induction machining region is basically uniform. In other words, the stable and ideal magnetic field is formed in machining region. When FMA was placed, the distributions of magnetic field have changed obviously (Figs. 4.30 and 4.31). The magnetic induction is obviously large near the two poles. Particularly, magnetic induction of the magnetic field generator 1 is larger between the two poles and the FMA. Two main machining regions are formed. It provides a good experimental foundation for the whole FMA finishing experiment.

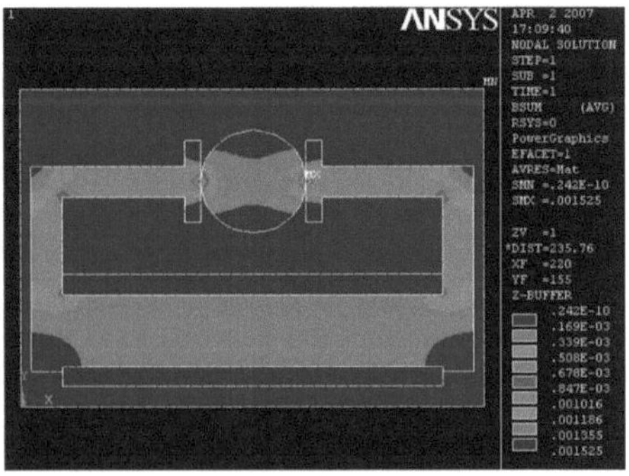


**Fig. 4.31** Distribution of magnetic induction of electromagnetic field device 2 with fluid magnetic abrasive

### 4.4.3 Device Design

The magnetic field generator is very important in the FMA finishing device. When designing the magnetic field, some factors need be mainly considered, such as magnetic field intensity, magnetic field direction, controllability of magnetic field, and the relative position of the magnetic field and the workpiece. In general, the magnetic induction of effective processing point should be greater than the saturation magnetic induction of FMA in order to be convenient for obtaining the maximum processing efficiency and the controlling region of the magnetic field. The magnetic field direction is perpendicular to the workpiece surface as far as possible. There are many abrasive particles near the processed surface, all of which contain a high shear force. Figure 4.32 is the photograph of the magnetic field generator for the finishing surface of rotational parts. Figure 4.33 is the structure diagram of the magnetic field generator.

**Fig. 4.32** Magnetic field generator for finishing surface of rotational parts

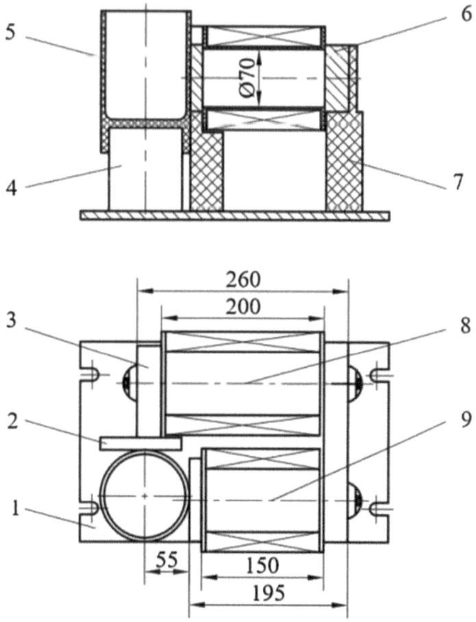


**Fig. 4.33** Structure diagram of magnetic field generator. 1—Floor; 2—Poles; 3—Small magnetic yoke; 4—Vessel base; 5—Vessel; 6—Big magnetic yoke; 7—Support; 8—Big iron core; 9—Coil

1. Design of coils and poles

FMA finishing is carried out between two poles. Figure 4.33 is the structure diagram of the magnetic field generator; the large and small electromagnetic coils are used as the source of magnetic field generator, and both the large and small coils are made of winding 0.9-mm enameled wire. Large coil has 3900 turns and small coil has 2900 turns. In order for the machining region to have sufficient magnetic induction between the two poles, the magnetic conductor, the magnetic yoke, and iron core must avoid magnetic saturation. At the same time, it is guaranteed that the magnetic flux goes through all unsaturated magnetic conductors, namely interlinking magnetic circuit [36].

According to the Eqs. (3.22), (3.23), (3.24), and (3.39) in the machining region, the magnetic flux gradually decreases with the increase of machining gap $\delta$ when the magnetic field generator is constant. As shown in Fig. 4.34, along the $X$-direction density of magnetic line becomes smaller and smaller.

The design of poles includes the design of magnetic pole, iron core, and magnetic yoke. Normally, the shape and size of the magnetic pole are determined according to the size of the workpiece, and the shape and size of the vessel. When the magnetic pole is determined, cross-sectional area of the magnetic pole $S_1$ can be calculated. When required magnetic induction $B$ and the machining gap $\delta$ of the finishing are determined, the permeance $\Lambda$ can be calculated according to the Eq. (3.23) in the magnetic field generator. Then, permeance $\Lambda$ substituted in Eq. (3.39) and Eq. (3.39) is used to solve for magnetic flux $\phi$. The size of cross-sectional area of the iron core and magnetic yoke is calculated by $B = \phi/S$, which can determine the shape and size of the iron core and magnetic yoke.

In the magnetic field generator of the FMA finishing, the arrangement and shape of the poles and the design of machining gap are shown in Fig. 4.34.

2. The analysis of magnetic induction

According to the analysis of the last section, it is demonstrated along the X-direction that the magnetic induction gradually decreases in the machining region of the magnetic field generator. Table 4.10 is the data obtained from actual measuring the magnetic field generator by the Tesla meter.

**Fig. 4.34** Arrangement of the poles and machining gap

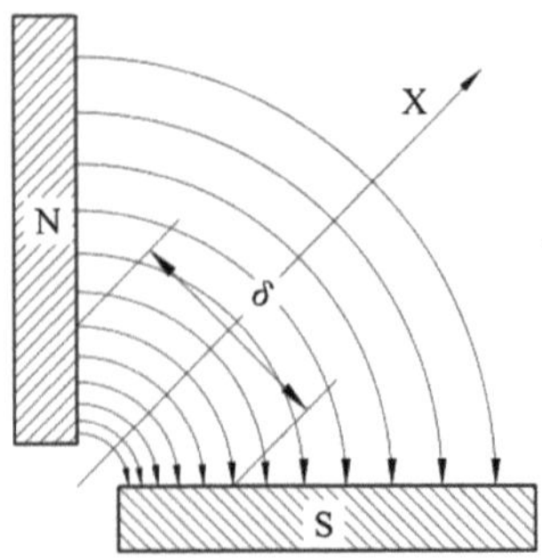

**Table 4.10** Measured values of magnetic induction

| | | | | | | |
|---|---|---|---|---|---|---|
| *Condition* Current 1 A; Voltage 60 V | | | | | | |
| *X*-axis distance/mm | 0 | 20 | 40 | 60 | 80 | 100 |
| Magnetic induction/T | 0.36 | 0.17 | 0.09 | 0.06 | 0.03 | 0.01 |
| *Condition* Current 2 A; Voltage 130 V | | | | | | |
| *X*-axis distance/mm | 0 | 20 | 40 | 60 | 80 | 100 |
| Magnetic induction/T | 0.56 | 0.28 | 0.16 | 0.10 | 0.05 | 0.03 |
| *Condition* Current 3 A; Voltage 150 V | | | | | | |
| *X*-axis distance/mm | 0 | 20 | 40 | 60 | 80 | 100 |
| Magnetic induction/T | 0.63 | 0.33 | 0.18 | 0.12 | 0.07 | 0.05 |

As shown in Fig. 4.35, distance is measured along the *X*-axis, starting from the origin. The above data was used to draw a curve as shown in Fig. 4.36.

From the above distribution curves of the magnetic induction, it can be seen that the magnetic induction increases gradually with the increase of current in the machining region and decreases rapidly along the *X*-axis. Figure 4.37 is a simulation image of the magnetic field intensity in the machining region. In the picture, different colors represent different density of magnetic line, namely the strength of the magnetic field intensity. When comparing Figs. 4.37 with 4.35, it can be seen that the region of track of the workpiece is located in the region of relatively large

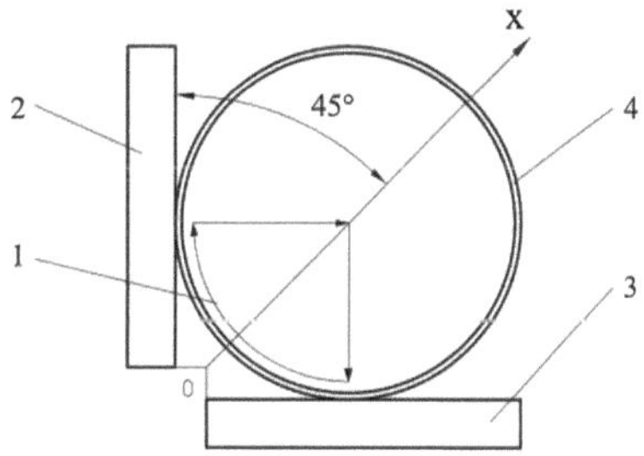


**Fig. 4.35** Track of workpiece and the arrangement of machining region. 1—Track of workpiece; 2—Pole I; 3—Pole II; 4—Vessel

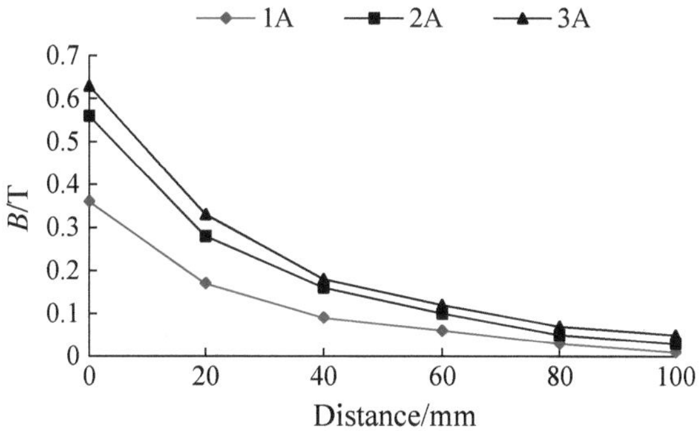


**Fig. 4.36** Distribution curves of magnetic induction along the *X*-axis

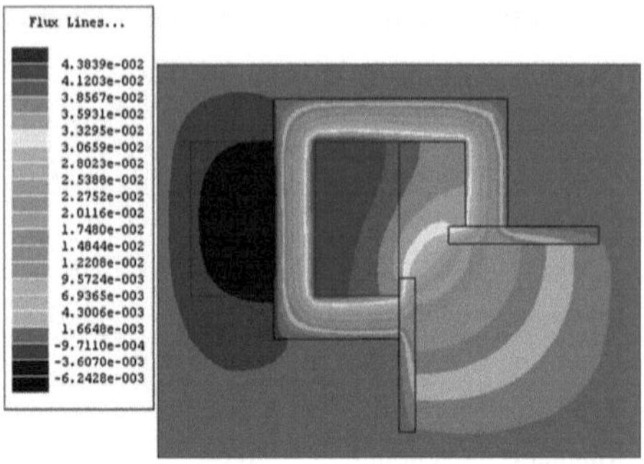


**Fig. 4.37** Renderings of the distribution of magnetic field intensity

magnetic field intensity. This ensures that the machining region has a sufficient magnetic field intensity in the process. Thus, on the surface of the workpiece, sufficient pressure is formed by the existence of FMA, and the structure of magnetic field generator becomes compact, which is convenient for manufacturing and installation.

## 4.5 Main Factors and Finishing Effects Affecting Fluid Magnetic Abrasive Performance

### *4.5.1 Finishing Process*

The process sequence of the fluid magnetic abrasives finishing is to prepare the processing device, the fluid magnetic abrasives, fix workpiece on the spindle of the machine, movement of the workpiece, application of the external magnetic field, closing the magnetic field after processing, stopping the movement of the workpiece, and removal of the workpiece.

The operation should be noted that the workpiece must be in high-speed running condition in the magnetic field. Generally speaking, the permeable material in the magnetic field is bound to be magnetized, which would influence the magnetic force line of the original magnetic field; the magnetic field strength changes correspondingly. The nature of the magnetization is the charged particles of the conductor under the influence of an external magnetic field, changed from the anisotropic to isotropic, and shows some magnetic properties. However, during the fluid magnetic abrasives machining process, the workpiece is always rotating at a high speed, the internal charged particles cannot always point to the same direction, so the magnetic materials will not be magnetized, which can be regarded as a non-magnetic conductive workpiece in the magnetic field, on condition that the workpiece must be in a rotational state. So long as the operation is reasonable, there is no essential difference to the magnetic and non-magnetic materials in the processing mechanism. The above conclusions are illustrated through some validation test, starting with the first one, which is to put the steel in a magnetic field and the workpiece, is held in stationary state. It is found that the workpiece is completely magnetized, and there is a layer of magnetic particles that are absorbing on the surface, which seriously damaged the structure of the fluid. Meanwhile, the isolation of the abrasive particles and the workpiece surface must be done, so that the finishing cannot continue normally. Instead, the rotation of the workpiece must take place first, and then the applying of the magnetic field. This way, there will not be any of the above phenomena, and the normal finishing will be achieved.

According to the mechanism analysis, the magnetic material workpiece in high-speed rotating will not be magnetized, even in a magnetic field, so that the steps should be noticed in the machining process. The first must be started from the CNC, to let the workpiece in motion; then apply a magnetic field; after finishing processing, turning off the power must be done first to remove the external magnetic field in order to ensure processing effect.

### 4.5.2 *Main Affecting Factors*

FMA is a kind of new technology used for precise finishing; the final finishing efficiency (generally represented by MRR) and the surface roughness value are the indicator of the processing performance. Finishing efficiency and final surface roughness values are affected by many parameters [19].

1. Abrasive type, size, and volume addition rate

There are many different types for abrasives; their selection principle has a direct relationship with the physical and mechanical properties of the workpiece material. In general, when the high tensile strength of the material, corundum abrasives, of a larger toughness is chosen, the lower tensile strength of the material, a kind of silicon carbide abrasive, is also chosen. The silicon type has a higher hardness but more brittle. Since the hardness of the silicon carbide abrasive (including black silicon carbide and green silicon carbide) is higher than the corundum abrasives, which shows the sharpened cutting edge and a good self-sharpening process, it can be see that this is more used in FMA. It should be noted that in the conventional grinding process, silicon carbide could have strong chemical reaction with steel material, due to thermal effect. Therefore, silicon carbide abrasives are generally not used in the finishing machining of steel materials. But in FMA, due to FMA having a good heat dissipation effect, the processing temperature is also not too high, it can be decided that the chosen material is the silicon carbide abrasive.

The selection of abrasive particle size mainly depends on the finishing efficiency and the surface roughness values. When the requirement of the surface roughness is lower, a finer size should be selected, so it can ensure that the residual traces of the workpiece surface are smaller with a low surface roughness. But the processing efficiency came out to below. If the initial roughness of the workpiece surface is higher, or in order to obtain higher processing efficiency, we can choose coarse grain abrasive, or rough grinding and fine grinding combined process.

The percentage of volume occupied by the abrasive up to a certain extent shows the number of the abrasive particles, which can take part in the grinding process in unit area. Generally, the higher the volume percentage of abrasives particles, the greater machining efficiency it is. However, since the proportion of solid particles in the liquid addition is certain, if the percentage of the abrasive particles is too high, it will affect the percentage of magnetic particles, resulting in the shear yield strength to reduce. At the same time, high proportion of the abrasive particles could potentially damage the stability of the FMA. According to the test, when the additive proportion of volume of abrasive particles is 30–40%, there will be a good comprehensive performance.

2. Magnetic field strength and direction

To better illustrate the influence of the following parameters on the FMA, a set of test data is combined to describe the situation. Test specimens are duralumin (YL12), non-magnetic materials, and 45# steel of the magnetic material. Test specimens are shown in Fig. 4.38, in which 20 mm diameter of the part is machined, and 12 mm diameter of the part is clipped into the chuck of the CNC milling machine drilling. Test processing device is the FMA device, which is designed in the fourth section of this chapter.

The compositions of the FMA include sodium dodecyl sulfate, pure water, brand for RXe carbonyl iron, 54# green silicon carbide, ethanol. Its mass percentage is as follows: the carbonyl iron, 35%; water, 25.6%; green silicon carbide, 38.5%; stabilizer, 0.5%; ethanol, 0.18%.

In the choice of the magnetic field direction, the common way is to let the direction of the magnetic field to be perpendicular to the workpiece surface. This way, it can improve the efficiency of finishing.

The workpiece surface roughness value decreased as it is shown in Fig. 4.39. Those of which is machined by different current values (the size of the current determine the magnetic field strength).

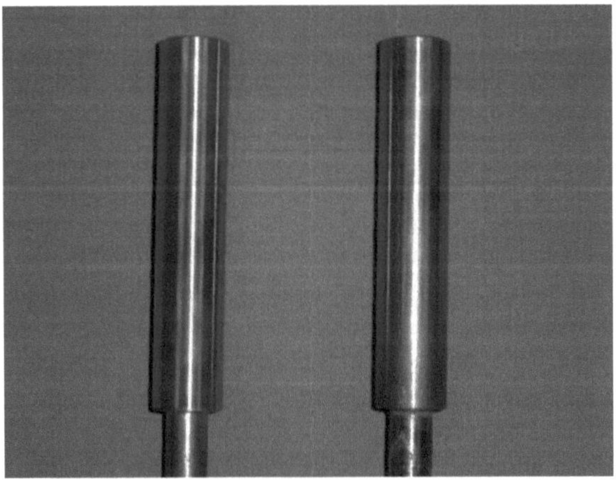

**Fig. 4.38** Test specimens

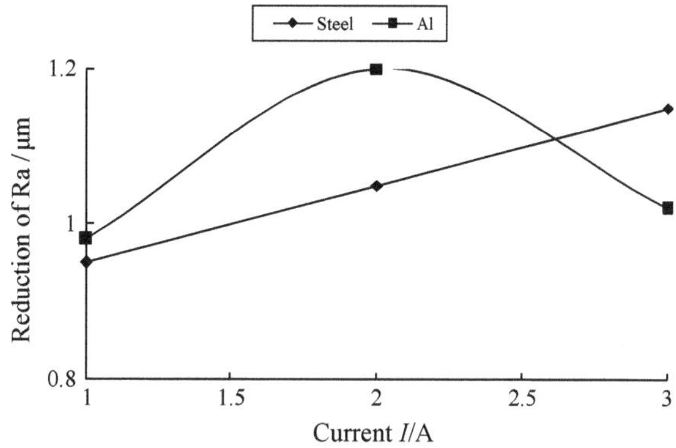


**Fig. 4.39** Relationship between surface roughness values reduce and the amount of the current

It can be found from Fig. 4.39 that the surface roughness of the steel reduced is proportional to the current size, as for aluminum, when the current, namely the magnetic field strength is less than a certain value, the roughness has a similar trend as the change of the steel, but when the magnetic field strength is over a certain range, the surface roughness decreasing amount is significantly reduced. In other word, when the surface roughness is decreased, these two different trends are caused due to the obvious gaps of steel and aluminum in the hardness. When the magnetic field strength is high, the grinding pressure and the shear stress will also increase, the surface, which is an original light and soft material surface, will appear in a darker corrugated manner, and have scratches under the score of abrasive. Thus, there will be a phenomenon that the workpiece surface roughness is increased. It should be noted that when the magnetic field strength is close to or exceeds the saturation flux density of MFA, then increasing the current would not affect the processing efficiency.

As shown from the above test data, the field strength should be appropriately selected depending on the different specimen material and the surface condition. The high magnetic field strength will improve the machining efficiency, but it can also cause an increase in final surface roughness value. Therefore, a strong magnetic field strength must be chosen in the roughing stage to improve the material removal rate. In the late stage of finishing, the magnetic field strength should be appropriately reduced, in order to achieve a lower surface roughness values.

3. Workpiece speed

Under the same conditions, the greater the relative speed between the abrasive and the workpiece surface, the more obvious the effect of slippery, scribing, and micro-cutting, although the high processing efficiency and the final surface roughness values are larger. After looking at the different speed of finishing, the decreased surface roughness values are shown in Fig. 4.40.

It can be found from Fig. 4.40 that the workpiece speed also has a great influence on the processing results, and the decreased roughness of steel and aluminum is similar to the trend of workpiece speed. When the speed is too fast and over a

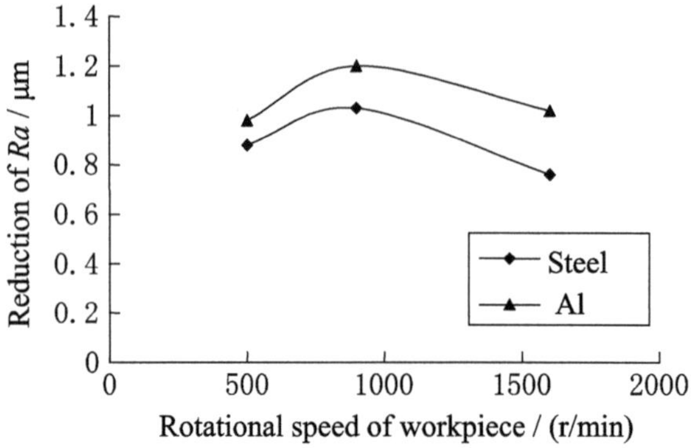


**Fig. 4.40** Relationship between the decreased surface roughness value and the rotational speed of workpiece

certain range (1000 r/min), the decreasing surface roughness values tend to be reduced. Therefore, the speed should not be too high. If the speed is too high, the metal removed will also reduce and cannot achieve the ideal surface roughness value. This is because when the speed is too high, there is the phenomenon of the shear thinning between the workpiece surface and FMA, leading to a decrease in the shear stress value, and the abrasive cannot achieve sufficient pressure on the surface, leaving the process ineffective.

4. Processing time

In general, with the growth of the processing time, there will be more and more material from the workpiece surface falling off, causing the surface roughness to be reduced. However, with the increase of processing time, the material removal begins to decrease, but the surface roughness is no longer declining after a certain time, and it even has a rising trend. There are many factors affecting the processing time, such as workpiece material performance, the shape of the workpiece, environmental factors. Normally, when the workpiece material is easy to split and the shape is thinner, the processing time should be less, on contrast, it is more.

The reduced surface roughness values are shown in Fig. 4.41, after different processing time.

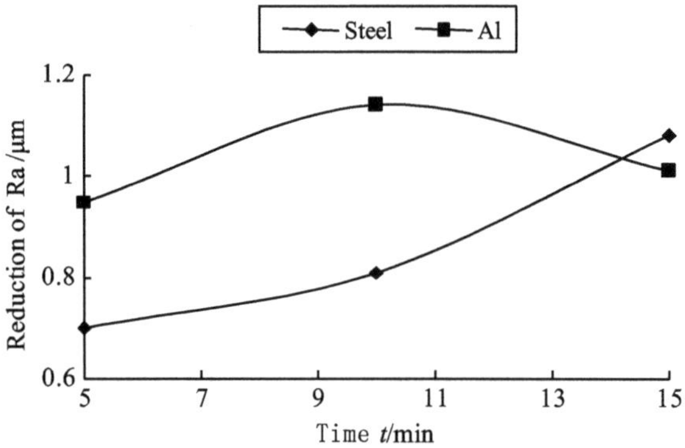


**Fig. 4.41** Relationship between the decreasing amount of surface roughness and the processing time

As can be seen from Fig. 4.41, with the extension of the processing time, the surface roughness of steel reduces more, and the material removal is growing. For aluminum, when the processing time is less than 10 min, the trend is similar to the steel's, but with the further extension of the processing time, the decreased surface roughness will evidently showing decreasing trend while the aluminum surface also has significant traces of pitting. This is because the hardness of abrasive is much higher than aluminum, while the removal is large, the removal efficiency is high, when the processing time is beyond a certain limit, because of the larger abrasive particle size (54#), the abrasive will wear off at the original uneven place and continue to carve, row, and scratch on the new surface. This causes an increase in the surface roughness; at the same time, because the FMA is existed for a long time in the deep scratches, the aluminum surface will be corroded, and pitting phenomenon occurred. The steel will not appear during this phenomenon, due to its high hardness and strong corrosion resistance. But with the further increase of processing time, the steel will also show this kind of phenomenon. In short, in the FMA finishing, the processing time should not be too long, for the lower hardness materials such as aluminum, copper [37].

Figure 4.42 is the curve diagram of the surface roughness *Ra* of 45# (the same condition as above, but the abrasive is green silicon carbide, 80# the current is 3A).

Under a certain condition, certain particle size of abrasive can only make the workpiece surface roughness values to reach to a certain degree. So it should be based on the original surface condition and the surface roughness values to select the polishing number. In the initial stage, bigger size should be chosen. In the finishing stage, smaller size should be chosen. In terms of material removal rate, factors need to be considered are material, size, shape, surface roughness, and others. In the roughing stage, the higher the productivity, the better the increasing in magnctic ficld strcngth, which hclps thc spccd of thc workpiccc; in thc finishing stage, the high surface quality is achieved by reducing the material removal rate.

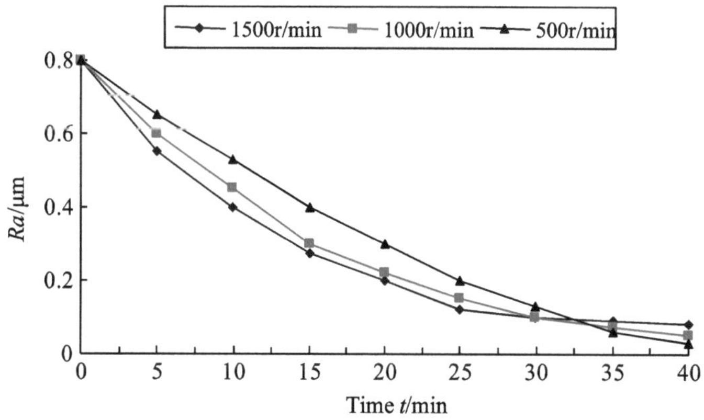


**Fig. 4.42** Relationship between the surface roughness *Ra* and processing time at different speeds

### 4.5.3 Finishing Effects

1. The outer surface of rotation

In order to test the processing capacity of the MFA on the outer surface of rotation, the inner hole surface shown in Fig. 4.43 has been machined, and the specimen material is aluminum (YL12). Under the condition that the test time and speed are at a certain value, the change of the hole surface roughness *Ra* before and after finishing is shown in Table 4.11 [38]. The processing time is 5 min, the rotating speed is 900 r/min, and the current is 2.5 A [39].

As shown from the data above, the FMA finishing has a good processing effect to the surface revolution, which can be used in the finishing of the curved face parts.

In order to analyze the workpiece surface morphology before and after processing, direct observation (such as the Mahr roughness tester and JSM-6700F field emission scanning electron microscopy) was used, respectively, to measure the experiment in the second part of this section. The 45 steel after the finishing is analyzed; the processing time is 20 min, the current is 3 A, and the speed of machine tool is 900 r/min.

By direct viewing (Fig. 4.44), the processed surface has improved obviously, its degree of surface is smooth, and the bright degree increases significantly.

The surface roughness of the workpiece contour is measured before and after the processing by using the roughness-measuring instrument of the Mahr Company from Germany (Figs. 4.45 and 4.46). It can be seen from the figure that the surface roughness value is decreased significantly, the flatness is improved, and the mean

**Fig. 4.43** Rotation specimen

**Table 4.11** Outer hole surface roughness *Ra* before and after finishing (μm)

| No. | Before finishing | After finishing | The average *Ra* (before finishing) | The average *Ra* (after finishing) |
|---|---|---|---|---|
| 1 | 1.120 | 0.910 | 1.199 | 0.917 |
| 2 | 1.273 | 0.935 | | |
| 3 | 1.090 | 0.872 | | |
| 4 | 1.275 | 0.942 | | |
| 5 | 1.235 | 0.926 | | |

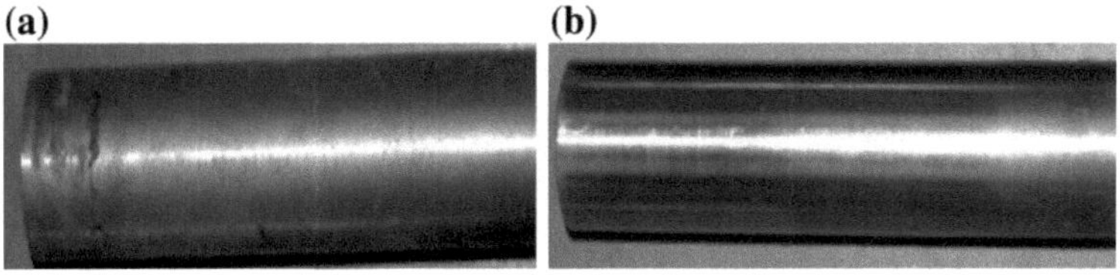


**Fig. 4.44** Comparison of the workpiece before and after processing. **a** Before finishing; **b** After finishing

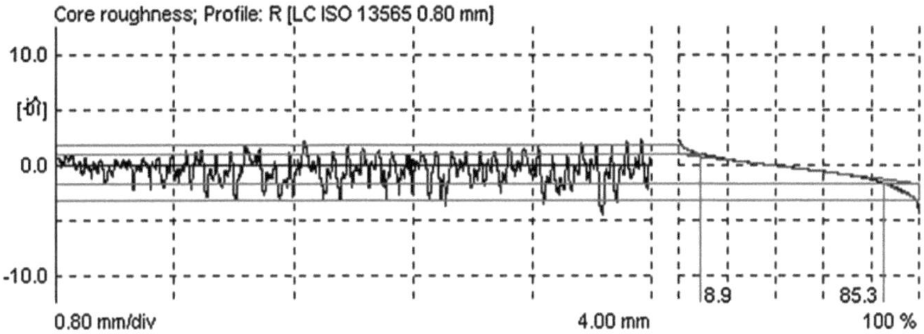


**Fig. 4.45** Surface roughness profile before processing

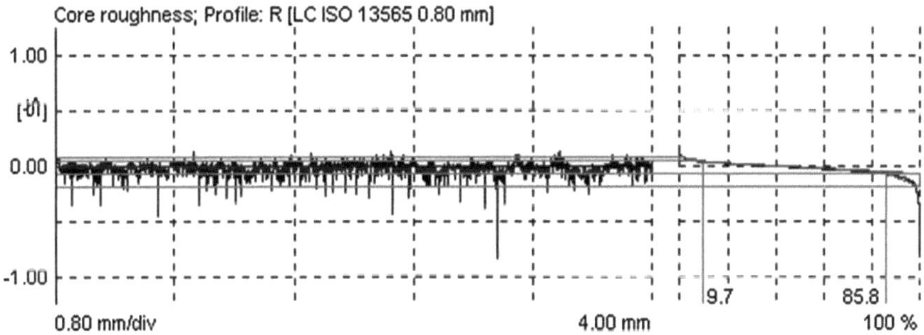


**Fig. 4.46** Surface roughness profile after processing

spacing of the profile irregularities is reduced. These can directly reflect the surface wear resistance and result the profile bearing ratio curve to become gentler. All of these instructions, after using liquid magnetic abrasive finishing process, impacted the workpiece surface quality and helped it to be improved significantly.

Through the JSM-6700 F field emission scanning of the electron microscope scanning, the 45 steel workpiece surface at 100× magnification micrograph was obtained (Fig. 4.47). Before finishing process was the turning processing, the surface morphology is consistent with the tip of the cyclical turning marks, which has a larger microscopic peaks and troughs. The surface textures have an obvious

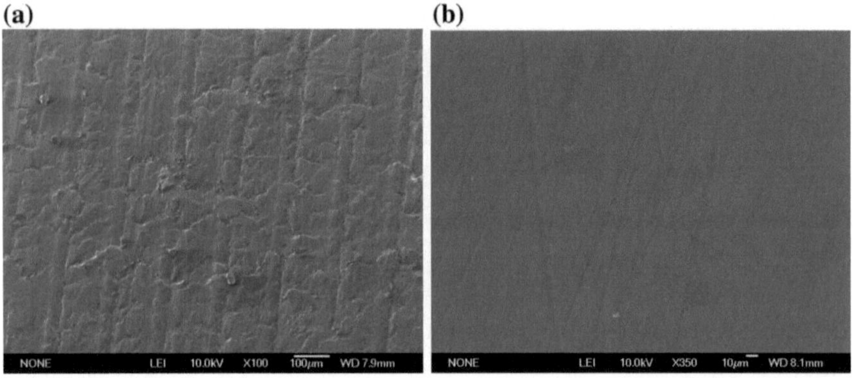


**Fig. 4.47** SEM of 45 steel surface before and after processing. **a** Before finishing; **b** After finishing

direction, and after the finishing, the cyclical turning trace disappeared entirely and the microscopic peaks and troughs are greatly reduced, resulting a uniform and smooth workpiece surface.

Therefore, the FMA is an ideal technology on precise surface finishing used in the finishing of the rotation. It not only helps obtain a lower surface roughness value, but also achieves a good surface morphology. It has obvious effects on improving the surface quality, especially for the surface part constituting the friction pair.

2. The inner surface of rotation

In order to test the processing capacity of the MFA on the inner surface of rotation, the inner hole surface in Fig. 4.48 has been machined, the specimen material is aluminum (YL12), the inner diameter is 30 mm, and the change of the hole surface roughness *Ra* before and after is shown in Table 4.12. The processing time is 5 min, the rotate speed is 900 r/min, and the current is 2.5 A [39].

As shown in the above data, the FMA has some effect on the inner surface of the rotary machining, but the processing efficiency is inferior to the outside surface. The generated magnetic field is not suitable for the finishing of the inner surface. In addition, due to the magnetic shielding effect, there is no field in the cavity of the workpiece; the aluminum workpiece was selected for the inner surface finishing.

**Fig. 4.48** Specimens of the inner hole surface

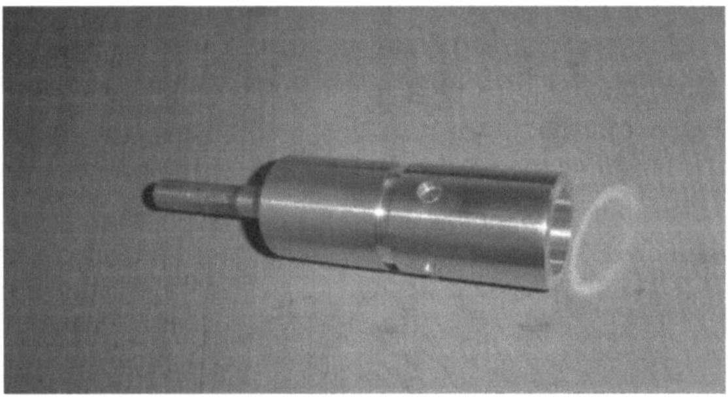

**Table 4.12** Inner hole surface roughness *Ra* before and after finishing

| No. | Before finishing | After finishing | The average *Ra* (before finishing) | The average *Ra* (after finishing) |
|---|---|---|---|---|
| 1 | 0.760 | 0.612 | 0.759 | 0.5545 |
| 2 | 0.791 | 0.572 | | |
| 3 | 0.783 | 0.442 | | |
| 4 | 0.790 | 0.607 | | |
| 5 | 0.720 | 0.602 | | |
| 6 | 0.729 | 0.587 | | |
| 7 | 0.753 | 0.489 | | |
| 8 | 0.746 | 0.525 | | |

## 4.6 The Surface Modification of Fluid Magnetic Abrasive

### *4.6.1 Existent Insufficiencies of Fluid Magnetic Abrasive*

The initial stage FMA was used for finishing, a good processing effect was obtained, but at the same time many problems were also exposed in which about itself property mainly shown in the following respects:

1. After the FMA standing, the magnetic particles are prone to reunite and subside;
2. Due to the use of sodium dodecyl sulfate as dispersing agent, during the process of finishing, with the increase of grinding temperature, a lot of bubbles are produced which definitely hinder the line of sight;
3. After finishing, the abrasive particles and magnetic abrasive particles subside and unable to restore the original steady state with stirring again, so the working life of the abrasives is shorter.

The first and the third questions involve the polymerization stability and sedimentation stability of fluid magnetic abrasives, and the follow-up study will adopt the way of surface modification to improve the stability of magnetic abrasive performance to solve the problem of agglomeration and sedimentation. The second problem is due to the nature of sodium dodecyl sulfate itself which will be solved by replacing.

### *4.6.2 The Surface Modification of Fluid Magnetic Abrasive*

The surface modification of FMA is to change the surface chemical properties of the particles through the methods of physical, chemical, and machinery. According to the need, the chemical and physical properties are changed in order to improve the anti-settlement behavior and the anti-agglomeration of the FMA.

In this paper, the settling properties of FMA are changed by adding surface modifiers, which improve the dispersibility of magnetic particles and decrease the behavior of the agglomeration of magnetic particles.

For solid particles, there are kinds of surface modifiers, which can be chosen according to the performance, characteristics, and application fields of the modified object. The base fluid is water, so the abrasive particles can be modified by the polyamine, poly acrylic acid salt of polyelectrolyte. In this paper, we take the sodium hexametaphosphate and polyacrylic acid as modifier for the modification of magnetic abrasive liquid, in order to compare the modified effect, and several copolymers containing different groups were chose, which are TH-3100, TH-613 (carboxyl, hydroxyl, and sulfo trimer), and TH-241 (carboxyl, hydroxyl, phosphonic acid, and sulfo tetramer). The main parameters of modified agents are shown in Table 4.13 [34, 40].

In addition, this paper introduces a new method to reduce the settling velocity and enhance the settling stability by adding a certain amount of non-magnetic nano-particles which is called performance additive. Some silica particles in 20–30 nm are selected as the performance additive. In addition, there are some other reagents in the modification test: deionized water, sodium nitrite (pure chemicals), sodium hydroxide (pure chemicals), hydrochloric acid (pure chemicals).

Under the use of the nano-silica, the comparison experiments of modification effect are carried out, which used sodium hexametaphosphate, PAA, TH-3100, TH-241, and TH-6131 as modifying agent at pH = 9, and the volume fraction of carbonyl iron particles settling over time is shown in the Fig. 4.49.

**Table 4.13** Chemical and physical properties of modifier

| The name of the modifier | Main component or group | Density (g/ml) | pH | Viscosity (mpa.s) | Solid content (%) | Appearance |
|---|---|---|---|---|---|---|
| Sodium hexametaphosphate | $(NaPO_3)$ 6 | 2.484 | | | | White powder |
| TH-3100 | Carboxyl, sulfonic group, non-ionic groups | 1.171 | 2.71 | 100–500 | 44.43 | Slightly yellow transparent liquid |
| PAA | Carboxyl | 1.09 | 3.0 | 200–600 | 30.0 | Slightly yellow transparent liquid |
| TH-613 | Carboxyl, hydroxyl, sulfonic group | 1.132 | 4.69 | 200–600 | 30.24 | Slightly yellow transparent liquid |
| TH-241 | Carboxyl, hydroxyl, phosphonic acid, and sulfonic group | 1.10 | 3.0–5.0 | 200–600 | >35 | Slightly yellow transparent liquid |

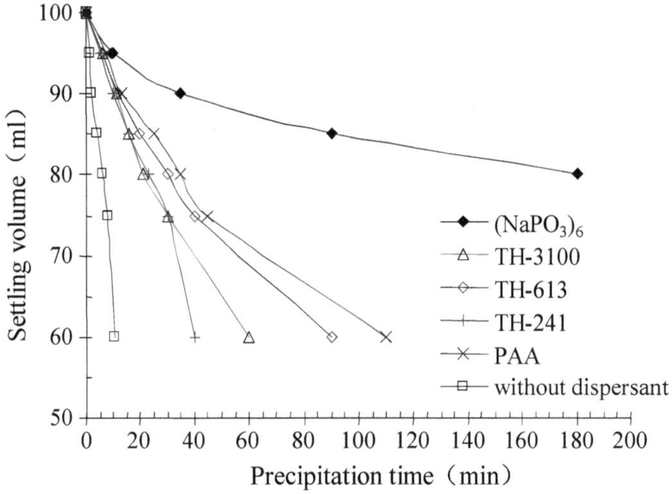


**Fig. 4.49** pH = 9 settlement curves

The results show that the dispersion effect of sodium hexametaphosphate and PAA has an obvious advantage over the rest of the modifier dispersion effect, so this article chooses sodium hexametaphosphate and PAA as modifier. After the experiments, the optimal modifying condition was found, and Table 4.14 shows two formulas of the FMA.

Preparation process is introduced as follows: (1) The 4 ml (4 g) PAA (or sodium hexametaphosphate) and the 70 ml of deionized water are mixed together to be stirred with powerful agitator; (2) the suitable amount of $NaNO_2$ is added to adjust pH value to 9, and 1.5 g nano-silicon dioxide is taken with electronic balance to prevent white dot in the mixture; (3) the mixture is scattered in the mixer for 30 min (700 RPM), then the right amount of antioxidants and 230 g carbonyl iron powder are added slowly to prevent the iron powder sink to the bottom of the container; (4) the mixture is again mixed with the suitable amount of $NaNO_2$ to adjust pH value to 9–11 and strongly scattered for more than 8 h in the mixer, and at the same time, the system temperature is put to 40–50 °C to make the distribution effect better; (5) the mixture is mixed with the 30 g green silicon carbide abrasive in the end, strongly spread for more than 30 min.

The processing experiment for the outside surface of cylindrical workpiece was carried out, which respectively use the FMA before and after modification, and the surface roughness value *Ra* was measured on the three different positions of cylindrical surface, and then took the average, the results were shown in Table 4.15.

As seen from Table 4.15, the material removal rate of modified FMA is high, and the final surface roughness value of workpiece is lower than before.

After finishing, the FMA can produce subsidence phenomenon, there-dispersant of before and after modification is shown in Table 4.16.

**Table 4.14** Two formulas of the FMA

| | Modifier | The amount of addition | Carbonyl iron powder g/100 ml | $NaNO_2$ g/100 ml | Nano-silica g/100 ml | SiC g/100 ml | Deionized water ml/100 ml | PH |
|---|---|---|---|---|---|---|---|---|
| 1 | PAA | 4 ml/100 ml | 230 | 2 | 1.5 | 30 | 70 | 9 |
| 2 | Sodium hexametaphosphate | 4 g/100 ml | 230 | 2 | 1.5 | 30 | 70 | 9 |

**Table 4.15** Finishing processing experimental data of modification before and after

| Workpiece | Surface roughness *Ra* (μm) | | | | The average value of *Ra* (μm) | The reduced value of *Ra* (μm) |
|---|---|---|---|---|---|---|
| Before modification | Before finishing | 1.420 | 1.339 | 1.284 | 1.373 | 0.573 |
| | After finishing | 0.777 | 0.717 | 0.821 | 0.800 | |
| After modification | Before finishing | 3.180 | 3.088 | 3.170 | 3.146 | 2.840 |
| | After finishing | 0.304 | 0.310 | 0.306 | 0.306 | |

**Table 4.16** Re-dispersion of FMA after settling

| | Before | PAA | Sodium hexametaphosphate |
|---|---|---|---|
| The ease or complexity of re-dispersible | No | Easier | Easier |
| Dispersion effect | Bad | Good | Good |
| Mellowness | Hard | Soft | Soft |
| The precipitates modality | Harden | Adhesion | Adhesion |

From Table 4.16, the modified FMA has good re-dispersibility. Before modification, the FMA is easy to harden and cannot be scattered again.

Compared with the FMA without modification, modified FMA has the following advantages: (1) the stability and re-dispersibility are obviously improved; (2) due to the increased use of carbonyl iron powder, the shear yield stress of the modified FMA is increased, so the workpiece material removability has also been increased.

## References


1. S.C. Yang, M.Z. Wang, Y.X. Zhang, *Surface Quality and Finishing* (China Machine Press, Beijing, 2000)
2. Z.J. Yuan, X.K. Wang, *Precision and Ultra-Precision Manufacturing Technology* (China Machine Press, Beijing, 2005)
3. C.H. Li, G.Q. Cai, Q. Li, et al., Generating mechanism of surface morphology finished by abrasive jet with grinding wheel as restraint. J. Northeast. Univ. (Nat. Sci.), **26**(6), 578–581 (2005)
4. J.D. Jiang, *Study on Magneto Rheological Transmission Technology and Devices* (Chongqing University, Chongqing, 2004)
5. Y. Yang, *The Design and Analysis of MRF Clutch* (Chongqing University, Chongqing, 2005)
6. H.T. Fu, F.J. Jiang, Research progress and application of magneto rheological fluids. Chem. Ind. Eng. Prog. **24**(2), 132–136 (2005)
7. D.C. Li, *Theory and Application of Magnetic Fluid* (Science Press, Beijing, 2003)

8. W.I. Kordonski, D. Golini, Fundamentals of magneto rheological fluid utilization in high precision finishing. J. Intell. Mater. Syst. Struct. **10**(9), 683–689 (2001)
9. N. Umehara, Magnetic fluid grinding-a new technique for finishing advanced ceramics. Annals CIRP **43**(1), 185–188 (1994)
10. W.W. You, *Research on Key Technology of Magneto Rheological Finishing* (National University of Defense Technology, Changsha, 2004)
11. W.D. Li, *Research and Development of the Liquid-Magnetic Abrasive Tool for Finishing Treatment* (Taiyuan University of Technology, Taiyuan, 2005)
12. W.D. Li, *Study on the Performance and Preparation Process of Fluid Magnetic Abrasive Finishing Technology* (Taiyuan University of Technology, Taiyuan, 2012)
13. S.C. Lu., Industrial Suspension-Performance Modulation and Process. Chemical Industry Press, (2003)
14. H. Si, *Mechanical Mechanism of Magneto Rheological Fluids* (Chongqing University, Chongqing, 2003)
15. W.F. Hughes, J.A. Brighton, *Schaum's Outline of Theory and Problems of Fluid Dynamics* (Science Press, Beijing, 2002)
16. H.T. Li, X.H. Peng, W.M. Chen, Yield shear-stress model of magneto rheological fluids. Chin. J. Chem. Phys. **18**(4), 505–509 (2005)
17. Y.S. Zhu, X.L. Gong, H. Li, Numerical analysis on shear yield stress of magneto rheological fluids. J. Chin. Univ. Min. Technol. **35**(4), 498–503 (2006)
18. H.J. Zhou, *Machining Principles and Experimental Study of Magneto Rheological Finishing for Super-Smooth Surface* (National University of Defense Technology, Changsha, 2002)
19. X.Q. Peng, *Study on the Key Techniques of Deterministic Magneto Rheological Finishing* (National University of Defense Technology, Changsha, 2004)
20. H.W. Sun, *Study on the Fluid Magnetic Abrasive Finishing Technology* (Taiyuan University of Technology, Taiyuan, 2008)
21. B.M. Li, B. Zhao, *Modern Grinding Technology* (China Machine Press, Beijing, 2003)
22. L.Y. Zhang, *Modern Processing Technology* (China Machine Press, Beijing, 2002)
23. T.H.C. Childs, S. Mahmood, H.J. Yoon, The material removal mechanism in magnetic fluid grinding of ceramic ball bearings. Proc. Inst. Mech. **208B**, 47–59 (1994)
24. T.H.C. Childs, D.J. Moss, Wear and cost issues in magnetic fluid grinding. Wear **249**, 509–516 (2001)
25. H.W. Sun, S.C. Yang, Revolved surface finishing with smart fluid abrasives. Key Eng. Mater. **304–305**, 579–583 (2006)
26. H.W. Sun, S.C. Yang, W.H. Li, Study on magneto rheological surface finishing. Key Eng. Mater. **259–260**, 653–656 (2004)
27. H.W. Sun, S.C. Yang., Fluid Magnetic Abrasives Based on Micron-Sized Carbonyl-Iron Particles and its Applications in the Precision Finishing Process, in *Proceedings of MNC2007* (2007)
28. X.M. Wang, *The Mechanism and Surface Characteristic of Pulse Electrochemical and its Compounded Finish Machining* (Dalian University of Technology, Dalian, 2002)
29. N. Umehara, K. Kato, A study on magnetic fluid grinding. Trans. Jpn. Soc. Mech. Eng., **61** (584), 1709–1714 (1988)
30. S. Jha, V.K. Jain, Modeling and simulation of surface roughness in magneto rheological abrasive flow finishing(MRAFF) process. Wear **261**, 856–866 (2006)
31. J.J. Hao, *Research and Development of Experimental Equipment in Liquid-Magnetic Abrasive Tool Finishing* (Taiyuan University of Technology, Taiyuan, 2005)
32. Q.S. Ru, *Research on Finishing Mechanism and Physical Property of Liquid-Magnetic Abrasive Tool* (Taiyuan University of Technology, Taiyuan, 2006)
33. S. Jha, V.K. Jain, Design and development of the magneto rheological abrasive flow finishing (MRAFF) process. Int. J. Mach. Tools Manuf. **44**, 1019–1029 (2004)
34. H. Lei, J.B. Luo, J.J. Ma, Advances and problems on chemical mechanical polishing. Lubr. Eng. **4**, 73–76 (2002)

35. R. Tao, X. Xu, Viscosity reduction in liquid suspensions by electric or magnetic fields. Int. J. Mod. Phys. **19**, 1283–1289 (2005)
36. X.W. Sun, *Research on Key CNC Technique of the Magneto Rheological Finishing Machine Tool* (Harbin Institute of Technology, Harbin, 2006)
37. S. Singh, H.S. Shan, P. Kumar, Wear behavior of materials in magnetically assisted abrasive flow machining. J. Mater. Process. Technol. **128**, 155–161 (2002)
38. F. Zhang, B.Z. Zhang, Surface roughness of optical elements fabricated by magnetic fluid-assisted polishing. Opt. Precis. Eng. **13**(1), 34–39 (2005)
39. N. Umehara, T. Kobayashi, T. Kato, Internal polishing of tube with magnetic fluid grinding. J. Magn. Magn. Mater. **149**, 188–191 (1995)
40. H.Y. Li, *Research on the Techniques of Pulse Electrochemical Finishing in Gears* (Dalian University of Technology, Dalian, 2003)

# Chapter 5
# Two-Phase Swirling Flow Hole Finishing

## 5.1 Introduction

### *5.1.1 Presentation of Two-Phase Swirling Flow Hole Finishing*

The traditional finishing methods such as honing, grinding, and rolling have some limitations owing to their process characteristics. The cavities, grooves, cross-holes, and the non-perfect rounded circular surfaces of workpiece are difficult to be processed through these methods [1]. For example, valve body as a common hydraulic part has many internal cross-holes. The hole fitted with valve core is processed through the fine boring or reaming. So the burrs would be produced at the corner of holes, which influenced the performance of the valve. The shedding burrs could lead to value core stuck, slide valve ineffective, performance degradation, and even leakage of hydraulic valve sealing when the hydraulic pressure increased or after a certain period of work. As another example, the burrs in gun barrels influenced the range and accuracy of gunshot seriously. In recent years, the nylon brush with abrasive has been used for deburring. However, the hand deburring is still the most common method, which leads to the lower productivity and unstable processing quality.

Since the 1970s, the free abrasive had been studied and applied in the hole finishing. The better fluidity of free abrasive was used to obtain satisfactory processing performance. The finishing methods such as magnetic abrasive finishing, barrel finishing, and abrasive flow machining have been widely used in many fields and have replaced the traditional finishing partly [2, 3]. However, the satisfied results were still not achieved by those finishing methods for some cases. The barrel finishing was used for the outer surfaces of the workpiece and the single hole with bigger dimension. But it was unsuitable for the finishing of grooves and smaller holes. The abrasive flow machining has better performance except for cross-holes

 
S. Yang and W. Li, *Surface Finishing Theory and New Technology*,
https://doi.org/10.1007/978-3-662-54133-3_5

and blind holes. The magnetic abrasive finishing was not applicable to some workpiece which could not form the necessary magnetic field [4–6].

Aiming at the current situation of the hole finishing, the two-phase swirling flow hole finishing was proposed, which inspired by the tornado, cyclone separator, and centrifugal pump. The process is commonly known as "flexible grinding wheel." The fluid with certain flow rate and pressure formed swirling flow through a specific device. The free abrasive was precedently loaded in holes of workpiece. Under the action of swirling flow, the free abrasive would be rolling and grinding against the surface. As a result, deburring and finishing were achieved, which improved the surface quality of workpiece [7, 8].

### 5.1.2 Characteristics of the Two-Phase Swirling Flow Hole Finishing

The two-phase swirling flow hole finishing has wide processing range, better adaptability, high processing efficiency, and better processing effect. The main characteristics are listed below:

(1) Owing to the sufficient kinetic energy produced by the swirling flow apparatus, this machining method has high processing efficiency. The surface roughness *R*a value after finishing of workpiece can be reduced by 1–2 level.
(2) The material removal amount is small in this method, which belongs to micro-processing. The removal thickness of workpiece is in the μm level, which means that workpiece can maintain the original machining accuracy.
(3) The length-to-diameter ratio of tube workpiece was not strictly demanded. Most of the straight tube workpiece with special requirements could be processed by this method.
(4) The media randomly distributed along the hole surface. Therefore, the part is uniformly forced in the circumferential direction. Then, the thin-walled workpiece could also be processed through the method.
(5) The processing was also suitable for special holes such as the cross-holes, stepped holes, blind holes.
(6) The processing could be used for deburring and finishing of the precision holes with small and medium size, such as holes in the hydraulic valve, oil cylinder, and gun barrel. It could also be used for cleaning and finishing of the large size holes, such as pipelines in the petroleum and natural gas industry.

## 5.2 Theory and Numerical Simulation of Swirling Flow

### 5.2.1 *Single-Phase Swirling Flow*

1. Basic definition

The axial velocity and the tangential velocity of media simultaneously exist in swirling flow finishing, and their magnitudes are quite high. The velocity distribution of media directly affects the flow characteristics and the application of swirling flow finishing.

In the case of constant flow rate, the swirling flow can be considered to be a constant flow when it reaches steady state. The isothermal continuum of the incompressible homogeneous fluid particle is selected as the research object. In addition, assume that the fluid density does not change with time and location. Continuity equation in cylindrical coordinate system can be obtained [9]:

$$\frac{1}{r}\frac{\partial}{\partial r}(ru_{\mathrm{r}})+\frac{1}{r}\frac{\partial u_{\theta}}{\partial \theta}+\frac{\partial u_{\mathrm{z}}}{\partial z}=0 \tag{5.1}$$

The momentum equation is:

$$\begin{cases} \rho\left(\frac{\partial u_{\mathrm{r}}}{\partial t}+u_{\mathrm{r}}\frac{\partial u_{\mathrm{r}}}{\partial r}+\frac{u_{\theta}}{r}\frac{\partial u_{\mathrm{r}}}{\partial \theta}+u_{z}\frac{\partial u_{z}}{\partial z}-\frac{u_{\theta}^{2}}{r}\right)= \\ \rho f_{\mathrm{r}}-\frac{\partial p}{\partial r}+\mu\left\{\frac{\partial}{\partial r}\left[\frac{1}{r}\frac{\partial}{\partial r}(ru_{\mathrm{r}})\right]+\frac{1}{r^{2}}\frac{\partial^{2}u_{\mathrm{r}}}{\partial \theta^{2}}-\frac{2}{r^{2}}\frac{\partial u_{\theta}}{\partial \theta}+\frac{\partial^{2}u_{\mathrm{r}}}{\partial z^{2}}\right\} \\ \rho\left(\frac{\partial u_{\theta}}{\partial t}+u_{r}\frac{\partial u_{\theta}}{\partial r}+\frac{u_{\theta}}{r}\frac{\partial u_{\theta}}{\partial \theta}+u_{\mathrm{z}}\frac{\partial u_{z}}{\partial z}+\frac{u_{\theta}u_{\mathrm{r}}}{r}\right)= \\ \rho f_{\theta}-\frac{1}{r}\frac{\partial p}{\partial \theta}+\mu\left\{\frac{\partial}{\partial r}\left[\frac{1}{r}\frac{\partial}{\partial r}(ru_{\theta})\right]+\frac{1}{r^{2}}\frac{\partial^{2}u_{\theta}}{\partial \theta^{2}}-\frac{2}{r^{2}}\frac{\partial u_{\mathrm{r}}}{\partial \theta}+\frac{\partial^{2}u_{\theta}}{\partial z^{2}}\right\} \\ \rho\left(\frac{\partial u_{\mathrm{z}}}{\partial t}+u_{\mathrm{r}}\frac{\partial u_{\mathrm{z}}}{\partial r}+\frac{u_{\theta}}{r}\frac{\partial u_{\mathrm{z}}}{\partial \theta}+u_{\mathrm{z}}\frac{\partial u_{\mathrm{z}}}{\partial z}\right)= \\ \rho f_{z}+\frac{\partial p}{\partial z}+\mu\left\{\frac{1}{r}\frac{\partial}{\partial r}\left(r\frac{\partial u_{\mathrm{z}}}{\partial r}\right)+\frac{1}{r^{2}}\frac{\partial^{2}u_{z}}{\partial \theta^{2}}+\frac{\partial^{2}u_{\mathrm{r}}}{\partial z^{2}}\right\} \end{cases} \tag{5.2}$$

where $\rho$ is fluid density (kg/m$^3$); $u_{\mathrm{r}}$, $u_{\theta}$, $u_{\mathrm{z}}$ and $f_{\mathrm{r}}, f_{\theta}, f_{\mathrm{z}}$ are the radial , tangential , and axial components of the fluid velocity (m/s) and unit mass force, respectively; $P$ is the pressure (Pa); $\mu$ is the fluid dynamic viscosity coefficient (Pa·s).

Swirling flow has the characteristics of high energy concentration, strong carrying capacity, and high transmission concentration. It is widely used in engineering, such as jet technology, combustion technology, pneumatic conveying, cyclone separation, and hydraulic flotation. In addition, swirling flow is often used in different types of separators, swirling water mist dryers, scrolls, agitators, and piping systems. In the combustion of engines or other furnaces, swirling flow is often used to stabilize the flame, to enhance the mixing of fuel and air, and at the same time to enhance the heat and mass exchange [10–12].

2. Numerical simulation and analysis

(1) Mathematical model

The swirling flow in pipe is the typical turbulent flow whose flow state is very complex. The mathematical model used to describe the turbulent flow has not been well developed [13]. The existing influential turbulence models include zero-equation model, one-equation model, two-equation model, algebraic stress model, Reynolds stress model, and other multi-equation models. The zero-equation model computes the turbulent transport term without increasing the partial differential equation. The one-equation model adds a partial differential equation for the turbulent function $k$. The two-equation model increases the partial differential equations for two turbulence terms: $k$ and $\varepsilon$. Algebraic stress model and the Reynolds stress model are two best models to describe the turbulence flow, but they are difficult to be computed.

The basic equations for describing turbulence include continuity equation, Navier–Stokes equation, Reynolds stress equation, turbulent kinetic energy equation, and turbulent kinetic energy dissipation rate equation [14, 15]. For constant temperature and incompressible flow, the continuity equation is:

$$\frac{\partial u_i}{\partial x_i} = 0 \tag{5.3}$$

Reynolds-averaged Navier–Stokes (RANS) equation is:

$$\frac{\partial}{\partial t}(\rho u_i) + \frac{\partial}{\partial x_j}(\rho u_i u_j) = -\frac{\partial p}{\partial x_i} + \frac{\partial}{\partial x_j}\left[\mu\left(\frac{\partial u_i}{\partial x_j} + \frac{\partial u_j}{\partial x_i} - \frac{2}{3}\delta_{ij}\frac{\partial u_l}{\partial u_l}\right) - \rho\overline{u_i' u_j'}\right] \tag{5.4}$$

where $\rho$ is the density of the fluid (kg/m$^3$), $u_i$ and $u_j$ are the time-averaged velocity component of the fluid, $i, j = 1, 2, 3$, $P$ is the pressure (Pa), $\mu$ is the dynamic viscous coefficient (Pa·s), $\delta_{ij}$ is the Kronecker constant (if $i = j$, $\delta_{ij} = 1$; else, $\delta_{ij} = 0$), and $-\rho\overline{u_i' u_j'}$ is the unknown Reynolds stress component (N/m$^2$).

Additional second-order correlation terms now appear in Eq. (5.4) that represents the effects of turbulence velocity, i.e., the Reynolds stress term. These Reynolds stresses, $-\rho\overline{u_i' u_j'}$, cannot be closed with the continuity equation, so it must be modeled in order to close equation. Different closure methods form different turbulence models.

(1) The Standard $k$-$\varepsilon$ Model

The model is based on the Boussinesq hypothesis of isotropic eddy viscosity, using the following Reynolds stress correlation:

$$-\rho\overline{u_i' u_j'} = \mu_T\left(\frac{\partial u_i}{\partial x_j} + \frac{\partial u_j}{\partial x_i}\right) - \frac{2}{3}\rho k\delta_{ij} \tag{5.5}$$

where $u_e = u + u_T$ is the effective viscous coefficient; $\mu_T = C_\mu \rho \frac{k^2}{\varepsilon}$ is the turbulent viscosity coefficient. The turbulence kinetic energy $k$ and its rate of dissipation $\varepsilon$ are obtained from the following transport equations:

$$\frac{\partial(\rho k)}{\partial t} + \frac{\partial(\rho u_j k)}{\partial x_j} = \frac{\partial}{\partial x_j}\left(\frac{\mu_e}{\sigma_k}\frac{\partial k}{\partial x_j}\right) + G_k - \rho\varepsilon \tag{5.6}$$

$$\frac{\partial(\rho \varepsilon)}{\partial t} + \frac{\partial(\rho u_j \varepsilon)}{\partial x_j} = \frac{\partial}{\partial x_j}\left(\frac{\mu_e}{\sigma_\varepsilon}\frac{\partial \varepsilon}{\partial x_j}\right) + \frac{\varepsilon}{k}(C_{\varepsilon 1} G_k - C_{\varepsilon 2}\rho\varepsilon) \tag{5.7}$$

where $G_k$ represents the generation of turbulence kinetic energy due to the mean velocity gradients; $\bar{S}_{ij} = \frac{1}{2}\left(\frac{\partial u_i}{\partial x_j} + \frac{\partial u_j}{\partial x_i}\right)$ is the strain rate tensor. The model constants $C_{\varepsilon 1}$, $C_{\varepsilon 2}$, $C_\mu$, $\sigma_k$, and $\sigma_\varepsilon$ have the following default values:

$$C_\mu = 0.09, C_{\varepsilon 1} = 1.45, C_{\varepsilon 2} = 1.92, \sigma_k = 1.0, \sigma_\varepsilon = 1.3.$$

Equation (5.5) and its parameter values apply to a high *Re* region turbulence at a distance from the wall, where Re is the square root of the turbulence kinetic energy. In the near-wall region, the turbulence Re is very low. In this case, the influence of molecular viscosities must be taken into account, and the wall function method is usually used.

(2) The RNG $k$-$\varepsilon$ Model

The RNG $k$-$\varepsilon$ model was derived using a rigorous statistical technique (called renormalization group theory). It is more accurate than the standard $k$-$\varepsilon$ model in solving the large curvature transition flow and the change of back-step boundary. It is similar in form to the standard $k$-$\varepsilon$ model, but includes an additional term in its $\varepsilon$ equation. The additional term increases significantly when the flow is rapidly distorted, and the empirical constant involved in the model is different from the standard $k$-$\varepsilon$ model.

The transport equations for $k$ and $\varepsilon$ in the RNG $k$-$\varepsilon$ model are:

$$\frac{\partial(\rho k)}{\partial t} + \frac{\partial(\rho u_j k)}{\partial x_j} = \frac{\partial}{\partial x_j}\left(\alpha_k \mu_e \frac{\partial k}{\partial x_j}\right) + G_k + \rho\varepsilon \tag{5.8}$$

$$\frac{\partial(\rho \varepsilon)}{\partial t} + \frac{\partial(\rho u_j \varepsilon)}{\partial x_j} = \frac{\partial}{\partial x_j}\left(\alpha_\varepsilon \mu_e \frac{\partial \varepsilon}{\partial x_j}\right) + \frac{\varepsilon}{k}(C^*_{\varepsilon 1} G_k - C_{\varepsilon 2}\rho\varepsilon) \tag{5.9}$$

where the model constants are as follows:

$$C_{\varepsilon 1}^{*} = C_{\varepsilon 1} - \frac{\eta(1-\eta/\eta_0)}{1+\beta\eta^3}\left(\eta = \frac{Sk}{\varepsilon}, S = \sqrt{2\overline{S_{ij}S_{ij}}}, \beta = 0.012, C_{\varepsilon 1} = 1.42,\right.$$
$$\left. C_{\varepsilon 2} = 1.68, \sigma_k = 0.7179, C_\mu = 0.0845, \sigma_\varepsilon = 0.7179\right)$$

Compared with the standard $k$-ε model, the constants in the RNG $k$-ε model are deduced by theory rather than experience; thus, they are more applicable. The coefficient of dissipation equation takes into account the effects of the average strain rate on the dissipation term and the turbulent swirl effects, so the calculation accuracy is also improved. The RNG $k$-ε model provides an analytical formula to calculate the turbulent Prandtl number for the low Reynolds number flow viscosity. These characteristics make the RNG $k$-ε model to have higher reliability and accuracy than the standard $k$-ε model.

(3) The Reynolds Stress Model (RSM)

The exact transport equations for the transport of the Reynolds stresses can be written as follows:

$$\frac{\partial}{\partial t}\left(\rho\overline{u_i'u_j'}\right) + \frac{\partial}{\partial x_k}\left(\rho u_k\overline{u_i'u_j'}\right) = D_{i,j} + P_{i,j} + G_{i,j} + \Phi_{i,j} + \varepsilon_{i,j} + F_{i,j} \tag{5.10}$$

where turbulent diffusion is $D_{i,j} = -\frac{\partial}{\partial x_k}\left[\rho\overline{u_i'u_j'u_k'} + \overline{p'u_i'}\delta_{kj} + \overline{p'u_j'}\delta_{ik} - \mu\frac{\partial}{\partial x_k}\left(\overline{u_i'u_j'}\right)\right]$; stress production is $P_{i,j} = -\rho\left(\overline{u_i'u_k'}\frac{\partial u_j}{\partial x_k} + \overline{u_j'u_k'}\frac{\partial u_i}{\partial x_k}\right)$; Buoyancy production is $G_{i,j} = -\rho\beta\left(g_i\overline{u_j'\theta} + g_j\overline{u_i'\theta}\right)$; pressure strain is $\Phi_{i,j} = \overline{p'\left(\frac{\partial u_i'}{\partial x_j} + \frac{\partial u_j'}{\partial x_i}\right)}$; dissipation is $\varepsilon_{i,j} = 2\mu\overline{\frac{\partial u_i'}{\partial x_k}\frac{\partial u_j'}{\partial x_k}}$; production by system rotation is $F_{i,j} = -2\rho\Omega_k\left(\overline{u_j'u_m'}e_{ikm} + \overline{u_i'u_m'}e_{jkm}\right)$ ($\Omega_k$ is the mean rate-of-rotation tensor; $e_{ijk}$ is the conversion symbol—if $i \neq j \neq k$ and meet the positive sequence, $e_{ijk}$ is equal to 1 and if $i, j, k$ meet reverse order, $e_{ijk}$ is equal to −1, else $e_{ijk}$ is equal to 0).

After modeling, the turbulent kinetic energy equation and the turbulent kinetic energy dissipation rate equation are [16]:

$$\frac{\partial}{\partial t}(\rho k) + \frac{\partial}{\partial x_i}(\rho k u_i) = \frac{\partial}{\partial x_j}\left[\left(\mu + \frac{\mu_i}{\sigma_k}\right)\frac{\partial k}{\partial x_j}\right] + \frac{1}{2}(P_{ii} + G_{ii}) - \rho\varepsilon \tag{5.11}$$

$$\frac{\partial}{\partial t}(\rho\varepsilon) + \frac{\partial}{\partial x_i}(\rho\varepsilon u_i) = \frac{\partial}{\partial x_j}\left[\left(\mu + \frac{\mu_i}{\sigma_\varepsilon}\right)\frac{\partial \varepsilon}{\partial x_j}\right] + C_{\varepsilon 1}\frac{1}{2}(P_{ii} + C_{\varepsilon 3}G_{ii})\frac{\varepsilon}{k} - C_{\varepsilon 2}\rho\frac{\varepsilon^2}{k} \tag{5.12}$$

where $C_\mu = 0.09$, $\sigma_k = 0.82$, $\sigma_\varepsilon = 1.0$, $C_{\varepsilon 1} = 1.44$, $C_{\varepsilon 2} = 1.92$, and $C_{\varepsilon 3}$ is a function of flow properties of the relative gravity of a point.

The time-averaged continuity Eq. (5.3), the Reynolds Eq. (5.4), the Reynolds stress transport Eq. (5.10), the turbulent kinetic energy Eq. (5.11), and the turbulent kinetic energy dissipation rate Eq. (5.12) constitute the basic closed-form equations of the Reynolds stress model (RSM) [17].

The Reynolds stress model (RSM) rejects the Boussinesq hypothesis based on isotropic viscousness. It takes into account the effects of multiple physical processes, including the effect of turbulent anisotropy, in particular the rotation effects, buoyancy effects, and curvature effects. In many cases, RSM can give results better than various $k$-$\varepsilon$ models. However, for the complex engineering applications and the three-dimensional problem, its calculation will be time-consuming and places high demands on the memory and CPU of computer. Secondly, the boundary conditions of each stress and flux component are not easy to be defined. In addition, there are a number of empirical coefficients to be determined, and the determination of these coefficients is even more difficult.

(2) Numerical calculation and simulation methods

Treating the dependent variables of the finite number positions (e.g., the mesh nodes) in the computational domain as basic unknowns, the numerical method provides algebraic equations and solutions method for these unknowns, i.e., substituting discrete values for the contact information contained in the exact solution of the differential equation, which can also be called the discretization method. In computational fluid dynamics, methods mainly used to solve these models include finite element method, finite volume method, finite analysis method, Lattice Boltzmann method [18].

(1) Finite difference method

The finite difference method is the earliest, the most classical, and commonly used numerical method. The basic idea is to establish the Taylor expansion of the discrete equation, to divide the solution region into differential grids, to replace the derivative of the governing equations with the quotient in each node of the grid, and to form an algebraic equation on each node. The finite difference equations with finite unknowns are obtained and solved to obtain the necessary numerical solutions. For different Taylor expansions, the different formats can be divided into the first order, second order, third order or higher order according to the approximation accuracy. They can also be divided into central and upwind format according to the format nature.

The advantages of finite difference method are simple form, easy understanding, and high processing capacity. The main drawback is the poor adaptability to complex regions. The precision and conservation of numerical solutions are difficult to be guaranteed.

(2) Finite element method

The finite element method (FEM) has been introduced in hydromechanics in the 1970s. FEM is based on the variational principle and the weighted residual method.

The basic solving principle is that the computational domain is divided into a finite number of non-overlapping units. Within each cell, selecting some appropriate nodes as solving the interpolation function, the differential equation of the variables would be rewritten by the variable or its derivative selected interpolation node value and the function of linear expression. With the aid of variational principle or weighted residual method, the discrete solutions of differential equation are obtained.

FEM can well adapt to the complex geometry, complex material properties, and complicated boundary conditions, but it involves large amount of computation. FEM is only used in solid mechanics simulation field. When solving the flow and heat transfer problems, it is less mature than the finite volume method for the discrete processing and incompressible primitive variable.

(3) Finite volume method

The finite volume method (FVM) is also termed as control volume method. Its basic idea is to divide the domain into a number of control volumes (cells, elements), where the variable of interest is located at the centroid of the control volume. The next step is to integrate the differential form of the governing equations over each control volume. As a result, a group of discretization equations, in which the unknown value is the dependent variable located in each node, are obtained. In order to calculate the integral of the control volume, the function to be solved is assumed as the first derivative of itself.

For FVM, the integral conservation of the dependent variable is satisfied for any control volume. So it is also satisfied for the whole computational domain, which is the advantage of the FVM [19]. FVM needs to assume the variation of the value between grid points (and the interpolation function) and to use the value as an approximate solution. The finite difference method considers only those values at the grid points, regardless of the change rule of the values between the grid points. In FVM, the interpolation function is only used to calculate the integral of the control volume. After the discrete equation is obtained, the interpolation function can be ignored. If necessary, different interpolation functions can be used for different terms in the differential equation.

The basic principle of the FVM is easy to understand. The coefficients of discrete equation have clear physical meaning. More importantly, FVM can be used to solve the complicated computational domain. Therefore, it is one of the most widely used numerical computation methods for engineering flow problems and is also the main discrete method in CFD commercial software [20, 21].

In order to better simulate the swirling flow field and to ensure the accuracy and stability of the calculation results, the gas-phase turbulence calculation model chosen is RSM model based on the anisotropic. SIMPLE algorithm, quadratic upwind interpolation, and body force-weighted discretization are selected in simulation. Through the simulation, the velocity field and pressure field can be obtained directly [22].

(3) Geometric model and boundary conditions

(1) Geometric model

Figure 5.1 shows the geometrical model with 40 mm pipe diameter and 400 mm pipe length. In addition, the nozzles' diameter and length are 4 mm and 16 mm, respectively. To form a spiral air flow and thus obtain better machining performance, the nozzles' angle $\alpha$ is equal to 10° [23], as shown in Fig. 5.2.

(2) Meshing

Software GAMBIT is used to generate meshes. The "bottom-up" meshing approach is used, which means first meshing lines, then meshing faces, and finally meshing volumes.

Due to the weak limit of the unstructured mesh to the grid nodes, it is easy to control the size, shape, and location of the grid cell, which makes the grid to have better adaptability in the complex geometric model. And the number of nodes and grid cells around a node is not fixed which is convenient to do adaptive computing. Therefore, unstructured grids are used for simulation. In the near-wall region, the GAMBIT is used to mesh the boundary layer. In the far-wall region, in order to adapt to the complicated 3D domain, tetrahedral grid cells or hexahedron grid cells are used. At the same time, the domain where the flow parameters change largely must be meshed by the grid encryption. The mesh of the geometrical model is shown in Fig. 5.3.

**Fig. 5.1** Geometrical model

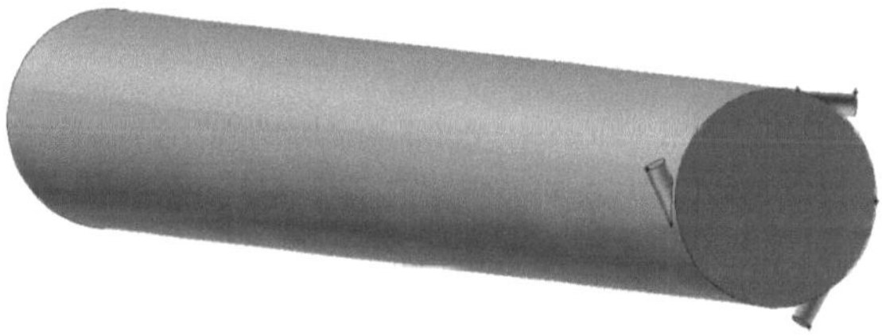

**Fig. 5.2** Nozzle distribution and angle $\alpha$

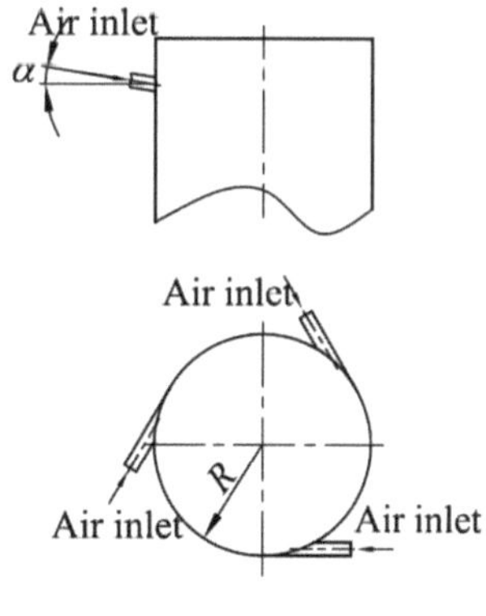

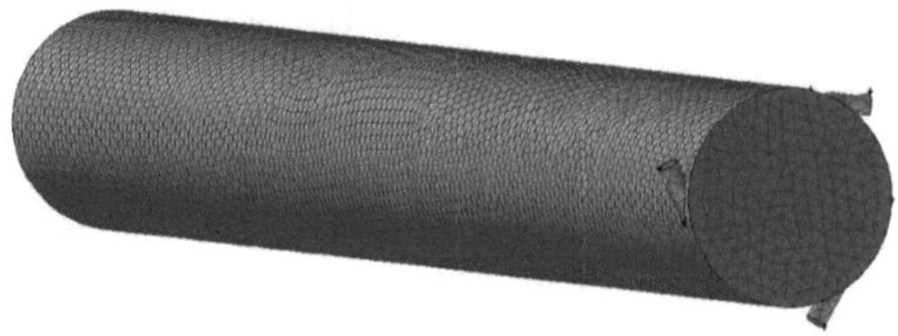

**Fig. 5.3** Mesh of the geometrical model

(3) Boundary conditions

Inlet boundary conditions: Velocity inlet boundary conditions are used to define the flow velocity when $Ma \leq 0.3$; gas is incompressible gas, and its density and viscosity are 1.205 kg/m$^3$ and $1.8 \times 105$ Pa.s, respectively. Pressure outlet boundary conditions are used when $Ma > 0.3$. Gas is compressible gas. Subsequent simulations used a velocity inlet boundary condition with inlet velocity of 100 m/s.

Outlet boundary conditions: The outlet is set as the "pressure outlet," and the operating pressure is set as 101325 Pa.

The wall boundary condition is a fixed heat flux of zero. By default, walls are not moving. For no-slip wall conditions, it indicates that the fluid sticks to the wall and moves with the same velocity as the wall. Wall effect is the main reason for the generation of the vortex and turbulence. Thus, the processing method of the near-wall domain has a significant effect on the simulation accuracy. For the simulation, the wall function method is used to deal with the first net nodes of the near wall.

(4) Simulation results and discussion

(1) Tangential velocity

Tangential velocity vector in different cross sections along the axial direction of the circular tube is shown in Fig. 5.4. The velocity contour in cross section $Y = 0$ is shown in Fig. 5.5.

As shown in Fig. 5.4, the tangential velocity and the vortex intensity of the gas decrease with the increase in the Z coordinate values. In addition, it is shown in Fig. 5.4 that the tangential velocity between the inside wall and the center has a large difference. Figure 5.4 also shows that the gas flows downstream and forms vortex flow near the wall after the gas flows into the pipe. Then, the gas flows upstream, forming an internal upward spiral movement, which means that reflux is produced and a stem is formed in the central region of the tube. In addition, the air flow inside the tube has good symmetry.

Figures 5.6 and 5.7 are the tangential velocity contour and variations of the tangential velocity in different lines ($Y = 0$, $Z = 0.05$; $Y = 0$, $Z = 0.1$; $Y = 0$, $Z = 0.2$; $Y = 0$, $Z = 0.3$), respectively.

As shown in Figs. 5.6 and 5.7, the tangential velocity of the gas shows a good axial symmetry, which reflects strong vortex characteristics. The tangential velocity profile shows "hump" distributions. Namely, on each side of the tangential velocity profile, there is a maximum tangential velocity point. In addition, the flow field

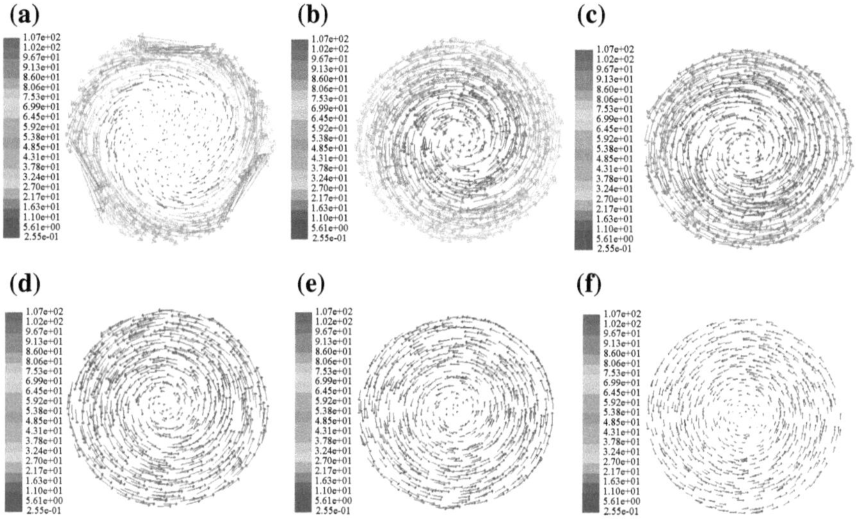


**Fig. 5.4** Tangential velocity vector in different cross sections. **a** $z = 0$ m; **b** $z = 0.05$ m; **c** $z = 0.1$ m; **d** $z = 0.15$ m; **e** $z = 0.2$ m; **f** $z = 0.3$ m

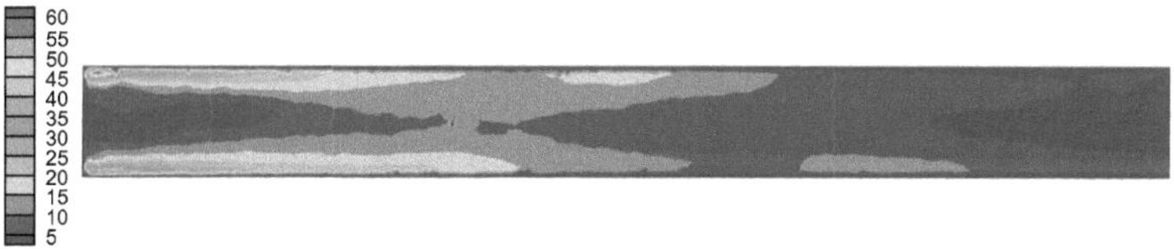


**Fig. 5.5** Velocity contour in cross section $Y = 0$

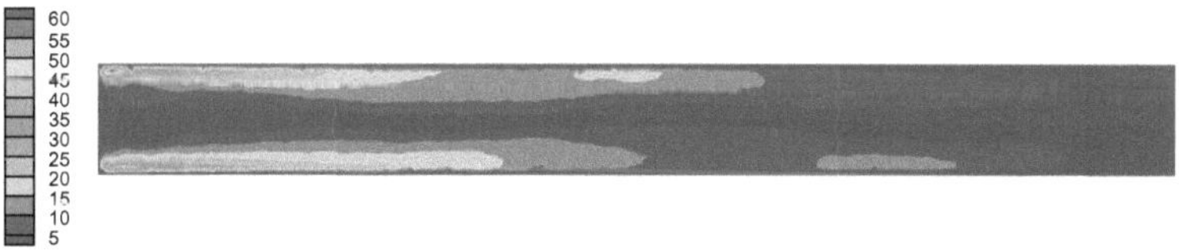


**Fig. 5.6** Tangential velocity contour of $Y = 0$

within the tube shows combined vortex features, all of the maximum tangential velocity points form an interface, and the interface divided the flow field into the quasi-forced vortex located inside of the interface and the quasi-free vortex located outside of the interface. The flow field characteristics are beneficial to the two-phase gas-particle finishing. The reason is that the centrifugal force caused by quasi-forced vortex makes the media move toward the inner wall surface of the pipe. On the other hand, due to the lower swirl intensity of the quasi-free vortex

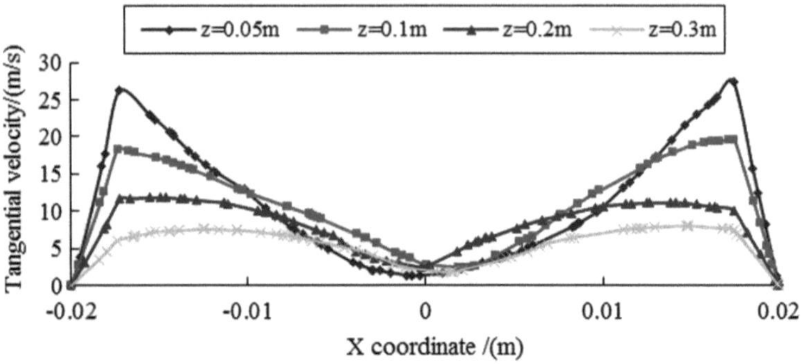


**Fig. 5.7** Variations of the tangential velocity in different lines. ($Y = 0$, $Z = 0.05$; $Y = 0$, $Z = 0.1$; $Y = 0$, $Z = 0.2$; $Y = 0$, $Z = 0.3$)

outside of the interface, the gas has a weaken carrying effects for media, which facilitates the movement of media in the vicinity of the inside wall.

In addition, the tangential velocity of most area is positive, but there is a very small part that has negative velocity, which is inconsistent with the mainstream, and it means the presence of the local secondary vortex. And the presence of local secondary vortex reflects the loss of rotational kinetic energy.

Figures 5.8 and 5.9 are the axial velocity contour and variations of the axial velocity in different lines ($Y = 0$, $Z = 0.05$; $Y = 0$, $Z = 0.1$; $Y = 0$, $Z = 0.2$; $Y = 0$, $Z = 0.3$), respectively.

As shown in Figs. 5.8 and 5.9, the axial velocity of the gas shows not only complicated radial distribution but also great change in the axial direction. It also shows that the axial velocity distribution has a well symmetry. It is also shown in Fig. 5.9 that the axial velocity is positive in the outside area of the interface and it is negative in the central area of the tube. As a result, the media near the inner wall surface of the pipe flow downstream.

Figure 5.9 shows that with the increase in the distance from the pipe centerline, the axial velocity increases firstly and then decreases to zero due to the wall effect. In addition, at the centerline, there is a minimum axial velocity point in different Z-planes. In $Z = 400$-mm plane, this minimum axial velocity closes to zero. In $Z = 0$-mm plane, the minimum axial velocity is negative which means that there has a stagnation or reflow phenomenon.

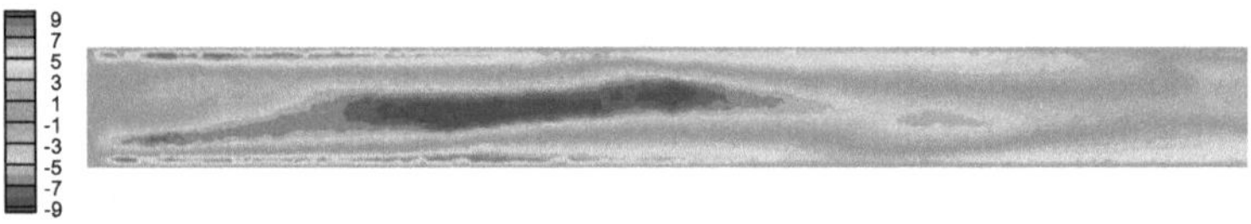


**Fig. 5.8** Axial velocity contour of $Y = 0$

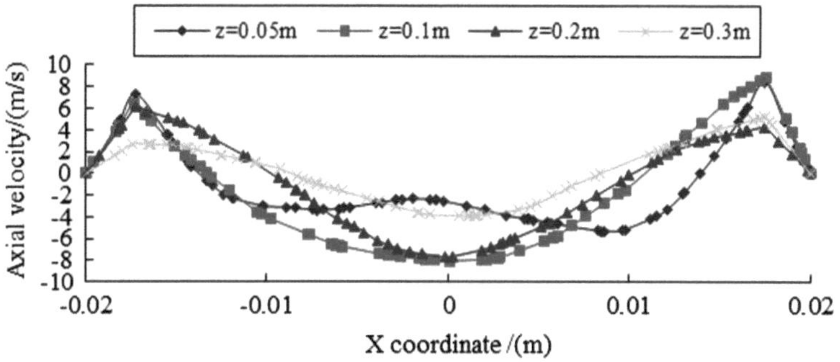


**Fig. 5.9** Variations of the axial velocity in different lines. ($Y = 0$, $Z = 0.05$; $Y = 0$, $Z = 0.1$; $Y = 0$, $Z = 0.2$; $Y = 0$, $Z = 0.3$)

(2) Turbulent kinetic energy and its dissipation rate

Turbulence occurs at high Reynolds number, and its basic characteristic is the pulsating of fluid particles. This irregular motion of fluid particles leads to the motion along the mainstream direction and the additional pulsating in all directions. Each fluid particle collides with each other, which leads to the fluid resistance during advancing and resulted in major energy consumption of turbulent kinetic energy.

Figures 5.10 and 5.11 show turbulent kinetic energy contour and variations of the radial turbulent kinetic energy in different lines ($Y = 0$, $Z = 0.05$; $Y = 0$, $Z = 0.1$; $Y = 0$, $Z = 0.2$; $Y = 0$, $Z = 0.3$), respectively; Figs. 5.12 and 5.13 show turbulent kinetic energy dissipation rate contour and variations of the turbulent kinetic energy dissipation rate in different lines ($Y = 0$, $Z = 0.05$; $Y = 0$, $Z = 0.1$; $Y = 0$, $Z = 0.2$; $Y = 0$, $Z = 0.3$), respectively. As shown in Figs. 5.10, 5.11, 5.12, and 5.13, the turbulent kinetic energy in the near-wall region of the tube close to the inlet is larger than that near the outlet and the inside of the pipe. So, the turbulent kinetic energy loss at the inner surface of the wall near the inlet is large. Both turbulent kinetic energy and turbulent kinetic energy dissipation rate show an axisymmetric distribution characteristic and reach the maximum at the inner surface of the wall near the inlet, which indicates that this domain has violently turbulence intensity and large energy consumption.

**Fig. 5.10** Turbulent kinetic energy contour

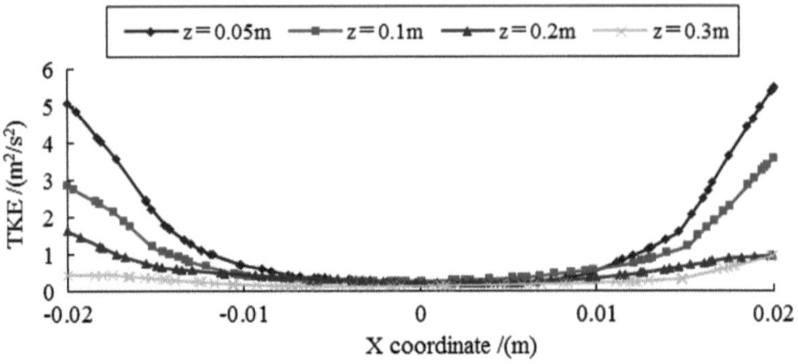


**Fig. 5.11** Variations of the radial turbulent kinetic energy in different lines ($Y = 0$, $Z = 0.05$; $Y = 0$, $Z = 0.1$; $Y = 0$, $Z = 0.2$; $Y = 0$, $Z = 0.3$)

**Fig. 5.12** Turbulent kinetic energy dissipation rate contour

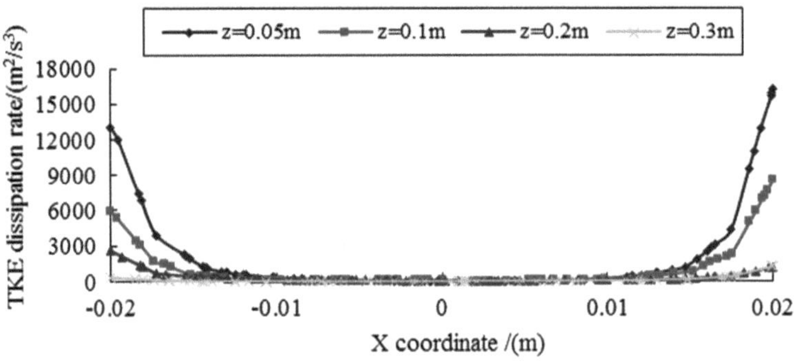


**Fig. 5.13** Variations of the turbulent kinetic energy dissipation rate in different lines. ($Y = 0$, $Z = 0.05$; $Y = 0$, $Z = 0.1$; $Y = 0$, $Z = 0.2$; $Y = 0$, $Z = 0.3$)

(3) Pressure

Main reasons for the change of pressure are: ① the kinetic energy loss caused by the gas rotation in the tube; ② the energy loss caused by the friction between the fluid particle and the wall.

Figures 5.14 and 5.15 show static pressure contour and variations of the static pressure in different lines ($Y = 0$, $Z = 0.05$; $Y = 0$, $Z = 0.1$; $Y = 0$, $Z = 0.2$; $Y = 0$, $Z = 0.3$), respectively; Figs. 5.16 and 5.17 show dynamic pressure contour and variations of the dynamic pressure in different lines ($Y = 0$, $Z = 0.05$; $Y = 0$, $Z = 0.1$; $Y = 0$, $Z = 0.2$; $Y = 0$, $Z = 0.3$), respectively.

As shown in Fig. 5.15, both the static pressure difference and the dynamic pressure difference decrease with the increase in the Z coordinate values in different lines. In addition, for each line, the dynamic pressure has a sharper decrease than static pressure. Both the static and dynamic pressures of the center vortex are lower than those of the inlet. It also can be shown that the static pressure and dynamic pressure change obviously at the inlet, while near the outlet, there is no obvious change for the static pressure and dynamic pressure. It is also shown in Fig. 5.14

**Fig. 5.14** Static pressure contour

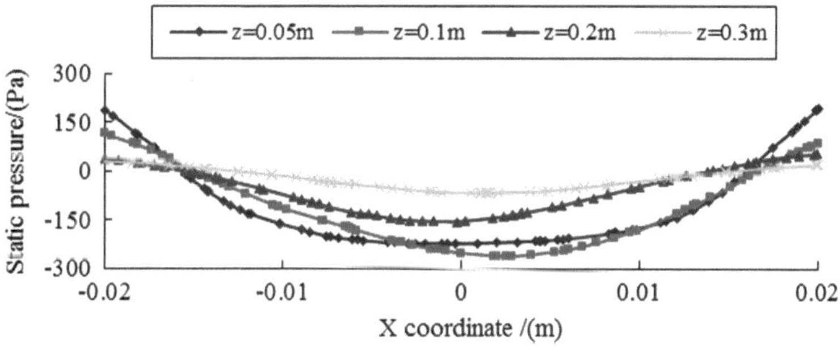


**Fig. 5.15** Variations of the static pressure in different lines. ($Y = 0$, $Z = 0.05$; $Y = 0$, $Z = 0.1$; $Y = 0$, $Z = 0.2$; $Y = 0$, $Z = 0.3$)

**Fig. 5.16** Dynamic pressure contour

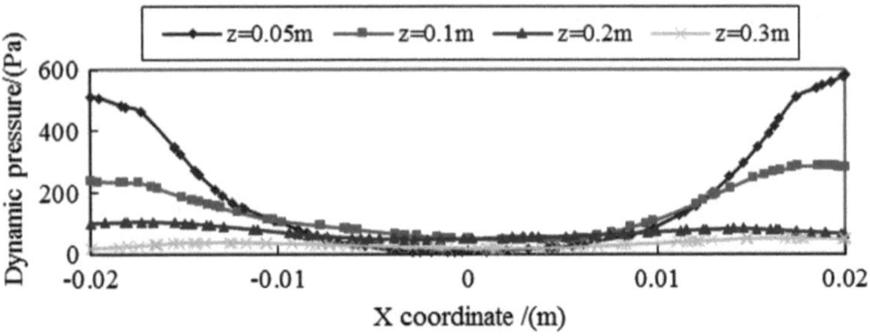


**Fig. 5.17** Variations of the dynamic pressure in different lines ($Y = 0$, $Z = 0.05$; $Y = 0$, $Z = 0.1$; $Y = 0$, $Z = 0.2$; $Y = 0$, $Z = 0.3$)

that the negative pressure occurs in the central part of the tube inlet, but the static pressure in a certain distance away from the inlet gradually increases until it reaches to outlet. Therefore, the reflux phenomenon is easy to appear.

The variations of the total pressure along the central axis and near the wall are shown in Fig. 5.18. It is shown in Fig. 5.18 that the total pressure near the wall has a sharper decrease at the inlet and then decreases slowly with the increase in the $Z$ coordinate value. Within the domain of $Z = 200$ mm to $Z = 400$ mm, the total pressure is close to zero and almost kept unchanged. In addition, the total pressure increases slowly with the increase in the $Z$ coordinate value along the central axis and reaches to 0 Pa at the outlet.

(4) Analysis of vortex intensity

Vortex intensity is used to characterize the flow field strength. The variations of the vortex intensity along the central axis are shown in Fig. 5.19. As shown in Fig. 5.19, the vortex intensity near the wall at the inlet is large and it decreases gradually along the axial direction. At the outlet, the outside gas has been inhaled and the vortex intensity increases. On the contrary, vortex intensity near the wall at the inlet is weaker. Reasons are as follows: The gas near the axis cannot be driven up when the gas flows into the tube, and when the gas flows into the entrance area, the vortex intensity decreases gradually.

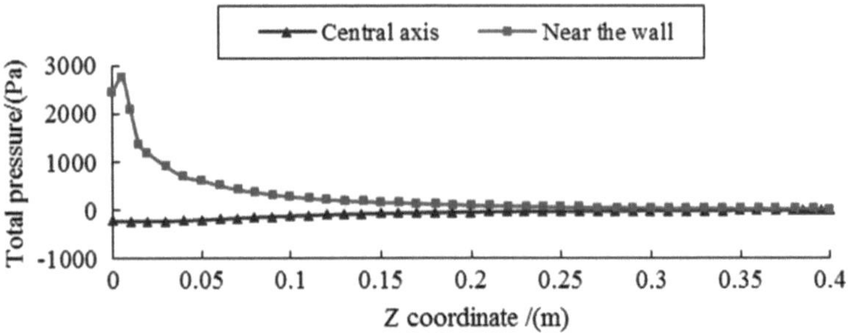


**Fig. 5.18** Variations of the total pressure in lines ($X = 0$, $Y = 0$; $X = 0$, $Y = 0.02$)

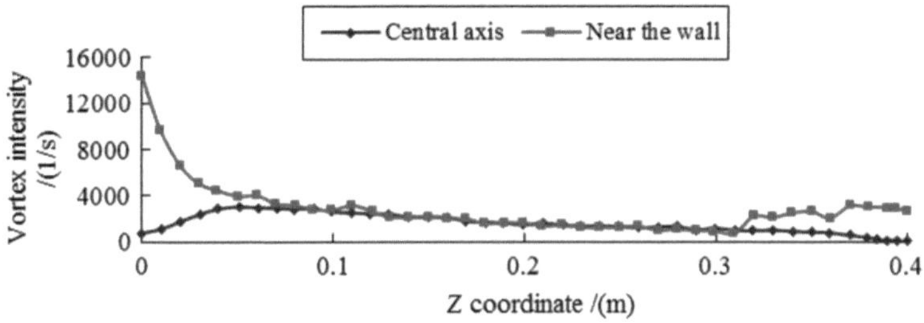


**Fig. 5.19** Variations of the vortex intensity along the central axis

(5) Comparison and analysis of the gas and the liquid

Under the same simulation conditions, it is found that the liquid and the gas have similar variation laws. Comparative analysis of the gas flow and liquid flow is carried out from the flow line, dynamic pressure, tangential velocity, and vortex intensity.

(1) Streamline

Figure 5.20 shows the swirling flow streamlines using gas and liquid as the working media.

As shown in Fig. 5.20, for both gas and liquid media, the abrasive particles can form a good swirling flow. The streamline pitch increases with the increase in the axial distance because of the fluid energy loss. The energy decreases with the increase in the axial distance. In addition, the time required for the fluid to complete a spiral is longer and the streamline pitch also becomes larger.

In addition, the streamline pitch of the liquid is smaller than that of the gas at the same position. For the tubular workpiece with the inner diameter of 40 mm, the energy generated by the liquid is larger than that of the gas at the same position and the time required for the fluid to complete one spiral is relatively short.

(2) Dynamic pressure

Figure 5.21 shows the dynamic pressure of the swirling flow using gas and liquid as the working media.

It is shown in Fig. 5.21 that the dynamic pressure using gas and liquid as the working media in the same cross section is basically symmetrical. When using liquid as the working medium, under the action of centrifugal force, most of the fluid is close to the wall and less fluid is located at the center of the tube, so the

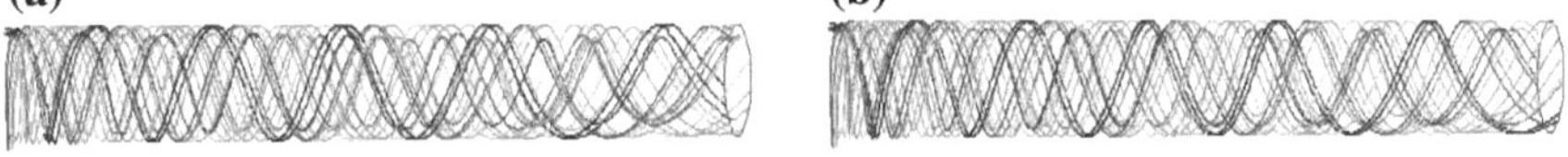


**Fig. 5.20** The swirling flow streamlines. **a** gas; **b** liquid

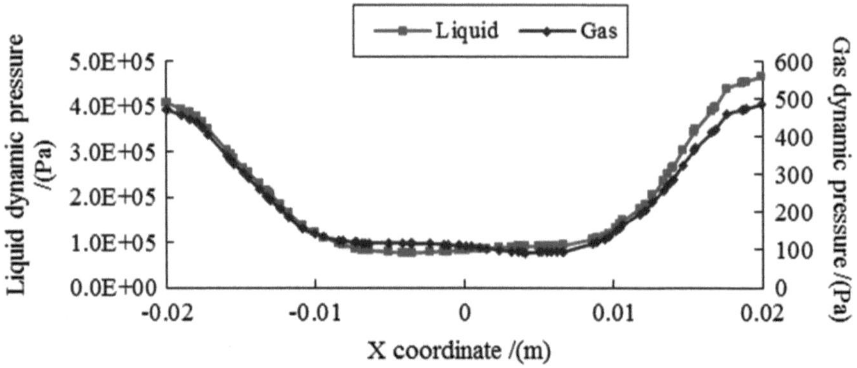


**Fig. 5.21** Dynamic pressure of the swirling flow

central dynamic pressure is very small. The energy of the liquid in the tube is larger than that of the gas. Although the energy loss of the liquid is much larger than that of the gas, the energy loss is not the main factor for the tube with an inner diameter of 40 mm. The energy of the liquid is larger than that of the gas at the same position. The dynamic pressure of the liquid is also greater than the dynamic pressure of the gas.

(3) Tangential velocity

Figure 5.22 shows the radial distribution of the tangential velocities using gas and liquid as working media.

It is shown in Fig. 5.22 that the tangential velocities of gas and liquid in the same cross section are almost symmetrical. The tangential velocities increase along the radial direction and rapidly reduce to zero at the wall. Due to the centrifugal force, the fluid mass increases along the radial direction; that is, the energy of the fluid increases along the radial direction, so the tangential velocity of the fluid increases. The tangential velocity near the inner hole surface of the tube is zero,

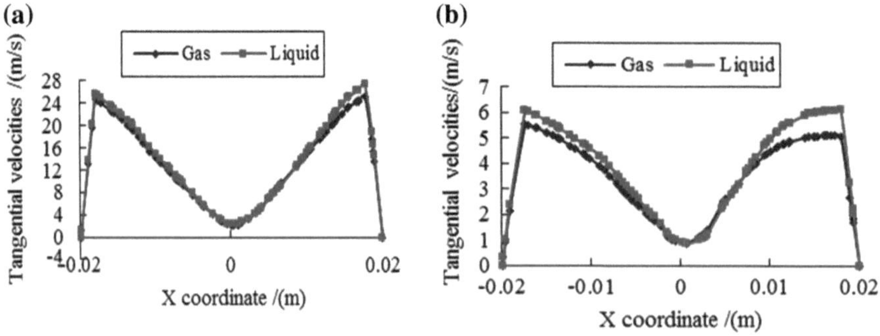


**Fig. 5.22** Radial distribution of tangential velocities. **a** $z = 0.05$ m; **b** $z = 0.3$ m

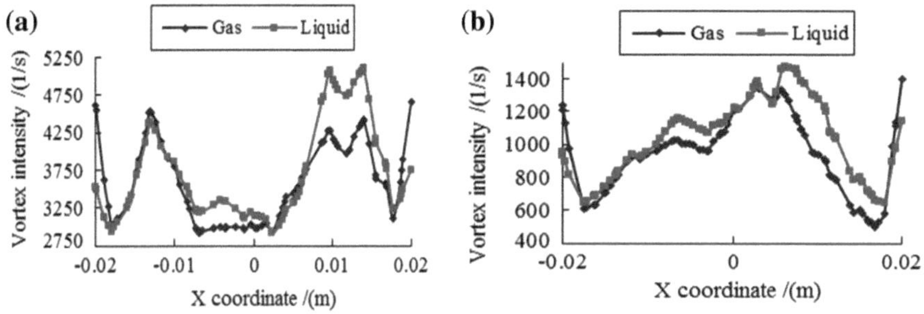


**Fig. 5.23** Radial distribution of the spiral vortex intensity. **a** $z = 0.05$ m; **b** $z = 0.3$ m

which is mainly due to the use of no-slip wall boundary conditions. In addition, the energy of the liquid at the same position is greater than that of the gas, and the energy loss is not the main factor. Therefore, the tangential velocity of the liquid is also greater than that of the gas.

(4) Vortex intensity

Figure 5.23 shows the radial distribution of the spiral vortex intensity of gas and liquid.

As shown in Fig. 5.23, the fluid in the tube forms a swirling flow, rather than a fixed axis rotation. Therefore, the vortex intensity is not completely symmetrical in the same cross section. As shown in Fig. 5.23a, b, the vortex intensity of gas and liquid changes more obviously at $z = 0.3$ m. In the same position, the tangential velocity of the liquid is larger than that of the gas; that is, the vortex effect of the liquid is bigger than that of the gas and the vortex intensity is slightly higher as well.

## 5.2.2 *Two-Phase Swirling Flow*

1. Introduction

Two-phase flow theory is the study of the movement of two phases that contain a complex mixture of states. The two-phase flow is generally defined as the movement which consists of two independent materials, and the two materials have changed interface. Two-phase flow may be plotted with three classes according to matter state: gas–solid two-phase flow; liquid–solid two-phase flow; gas-liquid two-phase flow [24].

An important feature of the two-phase flow is its non-equilibrium; namely, the two-phase interface has a velocity slip. Due to the velocity difference between the two phases, the two phases produce interaction to decrease the velocity difference. In the two-phase fluid dynamics, relaxation time is used to describe the non-equilibrium degree between the two phases. It also reflects the balancing process time [25].

This force balance equates the particle inertia with the forces acting on the particle and can be written (for the x direction in Cartesian coordinates) as:

$$\begin{cases} \frac{\mathrm{d}u_p}{\mathrm{d}t} = -\frac{u_p - u}{\tau_V} \\ \tau_V = \rho_p d_p^2 / 18\mu \end{cases} \tag{5.13}$$

where $u_\mathrm{p}$ is the particle velocity; $u$ is the fluid phase velocity; $\rho_\mathrm{p}$ is the density of the particle; $d_\mathrm{p}$ is the particle diameter; $\mu$ is the molecular viscosity of the fluid; $t$ is the time; $\tau_V$ is the velocity relaxation time.

If the initial particle velocity is zero, the particle velocity can be written as:

$$u_\mathrm{p} = u[1 - \exp(-t/\tau_\mathrm{V})] \tag{5.14}$$

It can be seen from Eq. 5.14 that the velocity difference $(v_p - v)$ decreases exponentially with the increase in the time. When $t = \tau_\mathrm{v}$, the velocity difference is equal to 1/e of the initial value; when $t = \tau_\mathrm{v}$, the velocity difference is equal to $1/\mathrm{e}^2$ of the initial value; when the time lean toward infinity, the velocity difference becomes zero. Relaxation time reflects the balancing process time between the particle velocity and the fluid velocity. Therefore, $\tau_\mathrm{v}$ is defined as a characteristic time to characterize the speed relaxation process, referred to the speed relaxation time.

2. Simulation and analysis

(1) Two-phase flow model

Based on the scale and the control body properties, two-phase flow models can be divided into three categories: continuum model, discrete particle model, and pseudoparticle model [26].

(1) Continuum Model

Continuum model is the most widely used method in the two-phase flow field. It treated the particles as fluid phase. Therefore, the continuum model is often referred to as "two-fluid model." In addition, fluids and particles are computed in Eulerian coordinates, so the corresponding mathematical methods are also termed as Eulerian Model. Typical mathematical models include k-ε model, Navier–Stokes promotion model, particle dynamics model, eddy kinetic model, coherent structure model. At present, the continuous medium model is usually used in the gas–solid two-phase study.

(2) Discrete Particle Model

In discrete particle model (DPM), particle is treated as discrete phase and gas is considered as a continuous phase. Not only the interactions between particles and gas, but also the interactions between the particles are taken into account. DPM can track the motion trajectory of the particles, so we also call it particle trajectory model. In addition, Lagrangian coordinate system and Euler coordinate system are used to descript the particle phase and the continuous phase, respectively, and the corresponding mathematics method for this model is called as Eulerian/Lagrangian model. The typical mathematical models also include the particle trajectory model, the particle stochastic model.

The substance of the particle trajectory model is to simulate the interparticle collision process by interparticle momentum exchange. Although the model is simple, it can reflect the real movement mechanism of the fluid–solid two-phase flow. In addition, the collision model theory is not complex. The resulting particle equation is ordinary differential equation and also easy to be solved. For the time-averaged and steady flow-solid two-phase flow coefficient, if only the fluid resistance acting on the particles is considered and other forces are neglected, the motion equation of the particle is:

$$m_{\mathrm{p}}\mathrm{d}u_{\mathrm{p}}/\mathrm{d}t = A_{\mathrm{p}}\rho_{\mathrm{p}}C_{\mathrm{D}}(u-u_{\mathrm{p}})\left|u-u_{\mathrm{p}}\right|/2 \tag{5.15}$$

$$\mathrm{d}x/\mathrm{d}t = u_{\mathrm{p}} \tag{5.16}$$

where $m_p$ is the mass of the particles (g); $\rho_p$ is the particles density (kg/m$^3$); $A_P$ is the particles windward area (m$^2$); $u$ *and* $u_{\mathrm{p}}$ are instantaneous velocities (m/s) of the fluid and particles, respectively; $C_{\mathrm{D}}$ is the resistance coefficient of particles.

(3) Pseudoparticle Model

The pseudoparticle model describes the motion characteristics in a single-particle scale. It not only treats macroscopically discrete particles as discrete phases, but also processes macro-continuous fluids using fluid micro-cells of the quasiparticle nature. By simulating the interaction between “gas particles” and solid particles, some classical phenomena and microscopic characteristics of two-phase flow can be described and reproduced. In the governing equations, both the movement of particles and the fluid are described in Lagrange coordinate system, so the corresponding mathematical method is “Lagrange method” (Lagrange model). At present, the more typical mathematical model is the gas “pseudoparticle” model.

Pseudoparticle model is still in the initial stage of development, and its numerical simulation for the flow field is also limited to some ideal cases, such as gas around a single cylinder, the gas around the double cylinder, and the drag between particles and fluid simulation. Therefore, this model needs to be further studied.

(2) Geometric model and boundary conditions

(1) Geometric model

Figure 5.1 shows the geometric model with pipe diameter of 40 mm and pipe length of 400 mm. In addition, the nozzles’ diameter and length are 4 mm and

16 mm, respectively. To form a swirling air flow and thus obtain better machining performance, the nozzles' angle $\alpha$ is equal to 10°, as shown in Fig. 5.2.

(2) Boundary Conditions

Gas-phase selection and parameter setting are the same as those in single-phase swirling flow simulation. The media material is corundum with density of 3900 kg/$m^3$ and diameter of 2 mm. Assuming that the media have the same diameter, the initial velocity of the media is set as zero and the media enter the tube under the drive of air flow. When the media move to the wall, the media are assumed to follow the principle of non-elastic collision to calculate the next position and velocity of media. As the media move to the outlet, the media escape from the outlet and its tracking stopped.

(3) Governing equation

Gas-phase selection and parameter setting are the same as those in single-phase swirling flow simulation. Stochastic trajectory model in the Lagrangian coordinate system was adopted to obtain the motion trajectory of the media.

When the media size is small and its concentration is low (the inlet media phase concentration is low, and it belongs to the dilute phase flow), the turbulence pulsation of the gas flow field has a significant effect on the media motion. And the media also have effects on the airflow, while the interaction between media could be neglected [27]. Therefore, stochastic trajectory model with phase coupling is used to consider the influence of media phase motion on gas-phase turbulence. In addition, the performance of swirling flow is analyzed by tracing the trajectory of media.

The stochastic trajectory model is a kind of particle orbital model, which takes into account the effects of particle motion on gas-phase turbulence and calculates the random trajectories and the orbital variations of the particles in the Lagrangian coordinate system. When the particle trajectory is complex, or when the particle phase concentration is very low, this model can better track the movement of particles.

(3) Simulation results and discussion

When the gas inlet velocity is 100 m/s, the simulation results are shown in Figs. 5.24, 5.25, 5.26, 5.27, 5.28, 5.29, 5.30, 5.31, 5.32, 5.33, 5.34, 5.35, 5.36, 5.37, 5.38, and 5.39. It is shown in Figs. 5.24, 5.25, 5.26, 5.27, 5.28, 5.29, 5.30, 5.31, 5.32, 5.33, 5.34, 5.35, 5.36, 5.37, 5.38, and 5.39 that under given conditions, particles can form a spiral movement under the driven spiral gas flow [28]. With the

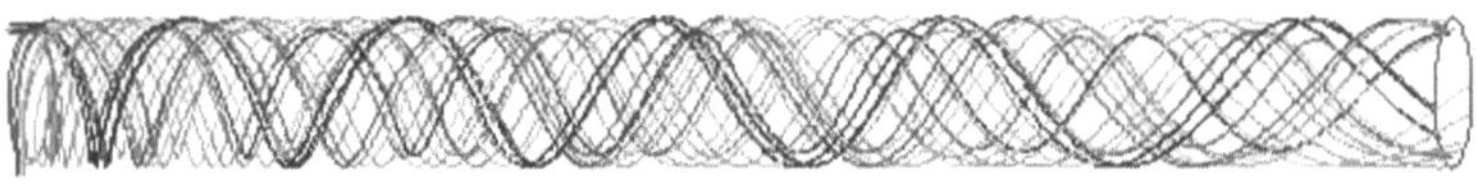

**Fig. 5.24** Streamline

**Fig. 5.25** Trajectory of abrasive particle

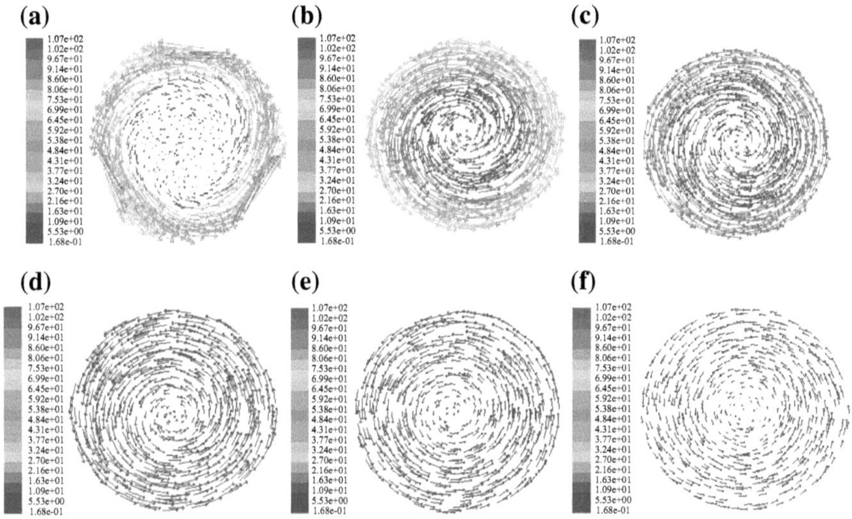


**Fig. 5.26** Vector of tangential velocity at different cross sections. **a** $z = 0$ m; **b** $z = 0.05$ m; **c** $z = 0.1$ m; **d** $z = 0.15$ m; **e** $z = 0.2$ m; **f** $z = 0.3$ m

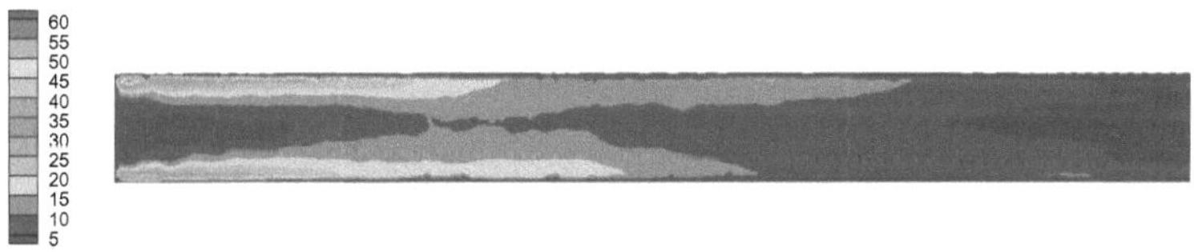


**Fig. 5.27** Velocity countour of longitudinal section ($Y = 0$)

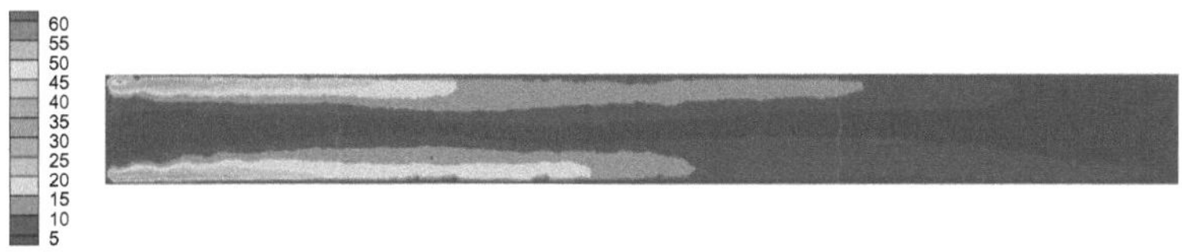


**Fig. 5.28** Tangential velocity countour

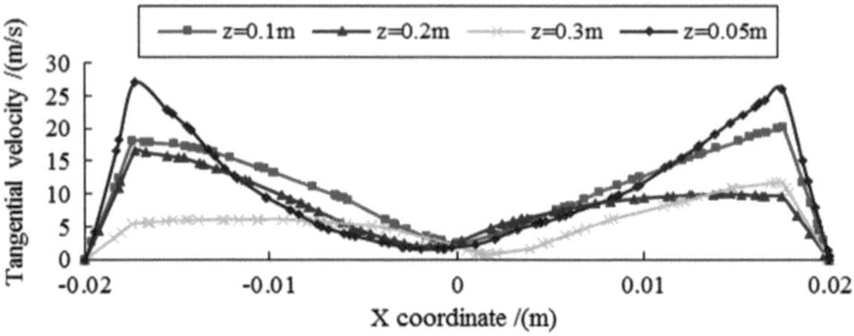


**Fig. 5.29** Variations of the tangential velocity in different lines ($Y = 0$, $Z = 0.05$; $Y = 0$, $Z = 0.1$; $Y = 0$, $Z = 0.2$; $Y = 0$, $Z = 0.3$)

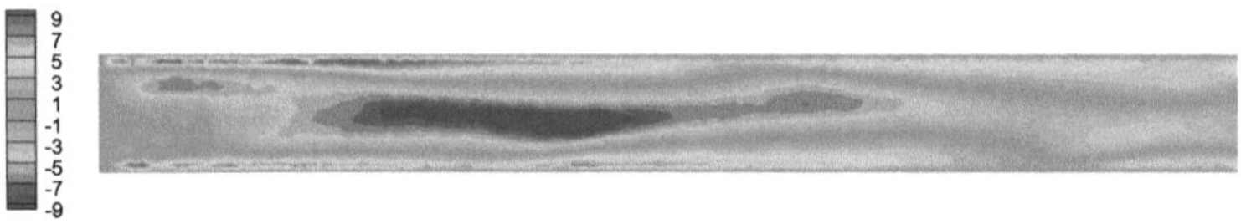


**Fig. 5.30** Axial velocity countour

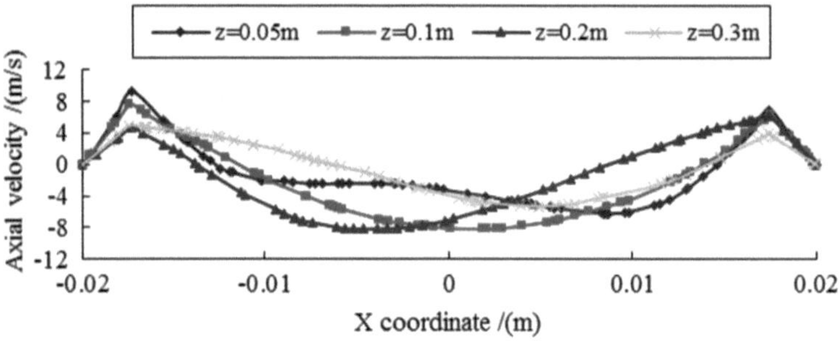


**Fig. 5.31** Variations of the axial velocity in different lines ($Y = 0$, $Z = 0.05$; $Y = 0$, $Z = 0.1$; $Y = 0$, $Z = 0.2$; $Y = 0$, $Z = 0.3$)

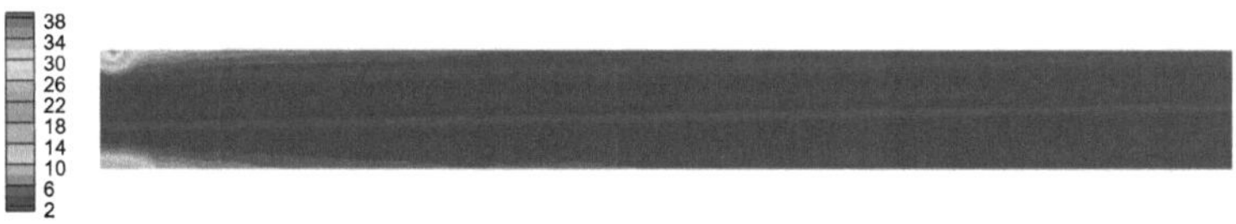


**Fig. 5.32** Turbulent kinetic energy contour

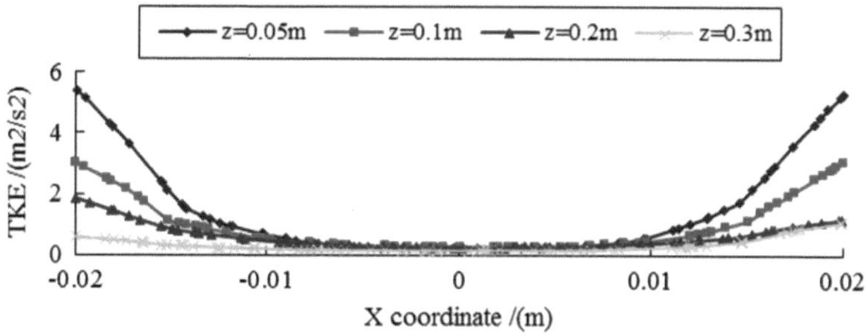


**Fig. 5.33** Variations of the radial turbulent kinetic energy in different lines ($Y = 0$, $Z = 0.05$; $Y = 0$, $Z = 0.1$; $Y = 0$, $Z = 0.2$; $Y = 0$, $Z = 0.3$)

**Fig. 5.34** Turbulent kinetic energy dissipation rate contour

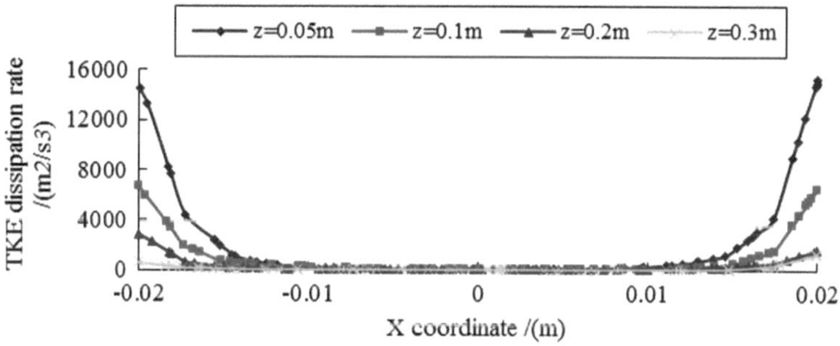


**Fig. 5.35** Variations of the turbulent kinetic energy dissipation rate in different lines ($Y = 0$, $Z = 0.05$; $Y = 0$, $Z = 0.1$; $Y = 0$, $Z = 0.2$; $Y = 0$, $Z = 0.3$)

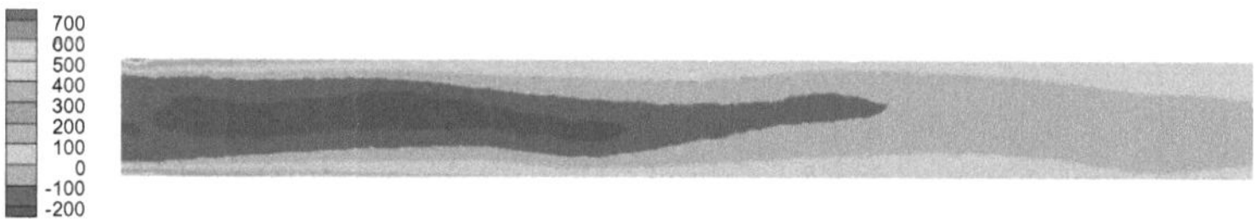


**Fig. 5.36** Static pressure contour

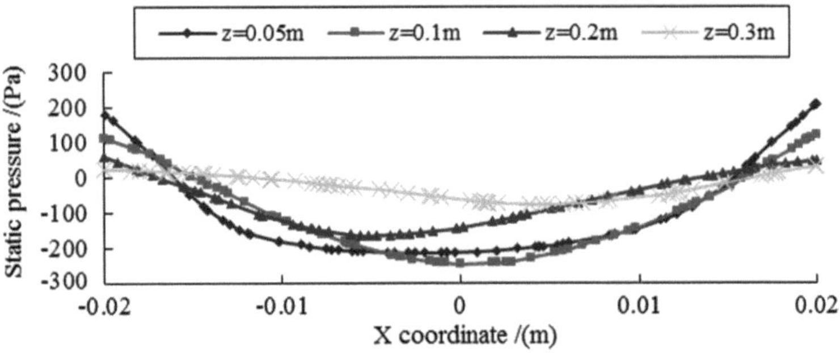


**Fig. 5.37** Variations of the static pressure in different lines ($Y = 0$, $Z = 0.05$; $Y = 0$, $Z = 0.1$; $Y = 0$, $Z = 0.2$; $Y = 0$, $Z = 0.3$)

**Fig. 5.38** Dynamic pressure contour

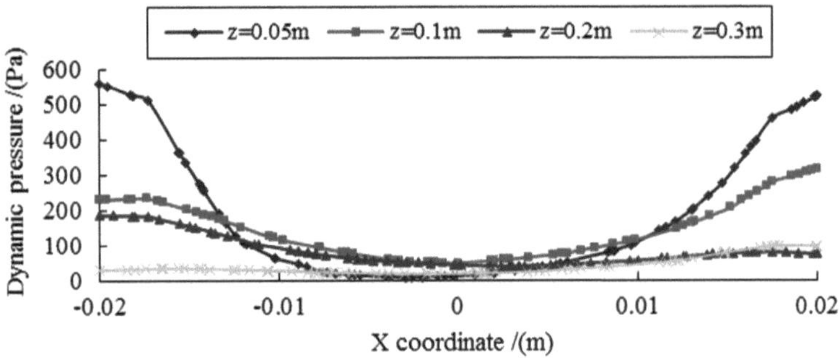


**Fig. 5.39** Variations of the dynamic pressure in different lines ($Y = 0$, $Z = 0.05$; $Y = 0$, $Z = 0.1$; $Y = 0$, $Z = 0.2$; $Y = 0$, $Z = 0.3$)

increases in axial distance, the tangential velocity, the axial velocity, and the vortex intensity decrease gradually. The velocity near the central axis and the velocity of the gas near the wall have a big difference. The tangential velocity and the axial velocity distributions along the radial direction show an obvious symmetry. The turbulent kinetic energy in the near-wall region at the tube inlet is large. And the turbulence energy at the outlet and in the tube is smaller, and the energy loss at the wall near the inlet is larger. The hydrostatic pressure and the dynamic pressure at the center of the vortex are lower than those at the inlet. Negative pressure is generated at the central region near the inlet; therefore, stagnation or reflow phenomenon occurs.

From the above analysis, it can be seen that the gas-particle two-phase swirling flow is consistent with that of gas single-phase swirling flow.

(4) Comparison of gas-phase swirling flow and gas-particle two-phase swirling flow

Under the same simulation conditions, dynamic pressure distribution along the axial direction, radial velocity distribution along the radial direction, and vortex intensity distribution along the axial direction are simulated for gas-phase swirling flow, and the gas-particle two-phase swirling flow. The simulation results are shown in Figs. 5.40, 5.41, and 5.42. It is shown in these figures that the tangential velocity along the radial direction is symmetrical and its magnitude is close for the two cases. The dynamic pressure and vortex intensity decrease gradually with the increase in the axial distance, and their change has the same tendency. The simulation results show that the gas-particle two-phase flow has the same flow field characteristics as the gas single-phase swirling flow. Under the gas swirling flow, the particles can form a certain helical motion. As the increases in the axial distance, the intensity of the swirling airflow gradually decreases.

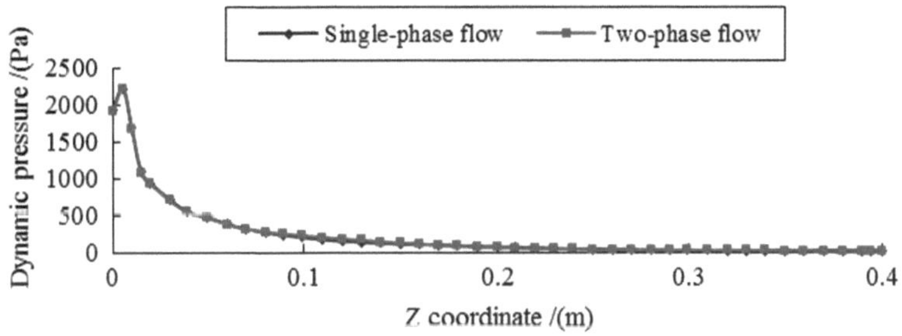


**Fig. 5.40** Comparison of dynamic pressure

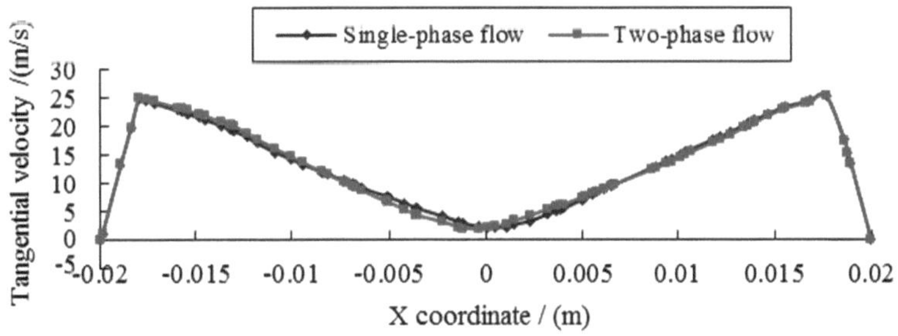


**Fig. 5.41** Comparison of tangential velocity

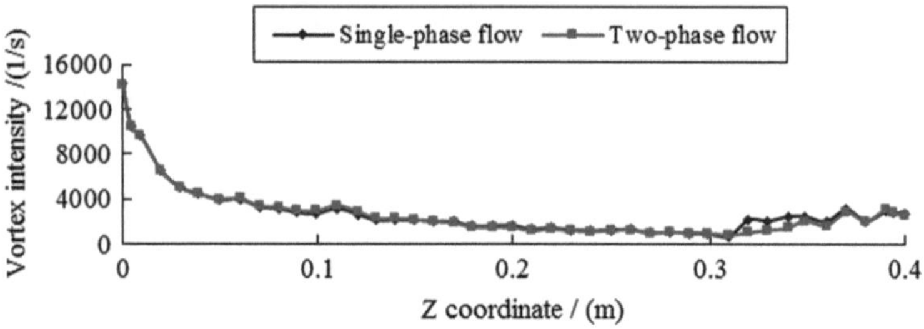


**Fig. 5.42** Comparison of vortex intensity

(5) Comparison of gas-particle and liquid-particle two-phase swirling flow

(1) Comparison of abrasive particle trajectory and streamline

Figures 5.43 and 5.44 show the particle trajectories and streamlines with the gas inlet velocity of 100 m/s, and the pressure outlet boundary, respectively. As shown in these figures, for both gas and liquid media, the abrasive particles can form a good swirling flow movement. When the gas is chosen as the working medium, the pitch of the particle trajectory is smaller than that of the particles with the liquid as the working medium under the same axial distance. In the liquid flow field, the spiral number of the particle trajectories is smaller than that in the gas flow field. This is mainly due to higher energy loss of liquid flow field.

(2) Tangential velocity comparison of the media

Figure 5.45 shows the tangential velocities of media for gas and liquid. It is shown in Fig. 5.45 that, at the inlet, liquid has a larger energy than gas. The tangential velocity of the media using liquid is larger than that using gas at the same position. But the energy loss is large for liquid due to the large viscosity of the liquid. The loss of the liquid's energy increases with the increase in the axial distance. When the axial distance increases to a certain value, the media tangential velocity in gas-particle two-phase swirling flow is bigger than that in liquid-particle two-phase swirling flow.

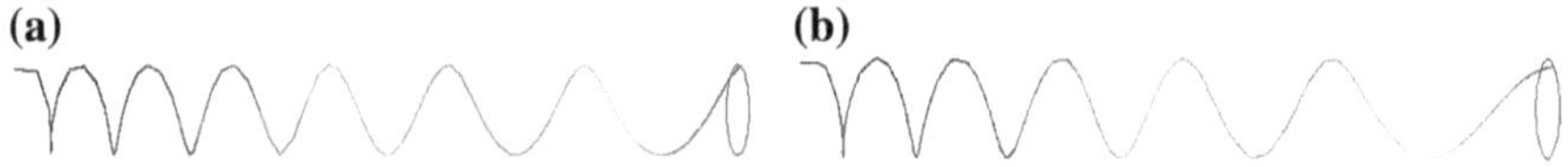


**Fig. 5.43** Particle trajectories. **a** Gas; **b** fluid

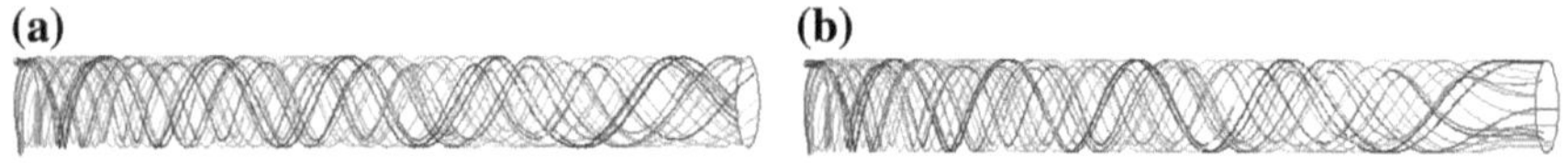


**Fig. 5.44** Particle streamlines. **a** Gas; **b** fluid

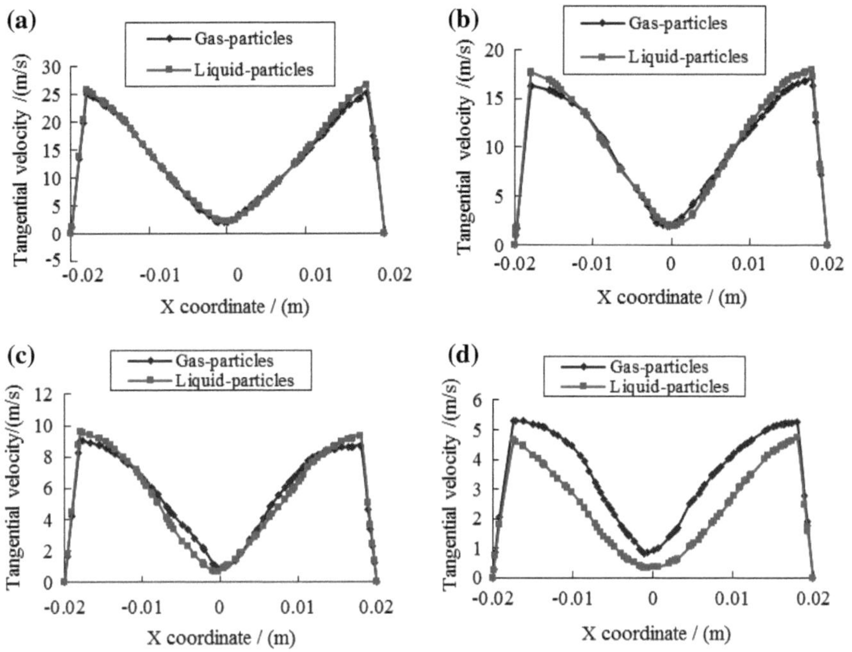


**Fig. 5.45** Variations of the media tangential velocity. **a** $z = 0.05$ m; **b** $z = 0.1$ m; **c** $z = 0.2$ m; **d** $z = 0.3$ m

(3) Comparison of dynamic pressure

Figure 5.46 shows the radial distributions of dynamic pressure for gas and liquid. As the liquid density is almost 1000 times of the gas, liquid dynamic pressure is significantly larger than the gas dynamic pressure. The simulation results show that the liquid dynamic pressure is approximately 1000 times of the gas dynamic pressure at the same location.

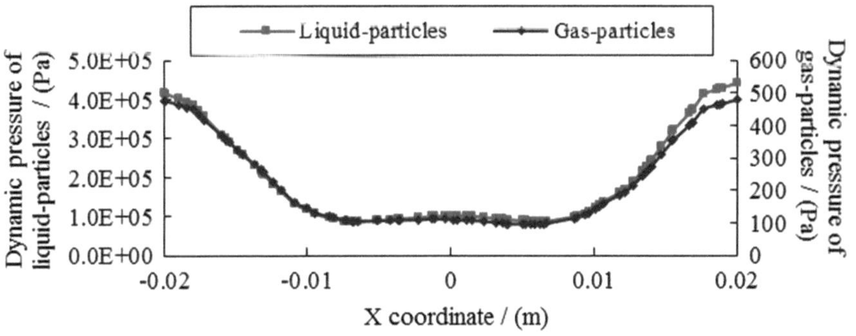


**Fig. 5.46** Variations of the dynamic pressure in line $Y = 0$, $Z = 0.05$ in gas and liquid

(4) Comparison of vortex intensity

Figure 5.47 shows the vortex intensity for gas and liquid media. It is shown in Fig. 5.47 that, at the inlet, liquid media have larger energy than gas media. The vortex intensity of the liquid media is larger than that of gas media at the same position. But the energy loss is large for liquid media due to the large viscosity of the liquid. With the increase in the axial distance, the energy loss of the liquid increases. When the axial distance increases to a certain extent, the vortex intensity of the gas is greater than that of the liquid.

In summary, whether liquid or gas is used as the working medium, stable spiral flow can be formed in the tubular workpiece. Therefore, the media can form spiral movement. The media and the workpiece have a relative motion and interaction force. Selecting reasonable processing parameters, surface finishing of the tubular workpiece can be achieved. However, the energy decay of the liquid-particle two-phase swirling flow is large. In addition, it requires the liquid to be filled in the tubular workpiece and complex experiment devices. So gas-particle two-phase swirling flow for finishing tubular workpiece is firstly studied.

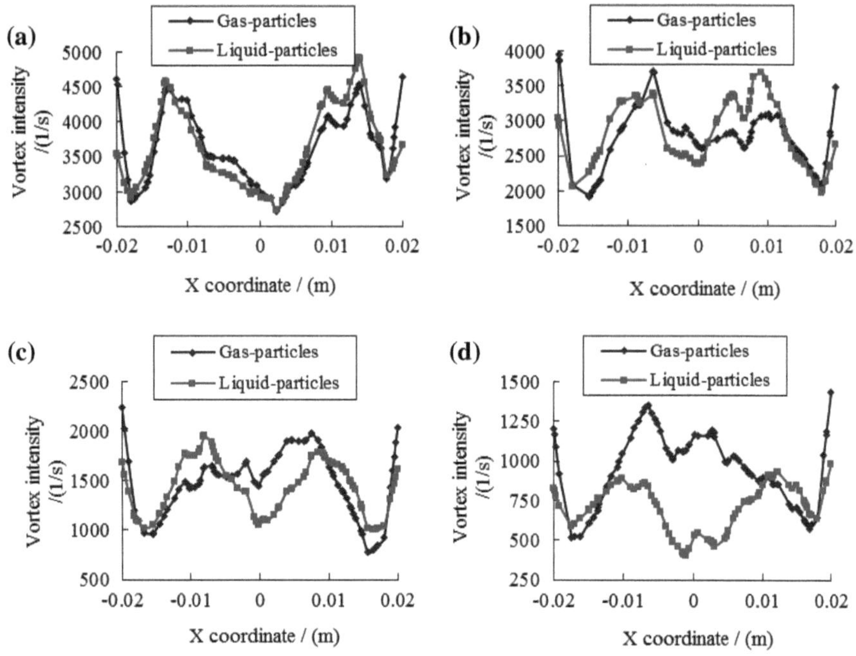


**Fig. 5.47** Variations of the vortex intensity using gas and liquid. **a** $z = 0.05$ m; **b** $z = 0.1$ m; **c** $z = 0.2$ m; **d** $z = 0.3$ m

### 5.2.3 Simulation of Gas-Particle Two-Phase Swirling Flow

For the workpiece with 32 mm hole diameter, the inlet velocity is 330 m/s and the outlet is set as a "pressure outlet." The effects of process parameters on the characteristics of the gas-particle two-phase swirling flow are obtained by numerical simulation.

1. Angle of nozzle

Figures 5.48, 5.49, 5.50, and 5.51 show simulation results of the media trajectories, dynamic pressure, tangential velocity, and vortex intensity under different nozzle angles.

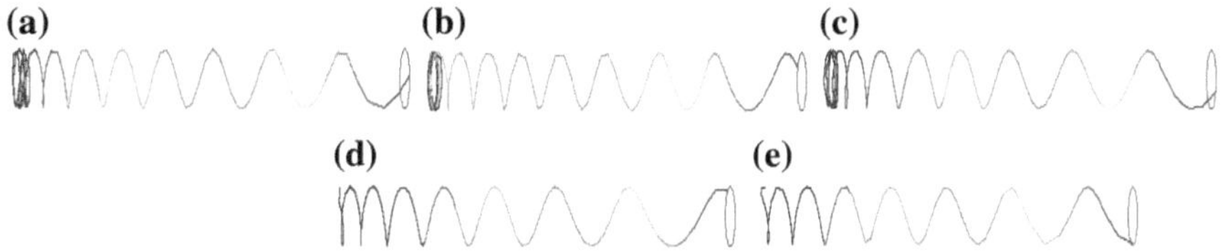


**Fig. 5.48** Media trajectories under different nozzle angles. **a** 5°; **b** 10°; **c** 15°; **d** 20°; **e** 30°

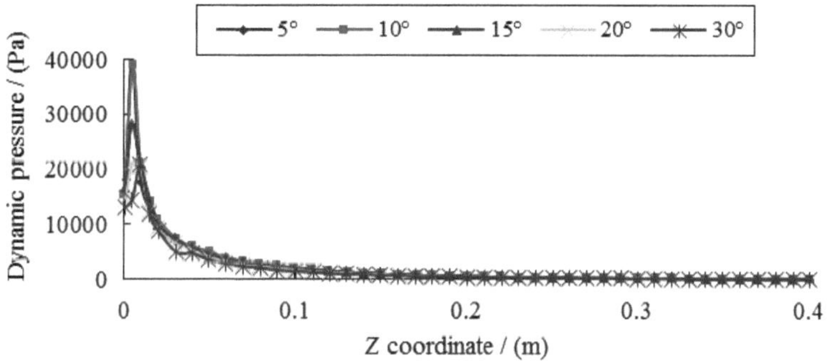


**Fig. 5.49** Dynamic pressure under different nozzle angles

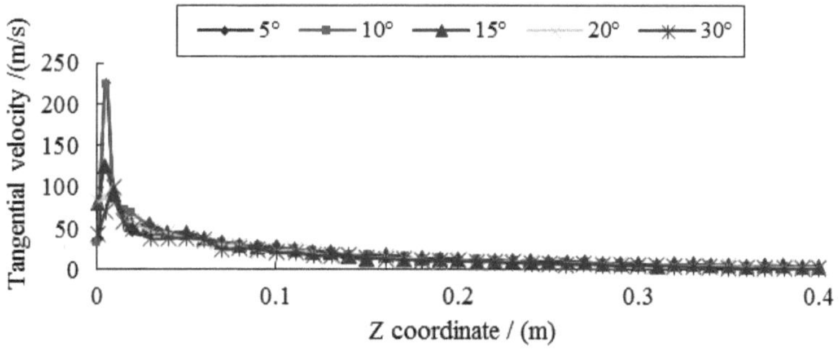


**Fig. 5.50** Tangential velocity under different nozzle angles

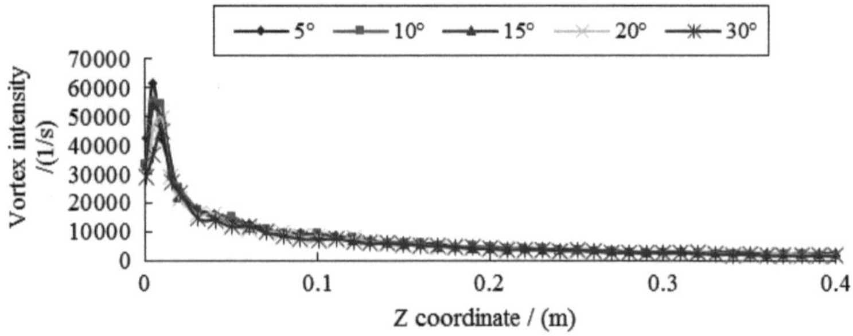


**Fig. 5.51** Vortex intensity under different nozzle angles

The media trajectories show that swirling flow is formed in the hole, and the pitch of the media trajectories increases with the increase in nozzle angle. Within about 120 mm of the axial distance from the inlet, dynamic pressure, tangential velocity, and vortex intensity along the axial direction decrease with the increase in the nozzle angle and the Z coordinate values.

The simulation results show that the gas-particle two-phase swirling flow finishing has processing ability in certain area of the flow field. In order to achieve better flow field characteristics and improve the processing ability, the desirable nozzle angle should be ranged from 5° to 10°.

2. Nozzle diameter

With the increase in nozzle diameter, the inlet velocity decreases under constant air source pressure. Figures 5.52, 5.53, 5.54, and 5.55 show simulation results of the particle trajectories, dynamic pressure, tangential velocity, and vortex intensity under different nozzle diameters.

Figures 5.56 and 5.57 show tangential velocity and vortex intensity under different nozzle diameters and 40 mm diameter hole.

As shown in Figs. 5.52, 5.53, 5.54, and 5.55, the pitch of the particle trajectories increases with the increase in the nozzle diameter. Within about 130 mm of the axial distance away from the inlet, dynamic pressure, tangential velocity, and vortex intensity along the axial direction decrease with the increase in the nozzle diameter. In addition, the particle trajectories, dynamic pressure, tangential velocity are relatively close under lower initial velocity, and 4 mm diameter nozzle, and 3 mm diameter nozzle. Due to higher energy attenuation, their variety is near and irregular as the axial distance increases.

**Fig. 5.52** Particle trajectories under different nozzle diameters. **a** 2 mm; **b** 3 mm; **c** 4 mm

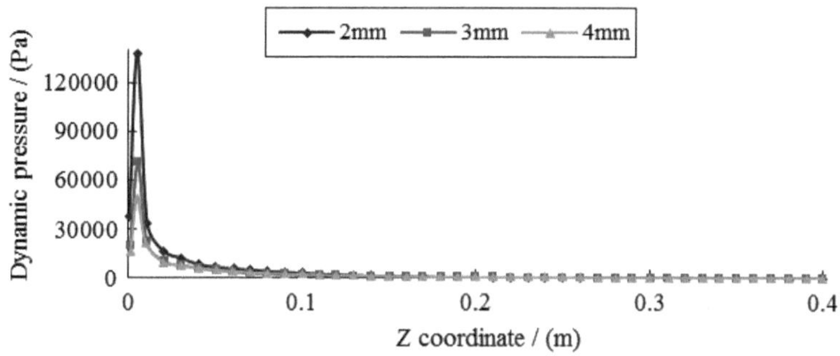


**Fig. 5.53** Variations of dynamic pressure under different nozzle diameters

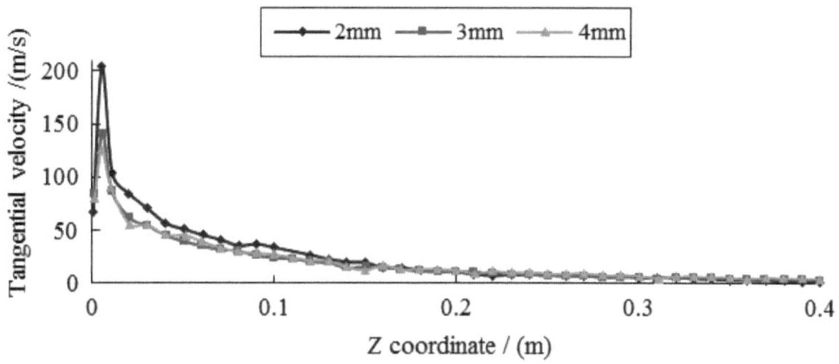


**Fig. 5.54** Variations of tangential velocity under different nozzle diameters

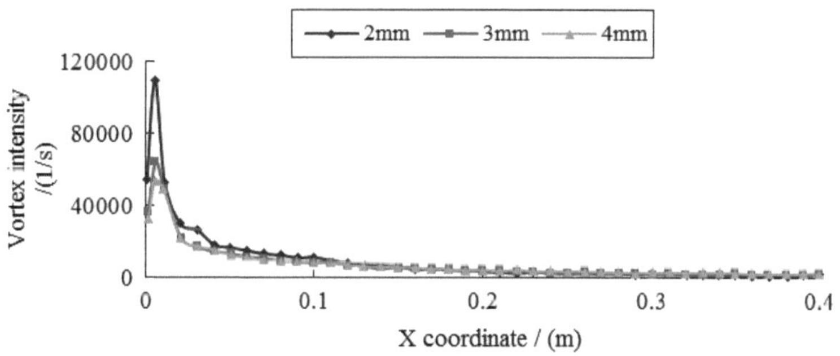


**Fig. 5.55** Variations of vortex intensity under different nozzle diameters

As shown in Figs. 5.56 and 5.57, tangential velocity and vortex intensity decrease with the increase in nozzle diameter. Therefore, the nozzle with small diameter should be selected for improving tangential velocity and vortex intensity. And when the diameter of the tube workpiece increases, smaller diameter nozzle should be used. In order to meet the requirement posed by the variation in hole

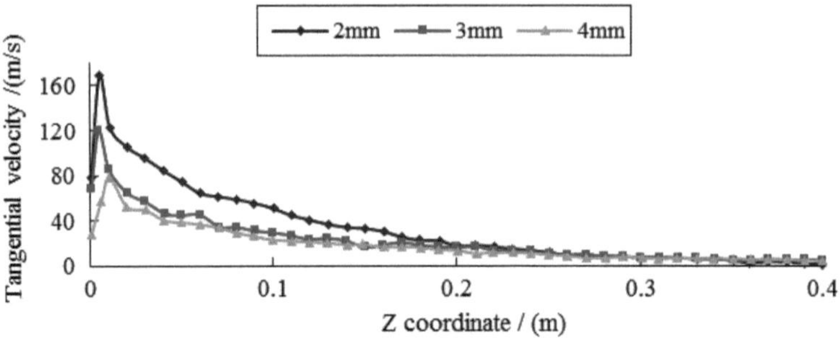


**Fig. 5.56** Variations of tangential velocity under different nozzle diameters

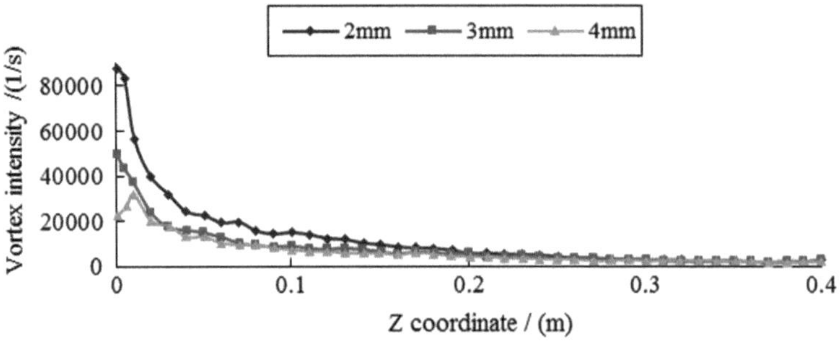


**Fig. 5.57** Variations of vortex intensity under different nozzle diameters

diameter of the workpiece, the nozzle diameter of about 2 mm is generally desirable for improving the flow field characteristics.

3. Nozzle quantity

The inlet velocity decreases with the increase in nozzle quantity under constant air source pressure. Simulation results show that the pitch of the particle trajectories increases and the inlet velocity decreases with the increase in the nozzle quantity [29]. Within about 150 mm of the axial distance, dynamic pressure, tangential velocity, and vortex intensity along the axial direction decrease with the increase in the nozzle quantity.

The smaller the nozzles number is, the larger the inlet kinetic energy will be. The large kinetic energy at the inlet leads to strong shock effect, but it has a sharper decrease in the axial direction. Therefore, it has little significance to internal hole surface finishing. In summary, three nozzles can be selected for industrial applications.

4. Particle diameter

The particle diameter is set to 1, 3, and 5 mm. Simulation results show that the increase in the particle diameter leads to decrease in the inlet velocities, dynamic pressure, vortex intensity, and tangential velocity, but this decrease is not obvious.

In addition, it also causes the increase in the pitch of the particle trajectories. For industrial applications, to improve the machining performance of the inner hole surface, the particle diameter is generally chosen from 3 to 5 mm.

5. Nozzle offset

The nozzle offset is the axis distance between the nozzle and the swirling head. When the distance is equal to the radius difference between the swirling head inner hole and the nozzle, the nozzle offset is zero.

The nozzle offset is set to 0, 1, 2, and 3 mm. Simulation results show that when the offset is 0 mm, the pitch of the particle trajectories is slightly bigger than those when the offsets are 1, 2, and 3 mm. This is because the gas flows from the nozzle directly reach to the inner surface of swirling head, which leads to high energy attenuation. When the offset is large, the spiral effect of the airflow is poor. When the distance is 3 mm, the dynamic pressure, vortex intensity, and tangential velocity of the particle decrease than those when the offsets are 1, 2, and 3 mm. So the nozzle offset is generally chosen from 1 to 2 mm in industrial applications.

6. Air source pressure

Under the different inlet velocities, simulation results show that with the increase in nozzle inlet velocity, the pitch of the particle trajectories decreases, and the dynamic pressure, vortex intensity, and tangential velocity of the particle increase correspondingly. To improve the vortex intensity and tangential velocity, the bigger inlet velocity should be selected. Therefore, the air source pressure is generally chosen from 0.4 to 0.6 MPa in industry application.

## 5.3 Finishing Mechanism of Gas-Particle Swirling Flow

Figure 5.58 shows the principle of the gas-particle two-phase swirling flow finishing. The high-pressure air from air pump is accelerated to subsonic airflow by convergent nozzle. The nozzles are tangent to the inside wall of swirling head and

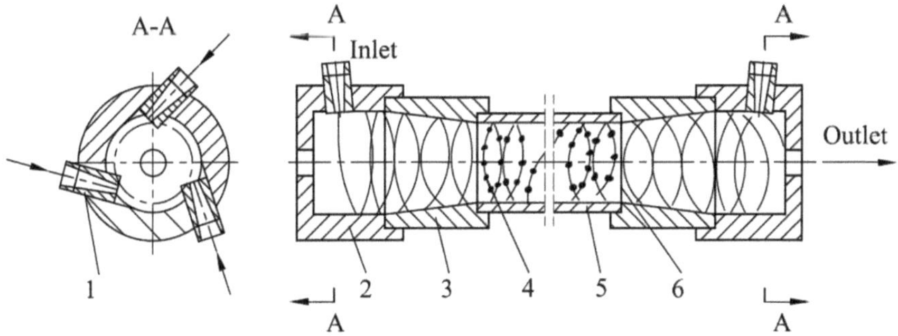


**Fig. 5.58** Schematic diagram of Gas-particles two-phase swirling flow finishing. 1—Nozzle; 2—swirling head; 3—union joint; 4—abrasive particles; 5—workpiece; 6—separation net

at an angle with the cross section of swirling head, so the air flow moves spirally in swirling head and the hole, which makes particles revolve at high speed, forming eccentric pressure. This eccentric pressure expels particles to make scraping and scoring the hole surface. Under the action of force and motion, the particle has rolling action, a minimum grinding, and scraping and scoring against the hole surface, and thereby, surface finishing is realized.

## *5.3.1 Formation of Flow Field*

1. Formation of swirling flow field

The high-pressure air from air pump is accelerated to subsonic airflow by convergent nozzle. The airflow moves spirally in swirling head to form the swirling flow field [30, 31].

Air with constant flow rate and constant pressure flows through the convergent nozzle and then becomes subsonic airflow. The direction of the subsonic airflow is not only along the tangential direction of the inside wall of swirling head but also has an angle $\alpha$ with the cross section of swirling head. The airflow velocity can be decomposed into radial velocity $u_{\mathrm{r}}$, tangential velocity $u_{\theta}$, and axial velocity $u_{\mathrm{z}}$.

The gas can rotate along the inner wall of the swirling head, and the centrifugal force can hinder the radial movement of the airflow. So it can be assumed that the airflow do not have radial direction movement, that is, $u_{\mathrm{r}} = 0$.

In the axial direction of the swirling head, the gas flow rotates along the inner wall of the swirling head. The circular motion of the gas flow is a free vortex motion, and the total water head is kept constant along the radial direction. Because of the axial velocity $u_{\mathrm{z}}$, the gas flow also advances along the axis direction; the synthesis of the axis direction and the circular motion makes the flow line in the vortex head a helix [32].

2. Velocity and pressure

According to the fluid dynamic theory, mathematical model of gas flow in the swirling head is established. To build the model, the gas flow is usually divided into two parts: the viscous flow located in the boundary layer and the ideal flow located out of the boundary layer [33–36]. The established model is only suitable for the ideal fluid flow located out of the boundary layer. A cylindrical coordinate system is built in the swirling head. The Z-axis of cylindrical coordinate system is the center axis of the tube. The cross section which is perpendicular to the Z-axis is selected as the polar coordinate. The continuity equation is:

$$\frac{\partial \rho}{\partial t} + \frac{1}{r}\frac{\partial}{\partial r}(\rho r u_{\mathrm{r}}) + \frac{1}{r}\frac{\partial(\rho u_{\theta})}{\partial \theta} + \frac{\partial(\rho u_{\mathrm{z}})}{\partial z} = 0 \tag{5.17}$$

Momentum equation is:

$$\begin{aligned}&\rho\left(\frac{\partial u_r}{\partial t}+u_r\frac{\partial u_r}{\partial r}+\frac{u_\theta}{r}\frac{\partial u_r}{\partial \theta}+u_z\frac{\partial u_z}{\partial z}-\frac{u_\theta^2}{r}\right)\\&=\rho f_r+\frac{1}{r}\left[\frac{\partial(r\tau_{rr})}{\partial r}+\frac{\partial \tau_{\theta r}}{\partial \theta}+\frac{\partial(r\tau_{zr})}{\partial z}\right]-\frac{\tau_{\theta\theta}}{r}\end{aligned}\tag{5.18}$$

$$\begin{aligned}&\rho\left(\frac{\partial u_\theta}{\partial t}+u_r\frac{\partial u_\theta}{\partial r}+\frac{u_\theta}{r}\frac{\partial u_\theta}{\partial \theta}+u_z\frac{\partial u_z}{\partial z}+\frac{u_\theta u_r}{r}\right)\\&=\rho f_\theta+\frac{1}{r}\left[\frac{\partial(r\tau_{r\theta})}{\partial r}+\frac{\partial \tau_{\theta\theta}}{\partial \theta}+\frac{\partial(r\tau_{z\theta})}{\partial z}\right]+\frac{\tau_{r\theta}}{r}\end{aligned}\tag{5.19}$$

$$\begin{aligned}&\rho\left(\frac{\partial u_z}{\partial t}+u_r\frac{\partial u_z}{\partial r}+\frac{u_\theta}{r}\frac{\partial u_z}{\partial \theta}+u_z\frac{\partial u_z}{\partial z}\right)\\&=\rho f_z+\frac{1}{r}\left[\frac{\partial(r\tau_{rz})}{\partial r}+\frac{\partial \tau_{\theta z}}{\partial \theta}+\frac{\partial(r\tau_{zz})}{\partial z}\right]\end{aligned}\tag{5.20}$$

where $\tau_{rr}$, $\tau_{\theta\theta}$, $\tau_{zz}$, $\tau_{r\theta}$, $\tau_{\theta z}$, $\tau_{zr}$ are stress components of the gas parcels.

When $\mu = 0$ and ignoring mass force that is $\upsilon_r = 0$, the airflow is steady flow. Therefore, $\tau_{rr} = \tau_{\theta\theta} = \tau_{zz} = p$, $\tau_{r\theta} = \tau_{\theta z} = \tau_{zr} = 0$, $f_r = f_\theta = f_z = 0$. Gas pressure and tangential velocity along the radial distribution are what we need to concern. According to the above conditions, from Eq. (5.18), Eq. (5.21) can be obtained:

$$\frac{\partial p}{\partial r}=\rho\frac{u_\theta^2}{r}\tag{5.21}$$

From Eq. (5.21) and gas state equation, Eq. (5.22) can be obtained:

$$\frac{p}{\rho}=RT\tag{5.22}$$

where $R$ is the gas constant (J/(kg K)); $T$ is absolute temperature (K).

From Eqs. (5.21) and (5.22), we can obtain:

$$\frac{\partial p}{\partial r}=\frac{p}{RT}\frac{u_\theta^2}{r}\tag{5.23}$$

Since the gas is assumed as ideal gas, the flow cross section is an isothermal surface. When the heat transfer between the gas and the airflow in the middle part of the swirling head reaches balance state, the flow field can be regarded as an isothermal field. According to the above analysis, it is assumed that the gas temperature in the airflow and the forced vortex is a constant value. Therefore, gas energy equation is:

$$H = h(T) + \frac{u_r^2 + u_\theta^2 + u_z^2}{2} \quad (5.24)$$

where $H$ is the energy of the gas (J) and $h$ is gas enthalpy (is the function of $T$ for the ideal gas).

When the flow line radius increases from $r$ to $r + dr$, the whole energy $H$ of the gas is:

$$\frac{dH}{dr} = \frac{dh(T)}{dr} + u_r \frac{du_r}{dr} + u_\theta \frac{du_\theta}{dr} + u_z \frac{du_z}{dr} \quad (5.25)$$

According to the characteristics of free vortex motion, the whole energy $H$ of the gas is constant along radial direction. So:

$$\frac{dH}{dr} = u_\theta \frac{du_\theta}{dr} = 0 \quad (5.26)$$

Because $u_\theta$ is not equal to zero, we can obtain:

$$\frac{du_\theta}{dr} = 0 \quad (5.27)$$

That is, tangential velocity $u_\theta$ of airflow is a constant value along the radial direction. From Eq. (5.23), pressure distribution of airflow along the radial direction is:

$$p = c_1 r^{u_\theta^2/RT} \quad (5.28)$$

where $c_1$ is the constant, which can be obtained from the boundary condition.

When the average pressure in the swirling head center is lower than 0.1 MPa, that is, the negative pressure formed in the center of the vortex head, the abrasive particles can go into the swirling head easily. When the average pressure in the swirling head center is larger than 0.1 MPa, it means that the swirling head center has a positive pressure, so supernumerary air pressure should be supplied to drive the abrasive particles into the swirling head.

### 5.3.2 Force Analysis

1. Force analysis of the abrasive particle

According to the force analysis of particles, forces acting on the abrasive particles can be divided into the following categories: Forces which have nothing to do with the relative motion between fluid and particles include the gravity of particles, the pressure difference force; forces which depend on the relative motion between fluid and particles and are in the direction along the relative motion include thrust, resistance, the additional body force, Basset force; forces which depend on the relative motion between fluid and particles and are in the direction of perpendicular to the direction of relative movement include lateral force, lift force, Magnus force, Saffman force [37, 38].

(1) Thrust force

$$F = \frac{1}{8}\pi C_{\mathrm{D}} d_{\mathrm{p}}^{2} \rho |u - u_{\mathrm{p}}|(u - u_{\mathrm{p}}) \tag{5.29}$$

(2) Pressure difference force

$$F_{\mathrm{D}} = -\frac{1}{6}\pi d_{\mathrm{p}}^{3} \frac{\mathrm{d}p}{\mathrm{d}x} \tag{5.30}$$

where d$p$/d$x$ is the pressure gradient.

(3) Gravity of particles

$$G = \frac{1}{6}\pi d_{\mathrm{p}}^{3} \rho_{\mathrm{p}} g \tag{5.31}$$

(4) Additional mass force

$$F_{\mathrm{III}} - -\frac{1}{12}\pi d_{\mathrm{p}}^{3} \rho \left( \frac{\mathrm{d}u_{\mathrm{p}}}{\mathrm{d}t} - \frac{\mathrm{d}u}{\mathrm{d}t} \right) \tag{5.32}$$

Unlike particle movements in vacuum, the accelerated particle in the fluid will accelerate the surrounding fluid. From fluid mechanics theory, this accelerating effect is equivalent to the additional mass of abrasive particle. For spherical particle, the additional mass is equal to half of fluid mass displaced by the spherical particle.

(5) Basset force

Basset force refers to a kind of resistance of abrasive particle. When the speed of the abrasive particle changes due to the fluid viscosity, the abrasive particle has relative acceleration. So the flow field around abrasive particle cannot achieve stability state. Therefore, fluid force applied against the abrasive particle not only depends on relative velocity and relative acceleration of the abrasive particles, but also depends on the former acceleration. The Basset force is important only at the early stage of accelerated motion. When the speed of the particle tends to stability, Basset force can be ignored.

(6) Magnus force

If the abrasive particle rotates with angular velocity $\omega$ and the rotation axis is perpendicular to the relative velocity $\left(u - u_{\mathrm{p}}\right)$, the abrasive particle is not only subjected to axis resistance, but also a lateral force, which is perpendicular to the relative velocity $\left(u - u_{\mathrm{p}}\right)$ and the rotation axis. This lateral force is called Magnus force. For the abrasive particles with lower rotation speed, the Magnus force is much smaller than the Stokes resistance, so it can be omitted.

(7) Saffman force

The non-uniform flow field leads to velocity gradient. Therefore, abrasive particle will be subject to an additional lateral force, which is called Saffman. Due to the larger diameter of the abrasive particle, velocity gradient only occurs at the area near the wall. Saffman has little effects on the abrasive particle and can be ignored.

(8) Lift force

$$F_{\mathrm{L}} = \frac{1}{8}\pi d_{\mathrm{p}}^{2}\rho C_{\mathrm{L}}\left(u - u_{\mathrm{p}}\right)^{2} \tag{5.33}$$

where $C_{\mathrm{L}}$ is lift coefficient of spherical abrasive particle. The lift force is zero because $C_{\mathrm{L}}$ is zero.

2. Normal force

When the abrasive particle enters into the swirling head, its velocity can reach a certain value in a very short time by the acceleration of the subsonic gas flow. The abrasive particle rotates along the inner hole surface and is subjected to 5 forces: the thrust which has same direction with the abrasive particle movement direction; the machining resistance; the normal force of the abrasive particle exerted by the inner hole surface; the centrifugal force generated by the high-speed rotation of the abrasive particle; the radial pressure difference, which is also called the buoyancy, caused by the pressure and the limited flow space. The resultant force of $F$ and $f_{\mathrm{N}}$ determines the movement state of the abrasive particle. When $F > f_{\mathrm{N}}$, the particle is accelerated; when $F = f_{\mathrm{N}}$, the abrasive particle moves at a constant velocity; when $F < f_{\mathrm{N}}$, the particle is decelerated.

When the abrasive particle is moving in a viscous gas, abrasive particle is subjected to forces which have the same direction with the gas flow direction. Customarily, these forces are referred to as thrust $F$ and can be calculated by Eq. (5.29).

The abrasive particle with a certain amount of kinetic energy will rotate along the inner surface by the centrifugal force and buoyancy force; as a result, the micro-grinding and rolling to workpiece can be realized. The resistance to the abrasive particle by cutting and sliding friction is:

$$f_N = F_N \cdot f \tag{5.34}$$

where $f$ is the resistance coefficient.

The abrasive particle rotates along the inner hole surface, and the normal force is balanced with the difference between the centrifugal force and the buoyancy force, That is:

$$F_N = F_C - F_D \tag{5.35}$$

$$F_C = \frac{4}{3}\pi\rho_p r_p^3 \frac{(u_p \cdot \cos\alpha)^2}{r_0 - r_p} \tag{5.36}$$

where $\rho_p$ is the density of the abrasive particle; $r_p$ is the radius of the abrasive particle; $r_0$ is the radius of inner hole; $\alpha$ is the lead angle of the abrasive particle helical motion.

The radial pressure differential force is made up of two parts: the force pointed to the workpiece axial center and generated by the pressure difference of the air flow; the other force pointed to the workpiece surface and generated by finite nature of air flow field. Because of high speed of air, the centrifugal acceleration is 100 times bigger than the gravity acceleration, but the centrifugal force and the gravity force are approximately equal. So the pressure difference force $F_D$ can be ignored. The normal force is:

$$F_N = F_C = \frac{4}{3}\pi\rho_p r_p^3 \frac{(u_p \cdot \cos\alpha)^2}{r_0 - r_p} \tag{5.37}$$

The abrasive particle moves along the inner hole surface of the workpiece. When the machining resistance force $f_N$ is equal to the gas thrust $F$, the abrasive particles will remain in an uniform motion state. Based on the expression $F = f_N$, the abrasive particles speed can be introduced:

$$u_p = \frac{\sqrt{3C_D\rho(r_0 - r_p)}}{\sqrt{8f\rho_p r_p}\cos\alpha + \sqrt{3C_D\rho(r_0 - r_p)}} \cdot u \tag{5.38}$$

In addition,

$$\rho = \frac{p}{RT} \tag{5.39}$$

Therefore, the normal force $F_N$ is:

$$F_{\mathrm{N}} = 4\pi\rho_{\mathrm{p}}r_{\mathrm{p}}^{3}\frac{C_{\mathrm{D}}pu^{2}}{\left(\sqrt{8f\rho_{\mathrm{p}}r_{\mathrm{p}}}+\sqrt{\frac{3C_{\mathrm{D}}p(r_{0}-r_{\mathrm{p}})}{RT\cos^{2}\alpha}}\right)^{2}RT} \tag{5.40}$$

It can be seen from Eq. (5.40) that $r_0$, $r_{\mathrm{p}}$, $C_{\mathrm{D}}$, $p$, $f$, $T$, $\alpha$, $u$ directly affect the size of the normal force $F_{\mathrm{N}}$. These parameters are independent, and their values have a certain range for engineering applications.

3. The effects of parameters on $F_{\mathrm{N}}$

Through the above analysis to the resistance coefficient, $C_{\mathrm{D}} \geq 0.44$, $r_{\mathrm{p}}$ should be less than the diameter of the gas beam. $P$ should not be less than 0.1 MPa. In addition, the designed gas inlet is the convergent nozzle, $v < 340$ m/s. The swirling flow cannot be formed for too bigger $\alpha$, so $\alpha$ value is less than or equal to 30°. Temperature $T$ has a smaller range. The effects of $r_0$, $r_{\mathrm{p}}$, $C_{\mathrm{D}}$, $p$, $f$, $T$, $\alpha$, $u$ on the normal force $F_{\mathrm{N}}$ are simulated using the software MATLAB. The simulation conditions are $r_0 = 20$ mm, $r_{\mathrm{p}} = 2$ mm, $C_{\mathrm{D}} = 0.44$, $f = 0.3$, $p = 0.4$ MPa, $T = 293$ K, $\alpha = 5°$, $u = 320$ m/s [39].

(1) The effects of $r_0$ on the $F_{\mathrm{N}}$

Figure 5.60 shows the effects of $r_0$ on the $F_{\mathrm{N}}$. It is observed in Fig. 5.60 that $F_{\mathrm{N}}$ decreases with the increase in the inner hole radius, but the change of $F_{\mathrm{N}}$ is in a small range, which demonstrates that the swirling flow finishing has a large machining range.

(2) The effects of $r_{\mathrm{p}}$ on the $F_{\mathrm{N}}$

Figure 5.61 shows the effects of $r_{\mathrm{p}}$ on the $F_{\mathrm{N}}$. It is observed in Fig. 5.61 that the $F_{\mathrm{N}}$ increases with the increase in the $r_{\mathrm{p}}$. So the abrasive particle radius can be chosen to control the $F_{\mathrm{N}}$. The abrasive particle radius should not be too large because of the effect of airflow.

(3) The effects of $C_{\mathrm{D}}$ on the $F_{\mathrm{N}}$

Figure 5.62 shows the effects of $C_{\mathrm{D}}$ on the $F_{\mathrm{N}}$. It is shown in Fig. 5.61 that $F_{\mathrm{N}}$ increases with the increase in the $C_D$.

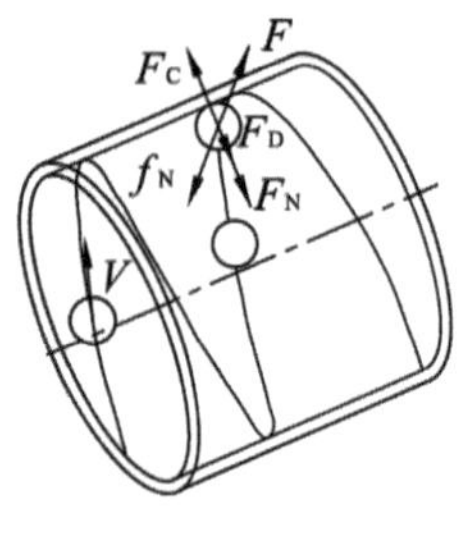


**Fig. 5.59** Force analysis of the abrasive particle

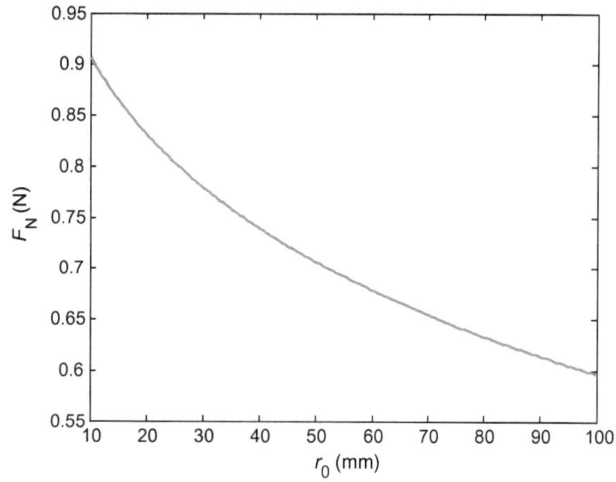


**Fig. 5.60** Effects of $r_0$ on the $F_N$

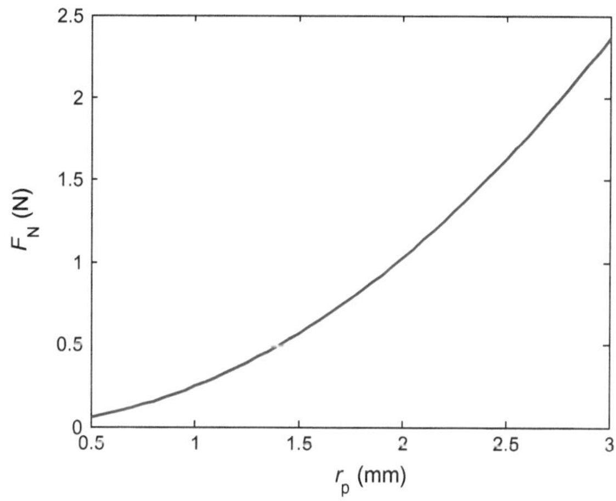


**Fig. 5.61** Effects of $r_p$ on the $F_N$

Considering the influence of the fluid compressibility, the surface roughness, and the effect of non-spherical of the abrasive particles on the resistance coefficient, the resistance coefficient is:

$$C_D = C_{Ds} \cdot f(Ma_r) \cdot S \tag{5.41}$$

where $C_{Ds}$ is the standard resistance coefficient; $S$ is the correction factor of the surface roughness and non-spherical nature of abrasive particles; $f$ (Mar) is the correction factor of the turbulence effect. According to the empirical formula, $f$ (Mar) is:

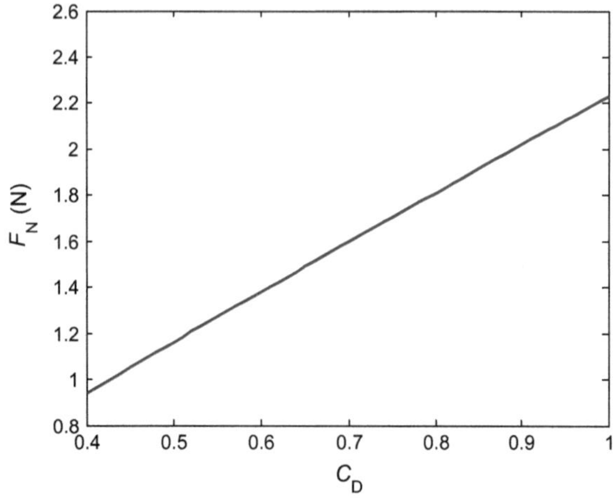


**Fig. 5.62** Effects of $C_D$ on the $F_N$

$$f(\mathrm{Mar}) = 1 + \exp(-\frac{0.427}{\mathrm{Mar}^{4.63}} - \frac{3}{\mathrm{Re}^{0.88}}) \tag{5.42}$$

where Mar is the relative Maher number. ($\mathrm{Mar} = |u - u_p|/c$, $c$ is the speed of sound.)

As can be seen from Eq. (5.42), in the two-phase flow of large Reynolds number, the correction factor of turbulence effect is determined by the Mar. In addition, $u_p < u$. So the Mar depends on $u$: The greater the $u$ is, the bigger the Maher number is, and the greater the resistance coefficient is. In certain $u$, the resistance coefficient is proportional to the non-spherical correction coefficient of the abrasive particles. The correct coefficients $S$ are listed in Table 5.1.

As shown in Table 5.1, the rougher the surface of the abrasive particles and the more irregular shape, the greater the resistance coefficient is, and the greater $F_N$ is.

(4) The effects of $f$, $\alpha$, $p$, $\upsilon$ on the $F_N$

According to the above-mentioned analysis methods, the following conclusion can be obtained. The normal force $F_N$ decreases with the increase in the friction coefficient $f$. When friction coefficient decreases, $F_N$ increases, and the abrasive particle has the rolling effect on the workpiece surface. Normal force $F_N$ can be changed by changing the $P$. In addition, the adjustment of the $P$ can be realized by changing the backpressure of the outlet. The normal force $F_N$ increases with the increase in the $u$. The speed of the airflow is determined by the ratio of the inlet pressure and the outlet pressure of the nozzle. The positive pressure $F_N$ decreases

**Table 5.1** Non-spherical correct coefficient of the abrasive particles

| Shape of abrasive particles | Spherical shape | Rough spherical shape | Elliptical shape | Irregular shape |
|---|---|---|---|---|
| $S$ | 1.0 | 2.42 | 3.08 | 2.75–3.5 |

with the increase in $\alpha$, but $\alpha$ has a narrow variation range. When $\alpha$ is changed from 5° to 20°, the normal force $F_N$ increased only by 0.006 N, so the effects of $\alpha$ on the $F_N$ can be neglected.

Through the above analysis, we can draw the following conclusions. The main factors that affect the process efficiency of the swirling flow are the material properties of the abrasive particles (such as material, shape, size), the abrasive particles' movement speed, and the force of the abrasive particles acting on the workpiece.

### 5.3.3 *Microscopic Motion Feature*

1. The trajectory of the spherical abrasive particles along the inner hole surface of the workpiece under the normal processing state

The trajectory of the abrasive particles is spiral line with a variable helix pitch; however, due to the inertia effect of the abrasive particles, the spiral angle of the trajectory using fluid as the working medium is less than that using gas working medium. Assuming that the helix angles of the abrasive particles and the vortex flow are equal, the parametric equation in the cylindrical coordinate system is:

$$\begin{cases} r = r_0 \\ \theta = \frac{u_p \cdot \cos\alpha}{r_0} \cdot t \\ z = u_p \cdot \sin\alpha \cdot t \end{cases} \tag{5.43}$$

where $u_p$ is the abrasive particle speed which can be obtained by Eq. (5.38) and $\alpha$ is the spiral angle of the abrasive particle trajectory.

$$\alpha = arctg \frac{\overline{u_z}}{\overline{u_\theta}} \tag{5.44}$$

where $\overline{u_\theta}$ is the average tangential velocity in the circle area which is the function of the time, and it decreases exponentially with the increase in the time; $u_z$ is the average axial velocity in the circle area. When the gas flow rate at the nozzle outlet is constant, the specific value of $u_z$ can be calculated according to the pipe diameter and the gas flow rate in unit time.

2. Micro-motion characteristics of the abrasive particle

Figure 5.59 shows the forces acting on the abrasive particle which moves at a high speed in the inner hole surface. According to the Dahlberg principle, the force balance equation of the abrasive particle is:

$$ma_{\tau} = F - f_1 \cdot F_{\mathrm{N}} - G\sin\alpha \tag{5.45}$$

$$ma_{\mathrm{n}} = F_{\mathrm{N}} - G\cos\alpha \tag{5.46}$$

$$J_0\varepsilon = f_1 \cdot F_{\mathrm{N}} \cdot r_{\mathrm{p}} - f_2 \cdot F_{\mathrm{N}} \cdot r_{\mathrm{p}} \tag{5.47}$$

where $m$ is the mass of the abrasive particle; $\alpha_{\tau}$ is tangential acceleration of the abrasive particle; $f_1$ is the sliding friction coefficient; $\alpha_{\mathrm{n}}$ is the normal acceleration of the abrasive particle; $f_2$ is the rolling friction coefficient; $J_0$ is the rotation inertia; $\varepsilon$ is the angular acceleration.

In addition,

$$a_{\tau} = \frac{\mathrm{d}u_{\mathrm{p}}}{\mathrm{d}t} \tag{5.48}$$

$$a_{\mathrm{n}} = \frac{u_{\mathrm{p}}^2}{r_0 - r_{\mathrm{p}}} \tag{5.49}$$

From Eqs. (5.45) to (5.49), we can obtain:

$$a_{\tau} \neq \varepsilon \times r_{\mathrm{p}} \tag{5.50}$$

So there are both rolling and sliding for the abrasive particle in the inner hole surface.

### 5.3.4 *Finishing Mechanism*

With the strong swirling flow of the air, the spherical abrasive particles with a certain speed exert normal force to the inner hole surface due to the action of the centrifugal force. Under the action of positive pressure and relative motion, the spherical abrasive particles produce grinding, rolling, carving, and scratching on the inner hole surface of the workpiece, so the finishing of the hole surface of the workpiece is achieved [37].

1. Micro-grinding

Under the action of the force, the free abrasive particles will glide along the inner hole surface with a certain relative speed. The abrasive particles liking the abrasive particles of the grinding wheel will carve, scratch, and cut the workpiece surface. Due to the relatively small force and the negative rake angle of the abrasive grains, the main role of the abrasive cutting edge on the workpiece surface is scrubbing and scoring. When the abrasive grains scratched one or more times, the superficial metal occurs one or more plastic deformation superposition. When the plastic deformation

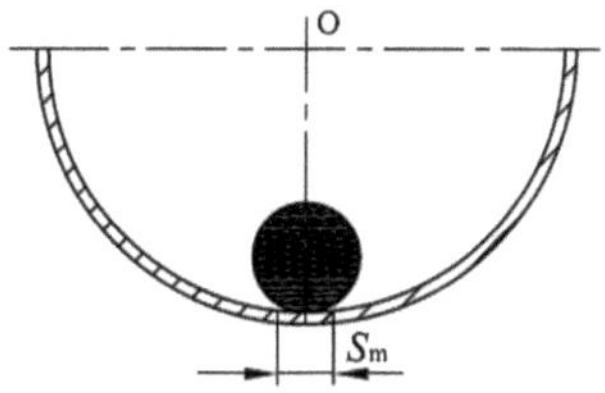


**Fig. 5.63** Contact arc length of the abrasive particles and the hole surface of the pipe wall during the finishing process

exceeds the limit of plastic deformation, microchip will fall from the workpiece. As a result, micro-grinding is achieved.

Assuming that micro-grinding is a micro-wear process, obviously, the trajectory of any point of the abrasive particles' surface is a cycloid relative to the inner hole surface. Figure 5.63 shows the contact arc length of the abrasive particles and the hole surface of the pipe wall during the finishing process. In Fig. 5.63, $S_m$ is the maximum contact length between the abrasive particle and the workpiece. The contact length is produced from the abrasive particle movement driven by the vortex flow. Due to the centrifugal force, the abrasive particle exerts normal force on the inner hole surface, which causes contact deformation in the inner hole surface. Under the action of the normal force and the relative sliding, micro-grinding of the abrasive particles on the workpiece is achieved.

The greater the normal force is, the longer the contact arc length is, and the greater the relative sliding speed is, the more significant the micro-grinding is. However, these grinding width and depth are very small, so wear debris is also very small.

2. Rolling

The free abrasive particles exert normal force on the inner hole surface and rolls relative to the workpiece surface with a certain angular velocity. During the relative motion, rolling process is realized. Under the rolling of the abrasive particles, micro-peak of the workpiece surface generates locally plastic deformation, which causes the decrease in the surface roughness. In addition, the workpiece surface is hardened and produces residual compressive stress. Due to the residual compressive stress, the wear resistance and fatigue strength of the workpiece are improved

There are two forms of the rolling action of the abrasive particles on the inner hole surface of the workpiece:

(1) Surface plastic deformation

In the finishing process, the rolling contact surface of the spherical abrasive particle with the workpiece is not a complete spherical crown, but irregular arc with multi-point contact. When the contact area is small, the compressive stress of the spherical grinding block on these contact points will reach the elastic limit. As a result, the plastic deformation of the local contact point occurs.

(2) Elastic deformation

When rolling the inner hole surface of the workpiece, the contact area between the abrasive particle and the wall surface may be large at one time. Due to the small normal force $F_N$, the micro-protrusion on the inner surface of the workpiece does not undergo plastic deformation, but only elastically deforms. However, under the alternating action of a large number of abrasive particles, these micro-protrusions will be fatigue-fractured due to the repeated elastic deformation. As a result, the machining chips are produced and the finishing of the inner surface is achieved.

3. Collision, carving, and scratching

(1) Collision

The finishing of the inner hole surface is conducted by a large number of abrasive particles. The single abrasive particle which may be away from the inner hole surface will not always rotate along the inner hole surface to realize the micro-cutting and rolling. Sometimes, a single abrasive particle will collide with other single abrasive particle. After repeatedly rebounding for a period time, the single abrasive particle then hits the surface, causing slight vibration, which results in collision. Although the collision effect will not change the whole surface, it can improve the mechanical properties of the workpiece inner surface.

(2) Carving and scratching

The abrasive particles have different sizes and shapes. When machining begins, some of the abrasive particles do not regularly rotate along the wall surface. But sharp edge, rough surface of the abrasive particles contacts with the inner surface; as a result, the abrasive particles have the carving and scratching effects on workpiece surface, which are also called plow wrinkle effect, as shown in Fig. 5.64.

The material at both sides, front and bottom of the grooves caused by carving and scratching, will be again pressed to smooth by the next time carving and scratching. The materials were deformed over and over again, which make the materials harden or materials strengthening. Finally, these materials peel off from the workpiece and form processing chips.

In summary, after the gas-particle swirling flow finishing, the micro-burr on the inner hole surface of the workpiece will be broken, grounded, and cut. So the micro-geometry of the workpiece surface is changed, and the surface roughness will decrease, and the initial running of the part will also be completed. At the same

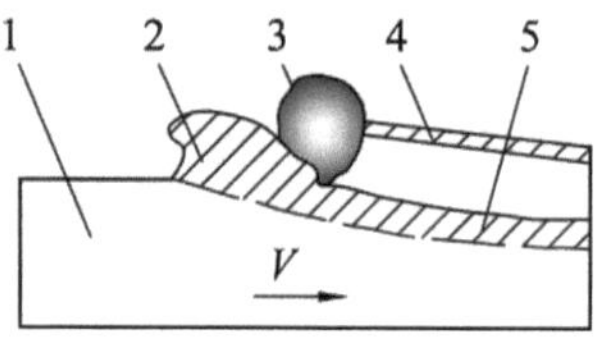


**Fig. 5.64** Micro-cutting of the abrasive particles. 1—Workpiece; 2—pushing deformation zone; 3—abrasive particle; 4—deformation ridge; 5—deformation layer

time, the material on the surface has plastic and elastic deformation by extruding and cutting. In addition, the finishing also changes the material organization, the physical and mechanical properties of the workpiece surface. In general, the gas-particle swirling flow finishing can improve the surface roughness and the integrity of surface.

## 5.4 Experimental Device of Gas Particle Two-Phase Swirling Flow Finishing

### *5.4.1 Overall Design*

The key for the gas-particle two-phase swirling flow finishing is how to produce high-speed airflow in the workpiece and make the media spiraling in a high speed along the surface of the internal hole. So, the core device, which can transfer the kinetic energy of the high-speed airflow into the kinetic energy of the grinding block, includes: nozzle producing the high-speed airflow, the swirling heads producing swirling airflow, and other components such as clamping fixture, pneumatic system, automatic control system.

In view of the non-uniformity of the finishing quality along the axis resulting from energy attenuation of the airflow, swirling heads are designed to provide for two-way airflow, which would result in the finishing quality along the axis tending to uniformity, and avoid the shortcoming of constantly adding and collecting the grinding block needed by the one-way airflow. Once the media are loaded, the whole finishing can be completed, resulting in the improvement of the efficiency.

In order to increase the level of automation, cylinder is used to clamp and loose the workpiece, and microcontroller is used to control the gas channel to realize automatic cycle. The media in the swirling head are accelerated by high-speed rotating airflow and spiral along the surface of the internal hole to achieve the finishing of the workpiece.

The schematic diagram of experimental device of the gas-particle two-phase swirling flow finishing is shown in Fig. 5.65. The sequence of operations is as follows: The workpiece is clamped in the swirling heads by pressing the "clamping" button. After pressing the "start" button, the high-pressure air flowing from the pump at the inlet of the nozzle would be accelerated due to the pressure decrease led by the structure of the nozzle, and become the high-speed rotating swirling airflow which enter the inner cavity of the swirling head along tangential direction, under the constraint of the cylindrical inner wall of the swirling head. The high-speed rotating swirling airflow drives the media loaded in the swirling head before spiraling in a high speed along the cylindrical inner wall of the swirling head. When the media reach the end point, the airflow direction is changed due to the automatic control assembly, and the high-speed rotating swirling airflow from another

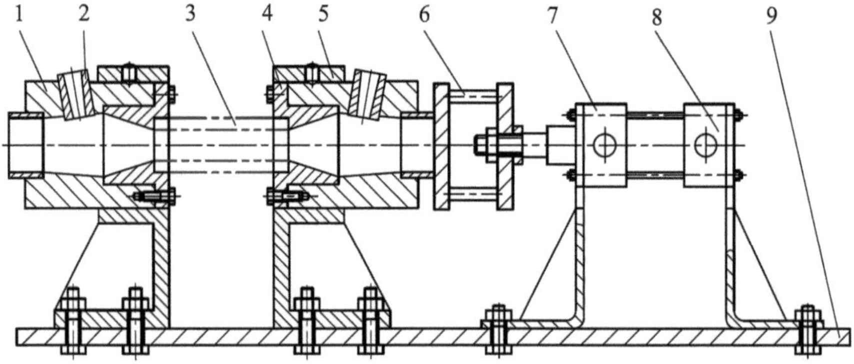


**Fig. 5.65** Experimental device schematic diagram of the gas-particle two-phase swirling flow finishing. 1—Left swirling head; 2—nozzle; 3—workpiece; 4—right swirling head; 5—support; 6—connector; 7—limit switch; 8—cylinder; 9—substructure

swirling head drives the grinding block spiraling reversely. The media reciprocate again and again until the workpiece is finished.

## *5.4.2 Formation Components of the Swirling Flow*

1. Nozzle

The nozzle is used in the device of the gas-particle swirling flow finishing; it is designed to convert the entropy of the gas into the kinetic energy of the airflow based on the jet wash theory of engineering thermodynamics. The nozzle is very short in length, and its cross section is converging or diverging. When the gas with a certain rate and pressure flows through a channel with variable section, it would generate the effect of pressure decreasing (or increasing) and flow speed increasing (or decreasing). So the rate and pressure of the airflow needed by finishing can be obtained by this effect. Figure 5.66 shows the schematic diagram of the structure of the converging nozzle.

(1) Basic theories

All devices generating high-speed airflow always utilize the nozzle to convert the entropy of the gas into the kinetic energy of the airflow based on the jet wash theory of engineering thermodynamics. That is:

**Fig. 5.66** Structure schematic diagram of the convergent nozzle

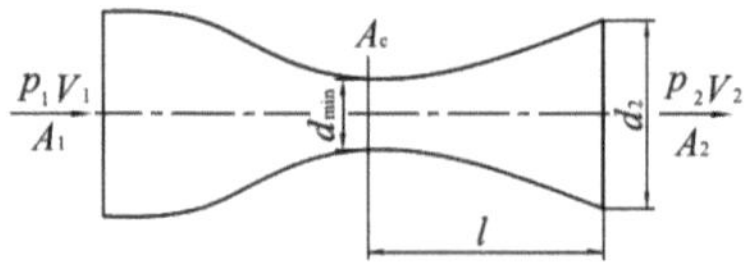

$$h = e + \frac{p}{\rho} = e + RT \tag{5.51}$$

where $h$ is the entropy of the gas (J) and $e$ is the internal energy of the gas (J).

Because the length of the nozzle is shorter and the speed of the flow is higher, the time needed by the gas flowing through the nozzle is lessen and the heat exchange between the gas and the nozzle can be neglected. Therefore, the airflow in the nozzle can be analyzed as a steady flow in isentropic process.

(1) The equation in isentropic process can be expressed as:

$$h_1 - h_2 = \frac{1}{2}\left(V_2^2 - V_1^2\right) \tag{5.52}$$

where $h_1$ and $h_2$ are the enthalpy of the inlet gas and the exit gas, respectively; $v_1$ and $v_2$ are the average speed of the inlet gas and the exit gas, respectively.

It can be seen from Eq. (5.52) that the enthalpy of the gas decreases and it is converted into kinetic energy of the airflow in isentropic process.

(2) Change equation between the flow speed and the pressure of gas is:

$$-\frac{\mathrm{d}p}{\rho} = v\mathrm{d}v \tag{5.53}$$

where $v$ is the average speed of the gas (m/s).

It can be seen from Eq. (5.53) that the speed of the gas increases when the pressure of the gas decreases.

(3) Relation equation between the section of nozzle and the flow speed of gas can be expressed as:

$$\frac{dA}{A} = \left(Ma_{\mathrm{r}}^2 - 1\right)\frac{\mathrm{d}v}{v} = \left(\frac{v^2}{c^2} - 1\right)\frac{\mathrm{d}v}{v} \tag{5.54}$$

where $A$ is the section of nozzle ($m^2$) and $Ma_{\mathrm{r}} = v/c$ is the Mach number.

When the speed of airflow is less than that of sound, the cross section of the nozzle should become smaller and smaller, that is, the converging nozzle. When the speed of the airflow is more than that of the sound, the cross section of the nozzle should increase gradually. If the gas is needed to accelerate to supersonic velocity, a diverging section must be added to a converging nozzle. The result is a converging–diverging nozzle often called Laval nozzle.

(4) Flow velocity at the nozzle exit can be calculated as:

$$v_2 = \sqrt{\frac{2k}{k-1}\frac{p_1}{\rho_1}\left[1 - \left(\frac{p_2}{p_1}\right)^{(k-1)/k}\right]} \tag{5.55}$$

where $k$ is the adiabatic exponent, and its value is 1.4 for diatomic molecule, namely $k = 1.4$; $\rho_1$ is the density of the gas (kg/m$^3$); $p_1$ is the gas pressure at the inlet of the nozzle depending on the actual flow value and the cross section area at the exit of the nozzle; $p_2$ is the gas pressure at the exit of the nozzle, which is often the environmental pressure.

(5) Cross-sectional area at the exit of the nozzle can be expressed as:

$$A_2 = \frac{\dot{m}}{\sqrt{\frac{2k}{k-1}\frac{p_1}{\rho_1}\left[1 - \left(\frac{p_2}{p_1}\right)^{(k-1)/k}\right]}} \tag{5.56}$$

where $\dot{m}$ is the mass flow rate of gas (kg/s).

(6) Critical pressure ratio of ideal gas in isentropic process can be expressed as:

$$K = \frac{p_2}{p_1} = \left(\frac{2}{k+1}\right)^{k/(k-1)} \tag{5.57}$$

(7) Throat area of converging–diverging nozzle can be expressed as:

$$A_c = \frac{\dot{m}}{\sqrt{kp_1\rho_1\left(\frac{2}{k+1}\right)^{(k\}1)/(k-1)}}} \tag{5.58}$$

(2) Basic design method

The major task of the nozzle design is to calculate the cross-sectional area at the exit of the nozzle and the throat area according to the given flow value. According to the principle that the flow pressure decreases and the flow velocity increases resulting from the gas expansion, the nozzle should be designed according to the pressure ratio of the outlet to the inlet and environmental pressure is often used as the outlet pressure.

The design procedures are as follows:

(1) The nozzle type

The critical pressure ration $K$ is equal to 0.528 for air ($k = 1.40$) according to Eq. (5.57). To obtain the maximum exit velocity, the converging nozzle should be selected when $K > 0.528$, and the converging–diverging nozzle should be selected when $K < 0.528$.

(2) Calculation of the main parameters includes the cross-sectional area at the exit of the nozzle, the throat area of the converging–diverging nozzle, and the converging length

For the converging and converging–diverging nozzle, the cross-sectional area at the outlet of the nozzle can be calculated from Eq. (5.56), and the inlet cross-sectional areas need not to be calculated. The cross-sectional area of the inlet is properly higher than that of the exit, and the converging section should keep a certain shape. The throat area $A$ can be calculated from Eq. 5.58 for the converging–diverging nozzle.

The cone angle of the converging section is general 10°–12° for the converging–diverging nozzle.

(3) Calculation of the possible maximum exit flow speed $v_2$

If the work condition is consistent with that of the first two steps, that is, to use the nozzle parameters calculated from the second step, and the flow rate, the inlet pressure, the exit pressure calculated from the first two steps, $v_2$ is the maximum speed of the nozzle under rated operation. If the flow rate and pressure are consistent with those calculated from the first two steps, the great change of the flow speed at the nozzle exit would happen.

For example, the converging–diverging nozzle should be selected according to the principle of the first step. But if the converging nozzle is actually used, the flow speed at the exit of the nozzle exit is the critical flow speed, which can be expressed as:

$$v_c = \sqrt{2\frac{k}{k+1}\frac{p_1}{\rho_1}} \tag{5.59}$$

(4) The actual flow speed at the nozzle outlet is determined according to the actual values of the flow rate and the inlet gas pressure

Because the working flow rate cannot often reach the designed flow rate in actual operation, the inlet pressure $p_1$ and the flow speed would decrease obviously.

**Table 5.2** Structure sizes of the nozzle

| No. | Inlet pressure $p_1$/MPa | Outlet pressure $p_2$/MPa | Predicted flow rate ($m^3$/min) | Nozzle types | Theoretical maximum velocity/(m/s) | Critical velocity/ (m/s) | Theoretical exit diameter $d_2$/mm | Throat diameter $d_{min}$/mm | Length of the diverging section $l$/mm |
|---|---|---|---|---|---|---|---|---|---|
| 1 | 0.2 | 0.1 | 0.16 | Converging | 313 | 313 | 3.009 | 3.006 | 0 |
| 2 | 0.4 | 0.1 | 0.16 | Converging and diverging | 438 | 313 | 2.346 | 2.126 | 1.257 |
| 3 | 0.6 | 0.1 | 0.16 | Converging and diverging | 485 | 313 | 2.104 | 1.735 | 2.108 |
| 4 | 0.2 | 0.1 | 0.25 | Converging | 313 | 313 | 3.761 | 3.758 | 0 |
| 5 | 0.2 | 0.1 | 0.3 | Converging | 313 | 313 | 4.120 | 4.117 | 0 |
| 6 | 0.2 | 0.1 | 0.4 | Converging | 313 | 313 | 4.758 | 4.753 | 0 |
| 7 | 0.4 | 0.1 | 0.3 | Converging and diverging | 438 | 313 | 3.213 | 2.911 | 1.714 |
| 8 | 0.4 | 0.1 | 0.4 | Converging and diverging | 438 | 313 | 3.710 | 3.361 | 1.994 |
| 9 | 0.6 | 0.1 | 0.3 | Converging and diverging | 485 | 313 | 2.882 | 2.376 | 2.891 |
| 10 | 0.6 | 0.1 | 0.4 | Converging and diverging | 485 | 313 | 3.328 | 2.744 | 3.337 |

(3) The design of the nozzle

The main function of the nozzle is to convert the high-pressure and low-speed airflow into the low-pressure and high-speed airflow, which can drive the media rotating at a high speed on the surface of the inner hole. The relation between the airflow speed and the velocity of the media needs to be analyzed.

(1) The determination of the flow rate $\dot{m}$, the inlet pressure $p_1$, and the exit pressure $p_2$

The maximum flow rate at the inlet of each nozzle is about 0.16 $m^3$/min in the experiment. The maximum output pressure of the pump is about 0.9 MPa. The inlet pressure $p_1$ is equal to 0.2, 0.4, and 0.6 MPa, respectively, according to the experimental conditions. The pressure at the exit of the nozzle lying in the inner cavity of the swirling head can be set as 0.1 MPa according to the test.

(2) Theoretical calculation of nozzle structure size

Theoretical calculation values of nozzle are listed in Table 5.2. It is shown in Table 5.2 that the velocity at the exit of the nozzle is larger for the converging nozzle. The difference between the exit and throat diameter of the same nozzle is very small (maximum value is 0.6 mm). Because the length of the diverging nozzle is only 3.4 mm, it is difficult to be processed. So the converging section is selected for manufacturing the nozzle.

(3) The size modification of the nozzle

Based on the engineering thermodynamics theory, the converging–diverging nozzle should be selected theoretically. But if the converging nozzle is actually used, the maximum flow velocity at the exit of the nozzle is the critical flow velocity at the throat of the converging–diverging nozzle, namely the local sound velocity (313 m/s). The diameter at the exit of the nozzle can only use the diameter at the throat of the converging–diverging nozzle. After the theoretical values of the nozzle throat are round-off, the actual diameter at the exit of the nozzle is 2 and 4 mm according to Table 5.2. For comparison, the diameter of 3 mm at the exit of the nozzle is selected. In addition, only after the jet flow pass through the pipe with a certain length, can the airflow keep steady stream. So the length of the nozzle exit is set as 20 mm.

(4) Calculating the actual working parameters

The actual gas flow rate at the exit under the different pressures can be calculated from Eq. (5.56) and (5.59). The relation between the exit velocity and the inlet pressure is listed in Table 5.3.

**Table 5.3** Velocity at the exit of the nozzle

| Inlet pressure/MPa | 0.4 | 0.35 | 0.3 | 0.25 | 0.23 | 0.2 | 0.19 | 0.18 | 0.17 | 0.16 | 0.15 | 0.14 | 0.13 |
|---|---|---|---|---|---|---|---|---|---|---|---|---|---|
| Exit pressure/MPa | 0.215 | 0.188 | 0.161 | 0.135 | 0.124 | 0.108 | 0.1 | 0.1 | 0.1 | 0.1 | 0.1 | 0.1 | 0.1 |
| Environmental pressure at the exit/MPa | 0.1 | 0.1 | 0.1 | 0.1 | 0.1 | 0.1 | 0.1 | 0.1 | 0.1 | 0.1 | 0.1 | 0.1 | 0.1 |
| Velocity at the exit/m/s | 313 | 313 | 313 | 313 | 313 | 313 | 313 | 301 | 287 | 272 | 254 | 232 | 206 |

According to Eqs. (5.46) and (5.48), the flow rate can be calculated as:

$$\dot{m} = A_2 \sqrt{2 \frac{k}{k-1} p_1 \rho_1 \left[ \left( \frac{p_2}{p_1} \right)^{2/k} - \left( \frac{p_2}{p_1} \right)^{(k+1)/k} \right]} \tag{5.60}$$

$$\dot{m} = A_c \sqrt{k p_1 \rho_1 \left( \frac{2}{k+1} \right)^{(k\}1)/(k-1)}} \tag{5.61}$$

Because the value of the environmental pressure at the exit is 0.1 MPa, the nozzle is under the critical working conditions. The flow rate passing through the nozzle can be calculated from Eq. (5.61) when $p_1 > 1/0.528$ and Eq. (5.60) when $p_1 < 1/0.528$.

The values of the flow rate at different inlet pressures and exit diameters of the nozzle are listed in Table 5.4. Figure 5.67 shows the relationships between the flow rate and the inlet pressure at different exit diameters.

When $p_1 > 1/0.528$, the formula for calculating flow rate can be simplified as:

$$\dot{m} = 0.0088511 \cdot d_2^2 \cdot p_1 \tag{5.62}$$

where $d_2$ is the exit diameter of the nozzle.

2. Formation components of the swirling flow

(1) The inner diameter of the swirling head

The swirling head is the supporter of the nozzles and the place where the swirling flow can be produced. When the gas energy is transferred to the spherical media, the spherical media would be accelerated and would obtain the initial

**Table 5.4** Flow rate at different inlet pressures and exit diameters of the nozzle

| Inlet pressure (MPa) | Exit diameter (mm) | | | |
|---|---|---|---|---|
| | 2 | 3 | 4 | 6 |
| 0.40 | 0.142 | 0.319 | 0.566 | 1.274 |
| 0.35 | 0.124 | 0.279 | 0.496 | 1.115 |
| 0.30 | 0.106 | 0.239 | 0.425 | 0.956 |
| 0.25 | 0.088 | 0.199 | 0.354 | 0.796 |
| 0.23 | 0.081 | 0.183 | 0.326 | 0.732 |
| 0.20 | 0.071 | 0.159 | 0.283 | 0.637 |
| 0.19 | 0.067 | 0.151 | 0.269 | 0.606 |
| 0.18 | 0.064 | 0.143 | 0.254 | 0.573 |
| 0.17 | 0.060 | 0.134 | 0.239 | 0.537 |
| 0.16 | 0.055 | 0.125 | 0.222 | 0.499 |
| 0.15 | 0.051 | 0.114 | 0.203 | 0.457 |
| 0.14 | 0.046 | 0.103 | 0.182 | 0.410 |
| 0.13 | 0.040 | 0.089 | 0.158 | 0.357 |

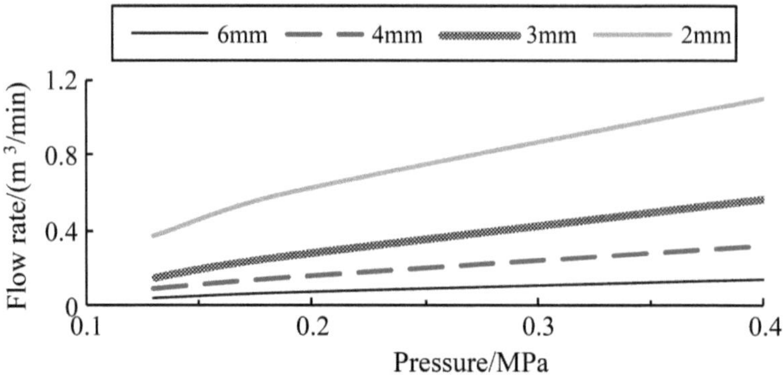


**Fig. 5.67** Relationships between the flow rate and the inlet pressure at different exit diameters

velocity. Then, the spherical media driven by airflow continue to move at a constant velocity. There are no other energy loss except from the loss of the field strength, because the gas flow is far away from the swirling head. The spherical media can obtain the maximum energy of the gas flow and reach maximum velocity.

Because of the limit of the actual size of the nozzle and a certain amount of media needed to be loaded in the swirling head, the diameter of the swirling head cannot to be too small for the workpiece with a small diameter; that is, the inner diameter of the swirling head is not the same as that of the workpiece. Excessive media relative to the accommodation space of the nozzle would block the motion of the airflow, especially when media are all concentrated in the swirling nozzle, and the gas flow field has not yet formed. This blockage would cause intensity turbulence and lead to a severe loss of gas energy.

(2) The number of the nozzle

When the number of nozzles is too small, the vortex effects would not be obvious and uniform. Insufficient gas flow rate and less total energy of the airflow as the power would lead to the velocity of the media decreasing, which would cause severe attenuation of the swirling flow field along the axis of the workpiece and influence the finishing effect.

Too many nozzles would increase the size of the swirling head and may cause the inner diameter of the swirling head not to be equal to that of the workpiece. So the structure sizes of the device increase.

Generally, three nozzles differing by 120° in peripheral are installed on the swirling head.

(3) Inclination angle of the nozzle

Only when there is a certain angle inclined to the workpiece between the axis of the nozzle and that of the swirling head, the swirling flow rotating at high speed can be produced to spiral at low velocity.

The inclination angle of the nozzle determines the spiral angle of the spherical media at initial time and influences the feed speed of the spherical media moving forward along the axis of workpiece. According to the simulation results, the big inclination angle of the nozzle would make the pitch of the media spiral to increase and lead to different finishing effects. If the inclination angle of the nozzle were too small, it would take too long time for the spherical media to pass through the workpiece, which can lead to the decrease in cycle numbers in a certain period of time. So the inclination angle of the nozzle is generally selected in the range of 5°–10°.

(4) Swirling head for one-way airflow and two-way airflow

(1) Swirling head for one-way airflow

When the gas from the nozzle enters into the swirling heads where the swirling flow field is formed, the negative pressure is produced at the center of the swirling head resulting from the swirling flow moving at high speed along the inner wall of the swirling head. The media in the container at the back of the swirling head enter into the swirling head, because of the action from the negative pressure and spiral driven by swirling flow, and then enter into the workpiece achieving the surface finishing of the inner hole. The inner hole of the workpiece is open, causing the spherical media to spiral away from the workpiece. Thus, they are collected by a recovery device.

When using the swirling head for one-way airflow, one end of the inner hole of the workpiece is connected to the swirling head, and another end is open. Because the airflow field does not have any blockage, the energy of the gas will not be lost. The spherical media make fast movements without any blockage. So, the required inlet pressure and gas flow do not need to be too large. The main motion control of the device is simple, and the control of the start–stop and the time are only needed.

But there are some disadvantages as follows: ① the utilization ratio of the spherical media is lower. The spherical media can only move along one-way, achieve single finishing, and then flow away from the workpiece. Therefore, plenty of spherical media are needed to be reserved. ② It is difficult to feed and control the spherical media. It is impossible for the feed inlet to be too large due to the limitation of the inner diameter. The spherical media may be blocked at the feed inlet. The feeding quantity is sometimes more, and sometimes less, and almost zero due to the blockage. The number of the spherical media is uneven. ③ The device is complex. The feeding and receiving device must be designed. For the feeding container, the pump is needed to make the spherical media entering into the swirling head evenly. The feeding number in unit time also needs to be controlled. Only are the spherical media recycled by the receiving device and then can be reused. ④ It is not suited for the workpiece whose diameter is smaller than the inner hole of the swirling head. The inner diameter of the swirling head cannot be too small resulting from the limitation of the various conditions. A conical interface between the swirling head and the workpiece must be added for the workpiece with a smaller diameter. As mentioned above, the backward airflow is produced on the

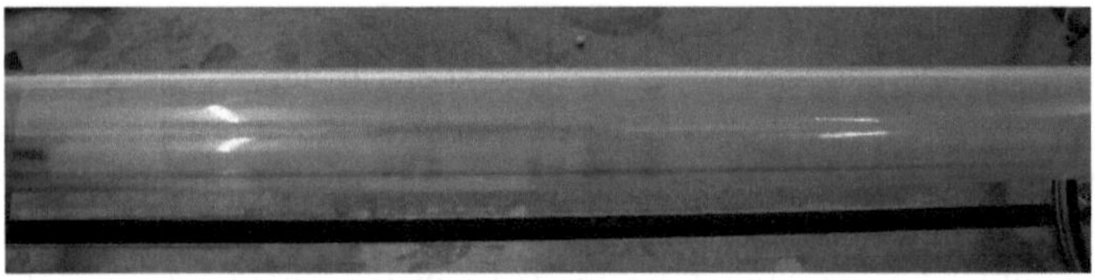

**Fig. 5.68** Photograph of finishing state using the swirling head for one-way airflow

conical surface of the workpiece interface, causing the negative pressure not to be produced at the central zone of the swirling head, and the internal pressure of the swirling head increases. Most spherical media flow out from the feeding inlet and cannot enter into the workpiece. If the feeding inlet is closed, the spherical media can easily enter into the workpiece, but cannot be added again, and the finishing effects are unsatisfactory. ⑤ The finishing quality is non-uniform. The one-way airflow and motion of the media would lead to the non-uniform finishing quality. The finishing effect of one end near swirling head is better.

Figure 5.68 is the photograph of finishing state using the swirling head for one-way airflow. As shown in Fig. 5.68, the pitch of the media spiraling changes which leads to the airflow attenuation and the finishing non-uniformity.

(2) The swirling heads for two-way airflow

The swirling heads for a two-way airflow are composed of the fixed and free swirling head. The vents of two ends are connected with two 2-position 2-way solenoid valves, respectively, by pipelines. The spherical media are loaded in the workpiece or fixed swirling head one time before finishing. The workpiece is tightly clamped by the free swirling head on the right. When the gas flow into the fixed swirling head on the left, the valve on the left connected with vents of the fixed swirling head is closed, and the valve on the right connected with vents of the free swirling head is opened to ensure the formation of the swirling flow field which drives the spherical media entering into the workpiece. After the spherical media from the fixed swirling head pass through the workpiece achieving the finishing one time and enter into the free swirling head, the direction of the gas flow changes, the valve on the right is closed, the valve on the left is opened, and the gas flows into the free swirling head on the right. The processing reciprocates again and again until the workpiece is finished.

There are some advantages using the swirling heads for two-way airflow as follows: ① The feeding of the media is simple, and the utilization ratio is high. After loaded appropriately in the swirling head one time, the spherical media need not to be added. The same media can be used repeatedly until the media lose the finishing capability. ② The master device is simple. The feeding and receiving device can be ignored, and the finishing efficiency improves greatly on the condition that the feeding is not blocked. ③ The finishing quality is uniform. According to the results of the simulation, the swirling airflow has a certain degree of attenuation along the axis. The media are driven by the gas from the swirling

**Fig. 5.69** Photograph of finishing state using the swirling head for two-way airflow

heads at the two ends to achieve the finishing of the workpiece. The surface texture of the workpiece shows reticular marks. The surface roughness values are relatively uniform, and the finishing uniformity of the surface is improved. ④ The finishing range increases. For a small diameter workpiece, the vent of one swirling head where gas enters must be closed, and the vent of the other swirling head must be opened. This blocks the backward airflow produced on the conical surface of the workpiece interface, and the swirling airflow field is formed, thereby driving the media entering into the workpiece to realize the finishing.

But there are some disadvantages as follows: ① The feeding of the media is uneven, not all the media are completely driven away from the swirling head, and redistributed uniformly inside the workpiece when the gas enter into the swirling head. The media are gradually driven away from the swirling head after a certain time. When the gas enters, the media flowing out of the swirling head becomes more, and it will decrease gradually along with the decrease in the media in the swirling head, which leads to the non-uniformity of the feeding and movement velocity of the media. ② The airflow is blocked. When the gas flows out of the narrow vent, the turbulent is produced inevitably, the swirling flow would be blocked which would lead to a loss of gas, and the velocity of media would be affected to some extent. ③ The control of the device is complicate. The regular reciprocating inlet and exhaust gas are needed.

Figure 5.69 is the photograph of finishing state using the swirling head for two-way airflow. Compared with Fig. 5.68, the distribution of the media is relatively uniform when using the swirling head for two-way airflow, which is beneficial to the finishing non-uniformity.

(5) Transformation interface and exhaust interface

Two swirling heads are placed at the two ends of the workpiece, respectively. When the nozzle at one end of the swirling heads works as the inlet, the nozzle at the other end of the swirling heads must be opened as the vent so as to keep the swirling flow spiral.

(1) Transformation interface

Because the inner diameter of the swirling heads is generally larger than that of the workpiece, the swirling head must be connected with the workpiece by a conical transformation interface out of which the swirling flow into the workpiece.

(2) Exhaust interface

The exhaust interface with a stepped adjustable diameter is connected with the back end of the swirling head in order to control the working state of the swirling head. So the inlet is open when the gas flows forward, and the vent is open when the exhaust begins.

Figure 5.70 shows the transformation interface at the front end of the workpiece and the exhaust interface at the back end of the workpiece.

(3) Negative function of the transformation interface

When the inner diameter of the swirling heads is equal to that of the workpiece, the transformation and exhaust interfaces need not to be used. The swirling airflow field forms without any resistance. If the swirling heads can connect well with the workpiece, the turbulence would not be produced and thus will not cause the loss of gas energy.

When the diameter of the vent at the back end of the swirling head is the maximum, that is, the exhaust interface is not in use, the strong airflow from the nozzle will spray onto the conical surface of the transformation interface and the force will be produced to drive the airflow backward. Because the back end of the swirling head is completely open, the resistance of the airflow moving backward is very small, and the airflow flowing out of the back end is more than those entering into the workpiece. Therefore, most media fly out of the back end of the swirling head. The swirling flow strength decreases greatly, and the velocity of the media is affected. The smaller the taper of the transformation interface is, the smaller the force driving the airflow moving backward. But the smaller taper would lead to an increase in the length of the transformation and a larger attenuation of the flow filed. The taper of the transformation interface is often set as 37° or so.

When the vent at the back end of the swirling head is completely closed, the gas will not flow and thus not flow out of the vent, and all the airflow would be forced to flow forward into the workpiece. But the turbulence zone is produced at the back end of the transformation interface, which would lead to the loss of airflow energy. In addition, when the gas flows into the swirling head at one end, the baffle of the swirling head at the other end must be removed to ensure the gas flowing out of the inner hole of the workpiece fluently. The control of the device becomes more complicated.

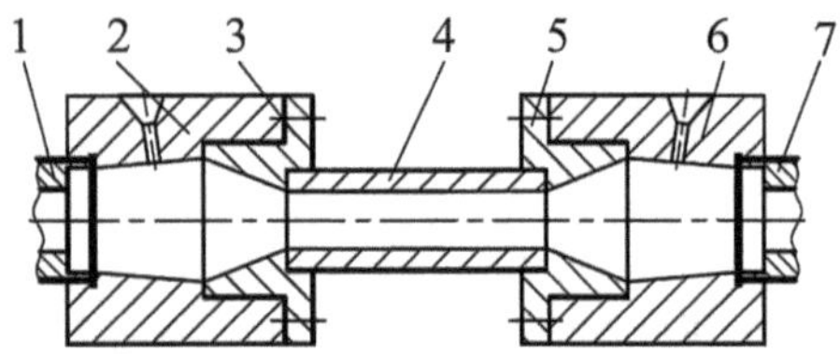


**Fig. 5.70** Transformation interface and exhaust interface

(4) Function and negative function of the exhaust interface

If the exhaust interface were not used, the gas would flow out of the vent from the three-position four-way valve at the non-working end, which led to the an excessive backpressure. The gas flow is disturbed, and the velocity of the media is smaller. The main reason is that the turbulent flow is produced resulting from the blockage of the narrow vent. The spiraling state of the swirling flow would be influenced resulting from the disturbance of the flow field. The energy of the gas loses greatly, and the energy transferred to the media decreases greatly.

When the exhaust interface is not used, the smaller pressure of the vent is the better and the backpressure is set as 0.1 MPa. At the same time, the inner diameter of the vent should be the same as that of the workpiece so as to make the airflow spiraling without any blockage. The swirling flow field would not form the disturbed zone causing the loss of the energy resulting from the influence of the vent. The energy transferred by the gas of the media increases greatly, the movement velocity of the media increases, and the fishing effect improves.

The vent is connected with a two-position two-way solenoid valve by the pipeline. When the gas begins to flow from the swirling head, the two-position two-way solenoid valve is closed and the vent stops the exhaust. When the gas flowing out of the swirling head needs to be exhausted, the two-position two-way solenoid valve is opened and the gas would flow out of the exhaust interface.

(6) The airflow spiraling direction and clamping force of the swirling heads

When the airflow from the swirling heads rotates in the opposite direction, the media can spiral in the opposite direction, and the reticular texture is formed at the surface of the workpiece, making the surface smoother. But it is difficult to deburr the cross-hole, because the media always contact with the same side of the edge. When the direction of the gas flow out of the two-way swirling head is the same, it is beneficial for the removal of burr, but it is difficult to form the reticular texture and make the surface smoother.

When finishing the workpiece, the fixed swirling head is stationary and the free swirling head is driven by the air cylinder with diameter of 40 mm, stroke of 20 mm or so, and working pressure of 0.1 MPa slides in the cylindrical slide realizing the clamping and loosening of the workpiece. When the length of the workpiece changes, the bracket of the free swirling head becomes stationary and the bracket of the fixed swirling head is fixed after sliding at an appropriate location.

### 5.4.3 Other Parts

1. Pneumatic system design.

The power source of the gas-particle two-phase swirling flow finishing device is the compressed air. According to the finishing requirements, the air source from the fixed and free swirling heads needs to be connected or cut off in real time, which can be controlled by the pneumatic system realizing the delivery of the airflow to the nozzle and ensuring the workpiece clamped and sealed firmly.

The pneumatic system is mainly used for the delivery of the airflow differing from the transfer of the gas pressure. The loss of the gas used for transferring the pressure is very small, but the gas will be exhausted outside for the gas-particle swirling flow finishing device. The gas pressure is determined by the load which is generally the cylinder for pneumatic motion. Whether the flow rate and energy of the pump is blocked and used for the finishing or not, the main load of the gas-particle two-phase swirling flow finishing is the nozzle and the media.

Figure 5.71 is the schematic of the pneumatic system using the gas transmission with the external diameter of 12 mm and the inner diameter of 8 mm based on the international direct terminal interface [40]. When the two-position two-way valve is opened, and the two-position four-way valve 7 is connected with the right, the gas flows through the flowmeter, the two-position two-way valve, the one-way valve, the air cleaner, the relief valve, and the two-position four-way valve arriving at the cylinder. The piston would drive the swirling head at the right end moving toward the left and clamping the workpiece. Otherwise, the two-position four-way valve is connected with the left and the piston would move toward the right loosening the workpiece. When the two-position two-way valve 14 is opened, and the corresponding two-position two-way valve 15 is closed, and the two-position two-way valve 9 is opened, and the corresponding two-position two-way valve 10 is closed, the gas flowing out of the swirling head can move from left to right, whereas the gas can move from right to left.

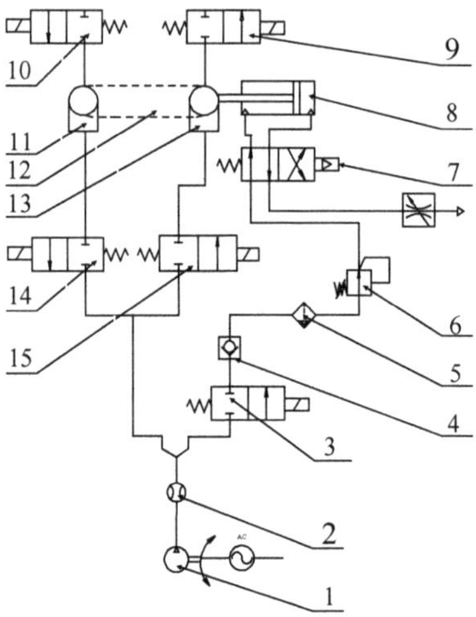


**Fig. 5.71** Schematic of pneumatic system. 1—Pump; 2—flowmeter; 3, 9, 10, 14, 15—two-position two-way valve; 4—one-way valve; 5—air cleaner; 6—relief valve; 7—two-position four-way valve; 8—cylinder; 11, 13—swirling head; 12—workpiece

Each pneumatic component has a choking effect. The airflow passes through the pipe, then the two-position two-way valve enters into the nozzle, and it also passes through the vent and the solenoid valve flowing out of the swirling head. There are the choking effect and the airflow blockage in the whole process leading to the flow rate decrease and the backpressure increase in the working airflow. The decrease in the flow rate means the energy of the gas decreasing, and the increase in the backpressure means that the energy transferring from the swirling field is influenced and disturbed. So the inlet pressure of the nozzle must be improved for the compensation. Compared with the gas not flowing through the nozzle, the flow rate of the gas flowing through the nozzle decreases to some extent and can be obtained by the theoretical calculation and the experimental measurement. The media have little influence on the flow rate, but the loaded quantity of the media is determined by the flow rate of the gas.

2. Circuit design of the relay

The relay control has wide application. The control of the circuit can be realized by the on or off of the trigger switch. According to the control requirements of the gas valve, the type of the relay is JZC-23FX5, the rated working voltage is 12 V, and the rated working current is 30 mA. The circuit diagram of the relay is shown in Fig. 5.72 [41].

The corresponding switch signal controller of the single-chip microcomputer is turned on as shown in Fig. 5.72, and the control circuit of the relay was triggered. The base electrode of the transistor would be in saturation making the collector and emitter of the transistor connected. The control coil is switched on, and the corresponding contact is made from controlling the electromagnet in the energized

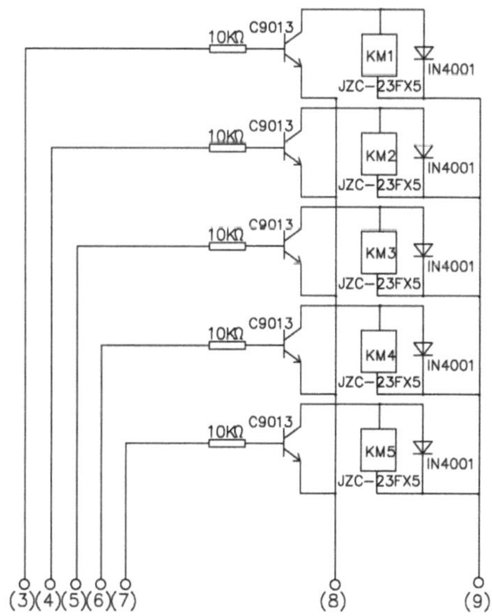


**Fig. 5.72** Circuit diagram of the relay

state. The corresponding gas valve is opened when the gas flows, and the finishing is completed.

In addition, the diode is connected with the control coils in parallel with the design preventing the transistor breakdown resulting from the high voltage at the moment of lose power.

3. Circuit design of the single-chip microcomputer

MCU control can realize the control function of the programmable controller. It has the characteristics of high reliability, strong expandability, and low control cost. Therefore, PIC16C74B was used to develop control system of swirling flow finishing device. For the program development of MCU, the main considerations are as follows:

(1) I/O ports of MCU are used to acquire switch states. For the development of MCU programs, the time delay should be considered so as to eliminate jitter problems of control buttons. When the microcontroller works, the real-time MCU program scans the I/O ports. When a button is pressed, the microcontroller's scanning program will perceive the corresponding information. When the operator presses the button, the button contact is a process from local contact to the overall contact; this will inevitably lead to changes in the resistance of the circuit, causing changes of the circuit current and the input voltage. In order to eliminate the influence of the access voltage on the control stability, a key without jitter program was prepared. That is, the microcontroller program scans the input I/O ports. When button pressing is scanned, the program read the port data by delaying 20 ms.
(2) Displaying of double-digit digital tube. The double-digit digital tube uses a common cathode connection, which determines that the displayed data are inconsistent with the data transmitted by DB bus. Therefore, before each display, it needs to convert the figures into single-byte 8-bit code and then transfers them to the display circuit through the DB bus. This transformation can be finished with subprogram. So, the figures are directly sent to the accumulator, and the transformation and the display can be achieved by the program.
(3) The I/O ports connected with the "machining/adjustment" button should have bidirectional input function; that is, when there are no inputs, it is in the processing state and when there are inputs, it is in the adjustment state.
(4) The I/O interface connected to the digital keypad is used as the input of the control program. It can set the initial value of the machining state from the program to determine the initial parameters of the machining. The set key is used to switch the machining to be a determined parameter.

4. The control panel design

The control panel should be designed of simple pattern and clear instruction meaning, so as to reduce the operator error.

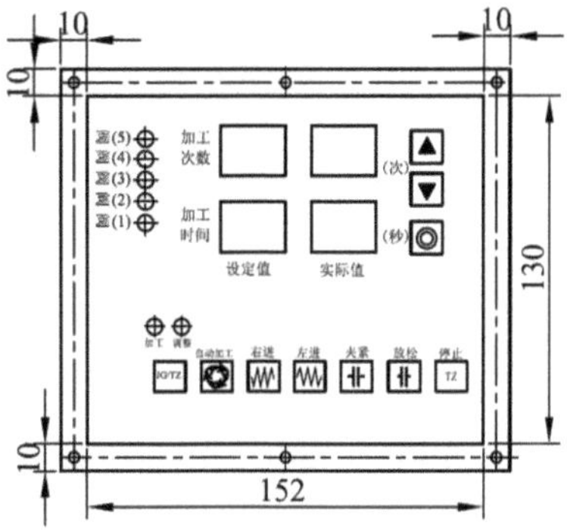


**Fig. 5.73** Appearance and related dimension diagram of the control panel

Light gray is used as the background color of the control panel. The blue color is applied to the control buttons ; black is served as the color of the buttons, symbols, and characters indicating the function. In order to warn and remind the operator, they should adjust processing parameters (processing time, the number of processing) and before the process, buttons used to adjust the parameters are painted as orange. Meanwhile, we are supposed to consider the position of the panel on the machine and its size; single-chip hardware circuit board's shape and size; installation; and other factors while designing the control panel. Figure 5.73 reveals the appearance of the control panel and the relevant dimensions.

## 5.5 Main Factors and Finishing Effects

### *5.5.1 Technical Process*

There are many factors affecting the finishing quality and efficiency in the gas-particle swirling flow hole finishing. The workpiece conditions, processing parameters, and processing media are the main factors. After the workpiece and requirements were determined, factors influencing the finishing effect and efficiency are: ① the parameters of equipment: nozzle angle, nozzle location, nozzle diameter, and nozzle number. The nozzle angle refers to the inclination angle between the axis of nozzle and the end face of the swirling head. The location of nozzle refers to the distance between the axis of nozzle and the axis of the swirling head; ② the dynamic parameter of equipment: power used to provide air supply. When the air supply is certain, the speed and the pressure distribution of the swirling flow are determined by the structure parameters; ③ the parameters of media: the abrasive grain, the media hardness, the shape and dimension, the load; ④ the processing parameters: the processing time is determined by the number of cycles and quantities.

The experimental device of gas-particle swirling flow finishing is shown in Fig. 5.74.

**Fig. 5.74** Experimental device of gas-particle swirling flow finishing

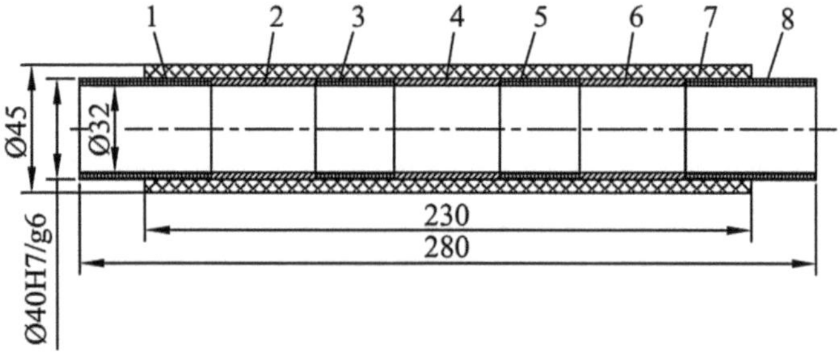


**Fig. 5.75** Assembly diagram of the workpiece. 1, 8—The sleeve connected with the swirling head; 2, 4, 6—workpiece; 3, 5—sleeves between the workpiece; 7—outside sleeve

The assembly diagram of the workpiece is shown in Fig. 5.75. The material of the workpiece is 20 steel. The surface roughness of the inner hole is about *Ra* 2.0 μm. The axial length of the workpiece is 40 mm. The material of the sleeve 1 and 8 connected with the swirling head is the organic glass. The axial length of the sleeve is 50 mm. The material of the sleeves 3 and 5 between the workpiece and the outside sleeve 7 is also an organic glass. The workpiece should be cleaned with kerosene and dried before and after the experiment.

Other experimental conditions are as follows: The offset distance of the nozzle is 0–3 mm. (It is assumed that distance between the axis of nozzle and the axis of swirling head equal to the radius of swirling head minus the radius of nozzle, meaning the offset distance is 0.) The diameter of nozzle is 2, 3, and 4 mm. The inclination angle of nozzle is 5°–30°. The gas flow rate of nozzle is 15 $m^3$/h. The material of media is corundum, and the diameter is 5 mm.

## 5.5.2 Main Factors

1. Parameters of device

When dynamic parameters (mainly refers to air supply) are given, the property of the swirling flow is determined by the parameters of the device structure such as the nozzle diameter, the nozzle number, and the nozzle location. These parameters affect the relative speed and force between the media and the surface of workpiece, resulting in different processing performances [42].

(1) The nozzle diameter

Table 5.5 shows the surface roughness *Ra* value with time using different nozzle diameters. Figure 5.76 shows the change of the surface roughness *Ra* with the time using different nozzle diameters.

The results show that the change of the surface roughness *Ra* value is very large in 1 min. The change curve tends to turn flat gradually with the increase in the processing time (over time), which indicates that there is an optimum processing time. When the flow rate of the gas remains unchanged, the processing effect of the nozzle with bigger diameter is inconspicuous. The changing of the diameter has little effect on the processing results. In the same condition, the processing effect is determined by the processing time and the force between the workpiece and the media. When the diameter of the nozzle increases, the tangential speed of the swirling flow and the speed of the media would decrease. The force of media acting on the workpiece also decreases. The contact time of the media and the surface of

**Table 5.5** Surface roughness *Ra* value with time using different nozzle diameters (μm)

| Diameter of nozzle/mm | Time *t*/min | | | | |
|---|---|---|---|---|---|
| | 0 | 1 | 3 | 5 | 8 |
| 4 | 1.77 | 0.84 | 0.60 | 0.71 | 0.47 |
| 3 | 1.72 | 0.82 | 0.61 | 0.56 | 0.50 |
| 2 | 1.65 | 0.81 | 0.60 | 0.57 | 0.56 |

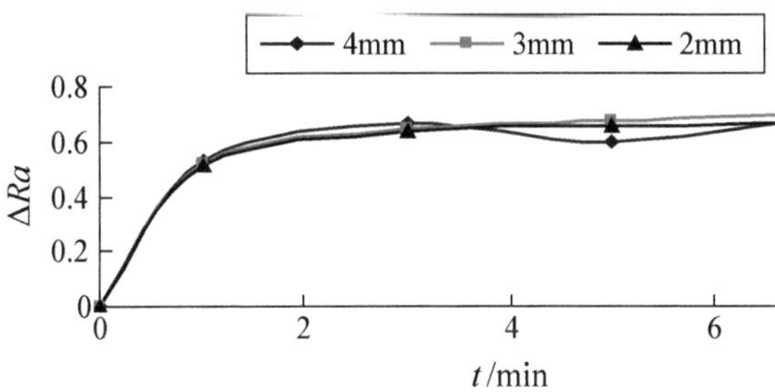


**Fig. 5.76** Change of the surface roughness *Ra* value with the time in different nozzle diameters

the workpiece become longer. In general, the change of diameter has relatively small effects on the processing performance. Therefore, the diameter of nozzle should not be too large in order to improve productivity.

(2) The inclination angle of the nozzle

Table 5.6 shows the surface roughness *Ra* value with time at different nozzle inclination angles. Figure 5.77 shows the change of the surface roughness *Ra* value with the time at different nozzle inclination angles.

The results show that the surface roughness *Ra* values could be reduced by 50% after 1 min in different inclination angles of the nozzle. It decreases slowly with time and reaches to about 0.5 μm at 8 min, which shows that the inclination angle of the nozzle has little effect on the processing. When the inclination angle of the nozzle increases, the tangential speed of the swirling flow would decrease, so did the force of the workpiece to the media. The axial speed of the swirling flow increased, which led to the increase in the contact number between the workpiece and media. The angle between the movement direction of media and the direction of the surface texture increased, which was beneficial to reduce the surface roughness. In general, the change of the inclination angle had relatively small

**Table 5.6** Surface roughness *Ra* value with time at different inclination angles of the nozzle (μm)

| Inclination angle of the nozzle | Time *t*/min | | | | |
|---|---|---|---|---|---|
| | 0 | 1 | 3 | 5 | 8 |
| 5° | 1.70 | 0.89 | 0.66 | 0.61 | 0.56 |
| 10° | 1.78 | 0.71 | 0.63 | 0.51 | 0.48 |
| 15° | 1.77 | 0.84 | 0.60 | 0.71 | 0.47 |
| 20° | 1.70 | 0.77 | 0.72 | 0.50 | 0.53 |
| 25° | 1.75 | 0.84 | 0.68 | 0.57 | 0.53 |
| 30° | 1.70 | 0.78 | 0.57 | 0.43 | 0.55 |

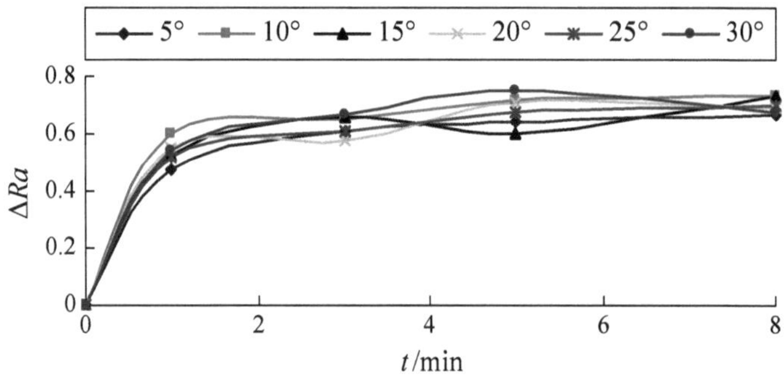


**Fig. 5.77** Change of the surface roughness *Ra* value with the time at different nozzle inclination angles

impact on the processing effect. The screw pitch would increase with the increase in inclination angle, which led to poor uniformity of surface in the axial direction. Therefore, inclination angle of the nozzle should not be too large.

(3) The nozzle location

Table 5.7 shows the surface roughness *Ra* value with time at different nozzle locations. Figure 5.78 shows the change of the surface roughness *Ra* with the time at different nozzle locations.

The results show that the surface roughness *Ra* increased with the increase in the nozzle offset distance. When the offset distance is 0 or negative, the gas flow from the nozzle would directly contact with the surface of the swirling head. The loss of the energy was greater. The offset distance should be considered in the design of the swirling head, and the most appropriate offset distance is about 2 mm.

2. Dynamic parameters of the device

Dynamic parameters of the device are mainly referred to the power used to provide air supply. From the simulation analysis, the inlet velocity of the nozzle would increase with the increase in air supply power. The increase in the power would also lead to the increase in the inlet speed of the nozzle and a smaller screw pitch. The dynamic pressure, the tangential speed, and the vortex strength would increase as well. So the larger the inlet speed of the nozzle is, the better the

**Table 5.7** Surface roughness *Ra* value with time at different nozzle locations (μm)

| Location of the nozzle/mm | Time *t*/min | | | | |
|---|---|---|---|---|---|
| | 0 | 1 | 3 | 5 | 8 |
| 0 | 1.78 | 0.71 | 0.63 | 0.51 | 0.48 |
| 1.5 | 1.87 | 0.80 | 0.77 | 0.67 | 0.52 |
| 3 | 2.13 | 0.87 | 0.55 | 0.49 | 0.50 |

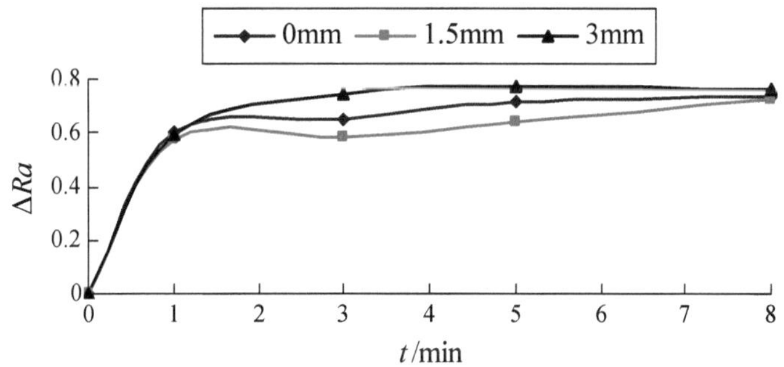


**Fig. 5.78** Change rate of the surface roughness *Ra* value with the time at different nozzle locations

processing effect and efficiency are, which is about 0.4–0.6 MPa considering the industrial application.

3. Processing parameters

(1) The processing time

In the case of gas flowing through the one-way swirling head, surface roughness was measured at the position of the axial distance of 29.4 mm. Figure 5.79 shows the surface roughness curve at different times. Figure 5.80 shows the changes of *Ra* and *Rmr*(*c*) with the machining time.

Figures 5.79 and 5.80 show that the *Ra* and *Rmr*(*c*) change with time. The surface roughness *Ra* value decreases significantly in 3 min and keeps at 0.42 μm after 3 min. This means that after a period of machining time, the increase in machining time has little improvement in surface roughness. As shown in Fig. 5.80b, the profile bearing ratio increases significantly with the machining time, which is beneficial to improving the load bearing ability and wear resistance of workpiece.

Figure 5.81 shows the variations of Δ*Ra* with the axial distance from the gas inlet after 1-min finishing for one-way swirling head.

Figure 5.81 shows that the Δ*Ra* decreases gradually with the increase in axial distance from the gas inlet. This phenomenon leads to the uniformity of the surface at the axial direction. The decrease was not obvious when the distance was about 150 mm. The process ability of the gas-particle swirling flow finishing reduced gradually with the increase in the axial distance, which was consistent with the simulation results. Therefore, the bidirectional spiral head structure should be used for processing in order to improve the uniformity.

(2) The parameters of media

Media is the “tool” used for surface finishing and its grain size, and hardness and size have a great effect on the finishing performance.

(1) The grain size of media

The larger grain size means that the media have finer abrasive. The media hardness determines how tightly the bond holds the abrasive. Table 5.8 and Fig. 5.82 show the variations of the surface roughness *Ra* with time under different abrasive grain sizes.

The experiment results show that the abrasive grain size of media has great influence on surface roughness *Ra*. The surface roughness *Ra* of the workpiece processed with $180^{\#}$ media is nearly half of the surface roughness *Ra* of the workpiece processed with $80^{\#}$ media. In surface finishing, media of larger abrasive grain provides better performance in decreasing surface roughness *Ra*. But, for the $120^{\#}$ media or larger does not show further improvement ability in decreasing surface roughness; therefore, the $120^{\#}$ media or larger should be selected to obtain a smaller *R*a.

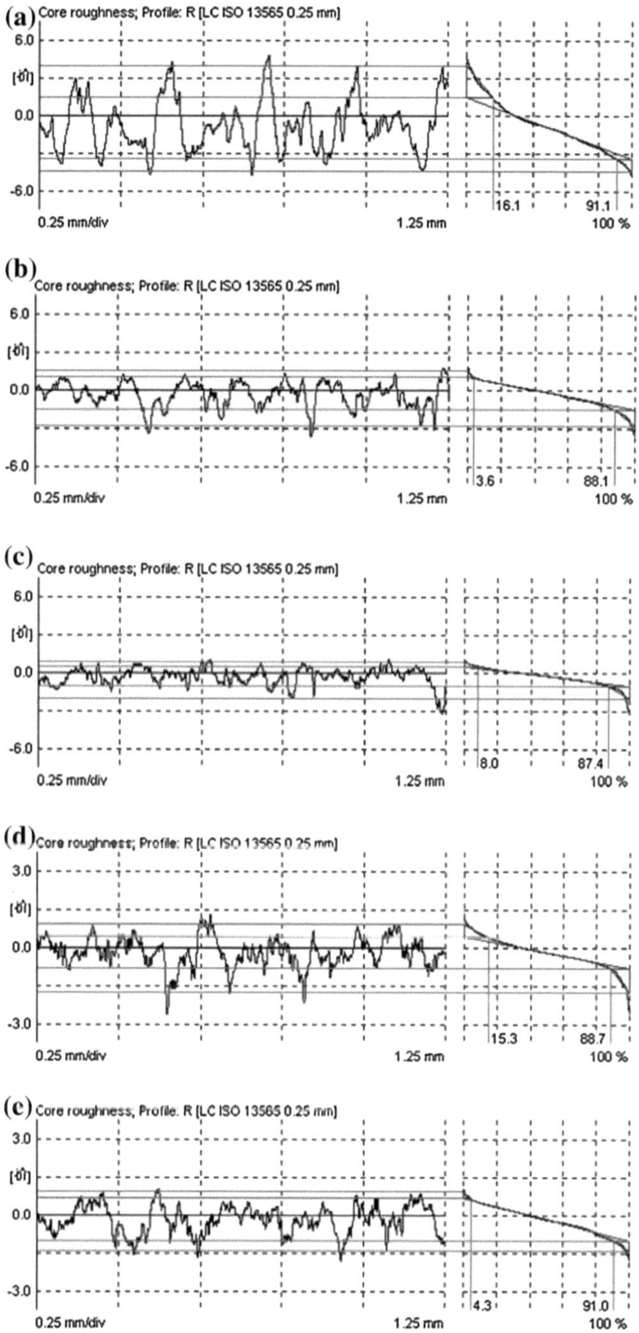


**Fig. 5.79** Change curve of surface roughness at different times. **a** Before processing; **b** 2 min; **c** 4 min; **d** 6 min; **e** 8 min

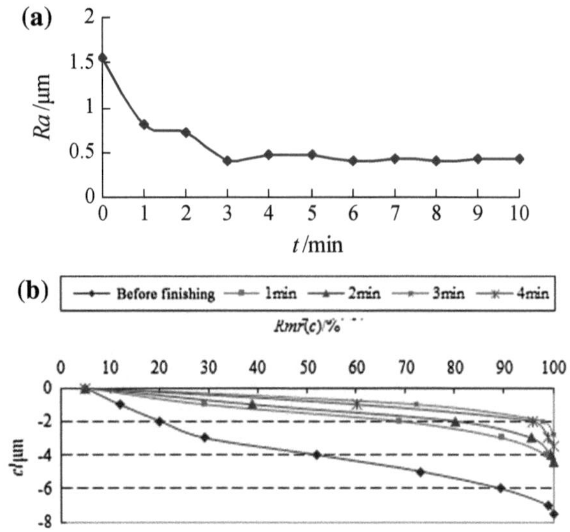


Fig. 5.80 Changes of *Ra* and *Rmr*(*c*) with the machining time. **a** *Ra*; **b** *Rmr*(*c*)

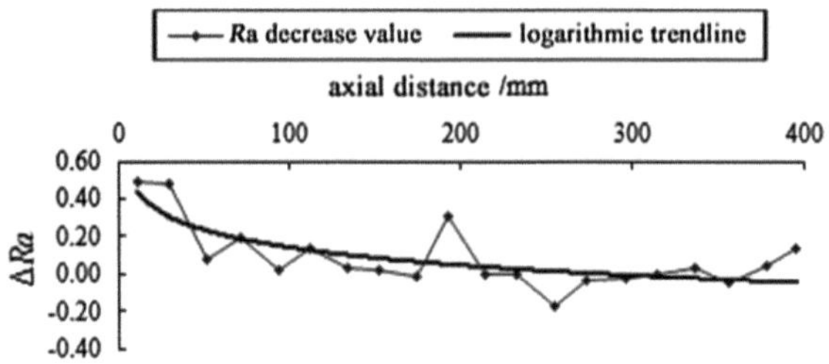


**Fig. 5.81** Variations of $\Delta Ra$ with the axial distance from the gas inlet

**Table 5.8** Surface roughness *Ra* with time under different abrasive grain sizes (μm)

| The abrasive grain of the media | Time *t*/min | | | | |
|---|---|---|---|---|---|
| | 0 | 1 | 3 | 5 | 8 |
| 180# | 1.72 | 0.82 | 0.61 | 0.56 | 0.50 |
| 80# | 1.70 | 1.26 | 1.09 | 1.05 | 1.00 |

(2) The media diameter

Table 5.9 shows the surface roughness *Ra* value with the time using different diameter media. Figure 5.82 shows the variations of surface roughness *Ra* with the time under different diameters.

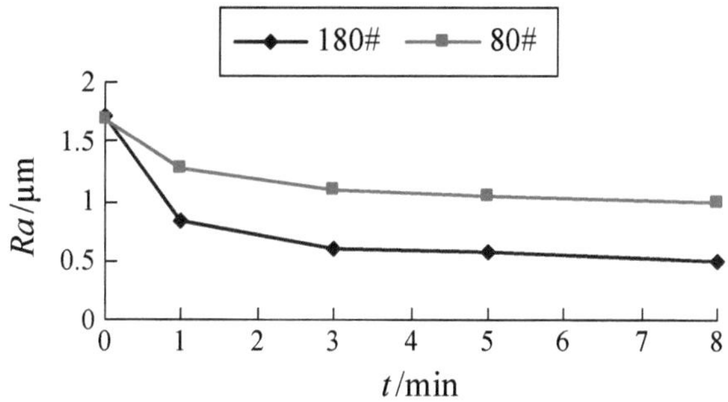


**Fig. 5.82** Variations of the surface roughness *Ra* value with time under different abrasive grain sizes

**Table 5.9** Surface roughness *Ra* with the time under different diameter media (μm)

| Diameter of the media/mm | Time *t*/min | | | | |
|---|---|---|---|---|---|
| | 0 | 1 | 3 | 5 | 8 |
| 1 | 1.65 | 1.18 | 0.99 | 0.85 | 0.89 |
| 3 | 1.68 | 0.99 | 0.92 | 0.93 | 0.88 |
| 5 | 1.72 | 0.82 | 0.61 | 0.56 | 0.50 |

The experiment results show that the media diameter has great influence on surface roughness *Ra*. The surface roughness of the workpiece decreases with the increase in media diameter. The *Ra* of the workpiece processed with 5 mm media diameter is nearly half of the surface roughness *Ra* of the workpiece processed with 5 mm media diameter. With other parameters are fixed, the media with bigger diameter leads to and thus smaller surface roughness *Ra*. However, the diameter of the media should not be too large, because the larger force produced by using larger diameter media leads poor surface roughness (Fig. 5.82).

(3) The media material

The workpiece material used in experiments is 45 steel. The nozzle diameter is 4 mm. The inclination angle of the nozzle is 10°, and the offset distance is 0 mm. Table 5.10 and Fig. 5.84 are the experiment results and variations of the surface roughness *Ra* with the time by using different material media.

As shown in Table 5.10 and Fig. 5.84, the surface roughness *Ra* does not have obvious improvement when the media material is corundum. The surface roughness

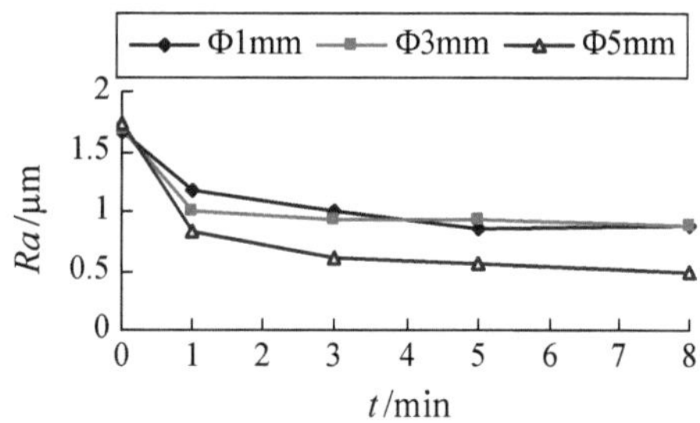


**Fig. 5.83** Variations of the surface roughness *Ra* with the time by using different diameter media

**Table 5.10** Surface roughness *Ra* with the time by using different materials (μm)

| Material of the media | Time *t*/min | | | |
|---|---|---|---|---|
| | 0 | 2 | 4 | 6 |
| Corundum | 0.43 | 0.41 | 0.40 | – |
| Steel Ball | 0.45 | 0.29 | 0.20 | 0.186 |

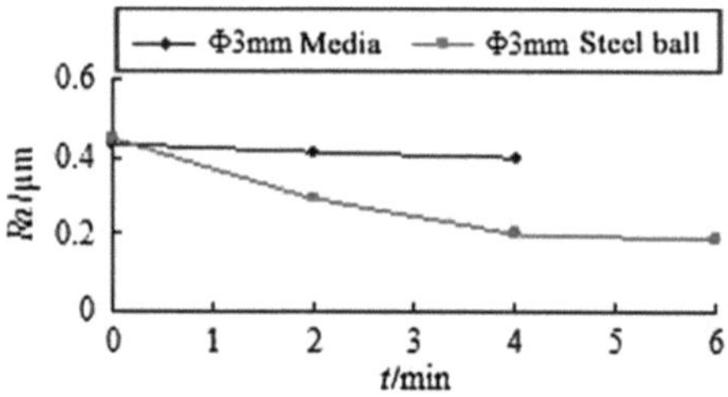


**Fig. 5.84** Variations of the surface roughness *Ra* with the time

*Ra* changed from 0.45 to 0.2 μm in 4 min firstly and then has little improvement when the media material is steel.

According to experimental studies, there is a limit value for surface roughness of part processed with a certain media material and shape. The above-conducted experiments show that the surface roughness would no longer decrease when the surface roughness value reached limit value. Therefore, considering the quality limit, the media made of corundum could not be used to finish when the surface roughness *Ra* value of part is lower than 0.4 μm.

(4) Loading capacity of media

As the spherical media are driven to flow into the workpiece, not all the spherical media flew fast out of the swirling head. The whole supply time can last 10 s or few tens of seconds according to the number of spherical media. The motion state of spherical media could be qualitatively or quantitatively described with the features: the stability of media supply, the number of media in stable supply, and the whole time of media moved from one side to the other side of the swirling head. Table 5.11 shows the motion state of the $Al_2O_3$ media.

Figure 5.85 shows the whole movement time and the number of media moving in 15 s or so at different masses of the media.

No matter what kind of nozzle is used, if the mass of the loaded media is 5 g, the media could rotate at high speed in the swirling head, but could not be normally pushed into the organic glass tube due to smaller mass of the loaded media. So the number of the media in simultaneously moving state in the tube is very little, and the media have lower axial velocity. Almost all the media do circular motion in the same position, and many media could not move out from the swirling head after 20 s.

Under the condition that the mass of loaded media is 10 g, the media could gradually move out from the swirling head in the one-way finishing. During the period from 5 to 10 s, the media show sparse uniform distribution state. After 10 s,

**Table 5.11** Motion state of the $Al_2O_3$ media

| Weight of media/g | Number of media | Number of media moved simultaneously | | | Time of 95% media/s | | | Remark |
|---|---|---|---|---|---|---|---|---|
| | | 2 mm | 4 mm | 6 mm | 2 mm | 4 mm | 6 mm | |
| 5 | 30 | – | – | – | – | – | – | unavailable |
| 10 | 60 | 15 | 15 | – | 20 | 20 | 15 | |
| 20 | 120 | 23 | 22 | – | 25 | 28 | 20 | |
| 30 | 180 | 28 | 22 | – | 30 | 40 | 25 | |
| 40 | 240 | 32 | 23 | – | 35 | 45 | 25 | |
| 50 | 300 | 28 | 28 | – | 40 | 50 | 20 | |
| 60 | 360 | 33 | 25 | – | 40 | 50 | 20 | unavailable |
| 70 | 420 | 30 | 30 | – | 35 | 45 | 20 | unavailable |
| 80 | 480 | 28 | 30 | – | 35 | 40 | 20 | unavailable |

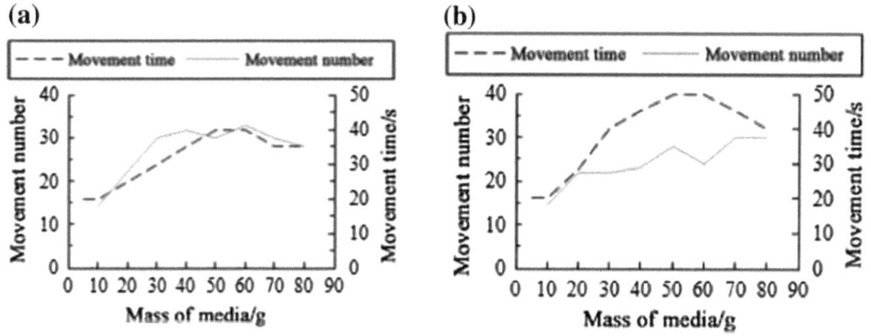


**Fig. 5.85** Whole movement time and the number of the media in simultaneously moving state in about 15 s. **a** Nozzle diameter is 2 mm; **b** nozzle diameter is 4 mm

the number of media rotating together gradually decreased. And its maximum value is about 15.

Under the condition that the mass of loaded media increases to 20 g, the working state shows better stability and uniformity. For the nozzle with the diameter of 4 mm, the speed of the media flowing out of the swirling head is faster and more uniform in 25 s. The number of the media moving together in the tube is about 20, and the whole movement time of media is about 28 s. The friction noise between the media and the pipe is relatively stable and loud, which meant that the speed of the media is high and stable.

Under the condition that the mass of loaded media increases to 50 g, the media moving in a straight line to the other end of the swirling head completely do not have the processing ability from 1 to 2 s. The number of the media flowing out of the swirling head was more. And it still increased along with the increase in media from 2 to 5 s. However, the axial speed of media is high, while the tangential speed is slow. Many media flow into the other side of the swirling head during the time and almost have not any processing capacity. During the period from 10 to 30 s, the number of media moving together decreases in the stable supply state. The speed of the media flowing out of the swirling head is uniform, and the axial speed of media is slower, while the tangential speed becomes faster, which is the state that the workpiece can be processed. The whole movement time for one cycle is 45 s for the media with mass of 30–50 g. When the movement time is at about 15 s, the number of the media moving together is about 25–30.

Under the condition that the mass of loaded media increases to 60–80 g, the excessive media almost occupy all the space of the swirling head, which leads to the difficulty in the swirling flow formation. The media only move from one swirling head to the other swirling head during from 2 to 5 s and have not any processing ability. The media speed is far less than that of the media with the load mass is 20 g. During the whole movement, the number of media is large and the speed of media is slow. So the processing ability of the media is weak. Because the formation of the swirling flow is blocked by too much media, the screw pitch of the media is large in the position ranging from $Z = 200$ mm to $Z = 400$ mm. And the tangential speed was slow. The whole movement time is about 40–45 s.

In a word, the number of the media moving together is about 25–30 for the nozzle with the diameter of 2 and 4 mm. For the same nozzle, because the certain flow rate could only support the certain intensity of vortex flow field, when the mass of the media is more than 30 g, the time of the media moving from one side to the other side almost kept the same. Because excessive media limits the formation of the swirling flow field, the media would be flow out of the swirling head or flow into another swirling head at a high speed by the airflow at the beginning of the movement. Thus, the number of media gradually decreases until the swirling flow field formed, and the media staying in the swirling head begin to spiral along the tube. The situation is similar to the nozzles with the diameters of 2 and 4 mm, but the motion state of the nozzle with the diameter of 4 mm is better. When using the nozzle with the diameter of 4 mm, the speed of the media flowing out of the swirling head is more uniform from the stable state to the end. There are not excessive media flowing out of the swirling head because there are more media in the swirling head, but there are always media continuously flowing out of the swirling head near the same rate of the speed until most media flow out. The stability is poor for the nozzle with the diameter of 2 mm. The more the media in the swirling head, the more the media flowing out of. The less the media in the swirling head, the less the media flowing out of. Although spherical media moving at the same time are less and less, the speed of spherical media could be obviously enhanced. The efficiency would decrease when the number of spherical media is large. The movement time of spherical media does not increase with the increase in the loading. The stage of abundant media flow from swirling head with fast axial velocity became longer, while the stage of uniform flowing with faster speed became shorter. That is to say, the effective processing time decreases. The load of the media is about 20–30 g for high efficiency. At the same time, for the nozzle with the diameter of 6 mm, the speed of the gas is very slow which resulted from the limitation of the gas source and pressure, and the flow field is weakening. During the whole one-way motion, the speed of the media is slow, and it is especially slow at the beginning of the motion causing the media having no processing ability [43].

Table 5.12 shows the motion state of the $Al_2O_3$ media with the diameter of 3.5 mm. When the mass of media is 5–10 g, the motion of media is fast and stable. When the mass of the media reaches to 30 g, there are more media flowing out of the swirling head, and the axial speed is high causing the media having the weaker processing capacity. This situation is more obvious with the increase in the mass of the media. The friction noise between the media and the tube is relatively smaller than that between the media with the diameter of 5 mm and the tube. That is because the mass of media with the diameter of 3.5 mm is relatively small leading to the small centrifugal force.

4. Other parameters

The material of the workpiece is 45 steel, the nozzle diameter is 4 mm, and the mass and the diameter of the media are 5 mm and 20 g, respectively. The one-way processing time is 15 s. Table 5.13 shows the *Ra* and removal thickness of the workpiece.

**Table 5.12** Motion state of the $Al_2O_3$ media with the diameter of 3.5 mm

| Quality of the media/g | Number of the media | Number of the media movement at the same time | | Whole time of 95% media/s | |
|---|---|---|---|---|---|
| | | 2 mm | 4 mm | 2 mm | 4 mm |
| 5 | 90 | 25 | 10 | 15 | 15 |
| 10 | 180 | 25 | 15 | 15 | 23 |
| 20 | 360 | 40 | 20 | 25 | 38 |
| 30 | 540 | 50 | 25 | 25 | 50 |
| 40 | 720 | 60 | 45 | 25 | 45 |
| 50 | 900 | 50 | 35 | 25 | 50 |
| 60 | 1100 | 50 | 35 | 25 | 50 |

**Table 5.13** Surface roughness *Ra* and the removal thickness of the workpiece

| Test item | Time *t*/min | | | | |
|---|---|---|---|---|---|
| | 0 | 5 | 10 | 15 | 25 |
| *Ra*/μm | 0.92 | 0.65 | 0.56 | 0.53 | 0.54 |
| The removal thickness of the workpiece/μm | 0 | 0.07 | 0.14 | 0.14 | 0.14 |

Table 5.13 shows that the surface roughness *Ra* and the removal thickness of the workpiece almost keeps unchanged when the machining time reaches to 5 min and it is also given in Table 5.13 that the maximum removal thickness is only 0.14 μm, which could keep the original surface accuracy. It also shows that the gas-particle two-phase swirling flow finishing is a microprocessing technology and it could be applied to the finishing and deburring of the hole surface.

### 5.5.3 Finishing Performance

1. The finishing of cross-holes

(1) The workpiece

The workpiece is shown in Fig. 5.86a. The material of the workpiece is 45 steel. There are one or two holes with the diameter of 3 mm punched on the arc surface. The burrs with the height of 0.02–0.1 mm occurred at the corner of the cross-holes. The distance from the workpiece to the swirling head is 20 mm.

(2) Media and the experiment setup

Burring and edge rounding could only use the different sizes of spherical media instead of the steel ball mainly used for rolling. So the $Al_2O_3$ media with the diameter of 5 mm and the mass of 20 g are selected. According to the analysis of the working state, the nozzle with the diameter of 4 mm is selected. The helical direction of the swirling head is contrary.

(3) The processing time

The workpiece is placed in the sleeve with 150 mm length. The time of the spherical media moving from one swirling head to the other swirling head is 12 s. One cycle consists of two one-way movements.

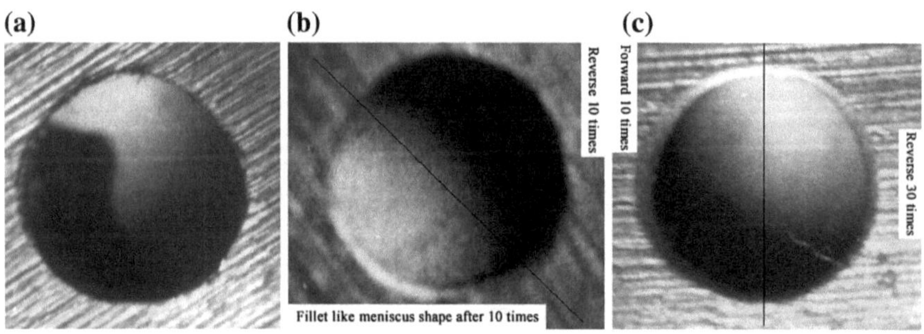


**Fig. 5.86** Processing performance of the cross-hole. **a** Before finishing; **b** forward finishing 5 min and reverse 5 min; **c** forward finishing 5 min and reverse 15 min

(4) The processing condition

The offset distance of the nozzle is 0 mm. The diameter of the nozzle is 4 mm. The inclination angle is 5°–30°. The flow rate of the nozzle is 15 $m^3$/h. The abrasive grain of corundum media is $180^{\#}$. The diameter of media is 5 mm.

(5) The processing performance

As shown in Fig. 5.86, the burr of the workpiece processed for ten times (5 min) was removed. And a slope with 0.1–0.3 mm wideness occurred on one side of the cross-holes [44]. The burrs locating at the other side of the cross-holes were reduced or bent. After the workpiece was placed reversely and processed for ten times, the burrs were partly removed. Then, the workpiece was continuously processed for ten times and almost all burrs were removed without a slope. The deburring principle diagram of the cross-hole is shown in Fig. 5.87.

From Fig. 5.87, the media driven by the gas flow would rotate along the hole surface when the gas flew through one side of the swirling head. It firstly reaches to the A point. The burrs were touched, collided, rolled, and cut slightly by the media. The burrs would be pushed down and bent or produced plastic deformation. Then, the burrs and edge located at the point B would be collided by the media along the tangential direction. Finally, the edge became curved and the burrs were removed. From experiments, the removal amount of burrs at the point B is larger and the edge becomes crescent shape. The processing is uneven for the burr located at the points A and B. So the direction of the swirling head should be changed in order to obtain better processing performance.

As can be seen from experiments, for the material with the lower hardness and larger plastic, burrs located in the bigger hole can be removed. But for the edge of the cross-hole, plastic deformation occurs and thus leads to the removal of burr. So the gas-particle swirling flow finishing can be used to finish brittle material workpiece.

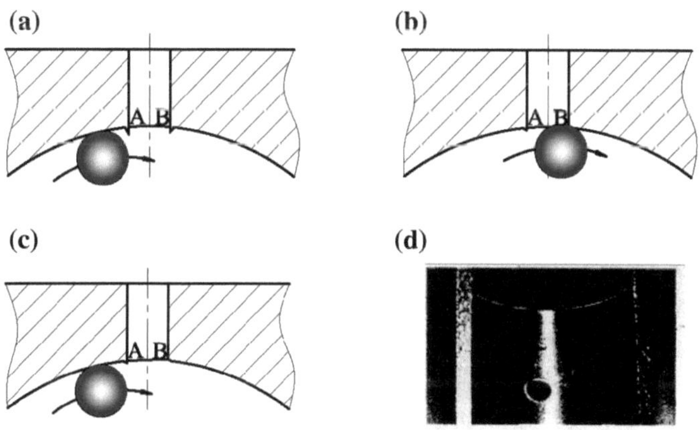


**Fig. 5.87** Deburring principle diagram of the cross-holes

2. The blind hole finishing

It was difficult to remove burrs located in the blind hole of workpiece such as the brake pump. At present, some tools are used to remove burrs by hand. The efficiency is low, and the processing quality is not stable. This kind of workpiece can only be used by single swirling head. To make the gas flow out of the blind hole, the pipe for discharging should be installed. The length of the pipe can be adjusted to reach the bottom of the blind hole to finish the whole process. After 2 min, burrs located in the cross-holes of the valve would be removed and crescent occurred. And the original surface roughness of the valve could be kept.

## 5.6 Two-Phase Compulsive Circulation Flow Finishing

### *5.6.1 Gas-Particle Two-Phase Compulsive Circulation Flow Finishing*

Gas-particle swirling flow can realize the hole surface finishing. Theoretical analysis and experimental results indicate that this process can decrease 1–2 level of surface roughness *Ra* value and have high efficiency. But its energy is much weakened along axial direction, and air initial pressure is low. So its application is limited. Therefore, as shown in Fig. 5.88, a new finishing process of gas-particle two-phase compulsive circulation flows is brought up. With specific component of circulation formation, the air, having certain pressure and flux, forms rotation motion along the hole surface. Particles are inhaled by the air or preloaded in the hole driven by the air, so the two-phase compulsive circulation flows emerges. High kinetic energy of the particle, obtained from compulsive circulation flows, is

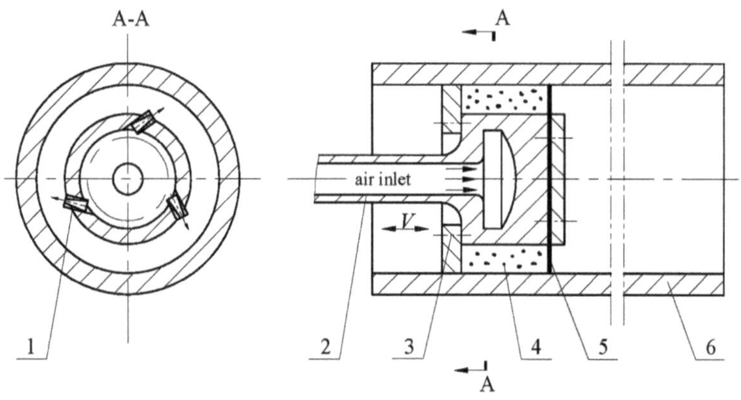


**Fig. 5.88** Basic principle diagram of gas-particle two-phase compulsive circulation flow finishing. 1—Nozzle; 2—circulation formation component; 3—oriented component; 4—particle; 5—isolated net; 6—workpiece

converted into finishing energy, which leads to high finishing efficiency. With the axial relative motion between hole surface and circulation formation component, surface finishing can be persisted. Under force and motion, the particle has the action of rolling and the minimum grinding with scraping and scoring against the hole surface. Finally, it realizes the hole surface finishing [45].

1. Numerical Simulation

For experimental research and improvement in foresight of practical application, numerical simulation of gas-particle two-phase compulsive circulation flow finishing is done through computational fluid dynamics(CFD) software(Fluent, Gambit), and it provides credible theoretical basis for thorough research. Figure 5.89 is its sketch model. Convergence curves of residual difference are shown in Fig. 5.90, and Fig. 5.91 shows the particles track at different inlet velocities. In Fig. 5.91, particles can form rotation motion along the hole surface with certain inlet velocity, so gas-particle two-phase compulsive circulation flow finishing is feasible [46].

2. Design of circulation formation component

Circulation formation component is the core part of gas-particle two-phase compulsive circulation flow finishing equipment, which has direct impact on finishing effect and finishing efficiency. Figure 5.92 is two different structures of component of circulation formation. Figure 5.92a is three convergent nozzles designed in light of the gas-particle two-phase swirling flow finishing principle, and it is easy to manufacture by milling. But this structure may have impact on follow-up air, and its sealing performance is not ideal. So the structure shown in Fig. 5.92b is adopted, which adopts whole structure, and its nozzles are machined by drilling.

**Fig. 5.89** Sketch model

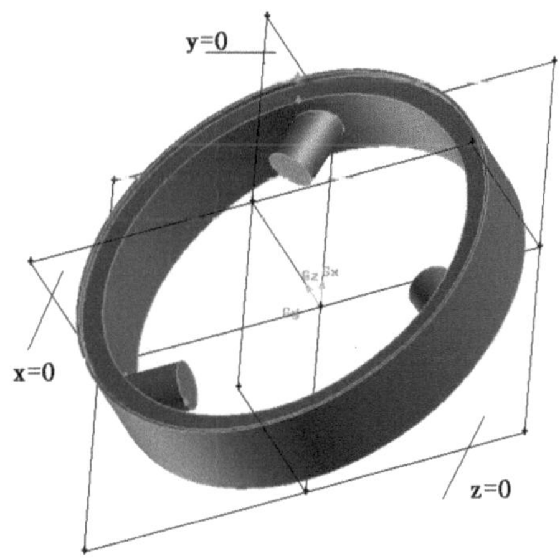

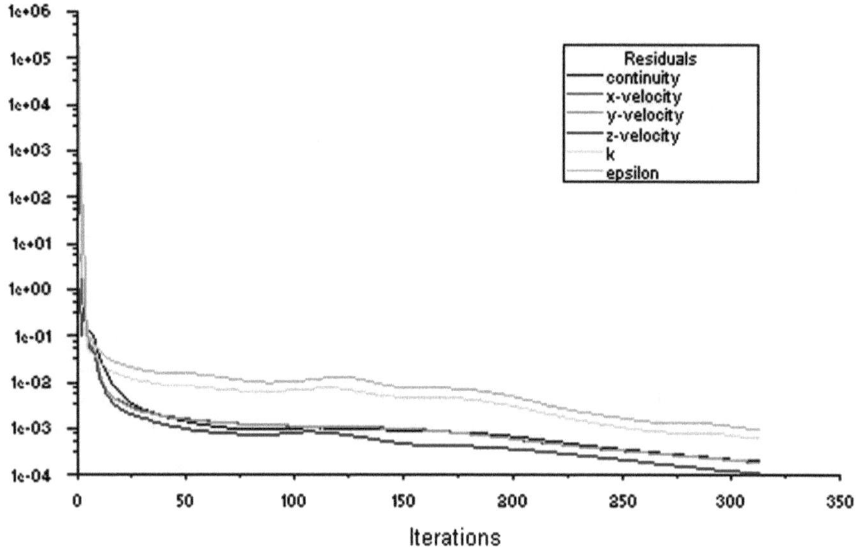


**Fig. 5.90** Convergence curves of residual difference

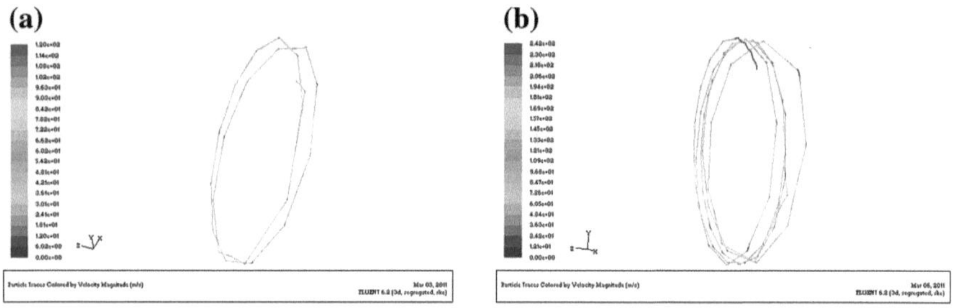


**Fig. 5.91** Particles track of different inlet velocities. **a** 100 m/s; **b** 200 m/s

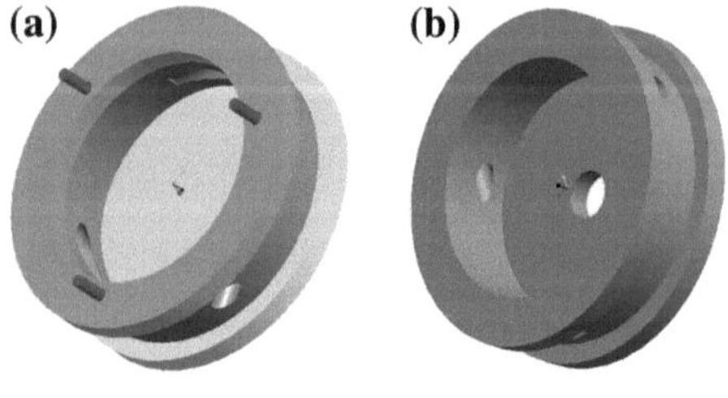


**Fig. 5.92** Two different structures of component of circulation formation. **a** Style 1; **b** Style 2

3. Experimental Analysis

(1) Gas flow

Figure 5.93 is motion status photographs for different flows with three nozzles. When air flow is large, as shown in Fig. 5.93, tangential velocity of the particle driven by the air increases and its centrifugal force also increases, which can have positive impact on the finishing effect and efficiency.

(2) Finishing effects

Workpiece is stainless steel tube, diameter is 140 mm, and length is 30 mm. The particle is spherical silicon carbide, and diameter is about 3 mm. Change curve of surface roughness value *Ra* with finishing time is shown in Fig. 5.94.

In Fig. 5.94, the surface roughness *Ra* value can decrease about two levels through gas-particle two-phase compulsive circulation flow finishing.

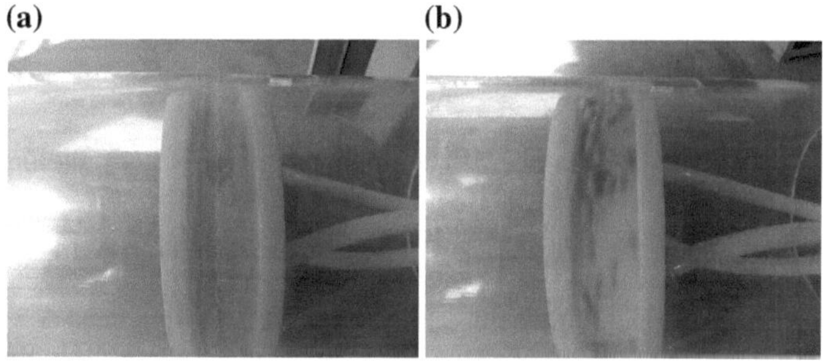


**Fig. 5.93** Photographs of different flows. **a** Large flow; **b** small flow

**Fig. 5.94** *Ra* curve with finishing time

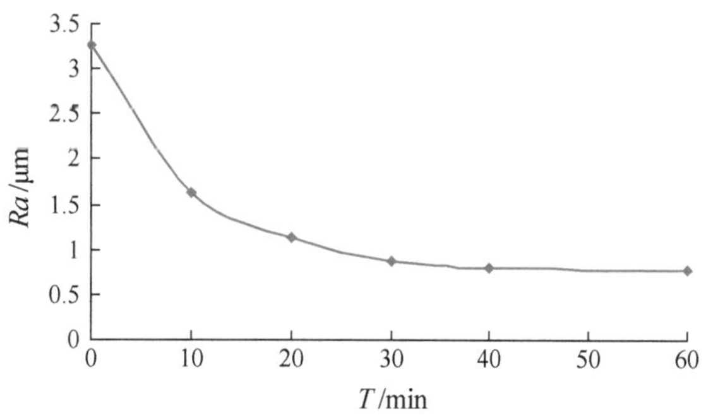

### 5.6.2 *Liquid-Particle Two-Phase Compulsive Circulation Flow Finishing*

The schematic diagram of liquid-particle two-phase compulsive circulation flow finishing is shown in Fig. 5.95. A stainless steel tube is connected through hoses and pump. When the pump starts, the liquid enters into component of circulation formation from the stainless steel tube and flows out from the nozzle of circulation formation component. Circulation region is formed by oriented component, component of circulation formation, and component controlling pressure. Particles are preloaded into circulation region and driven into spiral motion by the liquid out of nozzles. When the liquid flow is large enough, particles can have the action of rolling and the minimum grinding with scraping and scoring against the hole surface with the centrifugal force and relative motion. Finally, it realizes the hole surface finishing. In this region, liquid–particle two-phase flows can rotate continuously with continuous liquid inlet, and with axial relative motion, between oriented component and the finished hole, and the whole surface finishing can be realized [46, 47].

1. Flow field Simulation

(1) Geometry model

Figure 5.96 is the sketch model of two-phase compulsive circulation flow finishing project.

(2) Boundary condition and messing

Based on the $k$-$\varepsilon$ model, liquid flow in the nozzle of swirling head is regarded as boundary condition, and the liquid is supposed to be non-compression. Before CFD

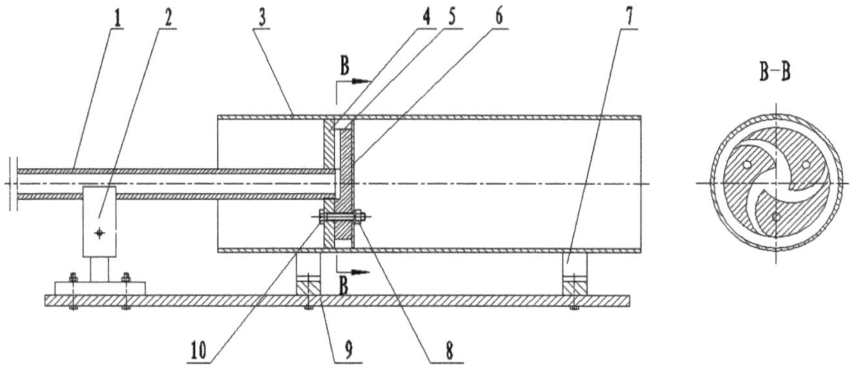


**Fig. 5.95** Schematic diagram of two-phase compulsive circulation flow finishing. 1—Stainless steel tube; 2—adjustable bearing; 3—plexiglass tube; 4—oriented component; 5—component of circulation formation; 6—component controlling pressure; 7—V-shaped block; 8-nut; 9—table; 10—bolt

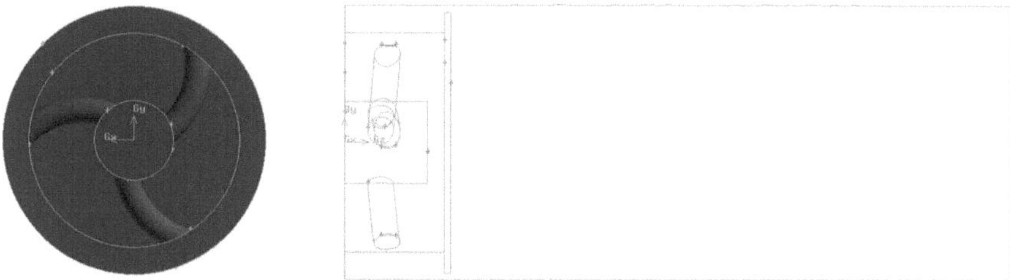

Fig. 5.96 Sketch model

computation on special problem, calculating area must disperse by messing. For geometry model, messing type is TGrid, which is obtained by Gambit program. Figure 5.97 is the grid model.

(3) Simulation results

Figures 5.98 and 5.99 are the velocity vectors and flow line graphs of the liquid. Figure 5.100 is the track graphs of particles. Figure 5.101 is the axial distribution graph of the liquid.

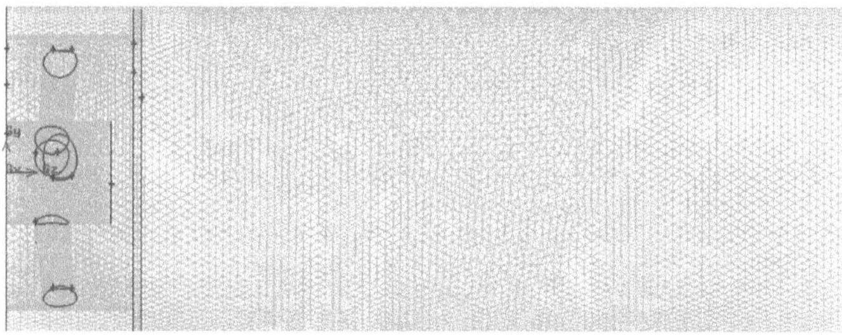

Fig. 5.97 Grid model

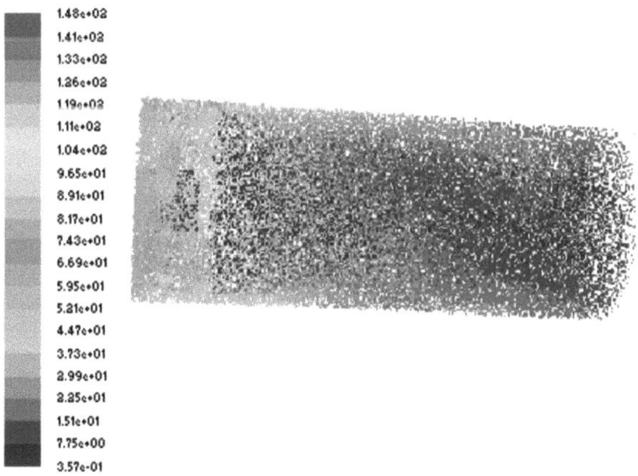


Fig. 5.98 Velocity vectors graphs

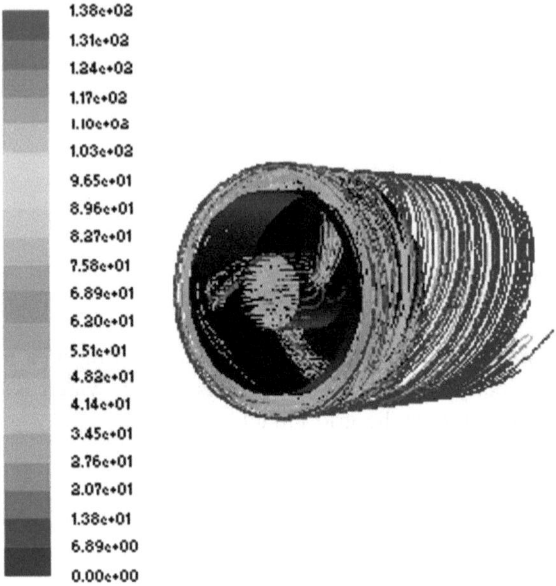


**Fig. 5.99** Flow line graphs

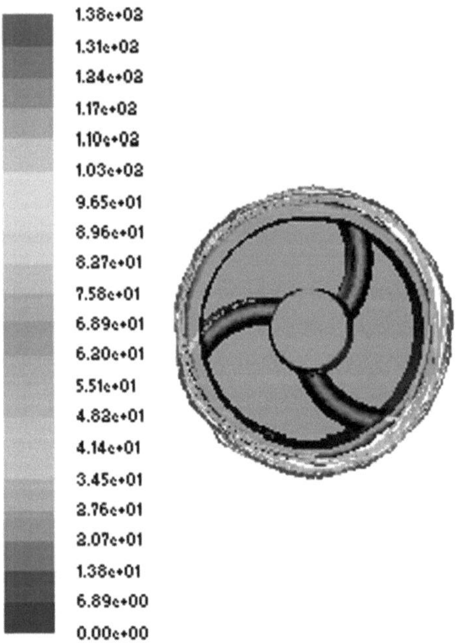


**Fig. 5.100** Track graphs of particles

In Figs. 5.98 and 5.99, velocity attention of the liquid is more along the axial position, and it has large velocity and spiral motion only in certain range of axial distance. In Fig. 5.100, the particle can form spiral motion in circulation region. In Fig. 5.101, when circulation region is in 35-mm axial position, velocity magnitude is beyond 30 m/s and tangential velocity is more than 20 m/s, so the particle can have larger kinetic energy in this region, which is conductive to good finishing effect.

(4) Simulation on different conditions

(1) Component of circulation formation

Figure 5.102 is the two different structures. Figure 5.103 is flow line graphs of component of circulation formation. As shown in Fig. 5.102a, three convergent nozzles are designed in light of gas-particle two-phase swirling flow finishing project, and it can form spiral motion in circulation region from simulation result as shown in Fig. 5.103a. But this structure has difficulty in machining, and it may have impact on follow-up liquid as nozzles are in the middle of circulation region. So the structure shown in Fig. 5.102b is adopted, which is easy to manufacture by milling. In Fig. 5.103b, flow lines can be clearly seen.

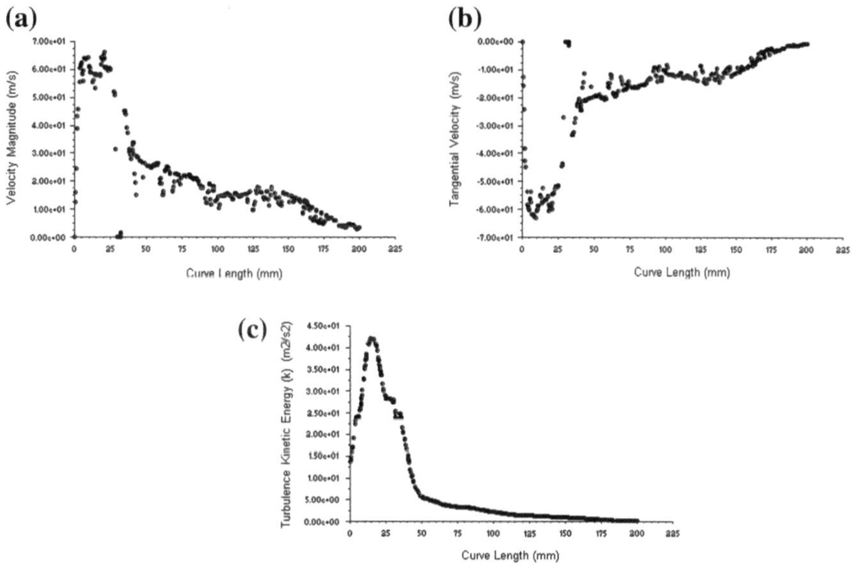


**Fig. 5.101** Axial distribution graph of the liquid. **a** Velocity magnitude distribution; **b** tangential velocity distribution; **c** turbulence kinetic energy

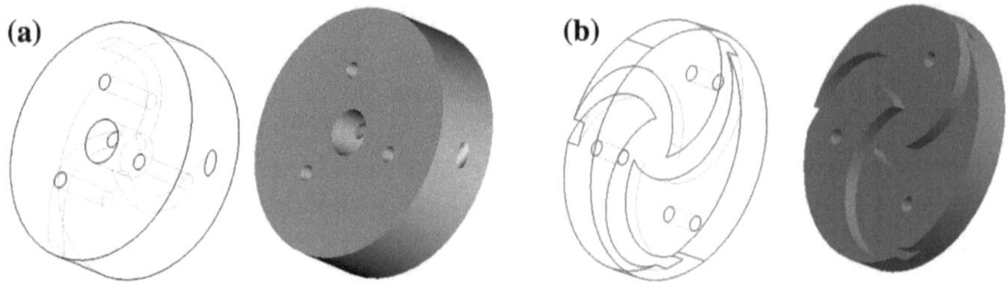


**Fig. 5.102** Two different structures of component of circulation formation. **a** Style 1; **b** Style 2

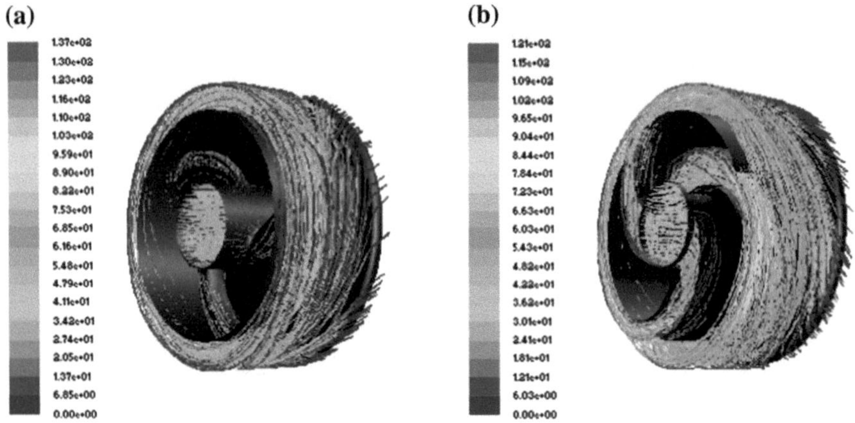


**Fig. 5.103** Flow line graphs. **a** Style 1; **b** Style 2

(2) Nozzle number

For two-phase compulsive circulation flow finishing, nozzle number and nozzle position are the main factors, which would directly influence the finishing effect and efficiency.

Numerical simulation of different nozzle numbers in circulation region is done through CFD software (Fluent, Gambit). Figure 5.104 is flow line graphs. As shown in Fig. 5.104, flow line of three nozzles is better and tangential velocity is larger, so it will have good finishing effect. However, the exit velocity is lower, and fluid energy loss through the nozzle also increases with the increase in the nozzle number. Therefore, nozzle number should have certain ranges and can define according to the diameter of circulation formation component and fluid flow.

(3) Nozzle position

When the liquid flow is larger, it has sufficient tangential kinetic energy, which can lead to good circulation motion of the particle. But when the liquid flow is not

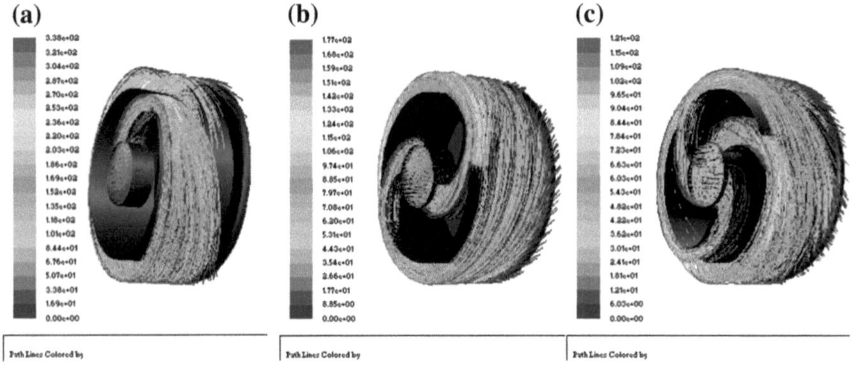


**Fig. 5.104** Flow line graphs. **a** One nozzle; **b** two nozzles; **c** three nozzles

larger, nozzle position would have influence on particle motion, so it cannot be ignored.

Figure 5.105 is two different positions of circulation formation component. Figure 5.106 is velocity vector graph of different nozzle positions. It can be seen from Fig. 5.106, when nozzles position of the circulation formation component is same as the position shown in Fig. 5.105a, the circulation formation component can just make the particle form circulation motion by adjusting the direction of liquid flow. In these conditions, nozzles position is adjusted into another position as shown in Fig. 5.105b, which cannot lead the particle to do full circulation motion.

(4) Liquid flow

Figure 5.107 is the velocity magnitude of the liquid along axial position with different inlet velocities.

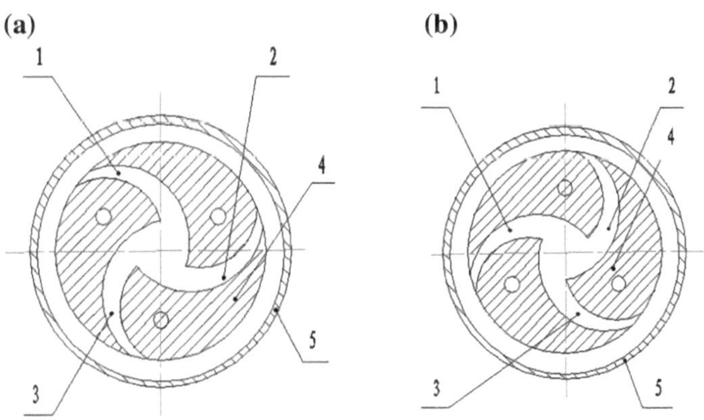


**Fig. 5.105** Different positions of circulation formation component. **a** One position; **b** another position

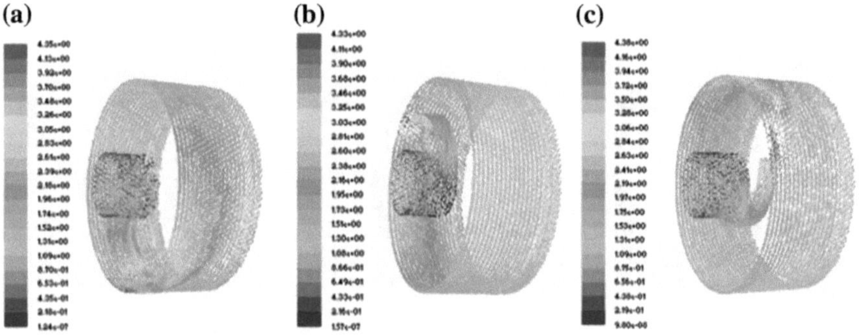


**Fig. 5.106** Velocity vector graph of different nozzle positions. **a** Nozzle 3; **b** Nozzle 2; **c** Nozzle 1

As shown in Fig. 5.107, velocity magnitude decreases constantly for inlet velocity is 17 m/s or 34 m/s, and velocity is larger than 17 m/s. Additionally, with certain width of circulation formation component, liquid flow should have certain range for good finishing effects.

2. Experimental Analysis

(1) Nozzle number

Figure 5.108 is the motion status photographs for different numbers of nozzles. For one nozzle, the exit velocity is greater than two and three nozzles. When the liquid from the nozzle turns around the hole surface once, most of the liquid discharges from circulation region with larger diameter of circulation formation component. So it may not form a complete compulsive circulation flow field, and total energy in circulation region becomes less. Finally, it greatly influences finishing effects, as shown in Fig. 5.108a. When there are two nozzles shown in Fig. 5.108b, another nozzle leads to the increase in the liquid flow continuity, which can be less for kinetic energy attenuation of the liquid, so the tangential kinetic energy of the particle increases. But negative force of the fluid to the hole surface will have impact on particle motion as the liquid ejects from the nozzle. For three nozzles, the distance between two nozzles is the smaller, which is benefit for circulation motion of the particle. Simulation results are similar to experimental phenomenon.

(2) Nozzle position

Figure 5.109 is the motion status photographs at different nozzle positions. Three-nozzle position shown in Fig. 5.109 corresponds to Fig. 5.105a. Experiments are done through two nozzles that is blocked up. In Fig. 5.109a, the liquid is only ejected from nozzle 3 and it just can lead to circulation motion of the particle by adjusting the liquid flow. At this time, it can be seen that complete circulation motion of the liquid cannot form better. But with the tangential kinetic energy of the particle, it can reach the limit of circulation area and arrive in front of the nozzle

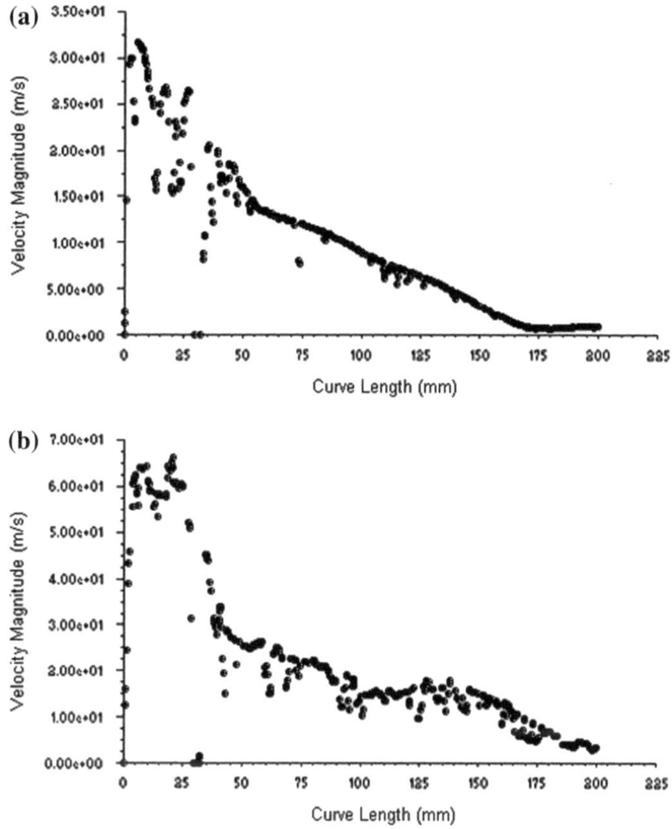


**Fig. 5.107** Velocity magnitude of the liquid along axial position with different inlet velocities. **a** 17 m/s; **b** 34 m/s

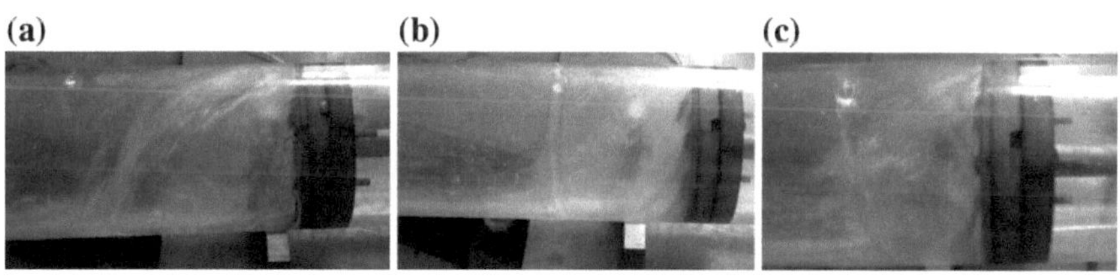


**Fig. 5.108** Motion status photographs of a different number of nozzles. **a** One nozzle; **b** two nozzles; **c** three nozzles

with the function of inertia force and gravity to overcome the impact of liquid, which leads to better circulation motion of the particle. Figure 5.109b, c cannot form better circulation motion of the particle. Experimental results are consistent with the numerical simulation.

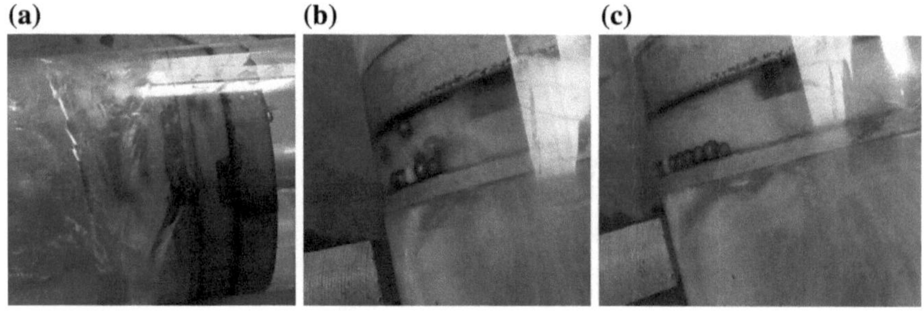

**Fig. 5.109** Motion status photographs at different nozzle positions. **a** Nozzle 3; **b** Nozzle 2; **c** Nozzle 1

(3) Liquid flow

Figure 5.110 is the motion status photographs for different flows as nozzle number is three. When liquid flow is large, as shown in Fig. 5.110a, tangential velocity of the particle driven by the liquid increases, and its centrifugal force also increases, which can have positive impact on finishing effect and efficiency.

(4) Finishing effect

Figure 5.111 is the experimental equipment photograph. Workpiece is stainless steel tube, diameter is 140 mm, and length is 30 mm. The particle is spherical silicon carbide, and diameter is about 3 mm. The liquid flow is 1.296 $m^3/h$. Nozzle is three, and the position is shown as Fig. 5.105a.

Figure 5.112 shows the change curves of surface roughness value *Ra* with finishing time at different positions. As shown in Fig. 5.112, the surface roughness degree value *Ra* of the workpiece can decrease about one level through two-phase compulsive circulation flow finishing. But *Ra* of the top and bottom is not uniform, so the workpiece or circulation formation component should rotate to improve uniformity.

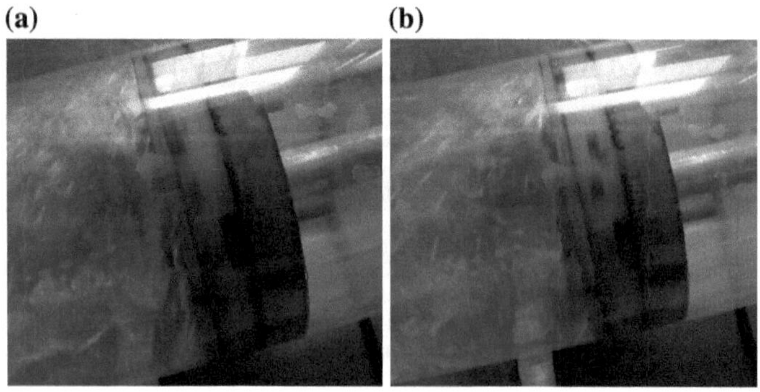

**Fig. 5.110** Motion status photographs of different flow. **a** Large flow; **b** small flow

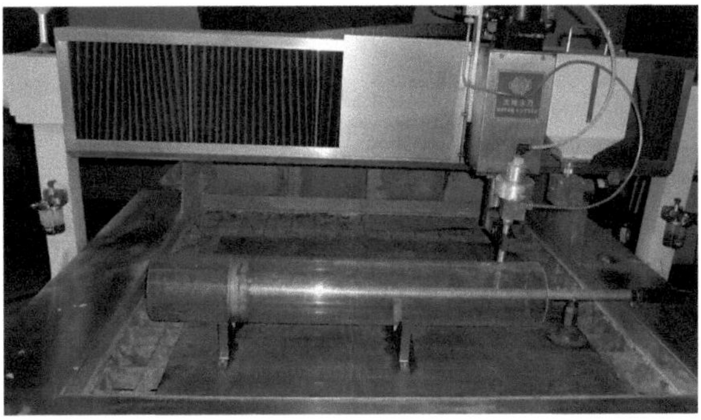

**Fig. 5.111** Experimental equipment photograph

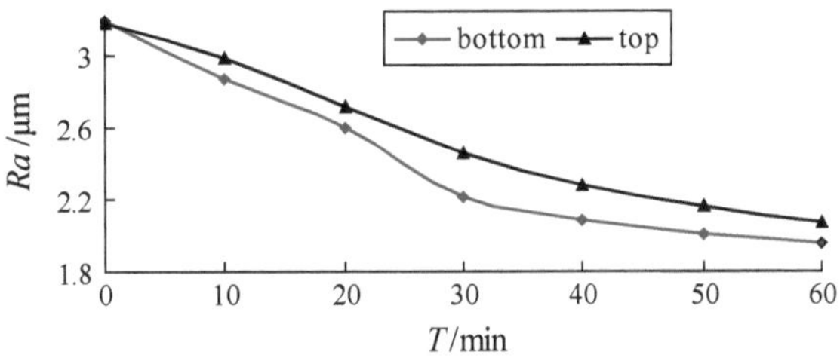


**Fig. 5.112** Change curves of surface roughness value *Ra* with finishing time at different positions

# References


1. S.C. Yang, M.Z. Wang, Y.X. Zhang, *Surface Quality and Finishing Technology* (Machinery Industry Press, Beijing, 2000)
2. X.H. Li, S.C. Yang, Mechanism research on the swirling air flow compounded with magnetic-field finishing. Adv. Mater. Res. **53–54**, 51–55 (2008)
3. W.H. Li, S.Q. Yang, X.H. Li, Theoretical analysis and experimental verification of hole surface finishing parts. Comput. Model. New Technol. **18**(10), 411–414 (2014)
4. B.M. Li, B. Zhao, *Modern Grinding Technology* (Mechanical Industry Press, Beijing, 2003)
5. L.Y. Zhang, *Modern Machining Technology* (Mechanical Industry Press, Beijing, 2002)
6. S.C. Yang, S.Q. Yang, The development direction of free abrasive finishing technology. Mach. Metal Cutting **1**, 11–12 (2002)
7. S.Q. Yang, *Theoretical Analysis and Experimental Study Two Phase Swirling Flow Hole Finishing Technology* (Taiyuan University of Technology, Taiyuan, 2009)
8. B.W. Xue, S.C. Yang, S.Q. Yang, Comparison and analysis of all kinds of swirling flow generator and the ways of generating swirling flow. Mech. Eng. Autom. **1**, 98–100 (2005)
9. H.X. Zhang, *Experimental investigation and numerical simulation of three dimensional turbulent swirling flow locally generatd in circular pipe* (Taiyuan University of Technology, Taiyuan, 2004)
10. J.R. Ni, G.Q. Wang, H.B. Zhang, *The Basic Theory and Latest Applications of Solid Liquid Two Phase Flow* (Science Press, Beijing, 1991)
11. S.Q. Yang, W.H. Li, S.C. Yang, Experimental research on improving performance of the engine by surface finishing technology. Trans. Chin. Soc. Agric. Mach. **37**(2), 17–20 (2006)
12. H.G. Du, L.S. Liu, W.B. Wang, Application and research of removal of inner hole burrs with abrasive water jet technology. Intern. Combust. Engine Powerplant **6**, 37–40 (2007)

13. J.O. Kitoh, Experimental study of turbulent swirling flow in a straight pipe. J. Fluid Mech. **225**, 445–479 (1991)
14. L.X. Zhou, *The Numerical Simulation of the Turbulent Two-Phase Flow and Combustion* (Tsinghua University Press, Beijing, 1991)
15. K.F. Cen, J.R. Fan, *Theory and Calculation of Gas-Solid Multiphase Flow in Engineering* (Zhejiang University Press, Hangzhou, 1990)
16. S.Q. Yang, W.H. Li, S.C. Yang, Numerical simulation of compulsive swirling flows field on two-phase swirling flows finishing. Comput. Appl. Technol. **29**(2–4), 205–207 (2007)
17. S.Q. Yang, W.H. Li, S.C. Yang, Simulation of flows field characteristics in the nozzle of two-phase compulsive circulation flows finishing. Adv. Mater. Res. **53–54**, 9–13 (2008)
18. F.J. Wang, *Computational Fluid Dynamics Analysis: Principle and Application of CFD Software* (Tsinghua University Press, Beijing, 2004)
19. Y.M. Lan, X.H. Sun, D.M. Huo, Numerical simulation of the 3D swirling flow in girallar pipe. J. Taiyuan Univ. Technol. **32**(3), 255–259 (2001)
20. S. Jha, V.K. Jain, Design and development of the magnetorheological abrasive flow finishing (MRAFF) process. Int. J. Mach. Tools Manuf. **44**, 1019–1029 (2004)
21. C.H. Zhang, *Study of Abrasive Water Jet Finishing and Experiment Equipment* (Nanjing University of Science and Technology, Nanjing, 2003)
22. S.Q. Yang, Y.N. Gao, W.H. Li et al., Numerical simulation and experiment study of swirling air stream finishing. New Technol. New Process **1**, 10–12 (2009)
23. W.H. Li, S.Q. Yang, S.C. Yang. Development and experimental research of machining medium in barrel finishing. DaLian: Proceedings of the International Conference on Surface Finishing Technology and Surface Engineering, Frontiers of Design and Manufacturing (2006)
24. Z.Y. Hou, S.C. Yang, S.Q. Yang, Comparison and analysis of the two-phase swirling flow field. Mech. Eng. Autom. **6**, 10–12 (2004)
25. J. Li, Y.Z. Zhang, K.F. Li et al., On the mathematical model of the particle motion in shear turbulent flow. J. Hydraul. Eng. **6**, 7–13 (1998)
26. W.F. Chen, Y. Chang, Three-dimensional supersonic gas-solid two-phase conduit flow with CFD/DSMC method. J. Propul. Technol. **26**(3), 239–241 (2005)
27. S.H. Huang, W. Li, L.J. Cheng, On numerical method of resolving discrete solid particles' motion equation and its applications. J. Hydrodyn. **14**(1), 51–61 (1999)
28. S.Q. Yang, W.H. Li, S.C. Yang, Flows field simulation of two-phase swirling flows finishing. Adv. Mater. Res. **24–25**, 17–22 (2007)
29. Y.M. Jia, X.H. Zhang, Q.X. Zhao, et al., The computer simulations and experiment studies on swirling flow finishing. J. Taiyuan Univ. Technol. **36**(6), 729–731 (2005)
30. A.K. Xiong, Q.D. Wei, Experiment study on a sort of strong swirling flow. Exp. Meas. Fluid Mech. **13**(4), 8–13 (1999)
31. G.W. He, S.C. Yang, S.Q. Yang, Analysis of the equipment of forming spiral liquid-flow in an annulus. Mech. Eng. Autom. **2**, 25–26 (2006)
32. S.Q. Yang, M.Z. Wang, S.C. Yang, Process research of swirling flow hole surface Finishing technology. Proceedings of the First International Conference on Mechanical Engineering, Shanghai, China (2000)
33. X.F. Wang, *Experimental Study and Analysis of Swirling Flow in Pipes* (Wuhan University of Technology, Wuhan, 2004)
34. A.K. Xiong, *Study on Strong Stable Swirling Flow and Application of Nonlinear Eddy Viscosity Model* (Peking University, Beijing, 1998)
35. Q.H. Wang, *The numerical simulation of solid/liquid two-phase flow in the screw centrifugal pump* (Lanzhou University of Technology, Lanzhou, 2005)
36. Y.L. Ma, *Numerical simulation of the dense two-phase flow* (Zhejiang University, Hangzhou, 2001)
37. S.Q. Yang, S.C. Yang, W.H. Li, Basic principle and theoretical research on the surface finishing of precise hole through gas-particle two-phase swirling flow. Key Eng. Mater. **259–260**, 549–552 (2004)

38. Y.M. Jia, X.Y. Ma, M.Z. Wang et al., The analysis of the machining mechanism and force suffered of swirling airflow finishing. China Mech. Eng. **13**(7), 600–602 (2002)
39. S.C. Yang, S.Q. Yang, T.M. Wang, Mechanism analysis of swirling air stream finishing. Electromachining Mould **2**, 13–15 (2002)
40. T.M. Wang, *The Theory Research, Experimental Equipment Design(Control Section) and Experiment Analysis of Whirling Air-Stream Finishing* (Taiyuan University of Technology, Taiyuan, 2001)
41. S.Q. Yang, S.C. Yang, W.H. Li, Experimental equipment for surface finishing precise hole through gas-particle two-phase swirling flow. Key Eng. Mater. **259–260**, 544–548 (2004)
42. S.Q. Yang, W.H. Li, S.C. Yang, Finishing effect and uniformity on gas-particle two-phase swirling flows finishing for hole surface. Adv. Mater. Res. **97–101**, 4124–4127 (2010)
43. Z.Y. Wu, *The Experiment Equipment Design and Process Study on the Swirling Air Stream Finishing* (Taiyuan University of Technology, Taiyuan, 2001)
44. W.H. Li, S.Q. Yang, X.C. Yang, Technique project and experimental study of gas-particle two-phase compulsive circulation flows finishing. Adv. Mater. Res. **305**, 438–441 (2011)
45. W.H. Li, S.Q. Yang, X.H. Li, Experimental study of two-phase compulsive circulation flows finishing. Adv. Mater. Res. **135**, 149–153 (2010)
46. S.Q. Yang, W.H. Li, S.C. Yang, Simulation and design of experimental equipment for two-phase compulsive circulation flows finishing. Adv. Mater. Res. **135**, 319–324 (2010)
47. X.C. Yang, *Numerical simulation and experimental study for gas-particle two-phase compulsive circulation flows finishing* (Taiyuan University of Technology, Taiyuan, 2011)

MIX
Papier aus verantwortungsvollen Quellen
Paper from responsible sources
FSC® C105338

If you have any concerns about our products,
you can contact us on
**ProductSafety@springernature.com**

In case Publisher is established outside the EU,
the EU authorized representative is:
**Springer Nature Customer Service Center GmbH**
**Europaplatz 3, 69115 Heidelberg, Germany**

Printed by Libri Plureos GmbH
in Hamburg, Germany